Current Topics in Membranes, Volume 55

Chemokines, Chemokine Receptors, and Disease

Current Topics in Membranes, Volume 55

Series Editors

Dale J. Benos
Department of Physiology and Biophysics
University of Alabama
Birmingham, Alabama

Sidney A. Simon
Department of Neurobiology
Duke University Medical Centre
Durham, North Carolina

Current Topics in Membranes, Volume 55

Chemokines, Chemokine Receptors, and Disease

Edited by

Lisa M. Schwiebert, Ph.D.
Department of Physiology and Biophysics and of Cell Biology and The Gregory Fleming James Cystic Fibrosis Research Center
University of Alabama at Birmingham
Birmingham, Alabama

ELSEVIER
ACADEMIC
PRESS

AMSTERDAM • BOSTON • HEIDELBERG • LONDON
NEW YORK • OXFORD • PARIS • SAN DIEGO
SAN FRANCISCO • SINGAPORE • SYDNEY • TOKYO

Elsevier Academic Press
525 B Street, Suite 1900, San Diego, California 92101-4495, USA
84 Theobald's Road, London WC1X 8RR, UK

This book is printed on acid-free paper. ♾

For all information on all Elsevier Academic Press publications visit our Web site at www.books.elsevier.com

ISBN-13: 978-0-12-153355-7
ISBN-10: 0-12-153355-7

Printed and bound in the United Kingdom
Transferred to Digital Printing, 2011

Contents

Contributors

Numbers in parentheses indicate the pages on which the author's contributions begin.

Nilo O. Arnaiz (255), Departments of Medicine, Pathology, and Pediatrics, Division of Pulmonary and Critical Care Medicine, School of Medicine, University of California, Los Angeles, Los Angeles, California 90024

John A. Belperio (255), Departments of Medicine, Pathology, and Pediatrics, Division of Pulmonary and Critical Care Medicine, School of Medicine, University of California, Los Angeles, Los Angeles, California 90024

Linda Bendall (331), Westmead Millennium Institute, University of Sydney, and Department of Haematology, Westmead Hospital Westmead, New South Wales, Australia 2145

Joan W. Berman (1), Department of Pathology and Department of Microbiology and Immunology, Albert Einstein College of Medicine, Bronx, New York 10461

J. Edwin Blalock (49), Department of Physiology and Biophysics, University of Alabama at Birmingham, Birmingham, Alabama 35294

Kenneth Bradstock (331), Westmead Millennium Institute, University of Sydney, and Department of Haematology, Westmead Hospital Westmead, New South Wales, Australia 2145

Marie D. Burdick (255), Departments of Medicine, Pathology, and Pediatrics, Division of Pulmonary and Critical Care Medicine, School of Medicine, University of California, Los Angeles, California 90024

Tina M. Calderon (1), Department of Pathology and Department of Microbiology and Immunology, Albert Einstein College of Medicine, Bronx, New York 10461

Keith M. Channon (223), Department of Cardiovascular Medicine, University of Oxford, Oxford OX1 3RE, United Kingdom

Karoll J. Cortez (289), Pediatric Oncology Branch, National Cancer Institute, National Institutes of Health, Bethesda, Maryland 20892

James M. Fox (73), Leukocyte Biology Section, Biomedical Sciences Division, Faculty of Medicine, Imperial College, London SW7 2AZ, United Kingdom

David R. Greaves (223), Sir William Dunn School of Pathology, University of Oxford, Oxford OX1 3RE, United Kingdom

Tadaatsu Imaizumi (103), Department of Vascular Biology, Hirosaki University School of Medicine, Hirosaki 036-8562, Japan

Julia O. Jackson (289), Laboratory of Immunoregulation, National Institute of Allergy and Infectious Diseases, National Institutes of Health, Bethesda, Maryland 20892

Alison E. John (223), Sir William Dunn School of Pathology, University of Oxford, Oxford OX1 3RE, United Kingdom

Michael P. Keane (255), Departments of Medicine, Pathology, and Pediatrics, Division of Pulmonary and Critical Care Medicine, School of Medicine, University of California, Los Angeles, Los Angeles, California 90024

Chang H. Kim (115), Laboratory of Immunology and Hematopoiesis, Department of Pathobiology; Purdue Cancer Center; Bindley Bioscience Center, Purdue University, West Lafayette, Indiana 47907

Robyn S. Klein (159), Division of Infectious Diseases and Departments of Molecular Microbiology, Pathology and Immunology, and Anatomy and Neurobiology, Washington University School of Medicine, St. Louis, Missouri 63110

Shyam Kottilil (289), Laboratory of Immunoregulation, National Institute of Allergy and Infectious Diseases, National Institutes of Health, Bethesda, Maryland 20892

Nicholas W. Lukacs (189), Department of Pathology, University of Michigan Medical School, Ann Arbor, Michigan 48109

Andrew D. Luster (159), Division of Rheumatology, Allergy, and Immunology, Massachusetts General Hospital, Boston, Massachusetts 02114

Allison L. Miller (189), Department of Pathology, University of Michigan Medical School, Ann Arbor, Michigan 48109

Jan Hendrik Niess (143), Gastrointestinal Unit and Center for the Study of Inflammatory Bowel Disease, Massachusetts General Hospital and Harvard Medical School, Boston, Massachusetts 02141

James E. Pease (73), Leukocyte Biology Section, Biomedical Sciences Division, Faculty of Medicine, Imperial College, London SW7 2AZ, United Kingdom

Hans-Christian Reinecker (143), Gastrointestinal Unit and Center for the Study of Inflammatory Bowel Disease, Massachusetts General Hospital and Harvard Medical School, Boston, Massachusetts 02141

Joshua B. Rubin (159), Division of Pediatric Oncology, Departments of Pediatrics and Anatomy and Neurobiology, Washington University School of Medicine, St. Louis, Missouri 63110

Ammar Sakkour (255), Departments of Medicine, Pathology, and Pediatrics, Division of Pulmonary and Critical Care Medicine, School of Medicine, University of California, Los Angeles, Los Angeles, California 90024

Kei Satoh (103), Department of Vascular Biology, Hirosaki University School of Medicine, Hirosaki 036-8562, Japan

Robert M. Strieter (255), Departments of Medicine, Pathology, and Pediatrics, Division of Pulmonary and Critical Care Medicine, School of Medicine, University of California, Los Angeles, Los Angeles, California 90024

Molly S. Thomas (189), Department of Pathology, University of Michigan Medical School, Ann Arbor, Michigan 48109

Nathaniel M. Weathington (49), Department of Physiology and Biophysics, University of Alabama at Birmingham, Birmingham, Alabama 35294

Hidemi Yoshida (103), Department of Vascular Biology, Hirosaki University School of Medicine, Hirosaki 036-8562, Japan

Foreword

Chemokines, Chemokine Receptors, and Disease: A comprehensive view of chemokine biology

Lisa M. Schwiebert, Ph.D.
Department of Physiology and Biophysics and of Cell Biology and
The Gregory Fleming James Cystic Fibrosis Research Center,
University of Alabama at Birmingham
Birmingham, Alabama 35294

Chemokines are small chemotactic proteins that mediate their function via binding to seven transmembrane-spanning G-protein-coupled receptors. Originally, chemokines were implicated in leukocyte trafficking and homing; recent work, however, demonstrates a role for chemokines and their receptors in a variety of disease-related pathologies including atherosclerosis, inflammation, allergy, infectious disease, and carcinogenesis. The chemokine receptor-mediated signaling pathways that trigger these pathologies are an intense area of study. In addition, chemokine receptor antagonists that block responses to these chemotactic proteins, and thereby attenuate or lessen associated tissue damage, are being actively developed.

The volume *Chemokine, Chemokine Receptors, and Disease* attempts to present these topics in an integrated manner that provides the reader with a comprehensive and up-to-date knowledge in both chemokine biology and related pathologies. We begin with introductory chapters that review, in detail, the history and biology of chemokine subfamilies, including CXC, CC, CX3C, and C chemokines, and their respective receptors. With that foundation in place, we next address chemokine-related topics in physiology and pathophysiology. Specifically, these topics discuss the role of chemokines and their receptors in such diverse areas as hematopoiesis, intestinal physiology, neurophysiology, pulmonary inflammation, vascular disease, carcinogenesis, and infectious disease. In the last chapter, we examine the emerging field of chemokine-related therapeutics and how such modalities may improve basic research as well as provide therapy to fight disease. Clearly, the topic of chemokine biology is a rapidly growing and changing field. As such, in every chapter, authors address the subject of future directions and unanswered questions in the area of chemokine biology in order to provide a road map for future chemokine-related research.

Acknowledgments

The Editor would like to thank all of the authors for their contributions and Academic Press for their expert publication of this volume for *Current Topics in Membranes*.

Previous Volumes in Series

Current Topics in Membranes and Transport

Volume 23 Genes and Membranes: Transport Proteins and Receptors* (1985)
Edited by Edward A. Adelberg and Carolyn W. Slayman

Volume 24 Membrane Protein Biosynthesis and Turnover (1985)
Edited by Philip A. Knauf and John S. Cook

Volume 25 Regulation of Calcium Transport across Muscle Membranes (1985)
Edited by Adil E. Shamoo

Volume 26 Na^+–H^+ Exchange, Intracellular pH, and Cell Function* (1986)
Edited by Peter S. Aronson and Walter F. Boron

Volume 27 The Role of Membranes in Cell Growth and Differentiation (1986)
Edited by Lazaro J. Mandel and Dale J. Benos

Volume 28 Potassium Transport: Physiology and Pathophysiology* (1987)
Edited by Gerhard Giebisch

Volume 29 Membrane Structure and Function (1987)
Edited by Richard D. Klausner, Christoph Kempf, and Jos van Renswoude

Volume 30 Cell Volume Control: Fundamental and Comparative Aspects in Animal Cells (1987)
Edited by R. Gilles, Arnost Kleinzeller, and L. Bolis

Volume 31 Molecular Neurobiology: Endocrine Approaches (1987)
Edited by Jerome F. Strauss, III, and Donald W. Pfaff

**Part of the series from the Yale Department of Cellular and Molecular Physiology*

Current Topics in Membranes

Volume 46 Potassium Ion Channels: Molecular Structure, Function, and Diseases (1999)
Edited by Yoshihisa Kurachi, Lily Yeh Jan, and Michel Lazdunski

Volume 47 AmilorideSensitive Sodium Channels: Physiology and Functional Diversity (1999)
Edited by Dale J. Benos

Volume 48 Membrane Permeability: 100 Years since Ernest Overton (1999)
Edited by David W. Deamer, Arnost Kleinzeller, and Douglas M. Fambrough

Volume 49 Gap Junctions: Molecular Basis of Cell Communication in Health and Disease
Edited by Camillo Peracchia

Volume 50 Gastrointestinal Transport: Molecular Physiology
Edited by Kim E. Barrett and Mark Donowitz

Volume 51 Aquaporins
Edited by Stefan Hohmann, Søren Nielsen and Peter Agre

Volume 52 Peptide–Lipid Interactions
Edited by Sidney A. Simon and Thomas J. McIntosh

Volume 53 CalciumActivated Chloride Channels
Edited by Catherine Mary Fuller

Volume 54 Extracellular Nucleotides and Nucleosides: Release, Receptors, and Physiological and Pathophysiological Effects
Edited by Erik M. Schwiebert

CHAPTER 1

Overview and History of Chemokines and Their Receptors

Tina M. Calderon and Joan W. Berman
Department of Pathology and Department of Microbiology and Immunology, Albert Einstein College of Medicine, Bronx, New York 10461

I. CHEMOKINES: GENERAL OVERVIEW

The large family of structurally related chemotactic cytokines involved in the migration and activation of specific leukocytes was designated "chemokines" by the Third International Symposium of Chemotactic Cytokines in 1992 (Taub and Oppenheim, 1993). This family of chemotactic proteins appears to be conserved evolutionarily, as evidenced by its high degree of homology (20 to 70% in amino acid sequences) across species from mammals to fish. Chemokines, small molecular weight proteins ranging in size from approximately 8 to 15 kDa, chemoattract specific subsets of leukocytes, in

Current Topics in Membranes, Volume 55

1063-5823/05 $35.00
DOI: 10.1016/S1063-5823(04)55001-9

addition to other cell types, and are characterized by the position of the first two of four cysteine residues in highly conserved positions within their amino (N)-terminal protein sequence (Rollins, 1997). Initially, chemokines were divided into four groups, α, β, γ, and δ, but a new classification nomenclature for chemokines and chemokine receptors was adopted to provide consistency and clarity in the rapidly expanding field of chemokine biology (Murphy *et al.*, 2000; Zlotnik and Yoshie, 2000). The chemokine family has now been subdivided into four major groups, designated CXC (formerly α chemokines), CC (formerly β chemokines), C (formerly γ chemokines), and CX_3C (formerly δ chemokines). X denotes amino acids positioned between the cysteine residues that are located at highly conserved positions in a chemokine domain at the N-terminus of these proteins. The addition of L to the chemokine designation (for example, in CCL2) indicates a chemokine ligand, and the addition of R (for example, in CCR2) indicates a chemokine receptor. Chemokines share structural motifs characterized by the presence of at least three β-pleated sheets, a carboxy(C)-terminal α-helix, and a flexible N-terminal region (Rollins, 1997).

The genes encoding CXC chemokines are generally located on human chromosome 4 (Modi and Chen, 1998); however, the gene for stromal cell–derived factor-1 (SDF-1/CXCL12) is located on chromosome 10 (Shirozu *et al.*, 1995). CC genes cluster on chromosome 17 (Naruse *et al.*, 1996), with a few exceptions, including macrophage inflammatory protein-3β (MIP-3β/CCL19), on chromosome 9 (Rossi *et al.*, 1997), and MIP-3α/CCL20, on chromosome 2 (Hieshima *et al.*, 1997). The gene that encodes lymphotactin/XCL1 maps to chromosome 1 (Kelner *et al.*, 1994; Kennedy *et al.*, 1995; Muller *et al.*, 1995; Yoshida *et al.*, 1995) and the gene encoding fractalkine/CX3CL1 maps to chromosome 16 (Bazan *et al.*, 1997; Harrison *et al.*, 1998; Nomiyama *et al.*, 1998).

CXC chemokines were originally characterized as neutrophil chemoattractants, although many family members, including SDF-1/CXCL12 and interleukin 8 (IL-8/CXCL8), are active on other classes of leukocytes, including monocytes (Larsen *et al.*, 1989; Bleul *et al.*, 1996b; Gerszten *et al.*, 1999). In addition to their ability to chemoattract leukocytes, CXC chemokines like IL-8/CXCL8, growth-related oncogene protein-α (GROα/CXCL1), and epithelial-cell-derived neutrophil-activating protein-78 (ENA-78/ CXCL5), which contain the highly conserved three amino acid ELR motif (Glu–Leu–Arg) immediately preceding the first cysteine in their N-terminal CXC motif (Strieter *et al.*, 1995; Luster, 1998), are also angiogenic (Koch *et al.*, 1992b; Strieter *et al.*, 1992). In general, CXC chemokines lacking an ELR motif, such as platelet factor 4 (PF4/CXCL4), interferon gamma–induced protein of MW10 (IP-10/CXCL10), and monokine induced by gamma interferon (MIG/CXCL9), possess anti-angiogenic properties.

However, the non-ELR CXC chemokine, SDF-1/CXCL12, induces neovascularization *in vivo* (Salcedo *et al.*, 1999). Most ELR-containing CXC chemokines chemoattract neutrophils but not monocytes, while non-ELR CXC chemokines are usually poor neutrophil chemoattractants but chemoattract lymphocytes and monocytes.

In general, CC chemokines chemoattract monocytes, basophils, eosinophils, and T-lymphocytes, but not neutrophils. Members of this family include monocyte chemoattractant protein-1 (MCP-1/CCL2), MIP-1α/CCL3, MIP-1β/CCL4, and, regulated upon activation, normal T-cell expressed and secreted (RANTES/CCL5). Lymphotactin/XCL1, a chemoattractant for T-lymphocytes (Kelner *et al.*, 1994; Kennedy *et al.*, 1995; Muller *et al.*, 1995; Yoshida *et al.*, 1995), and fractalkine/CX3CL1, a chemoattractant for T-lymphocytes and monocytes (Bazan *et al.*, 1997; Pan *et al.*, 1997), are members of the C and CX_3C subfamilies, respectively. Fractalkine/CX3CL1 has an unusual structure for a chemokine. It is a type-1 transmembrane glycoprotein with a chemokine domain located on top of an extended mucin-like stalk at its N-terminus. Fractalkine/CX3CL1 has the unique property of existing as a chemokine that is both soluble and membrane bound, and it is constitutively expressed in various tissues, including brain tissues (Bazan *et al.*, 1997; Pan *et al.*, 1997).

Although the regulation of leukocyte migration and activation during immune responses is the major function of chemokines, they are also thought to modulate additional biological activities, including hematopoiesis, apoptosis, angiogenesis, cell proliferation, and viral pathogenesis (Rollins, 1997; Bajetto *et al.*, 2001). A listing of the chemokines and chemokine receptors that have been identified as of this writing is shown in Table I. A more complete description of some of the major chemokines and chemokine receptors and their roles in inflammatory disorders and pathologies, as well as other immunological and developmental processes, is detailed in the next section.

A. ELR-Containing CXC Chemokines

Groα/CXCL1 functions as a neutrophil chemoattractant and is produced by many cell types, including endothelial cells, bronchial epithelial cells, macrophages, and neutrophils (Balentien *et al.*, 1990; Iida and Grotendorst, 1990; Moser *et al.*, 1990; Schroder *et al.*, 1990; Introna *et al.*, 1993; Becker *et al.*, 1994). The expression of this chemokine in multiple cell types, including epithelial cells, fibroblasts, and endothelial cells, is regulated both transcriptionally and posttranscriptionally by several factors, including tumor necrosis factor (TNF)-α, interleukin-1 beta (IL-1β), thrombin and oxidized

TABLE I
Chemokine and Receptor Families

Systematic nomenclature	Common nomenclature	Chemokine receptor(s)
CXC Family		
CXCL1	Groα/MGSA-α	CXCR2 > CXCR1
CXCL2	Groβ/MGSA-β	CXCR2
CXCL3	Groγ/MGSA-γ	CXCR2
CXCL4	PF4	Unknown
CXCL5	ENA-78	CXCR2
CXCL6	GCP-2	CXCR1, CXCR2
CXCL7	NAP-2	CXCR2
CXCL8	IL-8	CXCR1, CXCR2
CXCL9	Mig	CXCR3
CXCL10	IP-10	CXCR3
CXCL11	I-TAC	CXCR3
CXCL12	SDF-1α/β	CXCR4
CXCL13	BLC/BCA-1	CXCR5
CXCL14	BRAK	Unknown
CXCL15	Lungkine	Unknown
CXCL16	SR-PSOX	CXCR6
CC Family		
CCL1	I-309	CCR8
CCL2	MCP-1/MCAF/TDCF	CCR2
CCL3	MIP-1α/LD78α	CCR1, CCR5
CCL4	MIP-1β	CCR5
CCL5	RANTES	CCR1, CCR3, CCR5
CCL6	C10, MRP-1	CCR1?
CCL7	MCP-3	CCR1, CCR2, CCR3
CCL8	MCP-2	CCR1, CCR2, CCR3, CCR5
CCL9/10	MRP-2, CCF18, MIP-1γ	CCR1
CCL11	Eotaxin	CCR3 > CCR5
CCL12	MCP-5	CCR2
CCL13	MCP-4	CCR2, CCR3
CCL14	HCC-1	CCR1, CCR3, CCR5
CCL15	HCC-2/Lkn-1/MIP-1δ	CCR1, CCR3
CCL16	HCC-4/LEC/LCC-1	CCR1, CCR2, CCR5, CCR8
CCL17	TARC	CCR4
CCL18	DC-CK1/PARC/AMAC-1	Unknown

TABLE I (*Continued*)

Systematic Nomenclature	Common Nomenclature	Chemokine Receptor(s)
CCL19	MIP-3β/ELC/exodus-3	CCR7
CCL20	MIP-3α/LARC/exodus-1	CCR6
CCL21	6Ckine/SLC/exodus-2	CCR7
CCL22	MDC/STCP-1	CCR4
CCL23	MPIF-1/CKβ8/CKβ8-1	CCR1
CCL24	Eotaxin-2/MPIF-2	CCR3
CCL25	TECK	CCR9
CCL26	Eotaxin-3	CCR3
CCL27	CTACK/ALP/ILC	CCR10
CCL28	MEC	CCR3, CCR10
C Family		
XCL1	Lymphotactin/SCM-1α/ATAC	XCR1
XCL2	SCM-1β	XCR1
CX_3C Family		
CX3CL1	Fractalkine/Neurotactin	CX3CR1

low-density lipoprotein (LDL) cholesterol (Stoeckle, 1991, 1992; Introna *et al.*, 1993; Shattuck *et al.*, 1994; Murakami *et al.*, 1995; Lei *et al.*, 2001). In the central nervous system (CNS), Groα/CXCL1 is produced by subsets of astrocytes and neurons, and it is a potent inducer of oligodendrocyte precursor proliferation (Robinson *et al.*, 1998). The Groα/CXCL1 cDNA was initially cloned from Chinese hamster fibroblastic cell lines, and a homologous human cDNA was isolated from the T24 bladder carcinoma cell line (Anisowicz *et al.*, 1987). Message expression was regulated by the growth status of the Chinese hamster and human cell lines; thus, this protein was named Gro to signify that it is growth regulated. The murine homologue had been cloned previously as the platelet-derived growth factor (PDGF)-inducible KC gene (Cochran *et al.*, 1983; Oquendo *et al.*, 1989). The previously purified melanoma growth factor, melanoma growth stimulatory activity (MGSA), was identified as the human homologue of Groα/CXCL1 (Richmond *et al.*, 1988). Groβ/CXCL2 and Groγ/CXCL3 are similar to Groα/CXCL1 and are also neutrophil chemoattractants (Haskill *et al.*, 1990). These two proteins were initially identified as MIP-2α and MIP-2β in murine cells by the investigators who isolated the CC chemokines, MIP-1α/CCL3 and MIP-1β/CCL4 (Wolpe *et al.*, 1989; Tekamp-Olson *et al.*, 1990).

ENA-78/CXCL5, a neutrophil chemoattractant, was isolated from the human epithelial cell line A549 after TNF-α or IL-1β treatment and was found to have a greater amino-acid-sequence homology with the Gro proteins than with IL-8/CXCL8 (Walz *et al.*, 1991). ENA-78/CXCL5 is produced by other cell types than epithelial cells, including mast cells, monocytes, and activated endothelial cells (Imaizumi *et al.*, 1997; Walz *et al.*, 1997; Lukacs *et al.*, 1998). ENA-78/CXCL5 expression is increased in the synovial tissues and fluids of rheumatoid arthritis patients, and this chemokine contributes to neutrophil influx into arthritic synovial tissue (Koch *et al.*, 1994b). ENA-78/CXCL5 is also expressed by epithelial cells of the intestinal mucosa of patients with inflammatory bowel disease (IBS) (Z'Graggen *et al.*, 1997) and in the cerebrospinal fluid (CSF) of patients with bacterial meningitis (Zwijnenburg *et al.*, 2003).

Granulocyte chemotactic protein-2 (GCP-2)/CXCL6 was isolated from-cytokine-stimulated MG63 osteosarcoma cells and is chemotactic for neutrophilic granulocytes; it induces these cells to release matrix metalloproteinase (MMP) gelatinase B/MMP-9 (Proost *et al.*, 1993). Neutrophil accumulation *in vivo* is induced by the intradermal injection of GCP-2/CXCL6 into rabbit skin (Wuyts *et al.*, 1997), and human GCP-2/CXCL6 exhibits angiogenic activity (Strieter *et al.*, 1995). GCP–2/CXCL6 expression *in vitro* by fibroblasts, endothelial cells, chondrocytes, and macrophages is induced by IL-1β and lipopolysaccharide (LPS) treatment (Wuyts *et al.*, 2003). In contrast to ENA-78/CXCL5 and IL-8/CXCL8, GCP-2/CXCL6 is highly expressed by endothelial cells in inflamed intestinal tissue of individuals with IBS, which is characterized by abundant neutrophil infiltration (Gijsbers *et al.*, 2004). GCP-2/CXCL6 is also expressed by macrophages in the intestinal mucosa of people with Crohn's disease (Wuyts *et al.*, 2003).

Neutrophil-activating peptide-2 (NAP-2/CXCL7) is a neutrophil chemoattractant that is formed from a series of sequential proteolytic cleavages of a group of proteins named "β-thromboglobulins" that are predominantly expressed in platelet α-granules (Majumdar *et al.*, 1991). Platelet basic protein (PBP), which is synthesized in megakaryocytes (Holt *et al.*, 1988), is cleaved to form connective-tissue-activating peptide-III (CTAP-III) (Castor *et al.*, 1983), which is then proteolytically truncated to form the active chemokine, NAP-2/CXCL7, a process that occurs in the presence of neutrophils and appears to be catalyzed by a neutrophil enzyme (Walz and Baggiolini, 1989, 1990; Brandt *et al.*, 1991). In the early phase of inflammation, NAP-2/CXCL7 is rapidly produced in the vasculature after platelet degranulation and potently induces effector functions in neutrophils, such as chemotaxis and degranulation (Walz *et al.*, 1989).

IL-8/CXCL8 is secreted by many cell types, including endothelial cells, fibroblasts, epithelial cells, monocytes, and T-lymphocytes (Schroder *et al.*,

1987; Walz *et al.*, 1987; Yoshimura *et al.*, 1987; Schroder and Christophers, 1989; Strieter *et al.*, 1989a,b; Van Damme *et al.*, 1989; Standiford *et al.*, 1990), and was initially characterized as a neutrophil chemoattractant (Schroder *et al.*, 1987; Walz *et al.*, 1987; Yoshimura *et al.*, 1987). It was subsequently found to chemoattract T-lymphocytes and monocytes (Wilkinson and Newman, 1992; Bacon *et al.*, 1995; Xu *et al.*, 1995; Larsen *et al.*, 1989; Bishayi and Samanta, 1996; Taub *et al.*, 1996a; Lei *et al.*, 2002) and to induce the firm adhesion of monocytes to vascular endothelium under flow conditions (Gerszten *et al.*, 1999), as well as to possess angiogenic activity (Koch *et al.*, 1992b; Strieter *et al.*, 1992). IL-8/CXCL8 is also expressed by cells of the CNS, including activated and neoplastic astrocytes (Kasahara *et al.*, 1991; Aloisi *et al.*, 1992; Nitta *et al.*, 1992), as well as microglia (Ehrlich *et al.*, 1998; D'Aversa *et al.*, 2004). This chemokine enhances the survival of hippocampal neurons *in vitro* (Araujo and Cotman, 1993). The murine homologue of IL-8/CXCL8 has not yet been described, but two candidates are KC and macrophage inflammatory protein-2 (MIP-2), members of the murine CXC chemokine family (Huang *et al.*, 1992).

B. Non-ELR-Containing CXC Chemokines

PF4/CXCL4 is found in platelet α-granules along with PBP/CXCL7 (Files *et al.*, 1981) and was purified from human platelets and sequenced (Deuel *et al.*, 1977). PF4/CXCL4 is a poor neutrophil chemoattractant, but in the presence of TNF-α it stimulates neutrophil secondary granule exocytosis and enhances neutrophil adhesion to endothelial cells (Petersen *et al.*, 1996). PF4/CXCL4 inhibits angiogenesis (Maione *et al.*, 1990), which is likely to occur by direct inhibition of endothelial cell proliferation (Luster *et al.*, 1995). PF-4/CXCL4 also protects human monocytes from spontaneous apoptosis and induces differentiation of monocytes into a specific subtype of macrophages (Scheuerer *et al.*, 2000). The chemokine receptor specific to PF4/CXCL4 is unknown, but an integral chondroitin sulfate proteoglycan expressed on the surface of neutrophils is reported to mediate PF4/CXCL4 effects on this cell type (Petersen *et al.*, 1998).

Mig/CXCL9 was isolated from IFN-γ-treated macrophages and was found to have chemoattractant activity *in vitro* for activated T-lymphocytes (Farber, 1990, 1993, 1997; Liao *et al.*, 1995). Mig/CXCL9 is produced by IFN-γ-treated monocytes/macrophages, antigen-presenting cells, B-lymphocytes, and eosinophils (Farber, 1990, 1993; Park *et al.*, 2002; Dajotoy *et al.*, 2004). In humans, Mig/CXCL9 has been associated with T-lymphocyte infiltration in inflammatory diseases of the skin (Goebeler *et al.*, 1998; Flier *et al.*, 1999), joints (Konig *et al.*, 2000), and CNS (Sorensen *et al.*, 1999; Simpson *et al.*, 2000b), as well as in malignancy (Yoong *et al.*, 1999).

IP-10/CXCL10, cloned from IFN-γ-treated U937 cells and originally named IP-10 due to its molecular weight of ~10 kDa (Luster *et al.*, 1985), is chemotactic for activated T-lymphocytes (Taub *et al.*, 1993b; Farber, 1997). Its activity as a chemoattractant for monocytes is less clearly defined. It may be that contradictory data are a result of the need for monocytes to be activated, as studies demonstrate a low level of the IP-10/CXCL10 receptor, CXCR3, on unstimulated cells (Neville *et al.*, 1997). IP-10/CXCL10 induces the migration of specific subsets of stimulated T-lymphocytes (CD4+, CD29+) in *in vitro* chemotaxis assays (Taub *et al.*, 1993b). Data from *in vivo* studies also demonstrate IP10/CXCL10-induced T-lymphocyte migration. Subcutaneous administration of human IP-10/CXCL10 into severe combined immune deficiency (SCID) mice reconstituted with human peripheral blood mononuclear cells induced significant mononuclear cell migration to the injection site; T-lymphocytes comprised a large percentage of these cells (Taub *et al.*, 1996b). A variety of IFN-γ-treated cells express IP10/CXCL10, including mononuclear cells, keratinocytes, fibroblasts, endothelial cells, and T-lymphocytes (Luster *et al.*, 1985). IP-10/CXCL10 message and protein expression can also be induced in cultured astrocytes and microglia (Vanguri, 1995; Majumder *et al.*, 1996; D'Aversa *et al.*, 2004), and IP-10/CXCL10 protein is detected in a subpopulation of astrocytes in normal human brains (Xia *et al.*, 2000). IP-10/CXCL10 expression has been documented in lepromatous leprosy lesions, psoriasis, experimental autoimmune encephalomyelitis (EAE), an animal model of multiple sclerosis (MS), and standard human MS (Kaplan *et al.*, 1987; Gottlieb *et al.*, 1988; Ransohoff *et al.*, 1993; Sorensen *et al.*, 1999).

SDF-1/CXCL12 was originally purified from the supernatants of the murine bone marrow stroma cell line MS-5 and identified as a potent lymphocyte chemoattractant (Bleul *et al.*, 1996b). SDF-1/CXCL12 has unique properties when compared to other CXC chemokine family members. Differential splicing of the transcript from a single gene generates three isoforms: α, β, and γ (Nagasawa *et al.*, 1994; Shirozu *et al.*, 1995; Gleichmann *et al.*, 2000). SDF-1/CXCL12, unlike other CXC chemokines, is constitutively expressed in numerous tissues, including brain, heart, kidney, liver, lung, spleen, and bone marrow (Tashiro *et al.*, 1993; Shirozu *et al.*, 1995; McGrath *et al.*, 1999), by numerous cell types, including dermal fibroblasts (Fedyk *et al.*, 2001), endothelial cells (Imai *et al.*, 1999; Pablos *et al.*, 1999; Abi-Younes *et al.*, 2000), astrocytes, and neurons (Ohtani *et al.*, 1998; Bajetto *et al.*, 1999). Targeted deletion of SDF-1/CXCL12 or its receptor, CXCR4, is lethal in mice and causes profound defects in embryonic lymphopoiesis, vascularization, cardiac organogenesis, and CNS development (Nagasawa *et al.*, 1996; Ma *et al.*, 1998; Tachibana *et al.*, 1998;

Zou *et al.*, 1998). Though it lacks an ELR motif, SDF-1/CXCL12 is chemotactic for endothelial cells (Feil and Augustin, 1998; Gupta *et al.*, 1998; Volin *et al.*, 1998) and exhibits both angiogenic and angiostatic properties (Arenberg *et al.*, 1997; Salcedo *et al.*, 1999). SDF-1/CXCL12 is also the most highly conserved chemokine between species described to date (Shirozu *et al.*, 1995). All of these characteristics suggest that SDF-1/CXCL12 may play (an) important biological role(s) in embryogenesis, organogenesis, and vasculogenesis, in addition to its role as a chemoattractant for monocytes, resting and activated T-lymphocytes, pre-B-lymphocytes, dendritic cells, hematopoietic progenitor cells, endothelial cells, and astrocytes (Nagasawa *et al.*, 1994; Bleul *et al.*, 1996b; Aiuti *et al.*, 1997; Sozzani *et al.*, 1997b; Feil and Augustin, 1998; Gupta *et al.*, 1998; Volin *et al.*, 1998; Odemis *et al.*, 2002). Due to its constitutive expression in many organs, SDF-1/CXCL12 has been postulated to play a role in T-lymphocyte surveillance or baseline trafficking (Bleul *et al.*, 1996b). SDF-1/CXCL12 has also been shown to cause the rapid adhesion of T-lymphocytes to endothelium via ICAM-1 (Campbell *et al.*, 1998) and to induce MMP production by monocytes (Klier *et al.*, 2001), suggesting a possible mechanism by SDF-1/CXCL12 that facilitates monocyte transmigration across blood-vessel endothelium.

C. CC Chemokines

MCP-1/CCL2 was initially purified from the conditioned medium of baboon aortic smooth muscle cells based on its ability to chemoattract monocytes and inability to chemoattract neutrophils (Valente *et al.*, 1988). The human homologue was isolated from mononuclear and gliomal cells (Furutani *et al.*, 1989; Kuratsu *et al.*, 1989; Matsushima *et al.*, 1989; Yoshimura *et al.*, 1989a,b) and was found to be identical to the protein encoded by the JE gene, which was first cloned from PDGF-treated murine 3T3 cells (Cochran *et al.*, 1983; Rollins *et al.*, 1988; Rollins *et al.*, 1989). MCP-1/CCL2 is a potent monocyte chemoattractant, is also chemotactic for activated T-lymphocytes (Loetscher *et al.*, 1994c; Taub *et al.*, 1995a; Uguccioni *et al.*, 1995) and NK cells (Allavena *et al.*, 1994), and induces basophil and mast cell degranulation (Alam *et al.*, 1992b, 1994; Bischoff *et al.*, 1992; Kuna *et al.*, 1992a). MCP-1/CCL2 also plays a role in the differentiation of helper T-lymphocyte type 0 (Th0) cells to Th2 *in vitro* by a mechanism dependent on IL-4 (Karpus *et al.*, 1997). Monocytes, fibroblasts, and endothelial cells are major cellular sources of MCP-1/CCL2 (Strieter *et al.*, 1989c; Rollins *et al.*, 1990; Colotta *et al.*, 1992). Several other cell types synthesize MCP-1/CCL2, including astrocytes and microglia (Ransohoff *et al.*, 1993; Hayashi *et al.*, 1995; Hurwitz *et al.*, 1995; Ishizuka *et al.*, 1997; Peterson *et al.*, 1997;

McManus *et al.*, 1998a; D'Aversa *et al.*, 2002). MCP-1/CCL2 expression has been shown to be higher in many inflammatory diseases, including atherosclerosis (Nelken *et al.*, 1991; Yla-Herttuala *et al.*, 1991), rheumatoid arthritis (Koch *et al.*, 1992a; Villiger *et al.*, 1992), glomerulonephritis (Brown *et al.*, 1996), EAE (Hulkower *et al.*, 1993; Ransohoff *et al.*, 1993; Godiska *et al.*, 1995), MS (McManus *et al.*, 1998a; Simpson *et al.*, 1998; Sorensen *et al.*, 1999; Van Der Voorn *et al.*, 1999), and human immunodeficiency virus (HIV)-associated encephalitis and dementia (Sozzani *et al.*, 1997a; Cinque *et al.*, 1998; Conant *et al.*, 1998; Kelder *et al.*, 1998; Sanders *et al.*, 1998; McManus *et al.*, 2000). Elevated levels of MCP-1/CCL2 in the CNS are associated with monocyte infiltration; furthermore, in a coculture model of the human blood brain barrier, MCP-1/CCL2 was shown to induce monocyte transmigration (Eugenin and Berman, 2003; Weiss *et al.*, 1998). Several cell types within the CNS can be induced by inflammatory mediators and HIV-soluble proteins, as well as by infection with HIV-1, to express MCP-1/CCL2 (Sozzani *et al.*, 1997a; Conant *et al.*, 1998; McManus *et al.*, 2000; D'Aversa *et al.*, 2004).

Additional proteins homologous to MCP-1/CCL2 were isolated and named MCP-2/CCL8, MCP-3/CCL7, and MCP-4/CCL13. Human MCP-2/CCL8 has 62% identity with human MCP-1/CCL2, human MCP-3/CCL7 has 73% identity with human MCP-1/CCL2, and human MCP-4/CCL13 has 62% identity with human MCP-1/CCL2 (Van Damme *et al.*, 1992; Minty *et al.*, 1993; Opdenakker *et al.*, 1993; Garcia-Zepeda *et al.*, 1996a; Uguccioni *et al.*, 1996). In IL-1β-treated fibroblasts and mononuclear leukocytes, MCP-2/CCL8 is induced with MCP-1/CCL2, although at significantly lower levels (Van Damme *et al.*, 1994). After phytohemagglutinin (PHA) and IFN-γ treatment, MCP-3/CCL7 and MCP-1/CCL2 are both induced in mononuclear leukocytes (Minty *et al.*, 1993). IL-1β or TNF-α treatment of endothelial and epithelial cells induces MCP-4/CCL13 (Garcia-Zepeda *et al.*, 1996a). MCP-2/CCL8 is chemotactic for monocytes, T-lymphocytes, basophils, activated natural killer (NK) cells, and eosinophils. MCP-3/CCL7 is chemotactic for monocytes, T-lymphocytes, basophils, activated NK cells, eosinophils, and dendritic cells (Van Damme *et al.*, 1992; Minty *et al.*, 1993; Opdenakker *et al.*, 1993; Allavena *et al.*, 1994; Loetscher *et al.*, 1994c). MCP-2/CCL8 and MCP-3/CCL7 have been found to be expressed in EAE (Godiska *et al.*, 1995) and in MS brain tissue (McManus *et al.*, 1998a; Simpson *et al.*, 2000a). MCP-4/CCL13 is chemotactic for monocytes, eosinophils, and basophils, and it stimulates histamine release from basophils, suggesting that this chemokine plays a role in both Th1 and Th2 immune responses (Garcia-Zepeda *et al.*, 1996a; Uguccioni *et al.*, 1997). MCP-5/CCL12 is expressed only in the mouse and chemoattracts monocytes, eosinophils, and T-lymphocytes (Jia *et al.*, 1996; Sarafi *et al.*, 1997). MCP-5/CCL12 expression is elevated in the mouse lung during allergic inflammation, and neutralizing antibodies reduce eosinophil infiltration in this model (Jia *et al.*, 1996).

MIP-1α/CCL3 and MIP-1β/CCL4 were purified from LPS-treated monocytic cell lines (Davatelis *et al.*, 1988; Sherry *et al.*, 1988; Wolpe *et al.*, 1988). MIP-1α/CCL3 is a more potent monocyte chemoattractant and activator than MIP-1β/CCL4, but both chemokines are less active than MCP-1/CCL2 (Fahey *et al.*, 1992; Roth *et al.*, 1995; Uguccioni *et al.*, 1995). MIP-1α/CCL3 chemoattracts predominantly CD8+ T-lymphocytes, and MIP-1β/CCL4 chemoattracts predominantly CD4+ T-lymphocytes in chemotaxis-chamber assay systems (Schall *et al.*, 1993; Taub *et al.*, 1993a). However, in transendothelial migration assays, both chemokines chemoattract larger numbers of CD4+ T-lymphocytes across an endothelial cell monolayer than they do CD8+ T-lymphocytes; MIP-1α/CCL3 is more potent (Roth *et al.*, 1995). MIP-1α/CCL3 and MIP-1β/CCL4 are also chemoattractants for dendritic cells (Sozzani *et al.*, 1995, 1997b; Xu *et al.*, 1996) and NK cells (Maghazachi *et al.*, 1994; Taub *et al.*, 1995b; Loetscher *et al.*, 1996b), and basophil and eosinophil chemotaxis is induced by MIP-1α/CCL3 (Alam *et al.*, 1992a; Rot *et al.*, 1992; Dahinden *et al.*, 1994). MIP-1α/CCL3 and MIP-1β/CCL4 are expressed constitutively at low levels in most mature hematopoietic cells, in addition to other cell types, and expression can be induced by a variety of factors, including LPS and IL-1. Monocytes/macrophages (Ziegler *et al.*, 1991; Standiford *et al.*, 1993), neutrophils (Kasama *et al.*, 1993; Lapinet *et al.*, 2000), activated T-lymphocytes (Kim *et al.*, 1999), dendritic cells (Sallusto *et al.*, 1999), and fibroblasts (Bug *et al.*, 1998) are a few of the cellular sources of MIP-1α/CCL3 and MIP-1β/CCL4. These two chemokines are expressed at low levels by cultured human microglia, and expression is increased by LPS, TNF-α, or IL-1β treatment (Peterson *et al.*, 1997; McManus *et al.*, 1998b) as well as by treatment with the HIV transactivator protein tat (McManus *et al.*, 2000; D'Aversa *et al.*, 2004). Elevated expression of MIP-1α/CCL3 and MIP-1β/CCL4 is detected in multiple inflammatory diseases, including atherosclerosis (Wilcox *et al.*, 1994), rheumatoid arthritis (Koch *et al.*, 1994a), MS (Miyagishi *et al.*, 1995; Simpson *et al.*, 1998; Balashov *et al.*, 1999), and HIV and SIV encephalitis (Sasseville *et al.*, 1996; Schmidtmayerova *et al.*, 1996).

RANTES/CCL5 is chemotactic for T-lymphocytes of the memory/helper (CD4+/CD45RO+) phenotype (Schall *et al.*, 1990) as well as for monocytes (Schall *et al.*, 1990; Uguccioni *et al.*, 1995). In transendothelial migration assays, RANTES/CCL5 chemoattracts CD4+ and CD8+ T-lymphocytes (Roth *et al.*, 1995). RANTES/CCL5 also chemoattracts and activates NK cells (Taub *et al.*, 1995b; Maghazachi *et al.*, 1996), chemoattracts eosinophils (Rot *et al.*, 1992), and induces histamine release from basophils (Kuna *et al.*, 1992b). RANTES/CCL5 is expressed by activated T-lymphocytes (Schall *et al.*, 1988), cytokine-activated epithelial cells (Stellato *et al.*, 1995), fibroblasts (Rathanaswami *et al.*, 1993), human astrocytoma cells, astrocytes and microglia (Barnes *et al.*, 1996; Hu *et al.*, 1999; Janabi *et al.*, 1999; D'Aversa *et al.*, 2002, 2004), and thrombin-treated platelets (Kameyoshi *et al.*, 1992). The RANTES/CCL5 cDNA was cloned as a result of a T-lymphocyte-versus-B-lymphocyte differential screen and was found to be inducible in T-lymphocytes by mitogen or antigen (Schall *et al.*, 1988). Increased RANTES/CCL5 expression has been associated with a wide range of inflammatory pathologies, including rheumatoid arthritis (Robinson *et al.*, 1995), inflammatory airway disorders such as asthma (Lukacs *et al.*, 1996), glomerulonephritis (Lloyd *et al.*, 1997; Tang *et al.*, 1995), and HIV and SIV encephalitis (Sasseville *et al.*, 1996; Sanders *et al.*, 1998).

Eotaxin/CCL11 chemoattracts eosinophils (Rothenberg *et al.*, 1995, 1996; Garcia-Zepeda *et al.*, 1996b; Kitaura *et al.*, 1996; Ponath *et al.*, 1996b) and IL-2-treated T-lymphocytes (Uguccioni *et al.*, 1996), and it is induced by cytokine treatment of epithelial and endothelial cells (Rothenberg *et al.*, 1995; Garcia-Zepeda *et al.*, 1996b) and by IL-4 treatment of dermal fibroblasts (Mochizuki *et al.*, 1998). Eotaxin/CCL11 was first purified from bronchoalveolar lavage fluid of guinea pigs with eosinophil-rich pulmonary infiltrates (Jose *et al.*, 1994), and murine and human homologues were subsequently identified (Rothenberg *et al.*, 1995; Garcia-Zepeda *et al.*, 1996b; Kitaura *et al.*, 1996; Ponath *et al.*, 1996b). Eotaxin/CCL11 expression is increased in animal models of allergic inflammation (Jose *et al.*, 1994; Collins *et al.*, 1995) and at tissue sites of eosinophil accumulation in humans (Ponath *et al.*, 1996b).

D. C Chemokines

Lymphotactin/XCL1 chemoattracts CD4+ and CD8+ T-lymphocytes, NK cells (Kelner *et al.*, 1994; Kennedy *et al.*, 1995; Muller *et al.*, 1995; Giancarlo *et al.*, 1996; Hedrick *et al.*, 1997; Maghazachi *et al.*, 1997), neutrophils, and B-lymphocytes, but not monocytes (Huang *et al.*, 2001b). Lymphotactin/XCL1 is expressed predominantly by activated CD8+ T-lymphocytes and NK cells (Kelner *et al.*, 1994; Kennedy *et al.*, 1995; Muller *et al.*, 1995; Dorner *et al.*, 1997; Hedrick *et al.*, 1997), and expression of this chemokine in human tissues is mainly restricted to lymphoid tissues, including thymus and spleen (Kelner *et al.*, 1994; Kennedy *et al.*, 1995; Muller *et al.*, 1995; Yoshida *et al.*, 1995). Small intestine, colon, lung, ovary, and testis exhibit low levels of this chemokine (Kennedy *et al.*, 1995; Yoshida *et al.*, 1995). In an inducible brain-targeted tat transgenic mouse model, lymphotactin/XCL1 is upregulated in monocytes, astrocytes, and macrophages/microglia in the brains of tat-expressing mice (Kim *et al.*, 2004). The lymphotactin cDNA was cloned from an activated mouse pro-T-lymphocyte library (Kelner *et al.*, 1994) and from the human CD8+ T-cell clone A10 (Kennedy *et al.*, 1995), while independent investigators used PHA-stimulated human peripheral blood mononuclear cells to characterize a cDNA encoding for the SCM-1 protein, which was identical to human lymphotactin (Yoshida *et al.*, 1995). Two highly homologous SCM-1 genes, SCM-1α and SCM-1β, were subsequently identified, and the SCM-1α gene was identical to the lymphotactin gene. The amino-acid sequences of the coding regions of SCM-1α and SCM-1β differ by only two adjacent amino acids, at positions 28 and 29 (Yoshida *et al.*, 1996). There is little data as to the functional significance of these two chemokines.

E. CX_3C Chemokine

Fractalkine/CX3CL1 (also known as neurotactin) is a transmembrane protein that displays adhesion properties for T-lymphocytes and monocytes (Bazan *et al.*, 1997; Imai *et al.*, 1997; Fong *et al.*, 1998; Haskell *et al.*, 1999). The structural features of cell surface fractalkine/CX3CL1 include an extracellular domain consisting of a 76-amino-acid sequence that comprises the chemokine domain followed by a mucin-like stalk, a transmembrane region of 18 amino acids, and an intracellular region of 37 amino acids. The mucin-like stalk is thought to extend the chemokine domain away from the cell surface to enable the presentation of the CX_3C chemokine domain to leukocytes and to facilitate the adhesion of leukocytes to the membrane-bound chemokine in a selectin- and integrin-independent manner. Proteolytic cleavage of cell surface fractalkine/CX3CL1 is thought to produce a soluble form of this protein that is chemotactic for monocytes, T-lymphocytes (Bazan *et al.*, 1997; Imai *et al.*, 1997), neutrophils (Pan *et al.*, 1997), and microglia (Maciejewski-Lenoir *et al.*, 1999). Soluble fractalkine/CX3CL1 induced significant migration of CD16+ NK cells, CD3+CD8+ T-lymphocytes (both CD45RO+ and CD45RO–), and CD14+ monocytes across endothelial cell monolayers. CD3+CD4+ T-lymphocytes also transmigrated to soluble fractalkine/CX3CL1, but at much lower levels (Imai *et al.*, 1997). Significant neutrophil transmigration was not induced by soluble fractalkine/CX3CL1 (Beck *et al.*, 2003). Both anchored and soluble forms bind the only identified receptor for fractalkine/CX3CL1, CX3R1 (Imai *et al.*, 1997; Combadiere *et al.*, 1998; Dorf *et al.*, 2000). Fractalkine/CX3CL1 expression *in vitro* is detected in neurons, activated primary astrocytes (Harrison *et al.*, 1998; Maciejewski-Lenoir *et al.*, 1999) and endothelial cells (Bazan *et al.*, 1997). Low-level expression *in vivo* is detected in the colon, prostate, heart, lung, adrenal gland, skeletal muscle, and kidney (Bazan *et al.*, 1997; Harrison *et al.*, 1998). The highest levels of fractalkine/CX3CL1 expression *in vivo* are detected in the brain and localized to blood vessels, neurons, and microglia (Pan *et al.*, 1997; Harrison *et al.*, 1998). Fractalkine/CX3CL1 is the only chemokine expressed in the CNS in higher amounts than in the immune system and peripheral tissues. In astrocytes and endothelial cells, low constitutive fractalkine/CX3CL1 expression is upregulated by inflammatory mediators, including TNF-α and IL-1β (Bazan *et al.*, 1997; Maciejewski-Lenoir *et al.*, 1999). Fractalkine/CX3CL1 is markedly upregulated in brain tissue from pediatric patients with HIV encephalitis (Tong *et al.*, 2000) and in the brains of adults with HIV-associated dementia (Pereira *et al.*, 2001).

II. CHEMOKINE RECEPTORS: GENERAL OVERVIEW

Chemokines bind to cell surface proteins that belong to the family of seven-transmembrane domain receptors that signal through heterotrimeric GTP-binding proteins. The genes encoding chemokine receptors are generally clustered on chromosomes 2 and 3 (Rollins, 1997). GTP-binding or G-proteins are composed of three subunits, α, β, and γ. The α subunit family is divided into four major groups, $G_{\alpha i/o}$, $G_{\alpha q}$, $G_{\alpha s}$, and $G_{\alpha 12}$. The $G_{\alpha q}$ group includes $G_{\alpha q}$, $G_{\alpha 11}$, $G_{\alpha 14}$, $G_{\alpha 15}$, and $G_{\alpha 16}$. The $G_{\alpha i/o}$ group includes $G_{\alpha i1}$, $G_{\alpha i2}$, $G_{\alpha i3}$, $G_{\alpha o1}$, and $G_{\alpha o2}$ (Landry and Gies, 2002). Upon ligand binding, chemokine receptors activate G-proteins, particularly the *Bordetella pertussis* toxin (PTX)-sensitive subfamily of G proteins (G_i/G_o). However, other members of the G-protein family are also involved in chemokine receptor signaling, including the PTX-insensitive G_q (α14 and α16). Specifically, CCR2 (the receptor for MCP-1/CCL2) signaling involves $G_{\alpha i}$ and $G_{\alpha 16}$, in a cell-type-specific manner (Myers *et al.*, 1995; Arai and Charo, 1996). Chemokine receptor signaling is also mediated by G-protein $\beta\gamma$ subunits (Wu *et al.*, 1993; Kuang *et al.*, 1996).

In general, chemokine receptors range from 340 to 370 amino acids in length, with a relatively short extracellular acidic N-terminus as well as numerous serines and threonines in the short intracellular C-terminal tail that become phosphorylated upon ligand binding to the receptor. The seven-transmembrane domains are α-helical, and three intracellular and three extracellular loops exist between the transmembrane domains (Rollins, 1997). Some chemokine receptors form homodimers that may be involved in receptor signaling (Rodriguez-Frade *et al.*, 1999; Vila-Coro *et al.*, 1999). Chemokine binding to the appropriate receptor initiates signaling cascades that are involved in cytoskeletal reorganization, integrin activation, and other biochemical processes that result in increased cellular adhesion, migration, and activation. The majority of chemokine receptors are promiscuous in that they bind to more than one chemokine. In addition, many chemokines bind to more than one receptor. However, there are exclusive chemokine/receptor pairs, including SDF-1(CXCL12)/CXCR4, MIP-3α(CCL20)/CCR6, lymphotactin(XCL1)/XCR1 and fractalkine(CX3CL1)/CX3CR1 (Barker and Monk, 1997). Examples of receptors with multiple chemokine specificities include CXCR3, which binds IP-10/CXCL10 and MIG/CXCL9; CCR3, which binds eotaxin/CCL11 and RANTES/CCL5; and CCR5, which binds MIP-1α/CCL3, MIP-1β/CCL4, and RANTES/CCL5 (Premack and Schall, 1996).

A. CXC Chemokine Receptors

CXCR1 and CXCR2, which are 77% identical at the amino acid level, were both identified as receptors for IL-8/CXCL8 and were originally named IL8RA and IL8RB, respectively (Holmes *et al.*, 1991; Murphy and Tiffany, 1991). All of the other ELR-positive CXC chemokines are functional ligands of CXCR2. Groα/CXCL1, Groβ/CXCL2, Groγ/CXCL3, ENA-78/CXCL5, and NAP-2/CXCL7 bind with high affinity to CXCR2 (Moser *et al.*, 1991; LaRosa *et al.*, 1992; Lee *et al.*, 1992; Cerretti *et al.*, 1993; Geiser *et al.*, 1993; Ahuja and Murphy, 1996); IL–8/CXCL8 and GCP–2/CXCL6 bind to both CXCR1 and CXCR2 (Wuyts *et al.*, 1997; Wolf *et al.*, 1998). Groα/CXCL1, ENA-78/CXCL5, and NAP-2/CXCL7 also bind to CXCR1, but with low affinity (Loetscher *et al.*, 1994b; Ben-Baruch *et al.*, 1997; Wuyts *et al.*, 1998). CXCR1 and/or CXCR2 receptors are expressed on the surface of neutrophils, monocytes, mast cells, NK cells, and T-lymphocytes (Moser *et al.*, 1993; Chuntharapai *et al.*, 1994; Xu *et al.*, 1995; Bishayi and Samanta, 1996; Qin *et al.*, 1996; Lippert *et al.*, 1998; Tani *et al.*, 1998; Inngjerdingen *et al.*, 2001; Takata *et al.*, 2004). CXCR2 is also expressed by subsets of projection neurons in multiple regions of the brain and spinal cord (Horuk *et al.*, 1997) and by astrocytes and microglia (Lacy *et al.*, 1995; Muller-Ladner *et al.*, 1996).

CXCR3, the receptor for MIG/CXCL9, IP-10/CXCL10, and I-TAC/CXCL11 (Loetscher *et al.*, 1996a; Cole *et al.*, 1998), is highly expressed on memory and activated CD4+ and CD8+ T-lymphocytes, and on a small percentage of peripheral blood B-lymphocytes and NK cells (Loetscher *et al.*, 1996a; Qin *et al.*, 1998; Rabin *et al.*, 1999). *In vitro* studies indicate that CXCR3 is predominantly expressed by Th1 cells (Sallusto *et al.*, 1998). CXCR3 is also expressed in a subpopulation of neurons in various cortical and subcortical regions of the brain (Coughlan *et al.*, 2000; Xia *et al.*, 2000) and in astrocytes and microglia (Biber *et al.*, 2002).

CXCR4 was initially cloned by several laboratories (Federsppiel *et al.*, 1993; Herzog *et al.*, 1993; Jazin *et al.*, 1993; Nomura *et al.*, 1993) and identified as the orphan receptor LESTR by one group of investigators (Loetscher *et al.*, 1994a). This receptor was later identified as fusin, the co-receptor (with CD4) for syncytium-inducing laboratory strains of HIV-1 (Feng *et al.*, 1996). In subsequent studies, LESTR/fusin was characterized as the SDF-1/CXCL12 receptor and was renamed CXCR4 (Bleul *et al.*, 1996a; Oberlin *et al.*, 1996). CXCR4 is constitutively expressed by a wide variety of cell types, including naïve T-lymphocytes (Bonecchi *et al.*, 1998), endothelial cells (Feil and Augustin, 1998; Gupta *et al.*, 1998; Volin *et al.*, 1998; Molino *et al.*, 2000), microglia, astrocytes, and neurons (Lavi *et al.*, 1997; Tanabe *et al.*, 1997; Ohtani *et al.*, 1998; Vallat *et al.*, 1998; Westmoreland *et al.*, 1998; Albright *et al.*, 1999; Bajetto *et al.*, 1999; McManus *et al.*, 2000). The expression of this chemokine receptor is upregulated by vascular endothelial growth factor (VEGF), basic fibroblast growth factor (bFGF), TNF-α, and IL-1β, and it is downregulated by IFN-γ (Feil and Augustin, 1998; Gupta *et al.*, 1998; Salcedo *et al.*, 1999; Oh *et al.*, 2001).

B. CC Chemokine Receptors

CCR1, the receptor for MIP-1α/CCL3, RANTES/CCL5, and MCP-3/CCL7 (Gao *et al.*, 1993; Neote *et al.*, 1993; Combadiere *et al.*, 1995c), is expressed on peripheral blood lymphocytes and monocytes, but not on neutrophils. A majority of CD3+, CD4+, CD8+, and CD16+ peripheral blood lymphocytes, but not CD19+ lymphocytes, express CCR1. Among CD4+ peripheral blood lymphocytes, CD45RO+ cells express more CCR1 receptor compared with CD45RO– cells (Su *et al.*, 1996). CCR1 is also expressed by primary hippocampal neurons and cultured astrocytes (Meucci *et al.*, 1998; Dorf *et al.*, 2000).

CCR2 is expressed in two forms, CCR2A and CCR2B, as a result of alternative splicing (Charo *et al.*, 1994). These receptors differ only in their C-terminal intracellular tails. Both bind MCP-1/CCL2, MCP-3/CCL7, and MCP-4/CCL13, but not MCP-2/CCL8 (Ben-Baruch *et al.*, 1995; Combadiere *et al.*, 1995c; Franci *et al.*, 1995). CCR2 is expressed on monocytes, myeloid precursor cells, activated T-lymphocytes (Myers *et al.*, 1995; Qin *et al.*, 1996), basophils (Uguccioni *et al.*, 1997), and mast cells (Oliveira and Lukacs, 2001).

CCR3 was cloned from eosinophils on the basis of its enhanced expression on this cell type (Daugherty *et al.*, 1996; Ponath *et al.*, 1996a) and was also cloned from monocytes (Combadiere *et al.*, 1995a). This receptor binds eotaxin/CCL11, RANTES/CCL5, MCP-2/CCL8, MCP-3/CCL7, and MCP-4/CCL13 (Daugherty *et al.*, 1996; Uguccioni *et al.*, 1996), and it is expressed on basophils and eosinophils (Uguccioni *et al.*, 1997), Th2 cells, (Sallusto *et al.*, 1997) and microglia (Albright *et al.*, 1999).

CCR5 was cloned based on its homology to other CC chemokine receptors, and it binds MIP-1α/CCL3, MIP-1β/CCL4, and RANTES/CCL5 (Combadiere *et al.*, 1996; Raport *et al.*, 1996; Samson *et al.*, 1996a). CCR5 is expressed by monocytes (Combadiere *et al.*, 1996), IL-2-stimulated T-cells, a subset of effector/memory T-cells in blood, tissue macrophages (Wu *et al.*, 1997), and microglia (Albright *et al.*, 1999). CCR5 is a co-receptor (with CD4) for M-tropic strains of HIV-1 (Alkhatib *et al.*, 1996; Choe *et al.*, 1996; Dragic *et al.*, 1996).

C. The C Chemokine Receptor

The orphan receptor GPR5 (Heiber *et al.*, 1995) was identified as the lymphotactin/XCL1 receptor, XCR1 (Yoshida *et al.*, 1998). This receptor is expressed by CD8+ T-lymphocytes, NK cells (Yoshida *et al.*, 1999), neutrophils, and B-lymphocytes (Huang *et al.*, 2001b). In individuals with rheumatoid arthritis, significant XCR1 expression by infiltrating mononuclear cells and synoviocytes in synovial tissues is detected, and the XCR1 ligand, lymphotactin/XCL1, is also increased in arthritic joint fluid. Thus, XCR1 positive cells may play a role in the pathogenesis of arthritis (Blaschke *et al.*, 2003; Wang *et al.*, 2004).

D. The CX_3C Chemokine Receptor

The fractalkine/CX3CL1 receptor, CX3CR1, was first identified as the RBS11 orphan receptor in the rat (Harrison *et al.*, 1994), and the cDNA for the human homologue was subsequently cloned and named V28 (Combadiere *et al.*, 1995b; Raport *et al.*, 1995). In humans, CX3CR1 is expressed by a majority of CD16+ NK cells and CD14+ monocytes, as well as by a subpopulation of CD3+ T-lymphocytes that are mostly CD8+ (although some CD4+CD8+ T-lymphocytes also express this receptor) (Imai *et al.*, 1997). CX3CR1 is also expressed at high levels by microglia, which migrate in response to fractalkine/CX3CL1. This receptor is also expressed by astrocytes, but at much lower levels than by microglia, and astrocyte chemotaxis to fractalkine/CX3CL1 has not been detected (Harrison *et al.*, 1998; Maciejewski-Lenoir *et al.*, 1999). Functional expression of CX3CR1 by primary hippocampal neurons has also been reported, and fractalkine/CX3CL1 protects these cells from HIV-gp120 neurotoxicity (Meucci *et al.*, 2000).

III. CHEMOKINES AND CHEMOKINE RECEPTORS IN INFLAMMATION *IN VIVO*

Chemokines regulate the accumulation and activation of leukocytes at sites of inflammation. Alterations in chemokine expression are associated with the pathogenesis of a variety of acute and chronic inflammatory diseases. Methods of chemokine expression in a few representative inflammatory pathologies are detailed below.

MS is a putative autoimmune, inflammatory demyelinating disease characterized by focal destruction of myelin and by axonal loss in lesions within the CNS. The pathogenic mechanisms that contribute to MS include leukocyte chemotaxis into the CNS and production of inflammatory mediators, which result in augmented inflammation, oligodendrocyte damage, demyelination, and neuronal injury. Thus, factors that regulate leukocyte entry may contribute to early events in MS as well as to later stages of lesion pathogenesis. Increased expression of chemokines, including MCP-1/CCL2, RANTES/CCL5, MIP-1α/CCL3, MIP-1β/CCL4, and IP-10/CXCL10, by infiltrating leukocytes and resident CNS cells, has been documented in MS and in EAE (Hulkower *et al.*, 1993; Ransohoff *et al.*, 1993; Godiska *et al.*, 1995; McManus *et al.*, 1998a; Simpson *et al.*, 1998; Balashov *et al.*, 1999; Sorensen *et al.*, 1999; Van Der Voorn *et al.*, 1999). MIP-1α/CCL3 has been shown to play a key role in the initiation of EAE and MCP-1/CCL2 has been shown to play a role in relapsing EAE (Karpus and Kennedy, 1997). The administration of neutralizing polyclonal antibody for MIP-1α suppresses the severity of clinical EAE through partial inhibition of inflammatory mononuclear cell recruitment, including Th1 T-cells, into the CNS (Karpus *et al.*, 1995). Mice genetically deficient for MCP-1/CCL2 or its receptor CCR2 are resistant to EAE development due to impaired recruitment of monocytes into the CNS (Izikson *et al.*, 2000; Huang *et al.*, 2001a). MCP-1/CCL2 expression has also been demonstrated in human tissue from other inflammatory diseases, including atherosclerosis (Nelken *et al.*, 1991; Yla-Herttuala *et al.*, 1991), rheumatoid arthritis (Koch *et al.*, 1992a; Villiger *et al.*, 1992), and glomerulonephritis (Brown *et al.*, 1996). In transgenic mouse models, organ-specific expression of MCP-1/CCL2 results in a directed mononuclear cell infiltration into thymus, brain, pancreatic islets, and lung. In these transgenic mice, monocytes are not activated and there is no tissue damage. However, inflammatory responses to subsequent inflammatory stimuli are enhanced (Fuentes *et al.*, 1995; Rutledge *et al.*, 1995; Grewal *et al.*, 1997). IP-10/CXCL10 has also been implicated in the pathogenesis of EAE and MS, but conflicting results are obtained when IP10/CXCL10 is inhibited in the EAE model. Intrathecal administration of antisense oligonucleotides against IP-10/CXCL10 mRNA reduces the clinical severity of

EAE (Wojcik *et al.*, 1996), while administration of neutralizing monoclonal antibody to IP10/CXCL10 exacerbates clinical disease. Of interest is that enhanced expression of IL-12 has been shown in patients with progressive MS (Balashov *et al.*, 1997). IL-12 stimulates IFN-γ production, and IFN-γ is a potent inducer of IP-10/CXCL10 (Luster *et al.*, 1985).

It has also been postulated that chemokines play an important role in the pathogenesis of atherosclerosis, a chronic inflammatory disease of arterial blood vessels (Reape and Groot, 1999). The accumulation of leukocytes, smooth muscle cells, and lipids within the vessel wall promotes atherosclerotic plaque formation (Ross, 1993), and chemokine expression contributes to the directed migration of monocytes and T-lymphocytes to lesioned areas within the vasculature. Both MCP-1/CCL2 expression and IL-8/CXCL8 expression are increased in human atherosclerotic plaques (Nelken *et al.*, 1991; Yla-Herttuala *et al.*, 1991; Takeya *et al.*, 1993; Apostolopoulos *et al.*, 1996; Wang *et al.*, 1996), and genetically modified mouse strains have provided compelling evidence that MCP-1/CCL2 and its receptor CCR2 (Boring *et al.*, 1998; Gu *et al.*, 1998), as well as the murine IL-8 receptor (CXCR2) homologue mIL–8RH (Boisvert *et al.*, 1998), are directly involved in monocyte recruitment to atherosclerotic lesions. In addition to their effects on monocytes, IL-8/CXCL8 and MCP-1/CCL2 may also impact the contribution of smooth muscle cells to plaque development. IL-8/CXCL8 has been shown to induce proliferation and migration of vascular smooth muscle cells (Yue *et al.*, 1994). In addition, MCP-1/CCL2 has been shown to cnhance vascular smooth muscle cell proliferation induced by serotonin (Watanabe *et al.*, 2001), a vasoconstrictor released by aggregated platelets and expressed at high levels in the coronary sinus of patients with coronary artery disease (Rubanyl *et al.*, 1987). MCP-1/CCL2 also induces smooth muscle cell production of tissue factor, the initiator of coagulation, which may contribute to thrombus formation and plaque instability (Schecter *et al.*, 1997). SDF-1/CXCL12, an inducer of platelet aggregation and activation, is also highly expressed in human atherosclerotic plaques by smooth muscle cells, endothelial cells, and macrophages, and it is postulated to contribute to thrombus formation during plaque rupture and vessel occlusion (Abi-Younes *et al.*, 2000). In some human atherosclerotic lesions, fractalkine/CX3CL1 is detected in areas that contain significant numbers of macrophages (Greaves *et al.*, 2001). Another immunohistochemical study detected fractalkine/CX3CL1 in atherosclerotic, but not normal, coronary arteries (Wong *et al.*, 2002). In an epidemiological study, a genetic variation in the fractalkine/CX3CL1 receptor, CX3CR1, is associated with a lower incidence of coronary artery disease and a decreased number of fractalkine/CX3CL1 binding sites on monocytes (McDermott *et al.*, 2001, 2003; Moatti *et al.*, 2001). These data suggest that fractalkine/CX3CL1 expression may be involved in atherosclerotic plaque development. This hypothesis is supported by studies of CX3CL1$^{-/-}$ mice crossed with ApoE$^{-/-}$ mice, in which atherosclerotic lesion development was significantly further decreased in double knockout mice than in ApoE$^{-/-}$ mice (Combadiere *et al.*, 2003; Lesnik *et al.*, 2003).

Chemokines and their receptors are also involved in HIV pathophysiology. CCR3, CCR5, and CXCR4 are co-receptors for HIV-1 infection of CD4+ cells. Primary T-lymphocyte-tropic (T-tropic) HIV-1 viruses use the chemokine receptor CXCR4 as a co-receptor, whereas macrophage-tropic (M-tropic) viruses use CCR5 (Alkhatib *et al.*, 1996; Choe *et al.*, 1996; Deng *et al.*, 1996; Doranz *et al.*, 1996; Dragic *et al.*, 1996; Feng *et al.*, 1996). RANTES/CCL5, MIP-1α/CCL3, and MIP-1β/CCL4 are suppressive for HIV-1 infection of monocytes/macrophages, as they bind CCR5 (Cocchi *et al.*, 1995). In addition, individuals with a specific deletion mutation of CCR5 (CCR5 Δ32) are resistant to HIV infection (Dean *et al.*, 1996; Huang *et al.*, 1996; Liu *et al.*, 1996; Samson *et al.*, 1996b; Zimmerman *et al.*, 1997). The brain is a major tissue site for HIV infection, which leads to the development of HIV encephalitis and HIV-associated dementia in a significant percentage of HIV-infected individuals (Navia *et al.*, 1986; Wiley *et al.*, 1991; Gendelman *et al.*, 1994; Rappaport *et al.*, 1999; Gartner, 2000). CXCR4 and CCR5 are expressed on subpopulations of large hippocampal and neocortical pyramidal neurons, fetal neurons, microglia, and astrocytes in both normal and encephalitic brain tissue (He *et al.*, 1997; Lavi *et al.*, 1997; Tanabe *et al.*, 1997; Ohtani *et al.*, 1998; Vallat *et al.*, 1998; Westmoreland *et al.*, 1998; Albright *et al.*, 1999; Bajetto *et al.*, 1999; Klein *et al.*, 1999; Dorf *et al.*, 2000). However, the major cell type in the CNS infected by HIV is the microglial cell. Although astrocytes and neurons are not infected by HIV, they can respond to the chemokines produced in the CNS after HIV infection, and these interactions contribute to the cell activation, death, and/or dysfunction that is associated with HIV pathology in the brain. Elevated levels of IP-10/CXCL10 have been detected in the CSF of HIV-1-infected individuals, and the highest levels are detected in those with neurologic impairment (Kolb *et al.*, 1999). RANTES/CCL5 expression has been demonstrated in tissue from HIV encephalitic brains as well as in the CSF of HIV-infected individuals (Letendre *et al.*, 1999; Sanders *et al.*, 1998). IP-10/CXCL10 and RANTES/CCL5 have also been detected in SIV encephalitic brain tissue (Sasseville *et al.*, 1996). MCP-1/CCL2 expression is elevated in HIV-1 encephalitic brain tissue (Conant *et al.*, 1998; Sanders *et al.*, 1998; McManus *et al.*, 2000), and high levels of MCP-1/CCL2 have been reported in the CSF of individuals with AIDS dementia (Sozzani *et al.*, 1997a; Cinque *et al.*, 1998; Conant *et al.*, 1998; Kelder *et al.*, 1998). In SIV-infected macaque monkeys that developed encephalitis, elevated levels of MCP-1/CCL2 in the

CSF preceded the development of brain lesions (Zink *et al.*, 2001). Elevated expression of MIP-1α/CCL3 and MIP-1β/CCL4 is also detected in the brains of patients with HIV encephalitis (Schmidtmayerova *et al.*, 1996) and macaques with SIV encephalitis (Sasseville *et al.*, 1996). Fractalkine/CX3CL1 is markedly upregulated in neurons in brain tissue from pediatric patients with HIV encephalitis (Tong *et al.*, 2000). Fractalkine/CX3CL1 expression is also upregulated in the brains of adult AIDS patients with HIV-associated dementia; expression is mainly localized to astrocytes (Pereira *et al.*, 2001).

In addition to its role in inflammation, chemokine expression in the CNS is also postulated to regulate the migration of multipotent progenitor cells during development. In the case of mice genetically engineered to be deficient in SDF-1/CXCL12 or CXCR4, fetal cerebellar development shows marked changes, characterized by abnormal inward migration of external granule cells (Ma *et al.*, 1998; Tachibana *et al.*, 1998; Zou *et al.*, 1998). In addition, IL-8/CXCL8 and MCP-1/CCL2 are expressed transiently during CNS development, where they are thought to control migration of microglial precursors (Male and Rezaie, 2001).

IV. AN EXPANDING ROLE FOR CHEMOKINES

Chemokine and chemokine receptor expression by nonhematopoietic cells suggests that the function of this family of chemotactic proteins extends well beyond the realm of leukocyte and immune physiology. Chemokines have been shown to effect functions besides chemoattraction. In the CNS, chemokines act upon resident cells to induce cytokine and chemokine expression and to promote neuroprotection. MIP-1α/CCL3, MIP-1β/CCL4, and RANTES/CCL5 induce MCP-1/CCL2 expression by astrocytes (McManus *et al.*, 2000). MCP-1/CCL2 protects neurons and astrocytes from tat-induced apoptosis (Eugenin *et al.*, 2003). Chemokines also have been shown to induce tissue factor production by smooth muscle cells (Schecter *et al.*, 1997) and to induce cytokine production and MMPs (Locati *et al.*, 2002; Viedt *et al.*, 2002; Li *et al.*, 2003) by a variety of cell types. Thus, chemokines serve many significant biological functions.

References

Abi-Younes, S., Sauty, A., Mach, F., Sukhova, G. K., Libby, P., and Luster, A. D. (2000). The stromal cell-derived factor-1 chemokine is a potent platelet agonist highly expressed in atherosclerotic plaques. *Circ. Res.* **86,** 131–138.

Ahuja, S. K., and Murphy, P. M. (1996). The CXC chemokines growth-regulated oncogene (GRO) alpha, GRObeta, GROgamma, neutrophil-activating peptide-2, and epithelial cell-derived neutrophil-activating peptide-78 are potent agonists for the type B, but not the type A, human interleukin-8 receptor. *J. Biol. Chem.* **271,** 20545–20550.

Aiuti, A., Webb, I. J., Bleul, C., Springer, T., and Gutierrez-Ramos, J. C. (1997). The chemokine SDF-1 is a chemoattractant for human CD34+ hematopoietic progenitor cells and provides a new mechanism to explain the mobilization of CD34+ progenitors to peripheral blood. *J. Exp. Med.* **185,** 111–120.

Alam, R., Forsythe, P. A., Stafford, S., Lett-Brown, M. A., and Grant, J. A. (1992a). Macrophage inflammatory protein-1 alpha activates basophils and mast cells. *J. Exp. Med.* **176,** 781–786.

Alam, R., Lett-Brown, M. A., Forsythe, P. A., Anderson-Walters, D. J., Kenamore, C., Kormos, C., and Grant, J. A. (1992b). Monocyte chemotactic and activating factor is a potent histamine-releasing factor for basophils. *J. Clin. Invest.* **89,** 723–728.

Alam, R., Kumar, D., Anderson-Walters, D., and Forsythe, P. A. (1994). Macrophage inflammatory protein-1 alpha and monocyte chemoattractant peptide-1 elicit immediate and late cutaneous reactions and activate murine mast cells *in vivo*. *J. Immunol.* **152,** 1298–1303.

Albright, A. V., Shieh, J. T., Itoh, T., Lee, B., Pleasure, D., O'connor, M. J., Doms, R. W., and Gonzalez-Scarano, F. (1999). Microglia express CCR5, CXCR4, and CCR3, but of these, CCR5 is the principal coreceptor for human immunodeficiency virus type 1 dementia isolates. *J. Virol.* **73,** 205–213.

Alkhatib, G., Combadiere, C., Broder, C. C., Feng, Y., Kennedy, P. E., Murphy, P. M., and Berger, E. A. (1996). CC CKR5: A RANTES, MIP-1alpha, MIP-1beta receptor as a fusion cofactor for macrophage-tropic HIV-1. *Science* **272,** 1955–1958.

Allavena, P., Bianchi, G., Zhou, D., Van Damme, J., Jilek, P., Sozzani, S., and Mantovani, A. (1994). Induction of natural killer cell migration by monocyte chemotactic protein-1, -2 and -3. *Eur. J. Immunol.* **24,** 3233–3236.

Aloisi, F., Care, A., Borsellino, G., Gallo, P., Rosa, S., Bassani, A., Cabibbo, A., Testa, U., Levi, G., and Peschle, C. (1992). Production of hemolymphopoietic cytokines (IL-6, IL-8, colony-stimulating factors) by normal human astrocytes in response to IL-1 beta and tumor necrosis factor-alpha. *J. Immunol.* **149,** 2358–2366.

Anisowicz, A., Bardwell, L., and Sager, R. (1987). Constitutive overexpression of a growth-regulated gene in transformed Chinese hamster and human cells. *Proc. Natl. Acad. Sci. USA* **84,** 7188–7192.

Apostolopoulos, J., Davenport, P., and Tipping, P. G. (1996). Interleukin-8 production by macrophages from atheromatous plaques. *Arterioscler. Thromb. Vasc. Biol.* **16,** 1007–1012.

Arai, H., and Charo, I. F. (1996). Differential regulation of G-protein-mediated signaling by chemokine receptors. *J. Biol. Chem.* **271,** 21814–21819.

Araujo, D. M., and Cotman, C. W. (1993). Trophic effects of interleukin-4, -7 and -8 on hippocampal neuronal cultures: Potential involvement of glial-derived factors. *Brain Res.* **600,** 49–55.

Arenberg, D. A., Polverini, P. J., Kunkel, S. L., Shanafelt, A., Hesselgesser, J., Horuk, R., and Strieter, R. M. (1997). The role of CXC chemokines in the regulation of angiogenesis in non-small cell lung cancer. *J. Leukoc. Biol.* **62,** 554–562.

Bacon, K. B., Flores-Romo, L., Life, P. F., Taub, D. D., Premack, B. A., Arkinstall, S. J., Wells, T. N., Schall, T. J., and Power, C. A. (1995). IL-8-induced signal transduction in T lymphocytes involves receptor-mediated activation of phospholipases C and D. *J. Immunol.* **154,** 3654–3666.

Bajetto, A., Bonavia, R., Barbero, S., Piccioli, P., Costa, A., Florio, T., and Schettini, G. (1999). Glial and neuronal cells express functional chemokine receptor CXCR4 and its natural ligand stromal cell-derived factor 1. *J. Neurochem.* **73,** 2348–2357.

Bajetto, A., Bonavia, R., Barbero, S., Florio, T., and Schettini, G. (2001). Chemokines and their receptors in the central nervous system. *Front. Neuroendocrinol.* **22,** 147–184.

Balashov, K. E., Smith, D. R., Khoury, S. J., Hafler, D. A., and Weiner, H. L. (1997). Increased interleukin 12 production in progressive multiple sclerosis: induction by activated CD4+ T cells via CD40 ligand. *Proc. Natl. Acad. Sci. USA* **94,** 599–603.

Balashov, K. E., Rottman, J. B., Weiner, H. L., and Hancock, W. W. (1999). CCR5(+) and CXCR3(+) T cells are increased in multiple sclerosis and their ligands MIP-1alpha and IP-10 are expressed in demyelinating brain lesions. *Proc. Natl. Acad. Sci. USA* **96,** 6873–6878.

Balentien, E., Han, J. H., Thomas, H. G., Wen, D. Z., Samantha, A. K., Zachariae, C. O., Griffin, P. R., Brachmann, R., Wong, W. L., and Matsushima, K. (1990). Recombinant expression, biochemical characterization, and biological activities of the human MGSA/gro protein. *Biochemistry* **29,** 10225–10233.

Barker, M. D., and Monk, P. N. (1997). Structure-function relationships of leucocyte chemoattractant receptors. *Biochem. Soc. Trans.* **25,** 1027–1031.

Barnes, D. A., Huston, M., Holmes, R., Benveniste, E. N., Yong, V. W., Scholz, P., and Perez, H. D. (1996). Induction of RANTES expression by astrocytes and astrocytoma cell lines. *J. Neuroimmunol.* **71,** 207–214.

Bazan, J. F., Bacon, K. B., Hardiman, G., Wang, W., Soo, K., Rossi, D., Greaves, D. R., Zlotnik, A., and Schall, T. J. (1997). A new class of membrane-bound chemokine with a CX3C motif. *Nature* **385,** 640–644.

Beck, G., Ludwig, F., Schulte, J., Van Ackern, K., Van Der Woude, F. J., and Yard, B. A. (2003). Fractalkine is not a major chemoattractant for the migration of neutrophils across microvascular endothelium. *Scand. J. Immunol.* **58,** 180–187.

Becker, S., Quay, J., Koren, H. S., and Haskill, J. S. (1994). Constitutive and stimulated MCP-1, GRO alpha, beta, and gamma expression in human airway epithelium and bronchoalveolar macrophages. *Am. J. Physiol.* **266,** L278–L286.

Ben-Baruch, A., Xu, L., Young, P. R., Bengali, K., Oppenheim, J. J., and Wang, J. M. (1995). Monocyte chemotactic protein-3 (MCP3) interacts with multiple leukocyte receptors. C-C CKR1, a receptor for macrophage inflammatory protein-1 alpha/Rantes, is also a functional receptor for MCP3. *J. Biol. Chem.* **270,** 22123–22128.

Ben-Baruch, A., Bengali, K., Tani, K., Xu, L., Oppenheim, J. J., and Wang, J. M. (1997). IL-8 and NAP-2 differ in their capacities to bind and chemoattract 293 cells transfected with either IL-8 receptor type A or type B. *Cytokine* **9,** 37–45.

Biber, K., Dijkstra, I., Trebst, C., De Groot, C. J., Ransohoff, R. M., and Boddeke, H. W. (2002). Functional expression of CXCR3 in cultured mouse and human astrocytes and microglia. *Neuroscience* **112,** 487–497.

Bischoff, S. C., Krieger, M., Brunner, T., and Dahinden, C. A. (1992). Monocyte chemotactic protein 1 is a potent activator of human basophils. *J. Exp. Med.* **175,** 1271–1275.

Bishayi, B., and Samanta, A. K. (1996). Identification and characterization of specific receptor for interleukin-8 from the surface of human monocytes. *Scand. J. Immunol.* **43,** 531–536.

Blaschke, S., Middel, P., Dorner, B. G., Blaschke, V., Hummel, K. M., Kroczek, R. A., Reich, K., Benoehr, P., Koziolek, M., and Muller, G. A. (2003). Expression of activation-induced, T cell-derived, and chemokine-related cytokine/lymphotactin and its functional role in rheumatoid arthritis. *Arthritis Rheum.* **48,** 1858–1872.

Bleul, C. C., Farzan, M., Choe, H., Parolin, C., Clark-Lewis, I., Sodroski, J., and Springer, T. A. (1996a). The lymphocyte chemoattractant SDF-1 is a ligand for LESTR/fusin and blocks HIV-1 entry. *Nature* **382,** 829–833.

Bleul, C. C., Fuhlbrigge, R. C., Casasnovas, J. M., Aiuti, A., and Springer, T. A. (1996b). A highly efficacious lymphocyte chemoattractant, stromal cell-derived factor 1 (SDF-1). *J. Exp. Med.* **184,** 1101–1109.

Boisvert, W. A., Santiago, R., Curtiss, L. K., and Terkeltaub, R. A. (1998). A leukocyte homologue of the IL-8 receptor CXCR-2 mediates the accumulation of macrophages in atherosclerotic lesions of LDL receptor-deficient mice. *J. Clin. Invest.* **101,** 353–363.

Bonecchi, R., Bianchi, G., Bordignon, P. P., D'ambrosio, D., Lang, R., Borsatti, A., Sozzani, S., Allavena, P., Gray, P. A., Mantovani, A., and Sinigaglia, F. (1998). Differential expression of chemokine receptors and chemotactic responsiveness of type 1 T helper cells (Th1s) and Th2s. *J. Exp. Med.* **187,** 129–134.

Boring, L., Gosling, J., Cleary, M., and Charo, I. F. (1998). Decreased lesion formation in CCR2-/- mice reveals a role for chemokines in the initiation of atherosclerosis. *Nature* **394,** 894–897.

Brandt, E., Van Damme, J., and Flad, H. D. (1991). Neutrophils can generate their activator neutrophil-activating peptide 2 by proteolytic cleavage of platelet-derived connective tissue-activating peptide III. *Cytokine* **3,** 311–321.

Brown, Z., Robson, R. L., and Westwick, J. (1996). Regulation and expression of chemokines: Potential role in glomerulonephritis. *J. Leukoc. Biol.* **59,** 75–80.

Bug, G., Aman, M. J., Tretter, T., Huber, C., and Peschel, C. (1998). Induction of macrophage-inflammatory protein 1alpha (MIP-1alpha) by interferon-alpha. *Exp. Hematol.* **26,** 117–123.

Campbell, J. J., Hedrick, J., Zlotnik, A., Siani, M. A., Thompson, D. A., and Butcher, E. C. (1998). Chemokines and the arrest of lymphocytes rolling under flow conditions. *Science* **279,** 381–384.

Castor, C. W., Miller, J. W., and Walz, D. A. (1983). Structural and biological characteristics of connective tissue activating peptide (CTAP-III), a major human platelet-derived growth factor. *Proc. Natl. Acad. Sci. USA* **80,** 765–769.

Cerretti, D. P., Kozlosky, C. J., Vanden Bos, T., Nelson, N., Gearing, D. P., and Beckmann, M. P. (1993). Molecular characterization of receptors for human interleukin-8, GRO/melanoma growth-stimulatory activity and neutrophil activating peptide-2. *Mol. Immunol.* **30,** 359–367.

Charo, I. F., Myers, S. J., Herman, A., Franci, C., Connolly, A. J., and Coughlin, S. R. (1994). Molecular cloning and functional expression of two monocyte chemoattractant protein 1 receptors reveals alternative splicing of the carboxyl-terminal tails. *Proc. Natl. Acad. Sci. USA* **91,** 2752–2756.

Choe, H., Farzan, M., Sun, Y., Sullivan, N., Rollins, B., Ponath, P. D., Wu, L., Mackay, C. R., Larosa, G., Newman, W., Gerard, N., Gerard, C., and Sodroski, J. (1996). The beta-chemokine receptors CCR3 and CCR5 facilitate infection by primary HIV-1 isolates. *Cell* **85,** 1135–1148.

Chuntharapai, A., Lee, J., Hebert, C. A., and Kim, K. J. (1994). Monoclonal antibodies detect different distribution patterns of IL-8 receptor A and IL-8 receptor B on human peripheral blood leukocytes. *J. Immunol.* **153,** 5682–5688.

Cinque, P., Vago, L., Mengozzi, M., Torri, V., Ceresa, D., Vicenzi, E., Transidico, P., Vagani, A., Sozzani, S., Mantovani, A., Lazzarin, A., and Poli, G. (1998). Elevated cerebrospinal fluid levels of monocyte chemotactic protein-1 correlate with HIV-1 encephalitis and local viral replication. *AIDS* **12,** 1327–1332.

Cocchi, F., Devico, A. L., Garzino-Demo, A., Arya, S. K., Gallo, R. C., and Lusso, P. (1995). Identification of RANTES, MIP-1 alpha, and MIP-1 beta as the major HIV-suppressive factors produced by CD8+ T cells. *Science* **270,** 1811–1815.

Cochran, B. H., Reffel, A. C., and Stiles, C. D. (1983). Molecular cloning of gene sequences regulated by platelet-derived growth factor. *Cell* **33,** 939–947.

Cole, K. E., Strick, C. A., Paradis, T. J., Ogborne, K. T., Loetscher, M., Gladue, R. P., Lin, W., Boyd, J. G., Moser, B., Wood, D. E., Sahagan, B. G., and Neote, K. (1998). Interferon-inducible T cell alpha chemoattractant (I-TAC): A novel non-ELR CXC chemokine with potent activity on activated T cells through selective high affinity binding to CXCR3. *J. Exp. Med.* **187,** 2009–2021.

Collins, P. D., Marleau, S., Griffiths-Johnson, D. A., Jose, P. J., and Williams, T. J. (1995). Cooperation between interleukin-5 and the chemokine eotaxin to induce eosinophil accumulation *in vivo*. *J. Exp. Med.* **182,** 1169–1174.

Colotta, F., Borre, A., Wang, J. M., Tattanelli, M., Maddalena, F., Polentarutti, N., Peri, G., and Mantovani, A. (1992). Expression of a monocyte chemotactic cytokine by human mononuclear phagocytes. *J. Immunol.* **148,** 760–765.

Combadiere, C., Ahuja, S. K., and Murphy, P. M. (1995a). Cloning and functional expression of a human eosinophil CC chemokine receptor. *J. Biol. Chem.* **270,** 16491–16494.

Combadiere, C., Ahuja, S. K., and Murphy, P. M. (1995b). Cloning, chromosomal localization, and RNA expression of a human beta chemokine receptor-like gene. *DNA Cell Biol.* **14,** 673–680.

Combadiere, C., Ahuja, S. K., Van Damme, J., Tiffany, H. L., Gao, J. L., and Murphy, P. M. (1995c). Monocyte chemoattractant protein-3 is a functional ligand for CC chemokine receptors 1 and 2B. *J. Biol. Chem.* **270,** 29671–29675.

Combadiere, C., Ahuja, S. K., Tiffany, H. L., and Murphy, P. M. (1996). Cloning and functional expression of CC CKR5, a human monocyte CC chemokine receptor selective for MIP-1(alpha), MIP-1(beta), and RANTES. *J. Leukoc. Biol.* **60,** 147–152.

Combadiere, C., Salzwedel, K., Smith, E. D., Tiffany, H. L., Berger, E. A., and Murphy, P. M. (1998). Identification of CX3CR1. A chemotactic receptor for the human CX3C chemokine fractalkine and a fusion coreceptor for HIV-1. *J. Biol. Chem.* **273,** 23799–23804.

Combadiere, C., Potteaux, S., Gao, J. L., Esposito, B., Casanova, S., Lee, E. J., Debre, P., Tedgui, A., Murphy, P. M., and Mallat, Z. (2003). Decreased atherosclerotic lesion formation in CX3CR1/apolipoprotein E double knockout mice. *Circulation* **107,** 1009–1016.

Conant, K., Garzino-Demo, A., Nath, A., Mcarthur, J. C., Halliday, W., Power, C., Gallo, R. C., and Major, E. O. (1998). Induction of monocyte chemoattractant protein-1 in HIV-1 Tat-stimulated astrocytes and elevation in AIDS dementia. *Proc. Natl. Acad. Sci. USA* **95,** 3117–3121.

Coughlan, C. M., McManus, C. M., Sharron, M., Gao, Z., Murphy, D., Jaffer, S., Choe, W., Chen, W., Hesselgesser, J., Gaylord, H., Kalyuzhny, A., Lee, V. M., Wolf, B., Doms, R. W., and Kolson, D. L. (2000). Expression of multiple functional chemokine receptors and monocyte chemoattractant protein-1 in human neurons. *Neuroscience* **97,** 591–600.

Dahinden, C. A., Geiser, T., Brunner, T., Von Tscharner, V., Caput, D., Ferrara, P., Minty, A., and Baggiolini, M. (1994). Monocyte chemotactic protein 3 is a most effective basophil- and eosinophil-activating chemokine. *J. Exp. Med.* **179,** 751–756.

Dajotoy, T., Andersson, P., Bjartell, A., Lofdahl, C. G., Tapper, H., and Egesten, A. (2004). Human eosinophils produce the T cell-attracting chemokines MIG and IP-10 upon stimulation with IFN-gamma. *J. Leukoc. Biol.* **76,** 685–691.

Daugherty, B. L., Siciliano, S. J., Demartino, J. A., Malkowitz, L., Sirotina, A., and Springer, M. S. (1996). Cloning, expression, and characterization of the human eosinophil eotaxin receptor. *J. Exp. Med.* **183,** 2349–2354.

Davatelis, G., Tekamp-Olson, P., Wolpe, S. D., Hermsen, K., Luedke, C., Gallegos, C., Coit, D., Merryweather, J., and Cerami, A. (1988). Cloning and characterization of a cDNA for murine macrophage inflammatory protein (MIP), a novel monokine with inflammatory and chemokinetic properties. *J. Exp. Med.* **167,** 1939–1944.

D'Aversa, T. G., Weidenheim, K. M., and Berman, J. W. (2002). CD40-CD40L interactions induce chemokine expression by human microglia: Implications for human immunodeficiency virus encephalitis and multiple sclerosis. *Am. J. Pathol.* **160,** 559–567.

D'Aversa, T. G., Yu, K. O., and Berman, J. W. (2004). Expression of chemokines by human fetal microglia after treatment with the human immunodeficiency virus type 1 protein Tat. *J. Neurovirol.* **10,** 86–97.

Dean, M., Carrington, M., Winkler, C., Huttley, G. A., Smith, M. W., Allikmets, R., Goedert, J. J., Buchbinder, S. P., Vittinghoff, E., Gomperts, E., Donfield, S., Vlahov, D., Kaslow, R., Saah, A., Rinaldo, C., Detels, R., and O'brien, S. J. (1996). Genetic restriction of HIV-1 infection and progression to AIDS by a deletion allele of the CKR5 structural gene. Hemophilia Growth and Development Study, Multicenter AIDS Cohort Study, Multicenter Hemophilia Cohort Study, San Francisco City Cohort, ALIVE Study. *Science* **273,** 1856–1862.

Deng, H., Liu, R., Ellmeier, W., Choe, S., Unutmaz, D., Burkhart, M., Di Marzio, P., Marmon, S., Sutton, R. E., Hill, C. M., Davis, C. B., Peiper, S. C., Schall, T. J., Littman, D. R., and Landau, N. R. (1996). Identification of a major co-receptor for primary isolates of HIV-1. *Nature* **381,** 661–666.

Deuel, T. F., Keim, P. S., Farmer, M., and Heinrikson, R. L. (1977). Amino acid sequence of human platelet factor 4. *Proc. Natl. Acad. Sci. USA* **74,** 2256–2258.

Doranz, B. J., Rucker, J., Yi, Y., Smyth, R. J., Samson, M., Peiper, S. C., Parmentier, M., Collman, R. G., and Doms, R. W. (1996). A dual-tropic primary HIV-1 isolate that uses fusin and the beta-chemokine receptors CKR-5, CKR-3, and CKR-2b as fusion cofactors. *Cell* **85,** 1149–1158.

Dorf, M. E., Berman, M. A., Tanabe, S., Heesen, M., and Luo, Y. (2000). Astrocytes express functional chemokine receptors. *J. Neuroimmunol.* **111,** 109–121.

Dorner, B., Muller, S., Entschladen, F., Schroder, J. M., Franke, P., Kraft, R., Friedl, P., Clark-Lewis, I., and Kroczek, R. A. (1997). Purification, structural analysis, and function of natural ATAC, a cytokine secreted by CD8(+) T cells. *J. Biol. Chem.* **272,** 8817–8823.

Dragic, T., Litwin, V., Allaway, G. P., Martin, S. R., Huang, Y., Nagashima, K. A., Cayanan, C., Maddon, P. J., Koup, R. A., Moore, J. P., and Paxton, W. A. (1996). HIV-1 entry into CD4+ cells is mediated by the chemokine receptor CC-CKR-5. *Nature* **381,** 667–673.

Ehrlich, L. C., Hu, S., Sheng, W. S., Sutton, R. L., Rockswold, G. L., Peterson, P. K., and Chao, C. C. (1998). Cytokine regulation of human microglial cell IL-8 production. *J. Immunol.* **160,** 1944–1948.

Eugenin, E. A., and Berman, J. W. (2003). Chemokine-dependent mechanisms of leukocyte trafficking across a model of the blood-brain barrier. *Methods* **29,** 351–361.

Eugenin, E. A., D'Aversa, T. G., Lopez, L., Calderon, T. M., and Berman, J. W. (2003). MCP-1 (CCL2) protects human neurons and astrocytes from NMDA or HIV-tat-induced apoptosis. *J. Neurochem.* **85,** 1299–1311.

Fahey, T. J., 3rd, Tracey, K. J., Tekamp-Olson, P., Cousens, L. S., Jones, W. G., Shires, G. T., Cerami, A., and Sherry, B. (1992). Macrophage inflammatory protein 1 modulates macrophage function. *J. Immunol.* **148,** 2764–2769.

Farber, J. M. (1990). A macrophage mRNA selectively induced by gamma-interferon encodes a member of the platelet factor 4 family of cytokines. *Proc. Natl. Acad. Sci. USA* **87,** 5238–5242.

Farber, J. M. (1993). HuMig: A new human member of the chemokine family of cytokines. *Biochem. Biophys. Res. Commun.* **192,** 223–230.

Farber, J. M. (1997). Mig and IP-10: CXC chemokines that target lymphocytes. *J. Leukoc. Biol.* **61,** 246–257.

Federsppiel, B., Melhado, I. G., Duncan, A. M., Delaney, A., Schappert, K., Clark-Lewis, I., and Jirik, F. R. (1993). Molecular cloning of the cDNA and chromosomal localization of the gene for a putative seven-transmembrane segment (7-TMS) receptor isolated from human spleen. *Genomics* **16,** 707–712.

Fedyk, E. R., Jones, D., Critchley, H. O., Phipps, R. P., Blieden, T. M., and Springer, T. A. (2001). Expression of stromal-derived factor-1 is decreased by IL-1 and TNF and in dermal wound healing. *J. Immunol.* **166,** 5749–5754.

Feil, C., and Augustin, H. G. (1998). Endothelial cells differentially express functional CXC-chemokine receptor-4 (CXCR-4/fusin) under the control of autocrine activity and exogenous cytokines. *Biochem. Biophys. Res. Commun.* **247,** 38–45.

Feng, Y., Broder, C. C., Kennedy, P. E., and Berger, E. A. (1996). HIV-1 entry cofactor: Functional cDNA cloning of a seven-transmembrane, G protein-coupled receptor. *Science* **272,** 872–877.

Files, J. C., Malpass, T. W., Yee, E. K., Ritchie, J. L., and Harker, L. A. (1981). Studies of human plate alpha-granule release *in vivo*. *Blood* **58,** 607–618.

Flier, J., Boorsma, D. M., Bruynzeel, D. P., Van Beek, P. J., Stoof, T. J., Scheper, R. J., Willemze, R., and Tensen, C. P. (1999). The CXCR3 activating chemokines IP-10, Mig, and IP-9 are expressed in allergic but not in irritant patch test reactions. *J. Invest. Dermatol.* **113,** 574–578.

Fong, A. M., Robinson, L. A., Steeber, D. A., Tedder, T. F., Yoshie, O., Imai, T., and Patel, D. D. (1998). Fractalkine and CX3CR1 mediate a novel mechanism of leukocyte capture, firm adhesion, and activation under physiologic flow. *J. Exp. Med.* **188,** 1413–1419.

Franci, C., Wong, L. M., Van Damme, J., Proost, P., and Charo, I. F. (1995). Monocyte chemoattractant protein-3, but not monocyte chemoattractant protein-2, is a functional ligand of the human monocyte chemoattractant protein-1 receptor. *J. Immunol.* **154,** 6511–6517.

Fuentes, M. E., Durham, S. K., Swerdel, M. R., Lewin, A. C., Barton, D. S., Megill, J. R., Bravo, R., and Lira, S. A. (1995). Controlled recruitment of monocytes and macrophages to specific organs through transgenic expression of monocyte chemoattractant protein-1. *J. Immunol.* **155,** 5769–5776.

Furutani, Y., Nomura, H., Notake, M., Oyamada, Y., Fukui, T., Yamada, M., Larsen, C. G., Oppenheim, J. J., and Matsushima, K. (1989). Cloning and sequencing of the cDNA for human monocyte chemotactic and activating factor (MCAF). *Biochem. Biophys. Res. Commun.* **159,** 249–255.

Gao, J. L., Kuhns, D. B., Tiffany, H. L., Mcdermott, D., Li, X., Francke, U., and Murphy, P. M. (1993). Structure and functional expression of the human macrophage inflammatory protein 1 alpha/RANTES receptor. *J. Exp. Med.* **177,** 1421–1427.

Garcia-Zepeda, E. A., Combadiere, C., Rothenberg, M. E., Sarafi, M. N., Lavigne, F., Hamid, Q., Murphy, P. M., and Luster, A. D. (1996a). Human monocyte chemoattractant protein (MCP)-4 is a novel CC chemokine with activities on monocytes, eosinophils, and basophils induced in allergic and nonallergic inflammation that signals through the CC chemokine receptors (CCR)-2 and -3. *J. Immunol.* **157,** 5613–5626.

Garcia-Zepeda, E. A., Rothenberg, M. E., Ownbey, R. T., Celestin, J., Leder, P., and Luster, A. D. (1996b). Human eotaxin is a specific chemoattractant for eosinophil cells and provides a new mechanism to explain tissue eosinophilia. *Nat. Med.* **2,** 449–456.

Gartner, S. (2000). HIV infection and dementia. *Science* **287,** 602–604.

Geiser, T., Dewald, B., Ehrengruber, M. U., Clark-Lewis, I., and Baggiolini, M. (1993). The interleukin-8-related chemotactic cytokines GRO alpha, GRO beta, and GRO gamma activate human neutrophil and basophil leukocytes. *J. Biol. Chem.* **268,** 15419–15424.

Gendelman, H. E., Lipton, S. A., Tardieu, M., Bukrinsky, M. I., and Nottet, H. S. (1994). The neuropathogenesis of HIV-1 infection. *J. Leukoc. Biol.* **56,** 389–398.

Gerszten, R. E., Garcia-Zepeda, E. A., Lim, Y. C., Yoshida, M., Ding, H. A., Gimbrone, M. A., Jr., Luster, A. D., Luscinskas, F. W., and Rosenzweig, A. (1999). MCP-1 and IL-8 trigger firm adhesion of monocytes to vascular endothelium under flow conditions. *Nature* **398,** 718–723.

Giancarlo, B., Silvano, S., Albert, Z., Mantovani, A., and Allavena, P. (1996). Migratory response of human natural killer cells to lymphotactin. *Eur. J. Immunol.* **26,** 3238–3241.

Gijsbers, K., Van Assche, G., Joossens, S., Struyf, S., Proost, P., Rutgeerts, P., Geboes, K., and Van Damme, J. (2004). CXCR1-binding chemokines in inflammatory bowel diseases: Down-regulated IL-8/CXCL8 production by leukocytes in Crohn's disease and selective GCP-2/CXCL6 expression in inflamed intestinal tissue. *Eur. J. Immunol.* **34,** 1992–2000.

Gleichmann, M., Gillen, C., Czardybon, M., Bosse, F., Greiner-Petter, R., Auer, J., and Muller, H. W. (2000). Cloning and characterization of SDF-1gamma, a novel SDF-1 chemokine transcript with developmentally regulated expression in the nervous system. *Eur. J. Neurosci.* **12,** 1857–1866.

Godiska, R., Chantry, D., Dietsch, G. N., and Gray, P. W. (1995). Chemokine expression in murine experimental allergic encephalomyelitis. *J. Neuroimmunol.* **58,** 167–176.

Goebeler, M., Toksoy, A., Spandau, U., Engelhardt, E., Brocker, E. B., and Gillitzer, R. (1998). The C-X-C chemokine Mig is highly expressed in the papillae of psoriatic lesions. *J. Pathol.* **184,** 89–95.

Gottlieb, A. B., Luster, A. D., Posnett, D. N., and Carter, D. M. (1988). Detection of a gamma interferon-induced protein IP-10 in psoriatic plaques. *J. Exp. Med.* **168,** 941–948.

Greaves, D. R., Hakkinen, T., Lucas, A. D., Liddiard, K., Jones, E., Quinn, C. M., Senaratne, J., Green, F. R., Tyson, K., Boyle, J., Shanahan, C., Weissberg, P. L., Gordon, S., and Yla-Hertualla, S. (2001). Linked chromosome 16q13 chemokines, macrophage-derived chemokine, fractalkine, and thymus- and activation-regulated chemokine, are expressed in human atherosclerotic lesions. *Arterioscler. Thromb. Vasc. Biol.* **21,** 923–929.

Grewal, I. S., Rutledge, B. J., Fiorillo, J. A., Gu, L., Gladue, R. P., Flavell, R. A., and Rollins, B. J. (1997). Transgenic monocyte chemoattractant protein-1 (MCP-1) in pancreatic islets produces monocyte-rich insulitis without diabetes: Abrogation by a second transgene expressing systemic MCP-1. *J. Immunol.* **159,** 401–408.

Gu, L., Okada, Y., Clinton, S. K., Gerard, C., Sukhova, G. K., Libby, P., and Rollins, B. J. (1998). Absence of monocyte chemoattractant protein-1 reduces atherosclerosis in low density lipoprotein receptor-deficient mice. *Mol. Cell* **2,** 275–281.

Gupta, S. K., Lysko, P. G., Pillarisetti, K., Ohlstein, E., and Stadel, J. M. (1998). Chemokine receptors in human endothelial cells. Functional expression of CXCR4 and its transcriptional regulation by inflammatory cytokines. *J. Biol. Chem.* **273,** 4282–4287.

Harrison, J. K., Barber, C. M., and Lynch, K. R. (1994). cDNA cloning of a G-protein-coupled receptor expressed in rat spinal cord and brain related to chemokine receptors. *Neurosci. Lett.* **169,** 85–89.

Harrison, J. K., Jiang, Y., Chen, S., Xia, Y., Maciejewski, D., McNamara, R. K., Streit, W. J., Salafranca, M. N., Adhikari, S., Thompson, D. A., Botti, P., Bacon, K. B., and Feng, L. (1998). Role for neuronally derived fractalkine in mediating interactions between neurons and CX3CR1-expressing microglia. *Proc. Natl. Acad. Sci. USA* **95,** 10896–10901.

Haskell, C. A., Cleary, M. D., and Charo, I. F. (1999). Molecular uncoupling of fractalkine-mediated cell adhesion and signal transduction. Rapid flow arrest of CX3CR1-expressing cells is independent of G-protein activation. *J. Biol. Chem.* **274,** 10053–10058.

Haskill, S., Peace, A., Morris, J., Sporn, S. A., Anisowicz, A., Lee, S. W., Smith, T., Martin, G., Ralph, P., and Sager, R. (1990). Identification of three related human GRO genes encoding cytokine functions. *Proc. Natl. Acad. Sci. USA* **87,** 7732–7736.

Hayashi, M., Luo, Y., Laning, J., Strieter, R. M., and Dorf, M. E. (1995). Production and function of monocyte chemoattractant protein-1 and other beta-chemokines in murine glial cells. *J. Neuroimmunol.* **60,** 143–150.

He, J., Chen, Y., Farzan, M., Choe, H., Ohagen, A., Gartner, S., Busciglio, J., Yang, X., Hofmann, W., Newman, W., Mackay, C. R., Sodroski, J., and Gabuzda, D. (1997). CCR3 and CCR5 are co-receptors for HIV-1 infection of microglia. *Nature* **385,** 645–649.

Hedrick, J. A., Saylor, V., Figueroa, D., Mizoue, L., Xu, Y., Menon, S., Abrams, J., Handel, T., and Zlotnik, A. (1997). Lymphotactin is produced by NK cells and attracts both NK cells and T cells *in vivo*. *J. Immunol.* **158,** 1533–1540.

Heiber, M., Docherty, J. M., Shah, G., Nguyen, T., Cheng, R., Heng, H. H., Marchese, A., Tsui, L. C., Shi, X., George, S. R., *et al.* (1995). Isolation of three novel human genes encoding G protein-coupled receptors. *DNA Cell Biol.* **14,** 25–35.

Herzog, H., Hort, Y. J., Shine, J., and Selbie, L. A. (1993). Molecular cloning, characterization, and localization of the human homolog to the reported bovine NPY Y3 receptor: Lack of NPY binding and activation. *DNA Cell Biol.* **12,** 465–471.

Hieshima, K., Imai, T., Opdenakker, G., Van Damme, J., Kusuda, J., Tei, H., Sakaki, Y., Takatsuki, K., Miura, R., Yoshie, O., and Nomiyama, H. (1997). Molecular cloning of a novel human CC chemokine liver and activation-regulated chemokine (LARC) expressed in liver: Chemotactic activity for lymphocytes and gene localization on chromosome 2. *J. Biol. Chem.* **272,** 5846–5853.

Holmes, W. E., Lee, J., Kuang, W. J., Rice, G. C., and Wood, W. I. (1991). Structure and functional expression of a human interleukin-8 receptor. *Science* **253,** 1278–1280.

Holt, J. C., Rabellino, E. M., Gewirtz, A. M., Gunkel, L. M., Rucinski, B., and Niewiarowski, S. (1988). Occurrence of platelet basic protein, a precursor of low affinity platelet factor 4 and beta-thromboglobulin, in human platelets and megakaryocytes. *Exp. Hematol.* **16,** 302–306.

Horuk, R., Martin, A. W., Wang, Z., Schweitzer, L., Gerassimides, A., Guo, H., Lu, Z., Hesselgesser, J., Perez, H. D., Kim, J., Parker, J., Hadley, T. J., and Peiper, S. C. (1997). Expression of chemokine receptors by subsets of neurons in the central nervous system. *J. Immunol.* **158,** 2882–2890.

Hu, S., Chao, C. C., Ehrlich, L. C., Sheng, W. S., Sutton, R. L., Rockswold, G. L., and Peterson, P. K. (1999). Inhibition of microglial cell RANTES production by IL-10 and TGF-beta. *J. Leukoc. Biol.* **65,** 815–821.

Huang, D. R., Wang, J., Kivisakk, P., Rollins, B. J., and Ransohoff, R. M. (2001a). Absence of monocyte chemoattractant protein 1 in mice leads to decreased local macrophage recruitment and antigen-specific T helper cell type 1 immune response in experimental autoimmune encephalomyelitis. *J. Exp. Med.* **193,** 713–726.

Huang, H., Li, F., Cairns, C. M., Gordon, J. R., and Xiang, J. (2001b). Neutrophils and B cells express XCR1 receptor and chemotactically respond to lymphotactin. *Biochem. Biophys. Res. Commun.* **281,** 378–382.

Huang, S., Paulauskis, J. D., Godleski, J. J., and Kobzik, L. (1992). Expression of macrophage inflammatory protein-2 and KC mRNA in pulmonary inflammation. *Am. J. Pathol.* **141,** 981–988.

Huang, Y., Paxton, W. A., Wolinsky, S. M., Neumann, A. U., Zhang, L., He, T., Kang, S., Ceradini, D., Jin, Z., Yazdanbakhsh, K., Kunstman, K., Erickson, D., Dragon, E., Landau, N. R., Phair, J., Ho, D. D., and Koup, R. A. (1996). The role of a mutant CCR5 allele in HIV-1 transmission and disease progression. *Nat. Med.* **2,** 1240–1243.

Hulkower, K., Brosnan, C. F., Aquino, D. A., Cammer, W., Kulshrestha, S., Guida, M. P., Rapoport, D. A., and Berman, J. W. (1993). Expression of CSF-1, c-fms, and MCP-1 in the central nervous system of rats with experimental allergic encephalomyelitis. *J. Immunol.* **150,** 2525–2533.

Hurwitz, A. A., Lyman, W. D., and Berman, J. W. (1995). Tumor necrosis factor alpha and transforming growth factor beta upregulate astrocyte expression of monocyte chemoattractant protein-1. *J. Neuroimmunol.* **57,** 193–198.

Iida, N., and Grotendorst, G. R. (1990). Cloning and sequencing of a new gro transcript from activated human monocytes: Expression in leukocytes and wound tissue. *Mol. Cell. Biol.* **10,** 5596–5599.

Imai, K., Kobayashi, M., Wang, J., Shinobu, N., Yoshida, H., Hamada, J., Shindo, M., Higashino, F., Tanaka, J., Asaka, M., and Hosokawa, M. (1999). Selective secretion of chemoattractants for haemopoietic progenitor cells by bone marrow endothelial cells: A possible role in homing of haemopoietic progenitor cells to bone marrow. *Br. J. Haematol.* **106,** 905–911.

Imai, T., Hieshima, K., Haskell, C., Baba, M., Nagira, M., Nishimura, M., Kakizaki, M., Takagi, S., Nomiyama, H., Schall, T. J., and Yoshie, O. (1997). Identification and molecular characterization of fractalkine receptor CX3CR1, which mediates both leukocyte migration and adhesion. *Cell* **91,** 521–530.

Imaizumi, T., Albertine, K. H., Jicha, D. L., McIntyre, T. M., Prescott, S. M., and Zimmerman, G. A. (1997). Human endothelial cells synthesize ENA-78: Relationship to IL-8 and to signaling of PMN adhesion. *Am. J. Respir. Cell Mol. Biol.* **17,** 181–192.

Inngjerdingen, M., Damaj, B., and Maghazachi, A. A. (2001). Expression and regulation of chemokine receptors in human natural killer cells. *Blood* **97,** 367–375.

Introna, M., Breviario, F., D'Aniello, E. M., Golay, J., Dejana, E., and Mantovani, A. (1993). IL-1 inducible genes in human umbilical vein endothelial cells. *Eur. Heart J.* **14**(Suppl K), 78–81.

Ishizuka, K., Kimura, T., Igata-Yi, R., Katsuragi, S., Takamatsu, J., and Miyakawa, T. (1997). Identification of monocyte chemoattractant protein-1 in senile plaques and reactive microglia of Alzheimer's disease. *Psychiatry Clin. Neurosci.* **51,** 135–138.

Izikson, L., Klein, R. S., Charo, I. F., Weiner, H. L., and Luster, A. D. (2000). Resistance to experimental autoimmune encephalomyelitis in mice lacking the CC chemokine receptor (CCR)2. *J. Exp. Med.* **192,** 1075–1080.

Janabi, N., Hau, I., and Tardieu, M. (1999). Negative feedback between prostaglandin and alpha- and beta-chemokine synthesis in human microglial cells and astrocytes. *J. Immunol.* **162,** 1701–1706.

Jazin, E. E., Yoo, H., Blomqvist, A. G., Yee, F., Weng, G., Walker, M. W., Salon, J., Larhammar, D., and Wahlestedt, C. (1993). A proposed bovine neuropeptide Y (NPY) receptor cDNA clone, or its human homologue, confers neither NPY binding sites nor NPY responsiveness on transfected cells. *Regul. Pept.* **47,** 247–258.

Jia, G. Q., Gonzalo, J. A., Lloyd, C., Kremer, L., Lu, L., Martinez, A. C., Wershil, B. K., and Gutierrez-Ramos, J. C. (1996). Distinct expression and function of the novel mouse chemokine monocyte chemotactic protein-5 in lung allergic inflammation. *J. Exp. Med.* **184,** 1939–1951.

Jose, P. J., Griffiths-Johnson, D. A., Collins, P. D., Walsh, D. T., Moqbel, R., Totty, N. F., Truong, O., Hsuan, J. J., and Williams, T. J. (1994). Eotaxin: A potent eosinophil chemoattractant cytokine detected in a guinea pig model of allergic airways inflammation. *J. Exp. Med.* **179,** 881–887.

Kameyoshi, Y., Dorschner, A., Mallet, A. I., Christophers, E., and Schroder, J. M. (1992). Cytokine RANTES released by thrombin-stimulated platelets is a potent attractant for human eosinophils. *J. Exp. Med.* **176,** 587–592.

Kaplan, G., Luster, A. D., Hancock, G., and Cohn, Z. A. (1987). The expression of a gamma interferon-induced protein (IP-10) in delayed immune responses in human skin. *J. Exp. Med.* **166,** 1098–1108.

Karpus, W. J., and Kennedy, K. J. (1997). MIP-1alpha and MCP-1 differentially regulate acute and relapsing autoimmune encephalomyelitis as well as Th1/Th2 lymphocyte differentiation. *J. Leukoc. Biol.* **62,** 681–687.

Karpus, W. J., Lukacs, N. W., Kennedy, K. J., Smith, W. S., Hurst, S. D., and Barrett, T. A. (1997). Differential CC chemokine-induced enhancement of T helper cell cytokine production. *J. Immunol.* **158,** 4129–4136.

Karpus, W. J., Lukacs, N. W., McRae, B. L., Strieter, R. M., Kunkel, S. L., and Miller, S. D. (1995). An important role for the chemokine macrophage inflammatory protein-1 alpha in the pathogenesis of the T cell-mediated autoimmune disease, experimental autoimmune encephalomyelitis. *J. Immunol.* **155,** 5003–5010.

Kasahara, T., Mukaida, N., Yamashita, K., Yagisawa, H., Akahoshi, T., and Matsushima, K. (1991). IL-1 and TNF-alpha induction of IL-8 and monocyte chemotactic and activating factor (MCAF) mRNA expression in a human astrocytoma cell line. *Immunology* **74,** 60–67.

Kasama, T., Strieter, R. M., Standiford, T. J., Burdick, M. D., and Kunkel, S. L. (1993). Expression and regulation of human neutrophil-derived macrophage inflammatory protein 1 alpha. *J. Exp. Med.* **178,** 63–72.

Kelder, W., McArthur, J. C., Nance-Sproson, T., McClernon, D., and Griffin, D. E. (1998). Beta-chemokines MCP-1 and RANTES are selectively increased in cerebrospinal fluid of patients with human immunodeficiency virus-associated dementia. *Ann. Neurol.* **44,** 831–835.

Kelner, G. S., Kennedy, J., Bacon, K. B., Kleyensteuber, S., Largaespada, D. A., Jenkins, N. A., Copeland, N. G., Bazan, J. F., Moore, K. W., and Schall, T. J. (1994). Lymphotactin: A cytokine that represents a new class of chemokine. *Science* **266,** 1395–1399.

Kennedy, J., Kelner, G. S., Kleyensteuber, S., Schall, T. J., Weiss, M. C., Yssel, H., Schneider, P. V., Cocks, B. G., Bacon, K. B., and Zlotnik, A. (1995). Molecular cloning and functional characterization of human lymphotactin. *J. Immunol.* **155,** 203–209.

Kim, B. O., Liu, Y., Zhou, B. Y., and He, J. J. (2004). Induction of C chemokine XCL1 (lymphotactin/single C motif-1 alpha/activation-induced, T cell-derived and chemokine-related cytokine) expression by HIV-1 Tat protein. *J. Immunol.* **172,** 1888–1895.

Kim, J. J., Tsai, A., Nottingham, L. K., Morrison, L., Cunning, D. M., Oh, J., Lee, D. J., Dang, K., Dentchev, T., Chalian, A. A., Agadjanyan, M. G., and Weiner, D. B. (1999). Intracellular adhesion molecule-1 modulates beta-chemokines and directly costimulates T cells *in vivo*. *J. Clin. Invest.* **103,** 869–877.

Kitaura, M., Nakajima, T., Imai, T., Harada, S., Combadiere, C., Tiffany, H. L., Murphy, P. M., and Yoshie, O. (1996). Molecular cloning of human eotaxin, an eosinophil-selective CC chemokine, and identification of a specific eosinophil eotaxin receptor, CC chemokine receptor 3. *J. Biol. Chem.* **271,** 7725–7730.

Klein, R. S., Williams, K. C., Alvarez-Hernandez, X., Westmoreland, S., Force, T., Lackner, A. A., and Luster, A. D. (1999). Chemokine receptor expression and signaling in macaque and human fetal neurons and astrocytes: Implications for the neuropathogenesis of AIDS. *J. Immunol.* **163,** 1636–1646.

Klier, C. M., Nelson, E. L., Cohen, C. D., Horuk, R., Schlondorff, D., and Nelson, P. J. (2001). Chemokine-induced secretion of gelatinase B in primary human monocytes. *Biol. Chem.* **382,** 1405–1410.

Koch, A. E., Kunkel, S. L., Harlow, L. A., Johnson, B., Evanoff, H. L., Haines, G. K., Burdick, M. D., Pope, R. M., and Strieter, R. M. (1992a). Enhanced production of monocyte chemoattractant protein-1 in rheumatoid arthritis. *J. Clin. Invest.* **90,** 772–779.

Koch, A. E., Polverini, P. J., Kunkel, S. L., Harlow, L. A., Dipietro, L. A., Elner, V. M., Elner, S. G., and Strieter, R. M. (1992b). Interleukin-8 as a macrophage-derived mediator of angiogenesis. *Science* **258,** 1798–1801.

Koch, A. E., Kunkel, S. L., Harlow, L. A., Mazarakis, D. D., Haines, G. K., Burdick, M. D., Pope, R. M., and Strieter, R. M. (1994a). Macrophage inflammatory protein-1 alpha. A novel chemotactic cytokine for macrophages in rheumatoid arthritis. *J. Clin. Invest.* **93,** 921–928.

Koch, A. E., Kunkel, S. L., Harlow, L. A., Mazarakis, D. D., Haines, G. K., Burdick, M. D., Pope, R. M., Walz, A., and Strieter, R. M. (1994b). Epithelial neutrophil activating peptide-78: A novel chemotactic cytokine for neutrophils in arthritis. *J. Clin. Invest.* **94,** 1012–1018.

Kolb, S. A., Sporer, B., Lahrtz, F., Koedel, U., Pfister, H. W., and Fontana, A. (1999). Identification of a T cell chemotactic factor in the cerebrospinal fluid of HIV-1-infected individuals as interferon-gamma inducible protein 10. *J. Neuroimmunol.* **93,** 172–181.

Konig, A., Krenn, V., Toksoy, A., Gerhard, N., and Gillitzer, R. (2000). Mig, GRO alpha and RANTES messenger RNA expression in lining layer, infiltrates and different leucocyte populations of synovial tissue from patients with rheumatoid arthritis, psoriatic arthritis and osteoarthritis. *Virchows Arch.* **436,** 449–458.

Kuang, Y., Wu, Y., Jiang, H., and Wu, D. (1996). Selective G protein coupling by C-C chemokine receptors. *J. Biol. Chem.* **271,** 3975–3978.

Kuna, P., Reddigari, S. R., Rucinski, D., Oppenheim, J. J., and Kaplan, A. P. (1992a). Monocyte chemotactic and activating factor is a potent histamine-releasing factor for human basophils. *J. Exp. Med.* **175,** 489–493.

Kuna, P., Reddigari, S. R., Schall, T. J., Rucinski, D., Viksman, M. Y., and Kaplan, A. P. (1992b). RANTES, a monocyte and T lymphocyte chemotactic cytokine releases histamine from human basophils. *J. Immunol.* **149,** 636–642.

Kuratsu, J., Leonard, E. J., and Yoshimura, T. (1989). Production and characterization of human glioma cell-derived monocyte chemotactic factor. *J. Natl. Cancer Inst.* **81,** 347–351.

Lacy, M., Jones, J., Whittemore, S. R., Haviland, D. L., Wetsel, R. A., and Barnum, S. R. (1995). Expression of the receptors for the C5a anaphylatoxin, interleukin-8 and FMLP by human astrocytes and microglia. *J. Neuroimmunol.* **61,** 71–78.

Landry, Y., and Gies, J. P. (2002). Heterotrimeric G proteins control diverse pathways of transmembrane signaling, a base for drug discovery. *Mini Rev. Med. Chem.* **2,** 361–372.

Lapinet, J. A., Scapini, P., Calzetti, F., Perez, O., and Cassatella, M. A. (2000). Gene expression and production of tumor necrosis factor alpha, interleukin-1beta (IL-1beta), IL-8, macrophage inflammatory protein 1alpha (MIP-1alpha), MIP-1beta, and gamma interferon-inducible protein 10 by human neutrophils stimulated with group B meningococcal outer membrane vesicles. *Infect. Immun.* **68,** 6917–6923.

Larosa, G. J., Thomas, K. M., Kaufmann, M. E., Mark, R., White, M., Taylor, L., Gray, G., Witt, D., and Navarro, J. (1992). Amino terminus of the interleukin-8 receptor is a major determinant of receptor subtype specificity. *J. Biol. Chem.* **267,** 25402–25406.

Larsen, C. G., Anderson, A. O., Appella, E., Oppenheim, J. J., and Matsushima, K. (1989). The neutrophil-activating protein (NAP-1) is also chemotactic for T lymphocytes. *Science* **243,** 1464–1466.

Lavi, E., Strizki, J. M., Ulrich, A. M., Zhang, W., Fu, L., Wang, Q., O'Connor, M., Hoxie, J. A., and Gonzalez-Scarano, F. (1997). CXCR-4 (Fusin), a co-receptor for the type 1 human immunodeficiency virus (HIV-1), is expressed in the human brain in a variety of cell types, including microglia and neurons. *Am. J. Pathol.* **151,** 1035–1042.

Lee, J., Horuk, R., Rice, G. C., Bennett, G. L., Camerato, T., and Wood, W. I. (1992). Characterization of two high-affinity human interleukin-8 receptors. *J. Biol. Chem.* **267,** 16283–16287.

Lei, Z., Wen, C., Xu, J., Li, X., Qing, L., and Wei, M. (2001). OxLDL upregulates growth-regulation oncogene alpha expression in human endothelial cells. *Chin. Med. J. (Engl.)* **114,** 1240–1244.

Lei, Z. B., Zhang, Z., Jing, Q., Qin, Y. W., Pei, G., Cao, B. Z., and Li, X. Y. (2002). OxLDL upregulates CXCR2 expression in monocytes via scavenger receptors and activation of p38 mitogen-activated protein kinase. *Cardiovasc. Res.* **53,** 524–532.

Lesnik, P., Haskell, C. A., and Charo, I. F. (2003). Decreased atherosclerosis in CX3CR1-/- mice reveals a role for fractalkine in atherogenesis. *J. Clin. Invest.* **111,** 333–340.

Letendre, S. L., Lanier, E. R., and McCutchan, J. A. (1999). Cerebrospinal fluid beta chemokine concentrations in neurocognitively impaired individuals infected with human immunodeficiency virus type 1. *J. Infect. Dis.* **180,** 310–319.

Li, A., Dubey, S., Varney, M. L., Dave, B. J., and Singh, R. K. (2003). IL-8 directly enhanced endothelial cell survival, proliferation, and matrix metalloproteinases production and regulated angiogenesis. *J. Immunol.* **170,** 3369–3376.

Liao, F., Rabin, R. L., Yannelli, J. R., Koniaris, L. G., Vanguri, P., and Farber, J. M. (1995). Human Mig chemokine: biochemical and functional characterization. *J. Exp. Med.* **182,** 1301–1314.

Lippert, U., Artuc, M., Grutzkau, A., Moller, A., Kenderessy-Szabo, A., Schadendorf, D., Norgauer, J., Hartmann, K., Schweitzer-Stenner, R., Zuberbier, T., Henz, B. M., and Kruger-Krasagakes, S. (1998). Expression and functional activity of the IL-8 receptor type CXCR1 and CXCR2 on human mast cells. *J. Immunol.* **161,** 2600–2608.

Liu, R., Paxton, W. A., Choe, S., Ceradini, D., Martin, S. R., Horuk, R., Macdonald, M. E., Stuhlmann, H., Koup, R. A., and Landau, N. R. (1996). Homozygous defect in HIV-1 coreceptor accounts for resistance of some multiply-exposed individuals to HIV-1 infection. *Cell* **86,** 367–377.

Lloyd, C. M., Minto, A. W., Dorf, M. E., Proudfoot, A., Wells, T. N., Salant, D. J., and Gutierrez-Ramos, J. C. (1997). RANTES and monocyte chemoattractant protein-1 (MCP-1) play an important role in the inflammatory phase of crescentic nephritis, but only MCP-1 is involved in crescent formation and interstitial fibrosis. *J. Exp. Med.* **185,** 1371–1380.

Locati, M., Deuschle, U., Massardi, M. L., Martinez, F. O., Sironi, M., Sozzani, S., Bartfai, T., and Mantovani, A. (2002). Analysis of the gene expression profile activated by the CC chemokine ligand 5/RANTES and by lipopolysaccharide in human monocytes. *J. Immunol.* **168,** 3557–3562.

Loetscher, M., Geiser, T., O'Reilly, T., Zwahlen, R., Baggiolini, M., and Moser, B. (1994a). Cloning of a human seven-transmembrane domain receptor, LESTR, that is highly expressed in leukocytes. *J. Biol. Chem.* **269,** 232–237.

Loetscher, P., Seitz, M., Clark-Lewis, I., Baggiolini, M., and Moser, B. (1994b). Both interleukin-8 receptors independently mediate chemotaxis. Jurkat cells transfected with IL-8R1 or IL-8R2 migrate in response to IL-8, GRO alpha and NAP-2. *FEBS Lett.* **341,** 187–192.

Loetscher, P., Seitz, M., Clark-Lewis, I., Baggiolini, M., and Moser, B. (1994c). Monocyte chemotactic proteins MCP-1, MCP-2, and MCP-3 are major attractants for human CD4+ and CD8+ T lymphocytes. *Faseb J.* **8,** 1055–1060.

Loetscher, M., Gerber, B., Loetscher, P., Jones, S. A., Piali, L., Clark-Lewis, I., Baggiolini, M., and Moser, B. (1996a). Chemokine receptor specific for IP10 and mig: Structure, function, and expression in activated T-lymphocytes. *J. Exp. Med.* **184,** 963–969.

Loetscher, P., Seitz, M., Clark-Lewis, I., Baggiolini, M., and Moser, B. (1996b). Activation of NK cells by CC chemokines. Chemotaxis, Ca2+ mobilization, and enzyme release. *J. Immunol.* **156,** 322–327.

Lukacs, N. W., Strieter, R. M., Chensue, S. W., and Kunkel, S. L. (1996). Activation and regulation of chemokines in allergic airway inflammation. *J. Leukoc. Biol.* **59,** 13–17.

Lukacs, N. W., Hogaboam, C. M., Kunkel, S. L., Chensue, S. W., Burdick, M. D., Evanoff, H. L., and Strieter, R. M. (1998). Mast cells produce ENA-78, which can function as a potent neutrophil chemoattractant during allergic airway inflammation. *J. Leukoc. Biol.* **63,** 746–751.

Luster, A. D. (1998). Chemokines–chemotactic cytokines that mediate inflammation. *N. Engl. J. Med.* **338,** 436–445.

Luster, A. D., Unkeless, J. C., and Ravetch, J. V. (1985). Gamma-interferon transcriptionally regulates an early-response gene containing homology to platelet proteins. *Nature* **315,** 672–676.

Luster, A. D., Greenberg, S. M., and Leder, P. (1995). The IP-10 chemokine binds to a specific cell surface heparan sulfate site shared with platelet factor 4 and inhibits endothelial cell proliferation. *J. Exp. Med.* **182,** 219–231.

Ma, Q., Jones, D., Borghesani, P. R., Segal, R. A., Nagasawa, T., Kishimoto, T., Bronson, R. T., and Springer, T. A. (1998). Impaired B-lymphopoiesis, myelopoiesis, and derailed cerebellar neuron migration in CXCR4- and SDF-1-deficient mice. *Proc. Natl. Acad. Sci. USA* **95,** 9448–9453.

Maciejewski-Lenoir, D., Chen, S., Feng, L., Maki, R., and Bacon, K. B. (1999). Characterization of fractalkine in rat brain cells: migratory and activation signals for CX3CR-1-expressing microglia. *J. Immunol.* **163,** 1628–1635.

Maghazachi, A. A., Al-Aoukaty, A., and Schall, T. J. (1994). C-C chemokines induce the chemotaxis of NK and IL-2-activated NK cells. Role for G proteins. *J. Immunol.* **153,** 4969–4977.

Maghazachi, A. A., Al-Aoukaty, A., and Schall, T. J. (1996). CC chemokines induce the generation of killer cells from CD56+ cells. *Eur. J. Immunol.* **26,** 315–319.

Maghazachi, A. A., Skalhegg, B. S., Rolstad, B., and Al-Aoukaty, A. (1997). Interferon-inducible protein-10 and lymphotactin induce the chemotaxis and mobilization of intracellular calcium in natural killer cells through pertussis toxin-sensitive and -insensitive heterotrimeric G-proteins. *Faseb J.* **11,** 765–774.

Maione, T. E., Gray, G. S., Petro, J., Hunt, A. J., Donner, A. L., Bauer, S. I., Carson, H. F., and Sharpe, R. J. (1990). Inhibition of angiogenesis by recombinant human platelet factor-4 and related peptides. *Science* **247,** 77–79.

Majumdar, S., Gonder, D., Koutsis, B., and Poncz, M. (1991). Characterization of the human beta-thromboglobulin gene: Comparison with the gene for platelet factor 4. *J. Biol. Chem.* **266,** 5785–5789.

Majumder, S., Zhou, L. Z., and Ransohoff, R. M. (1996). Transcriptional regulation of chemokine gene expression in astrocytes. *J. Neurosci. Res.* **45,** 758–769.

Male, D., and Rezaie, P. (2001). Colonisation of the human central nervous system by microglia: The roles of chemokines and vascular adhesion molecules. *Prog. Brain Res.* **132,** 81–93.

Matsushima, K., Larsen, C. G., Dubois, G. C., and Oppenheim, J. J. (1989). Purification and characterization of a novel monocyte chemotactic and activating factor produced by a human myelomonocytic cell line. *J. Exp. Med.* **169,** 1485–1490.

McDermott, D. H., Halcox, J. P., Schenke, W. H., Waclawiw, M. A., Merrell, M. N., Epstein, N., Quyyumi, A. A., and Murphy, P. M. (2001). Association between polymorphism in the chemokine receptor CX3CR1 and coronary vascular endothelial dysfunction and atherosclerosis. *Circ. Res.* **89,** 401–407.

McDermott, D. H., Fong, A. M., Yang, Q., Sechler, J. M., Cupples, L. A., Merrell, M. N., Wilson, P. W., D'Agostino, R. B., O'Donnell, C. J., Patel, D. D., and Murphy, P. M. (2003). Chemokine receptor mutant CX3CR1-M280 has impaired adhesive function and correlates with protection from cardiovascular disease in humans. *J. Clin. Invest.* **111,** 1241–1250.

McGrath, K. E., Koniski, A. D., Maltby, K. M., McGann, J. K., and Palis, J. (1999). Embryonic expression and function of the chemokine SDF-1 and its receptor, CXCR4. *Dev. Biol.* **213,** 442–456.

McManus, C., Berman, J. W., Brett, F. M., Staunton, H., Farrell, M., and Brosnan, C. F. (1998a). MCP-1, MCP-2 and MCP-3 expression in multiple sclerosis lesions: An immunohistochemical and in situ hybridization study. *J. Neuroimmunol.* **86,** 20–29.

McManus, C. M., Brosnan, C. F., and Berman, J. W. (1998b). Cytokine induction of MIP-1 alpha and MIP-1 beta in human fetal microglia. *J. Immunol.* **160,** 1449–1455.

McManus, C. M., Weidenheim, K., Woodman, S. E., Nunez, J., Hesselgesser, J., Nath, A., and Berman, J. W. (2000). Chemokine and chemokine-receptor expression in human glial elements: induction by the HIV protein, Tat, and chemokine autoregulation. *Am. J. Pathol.* **156,** 1441–1453.

Meucci, O., Fatatis, A., Simen, A. A., Bushell, T. J., Gray, P. W., and Miller, R. J. (1998). Chemokines regulate hippocampal neuronal signaling and gp120 neurotoxicity. *Proc. Natl. Acad. Sci. USA* **95,** 14500–14505.

Meucci, O., Fatatis, A., Simen, A. A., and Miller, R. J. (2000). Expression of CX3CR1 chemokine receptors on neurons and their role in neuronal survival. *Proc. Natl. Acad. Sci. USA* **97,** 8075–8080.

Minty, A., Chalon, P., Guillemot, J. C., Kaghad, M., Liauzun, P., Magazin, M., Miloux, B., Minty, C., Ramond, P., and Vita, N. (1993). Molecular cloning of the MCP-3 chemokine gene and regulation of its expression. *Eur. Cytokine Netw.* **4,** 99–110.

Miyagishi, R., Kikuchi, S., Fukazawa, T., and Tashiro, K. (1995). Macrophage inflammatory protein-1 alpha in the cerebrospinal fluid of patients with multiple sclerosis and other inflammatory neurological diseases. *J. Neurol. Sci.* **129,** 223–227.

Moatti, D., Faure, S., Fumeron, F., Amara Mel, W., Seknadji, P., McDermott, D. H., Debre, P., Aumont, M. C., Murphy, P. M., De Prost, D., and Combadiere, C. (2001). Polymorphism in the fractalkine receptor CX3CR1 as a genetic risk factor for coronary artery disease. *Blood* **97,** 1925–1928.

Mochizuki, M., Bartels, J., Mallet, A. I., Christophers, E., and Schroder, J. M. (1998). IL-4 induces eotaxin: A possible mechanism of selective eosinophil recruitment in helminth infection and atopy. *J. Immunol.* **160,** 60–68.

Modi, W. S., and Chen, Z. Q. (1998). Localization of the human CXC chemokine subfamily on the long arm of chromosome 4 using radiation hybrids. *Genomics* **47,** 136–139.

Molino, M., Woolkalis, M. J., Prevost, N., Pratico, D., Barnathan, E. S., Taraboletti, G., Haggarty, B. S., Hesselgesser, J., Horuk, R., Hoxie, J. A., and Brass, L. F. (2000). CXCR4 on human endothelial cells can serve as both a mediator of biological responses and as a receptor for HIV-2. *Biochim. Biophys. Acta* **1500,** 227–240.

Moser, B., Clark-Lewis, I., Zwahlen, R., and Baggiolini, M. (1990). Neutrophil-activating properties of the melanoma growth-stimulatory activity. *J. Exp. Med.* **171,** 1797–1802.

Moser, B., Schumacher, C., Von Tscharner, V., Clark-Lewis, I., and Baggiolini, M. (1991). Neutrophil-activating peptide 2 and gro/melanoma growth-stimulatory activity interact with neutrophil-activating peptide 1/interleukin 8 receptors on human neutrophils. *J. Biol. Chem.* **266,** 10666–10671.

Moser, B., Barella, L., Mattei, S., Schumacher, C., Boulay, F., Colombo, M. P., and Baggiolini, M. (1993). Expression of transcripts for two interleukin 8 receptors in human phagocytes, lymphocytes and melanoma cells. *Biochem. J.* **294**(Pt. 1), 285–292.

Muller, S., Dorner, B., Korthauer, U., Mages, H. W., D'Apuzzo, M., Senger, G., and Kroczek, R. A. (1995). Cloning of ATAC, an activation-induced, chemokine-related molecule exclusively expressed in CD8+ T lymphocytes. *Eur. J. Immunol.* **25,** 1744–1748.

Muller-Ladner, U., Jones, J. L., Wetsel, R. A., Gay, S., Raine, C. S., and Barnum, S. R. (1996). Enhanced expression of chemotactic receptors in multiple sclerosis lesions. *J. Neurol. Sci.* **144,** 135–141.

Murakami, K., Ueno, A., Yamanouchi, K., and Kondo, T. (1995). Thrombin induces GRO alpha/MGSA production in human umbilical vein endothelial cells. *Thromb. Res.* **79,** 387–394.

Murphy, P. M., and Tiffany, H. L. (1991). Cloning of complementary DNA encoding a functional human interleukin-8 receptor. *Science* **253,** 1280–1283.

Murphy, P. M., Baggiolini, M., Charo, I. F., Hebert, C. A., Horuk, R., Matsushima, K., Miller, L. H., Oppenheim, J. J., and Power, C. A. (2000). International union of pharmacology. XXII. Nomenclature for chemokine receptors. *Pharmacol. Rev.* **52,** 145–176.

Myers, S. J., Wong, L. M., and Charo, I. F. (1995). Signal transduction and ligand specificity of the human monocyte chemoattractant protein-1 receptor in transfected embryonic kidney cells. *J. Biol. Chem.* **270,** 5786–5792.

Nagasawa, T., Hirota, S., Tachibana, K., Takakura, N., Nishikawa, S., Kitamura, Y., Yoshida, N., Kikutani, H., and Kishimoto, T. (1996). Defects of B-cell lymphopoiesis and bone-marrow myelopoiesis in mice lacking the CXC chemokine PBSF/SDF-1. *Nature* **382,** 635–638.

Nagasawa, T., Kikutani, H., and Kishimoto, T. (1994). Molecular cloning and structure of a pre-B-cell growth-stimulating factor. *Proc. Natl. Acad. Sci. USA* **91,** 2305–2309.

Naruse, K., Ueno, M., Satoh, T., Nomiyama, H., Tei, H., Takeda, M., Ledbetter, D. H., Coillie, E. V., Opdenakker, G., Gunge, N., Sakaki, Y., Iio, M., and Miura, R. (1996). A YAC contig of the human CC chemokine genes clustered on chromosome 17q11.2. *Genomics* **34,** 236–240.

Navia, B. A., Cho, E. S., Petito, C. K., and Price, R. W. (1986). The AIDS dementia complex: II. Neuropathology. *Ann. Neurol.* **19,** 525–535.

Nelken, N. A., Coughlin, S. R., Gordon, D., and Wilcox, J. N. (1991). Monocyte chemoattractant protein-1 in human atheromatous plaques. *J. Clin. Invest.* **88,** 1121–1127.

Neote, K., Digregorio, D., Mak, J. Y., Horuk, R., and Schall, T. J. (1993). Molecular cloning, functional expression, and signaling characteristics of a C-C chemokine receptor. *Cell* **72,** 415–425.

Neville, L. F., Mathiak, G., and Bagasra, O. (1997). The immunobiology of interferon-gamma inducible protein 10 kD (IP-10): A novel, pleiotropic member of the C-X-C chemokine superfamily. *Cytokine Growth Factor Rev.* **8,** 207–219.

Nitta, T., Allegretta, M., Okumura, K., Sato, K., and Steinman, L. (1992). Neoplastic and reactive human astrocytes express interleukin-8 gene. *Neurosurg. Rev.* **15,** 203–207.

Nomiyama, H., Imai, T., Kusuda, J., Miura, R., Callen, D. F., and Yoshie, O. (1998). Human chemokines fractalkine (SCYD1), MDC (SCYA22) and TARC (SCYA17) are clustered on chromosome 16q13. *Cytogenet. Cell Genet.* **81,** 10–11.

Nomura, H., Nielsen, B. W., and Matsushima, K. (1993). Molecular cloning of cDNAs encoding a LD78 receptor and putative leukocyte chemotactic peptide receptors. *Int. Immunol.* **5,** 1239–1249.

Oberlin, E., Amara, A., Bachelerie, F., Bessia, C., Virelizier, J. L., Arenzana-Seisdedos, F., Schwartz, O., Heard, J. M., Clark-Lewis, I., Legler, D. F., Loetscher, M., Baggiolini, M., and Moser, B. (1996). The CXC chemokine SDF-1 is the ligand for LESTR/fusin and prevents infection by T-cell-line-adapted HIV-1. *Nature* **382,** 833–835.

Odemis, V., Moepps, B., Gierschik, P., and Engele, J. (2002). Interleukin-6 and cAMP induce stromal cell-derived factor-1 chemotaxis in astroglia by up-regulating CXCR4 cell surface expression: Implications for brain inflammation. *J. Biol. Chem.* **277,** 39801–39808.

Oh, J. W., Drabik, K., Kutsch, O., Choi, C., Tousson, A., and Benveniste, E. N. (2001). CXC chemokine receptor 4 expression and function in human astroglioma cells. *J. Immunol.* **166,** 2695–2704.

Ohtani, Y., Minami, M., Kawaguchi, N., Nishiyori, A., Yamamoto, J., Takami, S., and Satoh, M. (1998). Expression of stromal cell-derived factor-1 and CXCR4 chemokine receptor mRNAs in cultured rat glial and neuronal cells. *Neurosci. Lett.* **249,** 163–166.

Oliveira, S. H., and Lukacs, N. W. (2001). Stem cell factor and igE-stimulated murine mast cells produce chemokines (CCL2, CCL17, CCL22) and express chemokine receptors. *Inflamm. Res.* **50,** 168–174.

Opdenakker, G., Froyen, G., Fiten, P., Proost, P., and Van Damme, J. (1993). Human monocyte chemotactic protein-3 (MCP-3): Molecular cloning of the cDNA and comparison with other chemokines. *Biochem. Biophys. Res. Commun.* **191,** 535–542.

Oquendo, P., Alberta, J., Wen, D. Z., Graycar, J. L., Derynck, R., and Stiles, C. D. (1989). The platelet-derived growth factor-inducible KC gene encodes a secretory protein related to platelet alpha-granule proteins. *J. Biol. Chem.* **264,** 4133–4137.

Pablos, J. L., Amara, A., Bouloc, A., Santiago, B., Caruz, A., Galindo, M., Delaunay, T., Virelizier, J. L., and Arenzana-Seisdedos, F. (1999). Stromal-cell derived factor is expressed by dendritic cells and endothelium in human skin. *Am. J. Pathol.* **155,** 1577–1586.

Pan, Y., Lloyd, C., Zhou, H., Dolich, S., Deeds, J., Gonzalo, J. A., Vath, J., Gosselin, M., Ma, J., Dussault, B., Woolf, E., Alperin, G., Culpepper, J., Gutierrez-Ramos, J. C., and Gearing, D. (1997). Neurotactin, a membrane-anchored chemokine upregulated in brain inflammation. *Nature* **387,** 611–617.

Park, M. K., Amichay, D., Love, P., Wick, E., Liao, F., Grinberg, A., Rabin, R. L., Zhang, H. H., Gebeyehu, S., Wright, T. M., Iwasaki, A., Weng, Y., Demartino, J. A., Elkins, K. L., and Farber, J. M. (2002). The CXC chemokine murine monokine induced by IFN-gamma (CXC chemokine ligand 9) is made by APCs, targets lymphocytes including activated B cells, and supports antibody responses to a bacterial pathogen *in vivo*. *J. Immunol.* **169,** 1433–1443.

Pereira, C. F., Middel, J., Jansen, G., Verhoef, J., and Nottet, H. S. (2001). Enhanced expression of fractalkine in HIV-1 associated dementia. *J. Neuroimmunol.* **115,** 168–175.

Petersen, F., Ludwig, A., Flad, H. D., and Brandt, E. (1996). TNF-alpha renders human neutrophils responsive to platelet factor 4. Comparison of PF-4 and IL-8 reveals different activity profiles of the two chemokines. *J. Immunol.* **156,** 1954–1962.

Petersen, F., Bock, L., Flad, H. D., and Brandt, E. (1998). A chondroitin sulfate proteoglycan on human neutrophils specifically binds platelet factor 4 and is involved in cell activation. *J. Immunol.* **161,** 4347–4355.

Peterson, P. K., Hu, S., Salak-Johnson, J., Molitor, T. W., and Chao, C. C. (1997). Differential production of and migratory response to beta chemokines by human microglia and astrocytes. *J. Infect. Dis.* **175,** 478–481.

Ponath, P. D., Qin, S., Post, T. W., Wang, J., Wu, L., Gerard, N. P., Newman, W., Gerard, C., and Mackay, C. R. (1996a). Molecular cloning and characterization of a human eotaxin receptor expressed selectively on eosinophils. *J. Exp. Med.* **183,** 2437–2448.

Ponath, P. D., Qin, S., Ringler, D. J., Clark-Lewis, I., Wang, J., Kassam, N., Smith, H., Shi, X., Gonzalo, J. A., Newman, W., Gutierrez-Ramos, J. C., and Mackay, C. R. (1996b). Cloning of the human eosinophil chemoattractant, eotaxin: Expression, receptor binding, and functional properties suggest a mechanism for the selective recruitment of eosinophils. *J. Clin. Invest.* **97,** 604–612.

Premack, B. A., and Schall, T. J. (1996). Chemokine receptors: Gateways to inflammation and infection. *Nat. Med.* **2,** 1174–1178.

Proost, P., De Wolf-Peeters, C., Conings, R., Opdenakker, G., Billiau, A., and Van Damme, J. (1993). Identification of a novel granulocyte chemotactic protein (GCP-2) from human tumor cells. *In vitro* and *in vivo* comparison with natural forms of GRO, IP-10, and IL-8. *J. Immunol.* **150,** 1000–1010.

Qin, S., Larosa, G., Campbell, J. J., Smith-Heath, H., Kassam, N., Shi, X., Zeng, L., Buthcher, E. C., and Mackay, C. R. (1996). Expression of monocyte chemoattractant protein-1 and interleukin-8 receptors on subsets of T cells: Correlation with transendothelial chemotactic potential. *Eur. J. Immunol.* **26,** 640–647.

Qin, S., Rottman, J. B., Myers, P., Kassam, N., Weinblatt, M., Loetscher, M., Koch, A. E., Moser, B., and Mackay, C. R. (1998). The chemokine receptors CXCR3 and CCR5 mark subsets of T cells associated with certain inflammatory reactions. *J. Clin. Invest.* **101,** 746–754.

Rabin, R. L., Park, M. K., Liao, F., Swofford, R., Stephany, D., and Farber, J. M. (1999). Chemokine receptor responses on T cells are achieved through regulation of both receptor expression and signaling. *J. Immunol.* **162,** 3840–3850.

Ransohoff, R. M., Hamilton, T. A., Tani, M., Stoler, M. H., Shick, H. E., Major, J. A., Estes, M. L., Thomas, D. M., and Tuohy, V. K. (1993). Astrocyte expression of mRNA encoding cytokines IP-10 and JE/MCP-1 in experimental autoimmune encephalomyelitis. *Faseb J.* **7,** 592–600.

Raport, C. J., Schweickart, V. L., Eddy, R. L., Jr., Shows, T. B., and Gray, P. W. (1995). The orphan G-protein-coupled receptor-encoding gene V28 is closely related to genes for chemokine receptors and is expressed in lymphoid and neural tissues. *Gene* **163,** 295–299.

Raport, C. J., Gosling, J., Schweickart, V. L., Gray, P. W., and Charo, I. F. (1996). Molecular cloning and functional characterization of a novel human CC chemokine receptor (CCR5) for RANTES, MIP-1beta, and MIP-1alpha. *J. Biol. Chem.* **271,** 17161–17166.

Rappaport, J., Joseph, J., Croul, S., Alexander, G., Del Valle, L., Amini, S., and Khalili, K. (1999). Molecular pathway involved in HIV-1-induced CNS pathology: Role of viral regulatory protein, Tat. *J. Leukoc. Biol.* **65,** 458–465.

Rathanaswami, P., Hachicha, M., Sadick, M., Schall, T. J., and Mccoll, S. R. (1993). Expression of the cytokine RANTES in human rheumatoid synovial fibroblasts: Differential regulation of RANTES and interleukin-8 genes by inflammatory cytokines. *J. Biol. Chem.* **268,** 5834–5839.

Reape, T. J., and Groot, P. H. (1999). Chemokines and atherosclerosis. *Atherosclerosis* **147,** 213–225.

Richmond, A., Balentien, E., Thomas, H. G., Flaggs, G., Barton, D. E., Spiess, J., Bordoni, R., Francke, U., and Derynck, R. (1988). Molecular characterization and chromosomal mapping of melanoma growth stimulatory activity, a growth factor structurally related to beta-thromboglobulin. *EMBO J.* **7,** 2025–2033.

Robinson, E., Keystone, E. C., Schall, T. J., Gillett, N., and Fish, E. N. (1995). Chemokine expression in rheumatoid arthritis (RA): Evidence of RANTES and macrophage inflammatory protein (MIP)-1 beta production by synovial T cells. *Clin. Exp. Immunol.* **101,** 398–407.

Robinson, S., Tani, M., Strieter, R. M., Ransohoff, R. M., and Miller, R. H. (1998). The chemokine growth-regulated oncogene-alpha promotes spinal cord oligodendrocyte precursor proliferation. *J. Neurosci.* **18,** 10457–10463.

Rodriguez-Frade, J. M., Vila-Coro, A. J., De Ana, A. M., Albar, J. P., Martinez, A. C., and Mellado, M. (1999). The chemokine monocyte chemoattractant protein-1 induces functional responses through dimerization of its receptor CCR2. *Proc. Natl. Acad. Sci. USA* **96,** 3628–3633.

Rollins, B. J. (1997). Chemokines. *Blood* **90,** 909–928.

Rollins, B. J., Morrison, E. D., and Stiles, C. D. (1988). Cloning and expression of JE, a gene inducible by platelet-derived growth factor and whose product has cytokine-like properties. *Proc. Natl. Acad. Sci. USA* **85,** 3738–3742.

Rollins, B. J., Stier, P., Ernst, T., and Wong, G. G. (1989). The human homolog of the JE gene encodes a monocyte secretory protein. *Mol. Cell. Biol.* **9,** 4687–4695.

Rollins, B. J., Yoshimura, T., Leonard, E. J., and Pober, J. S. (1990). Cytokine-activated human endothelial cells synthesize and secrete a monocyte chemoattractant, MCP-1/JE. *Am. J. Pathol.* **136,** 1229–1233.

Ross, R. (1993). The pathogenesis of atherosclerosis: A perspective for the 1990s. *Nature* **362,** 801–809.

Rossi, D. L., Vicari, A. P., Franz-Bacon, K., McClanahan, T. K., and Zlotnik, A. (1997). Identification through bioinformatics of two new macrophage proinflammatory human chemokines: MIP-3alpha and MIP-3beta. *J. Immunol.* **158,** 1033–1036.

Rot, A., Krieger, M., Brunner, T., Bischoff, S. C., Schall, T. J., and Dahinden, C. A. (1992). RANTES and macrophage inflammatory protein-1 alpha induce the migration and activation of normal human eosinophil granulocytes. *J. Exp. Med.* **176,** 1489–1495.

Roth, S. J., Carr, M. W., and Springer, T. A. (1995). C-C chemokines, but not the C-X-C chemokines interleukin-8 and interferon-gamma inducible protein-10, stimulate transendothelial chemotaxis of T lymphocytes. *Eur. J. Immunol.* **25,** 3482–3488.

Rothenberg, M. E., Luster, A. D., and Leder, P. (1995). Murine eotaxin: An eosinophil chemoattractant inducible in endothelial cells and in interleukin 4-induced tumor suppression. *Proc. Natl. Acad. Sci. USA* **92,** 8960–8964.

Rothenberg, M. E., Ownbey, R., Mehlhop, P. D., Loiselle, P. M., Van De Rijn, M., Bonventre, J. V., Oettgen, H. C., Leder, P., and Luster, A. D. (1996). Eotaxin triggers eosinophil-selective chemotaxis and calcium flux via a distinct receptor and induces pulmonary eosinophilia in the presence of interleukin-5 in mice. *Mol. Med.* **2,** 334–348.

Rubanyl, G. M., Frye, R. L., Holmes, D. R., Jr., and Vanhoutte, P. M. (1987). Vasoconstrictor activity of coronary sinus plasma from patients with coronary artery disease. *J. Am. Coll. Cardiol.* **9,** 1243–1249.

Rutledge, B. J., Rayburn, H., Rosenberg, R., North, R. J., Gladue, R. P., Corless, C. L., and Rollins, B. J. (1995). High level monocyte chemoattractant protein-1 expression in transgenic mice increases their susceptibility to intracellular pathogens. *J. Immunol.* **155,** 4838–4843.

Salcedo, R., Wasserman, K., Young, H. A., Grimm, M. C., Howard, O. M., Anver, M. R., Kleinman, H. K., Murphy, W. J., and Oppenheim, J. J. (1999). Vascular endothelial growth factor and basic fibroblast growth factor induce expression of CXCR4 on human endothelial cells: *In vivo* neovascularization induced by stromal derived factor-1alpha. *Am. J. Pathol.* **154,** 1125–1135.

Sallusto, F., Mackay, C. R., and Lanzavecchia, A. (1997). Selective expression of the eotaxin receptor CCR3 by human T helper 2 cells. *Science* **277,** 2005–2007.

Sallusto, F., Lenig, D., Mackay, C. R., and Lanzavecchia, A. (1998). Flexible programs of chemokine receptor expression on human polarized T helper 1 and 2 lymphocytes. *J. Exp. Med.* **187,** 875–883.

Sallusto, F., Palermo, B., Lenig, D., Miettinen, M., Matikainen, S., Julkunen, I., Forster, R., Burgstahler, R., Lipp, M., and Lanzavecchia, A. (1999). Distinct patterns and kinetics of chemokine production regulate dendritic cell function. *Eur. J. Immunol.* **29,** 1617–1625.

Samson, M., Labbe, O., Mollereau, C., Vassart, G., and Parmentier, M. (1996a). Molecular cloning and functional expression of a new human CC-chemokine receptor gene. *Biochemistry* **35,** 3362–3367.

Samson, M., Libert, F., Doranz, B. J., Rucker, J., Liesnard, C., Farber, C. M., Saragosti, S., Lapoumeroulie, C., Cognaux, J., Forceille, C., Muyldermans, G., Verhofstede, C., Burtonboy, G., Georges, M., Imai, T., Rana, S., Yi, Y., Smyth, R. J., Collman, R. G., Doms, R. W., Vassart, G., and Parmentier, M. (1996b). Resistance to HIV-1 infection in caucasian individuals bearing mutant alleles of the CCR-5 chemokine receptor gene. *Nature* **382,** 722–725.

Sanders, V. J., Pittman, C. A., White, M. G., Wang, G., Wiley, C. A., and Achim, C. L. (1998). Chemokines and receptors in HIV encephalitis. *Aids* **12,** 1021–1026.

Sarafi, M. N., Garcia-Zepeda, E. A., Maclean, J. A., Charo, I. F., and Luster, A. D. (1997). Murine monocyte chemoattractant protein (MCP)-5: A novel CC chemokine that is a structural and functional homologue of human MCP-1. *J. Exp. Med.* **185,** 99–109.

Sasseville, V. G., Smith, M. M., Mackay, C. R., Pauley, D. R., Mansfield, K. G., Ringler, D. J., and Lackner, A. A. (1996). Chemokine expression in simian immunodeficiency virus-induced AIDS encephalitis. *Am. J. Pathol.* **149,** 1459–1467.

Schall, T. J., Bacon, K., Camp, R. D., Kaspari, J. W., and Goeddel, D. V. (1993). Human macrophage inflammatory protein alpha (MIP-1 alpha) and MIP-1 beta chemokines attract distinct populations of lymphocytes. *J. Exp. Med.* **177,** 1821–1826.

Schall, T. J., Bacon, K., Toy, K. J., and Goeddel, D. V. (1990). Selective attraction of monocytes and T-lymphocytes of the memory phenotype by cytokine RANTES. *Nature* **347,** 669–671.

Schall, T. J., Jongstra, J., Dyer, B. J., Jorgensen, J., Clayberger, C., Davis, M. M., and Krensky, A. M. (1988). A human T cell-specific molecule is a member of a new gene family. *J. Immunol.* **141,** 1018–1025.

Schecter, A. D., Rollins, B. J., Zhang, Y. J., Charo, I. F., Fallon, J. T., Rossikhina, M., Giesen, P. L., Nemerson, Y., and Taubman, M. B. (1997). Tissue factor is induced by monocyte chemoattractant protein-1 in human aortic smooth muscle and THP-1 cells. *J. Biol. Chem.* **272,** 28568–28573.

Scheuerer, B., Ernst, M., Durrbaum-Landmann, I., Fleischer, J., Grage-Griebenow, E., Brandt, E., Flad, H. D., and Petersen, F. (2000). The CXC-chemokine platelet factor 4 promotes monocyte survival and induces monocyte differentiation into macrophages. *Blood* **95,** 1158–1166.

Schmidtmayerova, H., Nottet, H. S., Nuovo, G., Raabe, T., Flanagan, C. R., Dubrovsky, L., Gendelman, H. E., Cerami, A., Bukrinsky, M., and Sherry, B. (1996). Human immunodeficiency virus type 1 infection alters chemokine beta peptide expression in human monocytes: Implications for recruitment of leukocytes into brain and lymph nodes. *Proc. Natl. Acad. Sci. USA* **93,** 700–704.

Schroder, J. M., and Christophers, E. (1989). Secretion of novel and homologous neutrophil-activating peptides by LPS-stimulated human endothelial cells. *J. Immunol.* **142,** 244–251.

Schroder, J. M., Mrowietz, U., Morita, E., and Christophers, E. (1987). Purification and partial biochemical characterization of a human monocyte-derived, neutrophil-activating peptide that lacks interleukin-1 activity. *J. Immunol.* **139,** 3474–3483.

Schroder, J. M., Persoon, N. L., and Christophers, E. (1990). Lipopolysaccharide-stimulated human monocytes secrete, apart from neutrophil-activating peptide-1/interleukin-8, a second neutrophil-activating protein: NH2-terminal amino acid sequence identity with melanoma growth stimulatory activity. *J. Exp. Med.* **171,** 1091–1100.

Shattuck, R. L., Wood, L. D., Jaffe, G. J., and Richmond, A. (1994). MGSA/GRO transcription is differentially regulated in normal retinal pigment epithelial and melanoma cells. *Mol. Cell. Biol.* **14,** 791–802.

Sherry, B., Tekamp-Olson, P., Gallegos, C., Bauer, D., Davatelis, G., Wolpe, S. D., Masiarz, F., Coit, D., and Cerami, A. (1988). Resolution of the two components of macrophage inflammatory protein-1, and cloning and characterization of one of those components, macrophage inflammatory protein-1 beta. *J. Exp. Med.* **168,** 2251–2259.

Shirozu, M., Nakano, T., Inazawa, J., Tashiro, K., Tada, H., Shinohara, T., and Honjo, T. (1995). Structure and chromosomal localization of the human stromal cell-derived factor-1 (SDF1) gene. *Genomics* **28,** 495–500.

Simpson, J. E., Newcombe, J., Cuzner, M. L., and Woodroofe, M. N. (1998). Expression of monocyte chemoattractant protein-1 and other beta-chemokines by resident glia and inflammatory cells in multiple sclerosis lesions. *J. Neuroimmunol.* **84,** 238–249.

Simpson, J., Rezaie, P., Newcombe, J., Cuzner, M. L., Male, D., and Woodroofe, M. N. (2000a). Expression of the beta-chemokine receptors CCR2, CCR,3 and CCR5 in multiple sclerosis central nervous system tissue. *J. Neuroimmunol.* **108,** 192–200.

Simpson, J. E., Newcombe, J., Cuzner, M. L., and Woodroofe, M. N. (2000b). Expression of the interferon-gamma-inducible chemokines IP-10 and Mig and their receptor, CXCR3, in multiple sclerosis lesions. *Neuropathol. Appl. Neurobiol.* **26,** 133–142.

Sorensen, T. L., Tani, M., Jensen, J., Pierce, V., Lucchinetti, C., Folcik, V. A., Qin, S., Rottman, J., Sellebjerg, F., Strieter, R. M., Frederiksen, J. L., and Ransohoff, R. M. (1999). Expression of specific chemokines and chemokine receptors in the central nervous system of multiple sclerosis patients. *J. Clin. Invest.* **103,** 807–815.

Sozzani, S., Sallusto, F., Luini, W., Zhou, D., Piemonti, L., Allavena, P., Van Damme, J., Valitutti, S., Lanzavecchia, A., and Mantovani, A. (1995). Migration of dendritic cells in response to formyl peptides, C5a, and a distinct set of chemokines. *J. Immunol.* **155,** 3292–3295.

Sozzani, S., Introna, M., Bernasconi, S., Polentarutti, N., Cinque, P., Poli, G., Sica, A., and Mantovani, A. (1997a). MCP-1 and CCR2 in HIV infection: Regulation of agonist and receptor expression. *J. Leukoc. Biol.* **62,** 30–33.

Sozzani, S., Luini, W., Borsatti, A., Polentarutti, N., Zhou, D., Piemonti, L., D'amico, G., Power, C. A., Wells, T. N., Gobbi, M., Allavena, P., and Mantovani, A. (1997b). Receptor expression and responsiveness of human dendritic cells to a defined set of CC and CXC chemokines. *J. Immunol.* **159,** 1993–2000.

Standiford, T. J., Kunkel, S. L., Basha, M. A., Chensue, S. W., Lynch, J. P., III, Toews, G. B., Westwick, J., and Strieter, R. M. (1990). Interleukin-8 gene expression by a pulmonary epithelial cell line. A model for cytokine networks in the lung. *J. Clin. Invest.* **86,** 1945–1953.

Standiford, T. J., Kunkel, S. L., Liebler, J. M., Burdick, M. D., Gilbert, A. R., and Strieter, R. M. (1993). Gene expression of macrophage inflammatory protein-1 alpha from human blood monocytes and alveolar macrophages is inhibited by interleukin-4. *Am. J. Respir. Cell Mol. Biol.* **9,** 192–198.

Stellato, C., Beck, L. A., Gorgone, G. A., Proud, D., Schall, T. J., Ono, S. J., Lichtenstein, L. M., and Schleimer, R. P. (1995). Expression of the chemokine RANTES by a human bronchial epithelial cell line. Modulation by cytokines and glucocorticoids. *J. Immunol.* **155,** 410–418.

Stoeckle, M. Y. (1991). Post-transcriptional regulation of gro alpha, beta, gamma, and IL-8 mRNAs by IL-1 beta. *Nucleic Acids Res.* **19,** 917–920.

Stoeckle, M. Y. (1992). Removal of a 3' non-coding sequence is an initial step in degradation of gro alpha mRNA and is regulated by interleukin-1. *Nucleic Acids Res.* **20,** 1123–1127.

Strieter, R. M., Kunkel, S. L., Showell, H. J., Remick, D. G., Phan, S. H., Ward, P. A., and Marks, R. M. (1989a). Endothelial cell gene expression of a neutrophil chemotactic factor by TNF-alpha, LPS, and IL-1 beta. *Science* **243,** 1467–1469.

Strieter, R. M., Phan, S. H., Showell, H. J., Remick, D. G., Lynch, J. P., Genord, M., Raiford, C., Eskandari, M., Marks, R. M., and Kunkel, S. L. (1989b). Monokine-induced neutrophil chemotactic factor gene expression in human fibroblasts. *J. Biol. Chem.* **264,** 10621–10626.

Strieter, R. M., Wiggins, R., Phan, S. H., Wharram, B. L., Showell, H. J., Remick, D. G., Chensue, S. W., and Kunkel, S. L. (1989c). Monocyte chemotactic protein gene expression by cytokine-treated human fibroblasts and endothelial cells. *Biochem. Biophys. Res. Commun.* **162,** 694–700.

Strieter, R. M., Kunkel, S. L., Elner, V. M., Martonyi, C. L., Koch, A. E., Polverini, P. J., and Elner, S. G. (1992). Interleukin-8. A corneal factor that induces neovascularization. *Am. J. Pathol.* **141,** 1279–1284.

Strieter, R. M., Polverini, P. J., Kunkel, S. L., Arenberg, D. A., Burdick, M. D., Kasper, J., Dzuiba, J., Van Damme, J., Walz, A., Marriott, D., *et al.* (1995). The functional role of the ELR motif in CXC chemokine-mediated angiogenesis. *J. Biol. Chem.* **270,** 27348–27357.

Su, S. B., Mukaida, N., Wang, J., Nomura, H., and Matsushima, K. (1996). Preparation of specific polyclonal antibodies to a C-C chemokine receptor, CCR1, and determination of CCR1 expression on various types of leukocytes. *J. Leukoc. Biol.* **60,** 658–666.

Tachibana, K., Hirota, S., Iizasa, H., Yoshida, H., Kawabata, K., Kataoka, Y., Kitamura, Y., Matsushima, K., Yoshida, N., Nishikawa, S., Kishimoto, T., and Nagasawa, T. (1998). The chemokine receptor CXCR4 is essential for vascularization of the gastrointestinal tract. *Nature* **393,** 591–594.

Takata, H., Tomiyama, H., Fujiwara, M., Kobayashi, N., and Takiguchi, M. (2004). Cutting edge: Expression of chemokine receptor CXCR1 on human effector CD8+ T cells. *J. Immunol.* **173,** 2231–2235.

Takeya, M., Yoshimura, T., Leonard, E. J., and Takahashi, K. (1993). Detection of monocyte chemoattractant protein-1 in human atherosclerotic lesions by an anti-monocyte chemoattractant protein-1 monoclonal antibody. *Hum. Pathol.* **24,** 534–539.

Tanabe, S., Heesen, M., Yoshizawa, I., Berman, M. A., Luo, Y., Bleul, C. C., Springer, T. A., Okuda, K., Gerard, N., and Dorf, M. E. (1997). Functional expression of the CXC-chemokine receptor-4/fusin on mouse microglial cells and astrocytes. *J. Immunol.* **159,** 905–911.

Tang, W. W., Yin, S., Wittwer, A. J., and Qi, M. (1995). Chemokine gene expression in anti-glomerular basement membrane antibody glomerulonephritis. *Am. J. Physiol.* **269,** F323–F330.

Tani, K., Su, S. B., Utsunomiya, I., Oppenheim, J. J., and Wang, J. M. (1998). Interferon-gamma maintains the binding and functional capacity of receptors for IL-8 on cultured human T cells. *Eur. J. Immunol.* **28,** 502–507.

Tashiro, K., Tada, H., Heilker, R., Shirozu, M., Nakano, T., and Honjo, T. (1993). Signal sequence trap: A cloning strategy for secreted proteins and type I membrane proteins. *Science* **261,** 600–603.

Taub, D. D., and Oppenheim, J. J. (1993). Review of the chemokine meeting the Third International Symposium of Chemotactic Cytokines. *Cytokine* **5,** 175–179.

Taub, D. D., Conlon, K., Lloyd, A. R., Oppenheim, J. J., and Kelvin, D. J. (1993a). Preferential migration of activated CD4+ and CD8+ T cells in response to MIP-1 alpha and MIP-1 beta. *Science* **260,** 355–358.

Taub, D. D., Lloyd, A. R., Conlon, K., Wang, J. M., Ortaldo, J. R., Harada, A., Matsushima, K., Kelvin, D. J., and Oppenheim, J. J. (1993b). Recombinant human interferon-inducible protein-10 is a chemoattractant for human monocytes and T lymphocytes and promotes T cell adhesion to endothelial cells. *J. Exp. Med.* **177,** 1809–1814.

Taub, D. D., Proost, P., Murphy, W. J., Anver, M., Longo, D. L., Van Damme, J., and Oppenheim, J. J. (1995a). Monocyte chemotactic protein-1 (MCP-1), -2, and -3 are chemotactic for human T lymphocytes. *J. Clin. Invest.* **95,** 1370–1376.

Taub, D. D., Sayers, T. J., Carter, C. R., and Ortaldo, J. R. (1995b). Alpha and beta chemokines induce NK cell migration and enhance NK-mediated cytolysis. *J. Immunol.* **155,** 3877–3888.

Taub, D. D., Anver, M., Oppenheim, J. J., Longo, D. L., and Murphy, W. J. (1996a). T lymphocyte recruitment by interleukin-8 (IL-8). IL-8-induced degranulation of neutrophils releases potent chemoattractants for human T lymphocytes both *in vitro* and *in vivo*. *J. Clin. Invest.* **97,** 1931–1941.

Taub, D. D., Longo, D. L., and Murphy, W. J. (1996b). Human interferon-inducible protein-10 induces mononuclear cell infiltration in mice and promotes the migration of human T lymphocytes into the peripheral tissues and human peripheral blood lymphocytes-SCID mice. *Blood* **87,** 1423–1431.

Tekamp-Olson, P., Gallegos, C., Bauer, D., McClain, J., Sherry, B., Fabre, M., Van Deventer, S., and Cerami, A. (1990). Cloning and characterization of cDNAs for murine macrophage inflammatory protein-2 and its human homologues. *J. Exp. Med.* **172,** 911–919.

Tong, N., Perry, S. W., Zhang, Q., James, H. J., Guo, H., Brooks, A., Bal, H., Kinnear, S. A., Fine, S., Epstein, L. G., Dairaghi, D., Schall, T. J., Gendelman, H. E., Dewhurst, S., Sharer, L. R., and Gelbard, H. A. (2000). Neuronal fractalkine expression in HIV-1 encephalitis: Roles for macrophage recruitment and neuroprotection in the central nervous system. *J. Immunol.* **164,** 1333–1339.

Uguccioni, M., D'Apuzzo, M., Loetscher, M., Dewald, B., and Baggiolini, M. (1995). Actions of the chemotactic cytokines MCP-1, MCP-2, MCP-3, RANTES, MIP-1 alpha and MIP-1 beta on human monocytes. *Eur. J. Immunol.* **25,** 64–68.

Uguccioni, M., Loetscher, P., Forssmann, U., Dewald, B., Li, H., Lima, S. H., Li, Y., Kreider, B., Garotta, G., Thelen, M., and Baggiolini, M. (1996). Monocyte chemotactic protein-4 (MCP-4), a novel structural and functional analogue of MCP-3 and eotaxin. *J. Exp. Med.* **183,** 2379–2384.

Uguccioni, M., Mackay, C. R., Ochensberger, B., Loetscher, P., Rhis, S., Larosa, G. J., Rao, P., Ponath, P. D., Baggiolini, M., and Dahinden, C. A. (1997). High expression of the chemokine receptor CCR3 in human blood basophils: Role in activation by eotaxin, MCP-4, and other chemokines. *J. Clin. Invest.* **100,** 1137–1143.

Valente, A. J., Graves, D. T., Vialle-Valentin, C. E., Delgado, R., and Schwartz, C. J. (1988). Purification of a monocyte chemotactic factor secreted by nonhuman primate vascular cells in culture. *Biochemistry* **27,** 4162–4168.

Vallat, A. V., De Girolami, U., He, J., Mhashilkar, A., Marasco, W., Shi, B., Gray, F., Bell, J., Keohane, C., Smith, T. W., and Gabuzda, D. (1998). Localization of HIV-1 co-receptors CCR5 and CXCR4 in the brain of children with AIDS. *Am. J. Pathol.* **152,** 167–178.

Van Damme, J., Decock, B., Conings, R., Lenaerts, J. P., Opdenakker, G., and Billiau, A. (1989). The chemotactic activity for granulocytes produced by virally infected fibroblasts is identical to monocyte-derived interleukin-8. *Eur. J. Immunol.* **19,** 1189–1194.

Van Damme, J., Proost, P., Lenaerts, J. P., and Opdenakker, G. (1992). Structural and functional identification of two human, tumor-derived monocyte chemotactic proteins (MCP-2 and MCP-3) belonging to the chemokine family. *J. Exp. Med.* **176,** 59–65.

Van Damme, J., Proost, P., Put, W., Arens, S., Lenaerts, J. P., Conings, R., Opdenakker, G., Heremans, H., and Billiau, A. (1994). Induction of monocyte chemotactic proteins MCP-1 and MCP-2 in human fibroblasts and leukocytes by cytokines and cytokine inducers. Chemical synthesis of MCP-2 and development of a specific RIA. *J. Immunol.* **152,** 5495–5502.

Van Der Voorn, P., Tekstra, J., Beelen, R. H., Tensen, C. P., Van Der Valk, P., and De Groot, C. J. (1999). Expression of MCP-1 by reactive astrocytes in demyelinating multiple sclerosis lesions. *Am. J. Pathol.* **154,** 45–51.

Vanguri, P. (1995). Interferon-gamma-inducible genes in primary glial cells of the central nervous system. Comparisons of astrocytes with microglia and Lewis with brown Norway rats. *J. Neuroimmunol.* **56,** 35–43.

Viedt, C., Vogel, J., Athanasiou, T., Shen, W., Orth, S. R., Kubler, W., and Kreuzer, J. (2002). Monocyte chemoattractant protein-1 induces proliferation and interleukin-6 production in human smooth muscle cells by differential activation of nuclear factor-kappaB and activator protein-1. *Arterioscler. Thromb. Vasc. Biol.* **22,** 914–920.

Vila-Coro, A. J., Rodriguez-Frade, J. M., Martin De Ana, A., Moreno-Ortiz, M. C., Martinez, A. C., and Mellado, M. (1999). The chemokine SDF-1alpha triggers CXCR4 receptor dimerization and activates the JAK/STAT pathway. *Faseb J.* **13,** 1699–1710.

Villiger, P. M., Terkeltaub, R., and Lotz, M. (1992). Production of monocyte chemoattractant protein-1 by inflamed synovial tissue and cultured synoviocytes. *J. Immunol.* **149,** 722–727.

Volin, M. V., Joseph, L., Shockley, M. S., and Davies, P. F. (1998). Chemokine receptor CXCR4 expression in endothelium. *Biochem. Biophys. Res. Commun.* **242,** 46–53.

Walz, A., and Baggiolini, M. (1989). A novel cleavage product of beta-thromboglobulin formed in cultures of stimulated mononuclear cells activates human neutrophils. *Biochem. Biophys. Res. Commun.* **159,** 969–975.

Walz, A., Peveri, P., Aschauer, H., and Baggiolini, M. (1987). Purification and amino acid sequencing of NAF, a novel neutrophil-activating factor produced by monocytes. *Biochem. Biophys. Res. Commun.* **149,** 755–761.

Walz, A., Dewald, B., Von Tscharner, V., and Baggiolini, M. (1989). Effects of the neutrophil-activating peptide NAP-2, platelet basic protein, connective tissue-activating peptide III and platelet factor-4 on human neutrophils. *J. Exp. Med.* **170,** 1745–1750.

Walz, A., and Baggiolini, M. (1990). Generation of the neutrophil-activating peptide NAP-2 from platelet basic protein or connective tissue-activating peptide III through monocyte proteases. *J. Exp. Med.* **171,** 449–454.

Walz, A., Burgener, R., Car, B., Baggiolini, M., Kunkel, S. L., and Strieter, R. M. (1991). Structure and neutrophil-activating properties of a novel inflammatory peptide (ENA-78) with homology to interleukin-8. *J. Exp. Med.* **174,** 1355–1362.

Walz, A., Schmutz, P., Mueller, C., and Schnyder-Candrian, S. (1997). Regulation and function of the CXC chemokine ENA-78 in monocytes and its role in disease. *J. Leukoc. Biol.* **62,** 604–611.

Wang, C. R., Liu, M. F., Huang, Y. H., and Chen, H. C. (2004). Up-regulation of XCR1 expression in rheumatoid joints. *Rheumatology* (Oxford) **43,** 569–573.

Wang, N., Tabas, I., Winchester, R., Ravalli, S., Rabbani, L. E., and Tall, A. (1996). Interleukin-8 is induced by cholesterol loading of macrophages and expressed by macrophage foam cells in human atheroma. *J. Biol. Chem.* **271,** 8837–8842.

Watanabe, T., Pakala, R., Katagiri, T., and Benedict, C. R. (2001). Monocyte chemotactic protein-1 amplifies serotonin-induced vascular smooth muscle cell proliferation. *J. Vasc. Res.* **38,** 341–349.

Weiss, J. M., Downie, S. A., Lyman, W. D., and Berman, J. W. (1998). Astrocyte-derived monocyte-chemoattractant protein-1 directs the transmigration of leukocytes across a model of the human blood-brain barrier. *J. Immunol.* **161,** 6896–6903.

Westmoreland, S. V., Rottman, J. B., Williams, K. C., Lackner, A. A., and Sasseville, V. G. (1998). Chemokine receptor expression on resident and inflammatory cells in the brain of macaques with simian immunodeficiency virus encephalitis. *Am. J. Pathol.* **152,** 659–665.

Wilcox, J. N., Nelken, N. A., Coughlin, S. R., Gordon, D., and Schall, T. J. (1994). Local expression of inflammatory cytokines in human atherosclerotic plaques. *J. Atheroscler. Thromb.* **1**(Suppl. 1), S10–S13.

Wiley, C. A., Masliah, E., Morey, M., Lemere, C., Deteresa, R., Grafe, M., Hansen, L., and Terry, R. (1991). Neocortical damage during HIV infection. *Ann. Neurol.* **29,** 651–657.

Wilkinson, P. C., and Newman, I. (1992). Identification of IL-8 as a locomotor attractant for activated human lymphocytes in mononuclear cell cultures with anti-CD3 or purified protein derivative of Mycobacterium tuberculosis. *J. Immunol.* **149,** 2689–2694.

Wojcik, W. J., Swoveland, P., Zhang, X., and Vanguri, P. (1996). Chronic intrathecal infusion of phosphorothioate or phosphodiester antisense oligonucleotides against cytokine responsive gene-2/IP-10 in experimental allergic encephalomyelitis of lewis rat. *J. Pharmacol. Exp. Ther.* **278,** 404–410.

Wolf, M., Delgado, M. B., Jones, S. A., Dewald, B., Clark-Lewis, I., and Baggiolini, M. (1998). Granulocyte chemotactic protein-2 acts via both IL-8 receptors, CXCR1 and CXCR2. *Eur. J. Immunol.* **28,** 164–170.

Wolpe, S. D., Davatelis, G., Sherry, B., Beutler, B., Hesse, D. G., Nguyen, H. T., Moldawer, L. L., Nathan, C. F., Lowry, S. F., and Cerami, A. (1988). Macrophages secrete a novel heparin-binding protein with inflammatory and neutrophil chemokinetic properties. *J. Exp. Med.* **167,** 570–581.

Wolpe, S. D., Sherry, B., Juers, D., Davatelis, G., Yurt, R. W., and Cerami, A. (1989). Identification and characterization of macrophage inflammatory protein-2. *Proc. Natl. Acad. Sci. USA* **86,** 612–616.

Wong, B. W., Wong, D., and Mcmanus, B. M. (2002). Characterization of fractalkine (CX3CL1) and CX3CR1 in human coronary arteries with native atherosclerosis, diabetes mellitus, and transplant vascular disease. *Cardiovasc. Pathol.* **11,** 332–338.

Wu, D., Larosa, G. J., and Simon, M. I. (1993). G protein-coupled signal transduction pathways for interleukin-8. *Science* **261,** 101–103.

Wu, L., Paxton, W. A., Kassam, N., Ruffing, N., Rottman, J. B., Sullivan, N., Choe, H., Sodroski, J., Newman, W., Koup, R. A., and Mackay, C. R. (1997). CCR5 levels and expression pattern correlate with infectability by macrophage-tropic HIV-1, *in vitro*. *J. Exp. Med.* **185,** 1681–1691.

Wuyts, A., Van Osselaer, N., Haelens, A., Samson, I., Herdewijn, P., Ben-Baruch, A., Oppenheim, J. J., Proost, P., and Van Damme, J. (1997). Characterization of synthetic human granulocyte chemotactic protein-2: Usage of chemokine receptors CXCR1 and CXCR2 and *in vivo* inflammatory properties. *Biochemistry* **36,** 2716–2723.

Wuyts, A., Proost, P., Lenaerts, J. P., Ben-Baruch, A., Van Damme, J., and Wang, J. M. (1998). Differential usage of the CXC chemokine receptors 1 and 2 by interleukin-8, granulocyte chemotactic protein-2 and epithelial-cell-derived neutrophil attractant-78. *Eur. J. Biochem.* **255,** 67–73.

Wuyts, A., Struyf, S., Gijsbers, K., Schutyser, E., Put, W., Conings, R., Lenaerts, J. P., Geboes, K., Opdenakker, G., Menten, P., Proost, P., and Van Damme, J. (2003). The CXC chemokine GCP-2/CXCL6 is predominantly induced in mesenchymal cells by interleukin-1beta and is down-regulated by interferon-gamma: Comparison with interleukin-8/CXCL8. *Lab. Invest.* **83,** 23–34.

Xia, M. Q., Bacskai, B. J., Knowles, R. B., Qin, S. X., and Hyman, B. T. (2000). Expression of the chemokine receptor CXCR3 on neurons and the elevated expression of its ligand IP-10 in reactive astrocytes: *In vitro* ERK1/2 activation and role in Alzheimer's disease. *J. Neuroimmunol.* **108,** 227–235.

Xu, L., Kelvin, D. J., Ye, G. Q., Taub, D. D., Ben-Baruch, A., Oppenheim, J. J., and Wang, J. M. (1995). Modulation of IL-8 receptor expression on purified human T lymphocytes is associated with changed chemotactic responses to IL-8. *J. Leukoc. Biol.* **57,** 335–342.

Xu, L. L., Warren, M. K., Rose, W. L., Gong, W., and Wang, J. M. (1996). Human recombinant monocyte chemotactic protein and other C-C chemokines bind and induce directional migration of dendritic cells in vitro. *J. Leukoc. Biol.* **60,** 365–371.

Yla-Herttuala, S., Lipton, B. A., Rosenfeld, M. E., Sarkioja, T., Yoshimura, T., Leonard, E. J., Witztum, J. L., and Steinberg, D. (1991). Expression of monocyte chemoattractant protein-1 in macrophage-rich areas of human and rabbit atherosclerotic lesions. *Proc. Natl. Acad. Sci. USA* **88,** 5252–5256.

Yoong, K. F., Afford, S. C., Jones, R., Aujla, P., Qin, S., Price, K., Hubscher, S. G., and Adams, D. H. (1999). Expression and function of CXC and CC chemokines in human malignant liver tumors: A role for human monokine induced by gamma-interferon in lymphocyte recruitment to hepatocellular carcinoma. *Hepatology* **30,** 100–111.

Yoshida, T., Imai, T., Kakizaki, M., Nishimura, M., and Yoshie, O. (1995). Molecular cloning of a novel C or gamma type chemokine, SCM-1. *FEBS Lett.* **360,** 155–159.

Yoshida, T., Imai, T., Takagi, S., Nishimura, M., Ishikawa, I., Yaoi, T., and Yoshie, O. (1996). Structure and expression of two highly related genes encoding SCM-1/human lymphotactin. *FEBS Lett.* **395,** 82–88.

Yoshida, T., Imai, T., Kakizaki, M., Nishimura, M., Takagi, S., and Yoshie, O. (1998). Identification of single C motif-1/lymphotactin receptor XCR1. *J. Biol. Chem.* **273,** 16551–16554.

Yoshida, T., Izawa, D., Nakayama, T., Nakahara, K., Kakizaki, M., Imai, T., Suzuki, R., Miyasaka, M., and Yoshie, O. (1999). Molecular cloning of mXCR1, the murine SCM-1/lymphotactin receptor. *FEBS Lett.* **458,** 37–40.

Yoshimura, T., Matsushima, K., Oppenheim, J. J., and Leonard, E. J. (1987). Neutrophil chemotactic factor produced by lipopolysaccharide (LPS)-stimulated human blood mononuclear leukocytes: partial characterization and separation from interleukin-1 (IL 1). *J. Immunol.* **139,** 788–793.

Yoshimura, T., Robinson, E. A., Tanaka, S., Appella, E., Kuratsu, J., and Leonard, E. J. (1989a). Purification and amino acid analysis of two human glioma-derived monocyte chemoattractants. *J. Exp. Med.* **169,** 1449–1459.

Yoshimura, T., Robinson, E. A., Tanaka, S., Appella, E., and Leonard, E. J. (1989b). Purification and amino acid analysis of two human monocyte chemoattractants produced by phytohemagglutinin-stimulated human blood mononuclear leukocytes. *J. Immunol.* **142,** 1956–1962.

Yue, T. L., Wang, X., Sung, C. P., Olson, B., Mckenna, P. J., Gu, J. L., and Feuerstein, G. Z. (1994). Interleukin-8: A mitogen and chemoattractant for vascular smooth muscle cells. *Circ. Res.* **75,** 1–7.

Z'Graggen, K., Walz, A., Mazzucchelli, L., Strieter, R. M., and Mueller, C. (1997). The C-X-C chemokine ENA-78 is preferentially expressed in intestinal epithelium in inflammatory bowel disease. *Gastroenterology* **113,** 808–816.

Ziegler, S. F., Tough, T. W., Franklin, T. L., Armitage, R. J., and Alderson, M. R. (1991). Induction of macrophage inflammatory protein-1 beta gene expression in human monocytes by lipopolysaccharide and IL-7. *J. Immunol.* **147,** 2234–2239.

Zimmerman, P. A., Buckler-White, A., Alkhatib, G., Spalding, T., Kubofcik, J., Combadiere, C., Weissman, D., Cohen, O., Rubbert, A., Lam, G., Vaccarezza, M., Kennedy, P. E., Kumaraswami, V., Giorgi, J. V., Detels, R., Hunter, J., Chopek, M., Berger, E. A., Fauci, A. S., Nutman, T. B., and Murphy, P. M. (1997). Inherited resistance to HIV-1 conferred by an inactivating mutation in CC chemokine receptor-5: Studies in populations with contrasting clinical phenotypes, defined racial background, and quantified risk. *Mol. Med.* **3,** 23–36.

Zink, M. C., Coleman, G. D., Mankowski, J. L., Adams, R. J., Tarwater, P. M., Fox, K., and Clements, J. E. (2001). Increased macrophage chemoattractant protein-1 in cerebrospinal fluid precedes and predicts simian immunodeficiency virus encephalitis. *J. Infect. Dis.* **184,** 1015–1021.

Zlotnik, A., and Yoshie, O. (2000). Chemokines: A new classification system and their role in immunity. *Immunity* **12,** 121–127.

Zou, Y. R., Kottmann, A. H., Kuroda, M., Taniuchi, I., and Littman, D. R. (1998). Function of the chemokine receptor CXCR4 in hematopoiesis and in cerebellar development. *Nature* **393,** 595–599.

Zwijnenburg, P. J., De Bie, H. M., Roord, J. J., Van Der Poll, T., and Van Furth, A. M. (2003). Chemotactic activity of CXCL5 in cerebrospinal fluid of children with bacterial meningitis. *J. Neuroimmunol.* **145,** 148–153.

CHAPTER 2

The Biology of CXC Chemokines and Their Receptors

Nathaniel M. Weathington and J. Edwin Blalock
Department of Physiology and Biophysics,
University of Alabama at Birmingham,
Birmingham, Alabama 35294

I. INTRODUCTION

Chemokine biology has seen a renaissance in the almost two decades since the first discovery and characterization of a chemokine molecule, known as monocyte derived neutrophil chemotactic factor (Yoshimura *et al.*, 1987) or neutrophil activating factor (Walz *et al.*, 1987) and later renamed IL-8 and then CXCL-8. It is doubtful that those who originally described this novel factor had any idea of the explosion in immunology they had ignited. In short order, the two receptors for this factor were identified and cloned (Holmes *et al.*, 1991; Murphy and Tiffany, 1991), and then similar molecules were discovered and characterized until, by 1997, an extensive constellation of ligands and receptors (key signaling molecules in the immune system) had been recognized (Baggiolini *et al.*, 1997). Generally speaking, chemokines are released by cells to induce the migration of other cells that possess

1063-5823/05 $35.00
DOI: 10.1016/S1063-5823(04)55002-0

appropriate G-protein-coupled receptors towards higher chemokine concentrations (a process called chemotaxis). Chemokine activity is now understood to be the immune system's predominant type of internal signal for specific recruitment of particular cell types to important sites. This cell recruitment is essential for the innate immune system's inflammatory responses, for proper tissue development in organs of the cardiovascular, nervous, and immune systems, and for the activation as well as the specialized regional development of memory and effector lymphocytes in the adaptive immune system of the adult animal (Rot and von Andrian, 2004). Consequently, for some disorders of immunity the activities of various chemokines have become recognized as major potential therapeutic targets for diseases of many organ and tissue systems (Wang *et al.*, 1998; Howard *et al.*, 1999; Nickel *et al.*, 1999), because their targeted disruption of leukocyte trafficking and intercellular communication would allow, by today's standards, specific and elegant modulation of the immune response.

We will explore the biology of the CXC family of chemokines and receptors in three sections. The first section will discuss the structure, function, and history of the CXC chemokine ligands and receptors. In the second section we will discuss the biochemistry and cell biology underlying the signaling of these molecules. We will finish with a discussion of cellular trafficking and the roles these chemokines play in normal and pathological states, addressing potential therapeutic targets of this system. By the end of the article, the reader should have gained a solid understanding of the basic mechanisms of the CXC chemokine system, as well as some insight into the contributions of this signalling activity at the level of whole animal physiology.

II. GENETIC AND PROTEIN STRUCTURE OF THE CXC CHEMOKINE FAMILY

We will begin with a note on the nomenclature of the CXC chemokines: Because of the numerous chemokine ligands and receptors described since 1990, a taxonomy system has been developed (Murphy *et al.*, 2000). This system is based on the primary sequences of the chemokine ligands, which have four bridging cysteines. The CXC chemokine ligands, called CXCLs and numbered CXCL1, CXCL2, etc., each have a nonconserved amino acid separating the first two cysteines, but the initial cysteines of the CCL chemokine ligands have no such intervening residue. Likewise, the CX3CL fractalkine possesses three amino acids between these cysteines. The chemokine receptors are designated CXCRs, CCRs, or CX3CRs based on their cognate ligands, and they are also numbered CXCR2, CCR7, etc.

Though other names for these factors may be mentioned, we will adhere to the nomenclature laid out here.

A. CXC Ligands

All of the CXC chemokines, with the exception of CXCLs 12, 14, and 16, share chromosomal locations: 4q21.1 in humans and 5E in mice. This suggests that these chemokines have diverged from one another rather recently. Phylogenetic analysis reveals that these chemokines (CXCLs 1–11 and CXCL13) are all derived from a common progenitor CXC chemokine (perhaps closely related to CXCL13), which likely underwent a series of gene duplications at this chromosomal location some time after the tetrapods diverged from the fish. CXCLs 12, 14, and 16 are the more primordial isoforms of the ligands; CXCL12 is apparently the earliest of the CXC ligands to have come about, as homologs of this ligand trace back as far as the early boned fish, pufferfish and catfish. Figure 1 displays the impressive phylogenetic analysis of the CXC ligands and receptors recently performed by Huising *et al.* (2003).

This phylogenetic study, along with tissue-specific expression screens for the different members of the CXC ligand family, reveals that these ligands (and their receptors) are not exclusive to the immune system in which they have been so well characterized. In fact, it is now theorized that CXCL12 has the origins of its expression in the central nervous system (CNS). It has been shown as a determinant of proper cellular migration during the early development of the brain in most higher vertebrates: CXCL12 knockout mice die *in utero* with underdeveloped cerebella (Zou *et al.*, 1998). Once the signaling machinery for cell-specific chemotaxis was set up, this activity of CXCL12 was adapted for the development of other organ systems, particularly the cardiovascular and lymphatic systems. As vertebrates evolved, other paralogs of the CXC genes came about as more highly specialized leukocyte traffic signals essential for the appropriate functioning of the complex immune systems present in amphibians, birds, and mammals.

All of the CXC ligand proteins are active as chemoattractant factors. As mentioned earlier, the first chemokine identified was CXCL8, which has now been extensively characterized and will thus be used as a representative for this discussion on chemokine structures and functions. The amino-acid alignment of all the human CXC chemokines shown in Fig. 2 is therefore arranged with respect to CXCL8, and references to specific sections of CXCLs will be numbered according to the CXCL8 primary sequence.

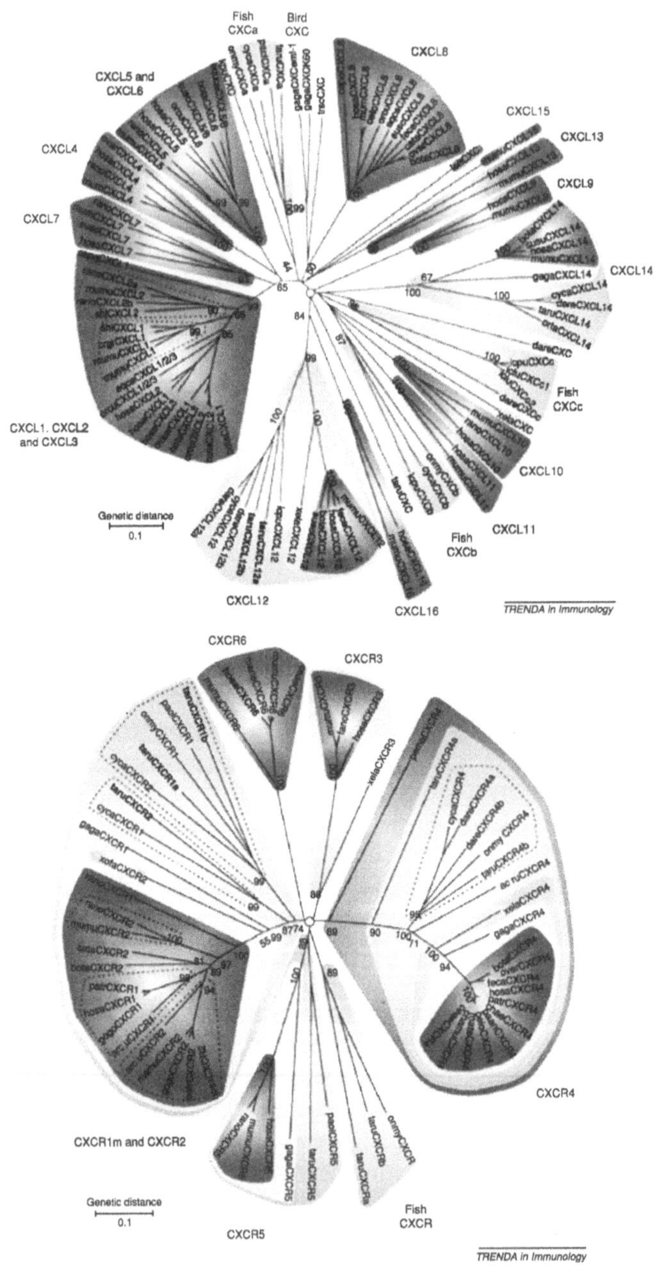

FIGURE 1 (*For legend, see opposite page.*)

The gross features of the ligands' macromolecular structures are determined by the disulfide linkages of their four cysteines; the first bridges to the third and the second to the fourth. This arrangement of bonds causes all chemokines to have a similar degenerate structure (Fig. 3A and a flexible N-terminal region preceding the first pair of bridged cysteines (Fernandez and Lolis, 2002). A stretch of ~6 amino acids following the second cysteine of the molecule constitutes the 'N-loop', whose carboxyl end is the first of three β-strands. The C-terminus of the molecule is an alpha helix; between each of the β-strands and between $\beta3$ and the C terminal helix are the 30s, 40s, and 50s loops (from N to C). Disulfides tie the N terminal region to the β-sheet region of the molecule at the 30s and 50s loops (Fig. 3A). Thus configured, most of the CXC ligands typically homodimerize along the first β-strand in crystallography and NMR studies (Clore *et al.*, 1990), as shown in Fig. 3B, though monomers of CXCLs 7 and 12 have been seen, and tetramers (or paired dimers) of CXCL4 have also been observed (Fernandez and Lolis, 2002). Notably, though, such homodimers (at least those of CXCL8) dissociate at micromolar concentrations, suggesting that they are physiologically active only as monomers since their *in vivo* concentrations rarely reach this range (Burrows *et al.*, 1994).

Although the CXC chemokines are all very similar three-dimensionally, each specifically ligates only its target receptor(s), depending on fine structural motifs attributable to its primary sequence. The main subdividing determinant of the family is the presence of a Glu–Leu–Arg, or ELR, motif (Fig. 3A) at residues 4–6, immediately preceding the CXC sequence in the

---

FIGURE 1 Phylogenetic analysis of the CXC chemokine ligands (A) and receptors (B). Colors represent major vertebrate divisions: red, mammalian; yellow, avian; green, amphibian; lilac, teleost fish; and grey, agnathan. Human and pufferfish sequences are indicated in bold Subclusters (statistically reliable clusters within clusters) are indicated by broken lines. The abbreviations for the organisms are as follows: bota, *Bos taurus* (cow); cafa, *Canis familiaris* (dog); ceae, *Cercopithecus aethiops* (vervet monkey); capo, *Cavia porcellus* (guinea pig); ceto, *Cercocebus torquatus atys* (mangabey); chae, *Chlorocebus aethiops* (African green monkey); crgr, *Cricetulus griseus* (Chinese hamster); cyca, *Cyprinus carpio* (carp); dare, *Danio rerio* (zebrafish); eqca, *Equus caballus* (horse); feca, *Felis cattus* (cat); gaga, *Gallus gallus* (chicken); gogo, *Gorilla gorilla* (gorilla); hosa, *Homo sapiens* (human); icfu, *Ictalurus furcatus* (blue catfish); icpu, *Ictalurus punctatus* (channel catfish); lafl, *Lampetra fluviatilis* (lamprey); mamu, *Macaca mulatta Macaca nemestrina* (pig-tailed macaque); mumu, *Mus musculus* (mouse); onmy, *Oncorhynchus mykiss* (rainbow trout); orcu, *Oryctolagus cuniculus* (rabbit); orla, *Oryzias latipes* (medaka); ovar, *Ovis aries* (sheep); paan, *Papio anubis* (olive baboon); paol, *Paralichtys olivaceous* (flounder); patr, *Pan troglodytes* (chimpanzee); pema, *Petromyzon marinus* (sea lamprey); rano, *Rattus norvegicus* (rat); sihi, *Sigmodon hispidus* (hispid cotton rat); susc, *Sus scrofa* (pig); taru, *Takifugu rubripes* (pufferfish); trsc, *Triakis scyllium* (shark); xela, *Xenopus laevis* (African clawed toad). Figure and parts of legend reprinted from Huising *et al.* (2003) with permission. (See Color Insert.)

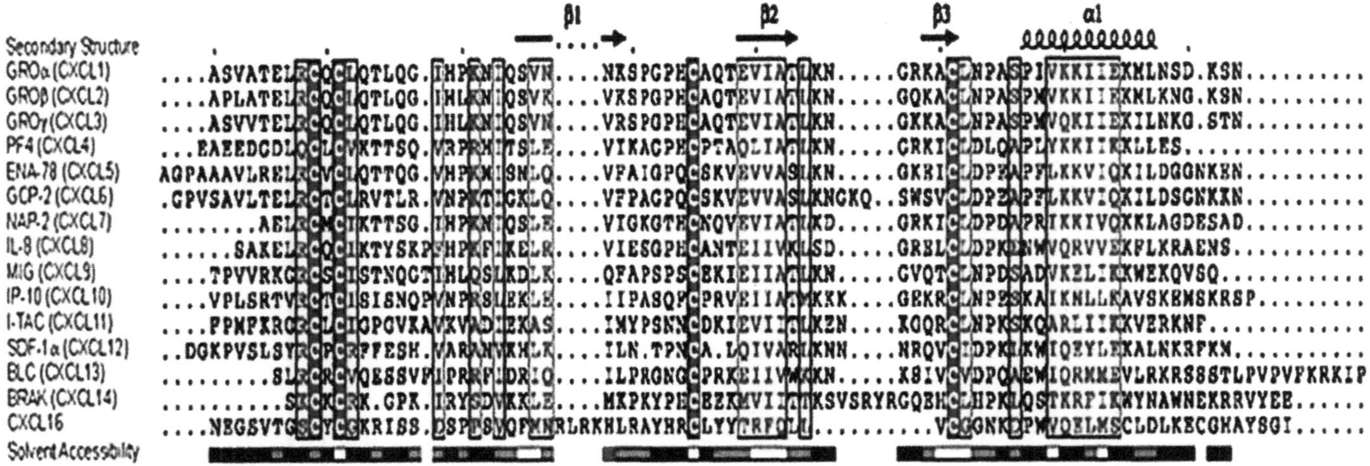

FIGURE 2 Protein sequence alignment of the human CXC chemokine family. Structural cysteines are shown with a red background, and homologous residues in the family are shown in red. Secondary structural character is designated at the top of the diagram, while solvent accessibility is shown at the bottom. Dark blue indicates completely solvent exposed side chains, while light blue indicates partially exposed residues, and white amino acids are buried. Reprinted from Fernandez and Lolis (2002) with permission. (See Color Insert.)

CXCLs 1–8 molecules (except CXCL4). All of the ELR^+ ligands activate the CXCR2 receptor, but only CXCL8 can activate both CXCR1 and CXCR2 (Ahuja and Murphy, 1996). These receptors are abundantly expressed on neutrophils, and the ELR^+ CXC chemokines are primarily considered neutrophil chemoattractants.

The binding and activation of receptors by the chemokine ligands has been modeled with two steps, indicating a cooperative process in which multiple sites on the ligand bind to multiple sites on the receptor and eventually activate it. The initial binding events and determination of receptor specificity for the chemokines in general seem to be largely conferred by the N-loop sequence. For example, when this region of CXCL1 is exchanged with that of CXCL8, the mutant CXCL8 fails to bind CXCR1, while the CXCL1 chimera is active on both CXCR1 and CXCR2 (Lowman *et al.*, 1996). Likewise, the substitution of the N-loop from CXCL12 into the CXCL1 molecule results in a chemokine active on CXCR4, whose cognate ligand is CXCL12 (Crump *et al.*, 1997).

In the ELR^+ chemokines, this namesake sequence is required for appropriate functioning of the molecule, as revealed by serial mutagenesis, and mutation of all three residues to alanines abolishes the activity of the molecule (Hebert *et al.*, 1991). However, mutation of the leucine in this region to cysteine concurrent with a His-33 mutation to cysteine in the 30s loop creates an engineered disulfide and tethers the ELR to the nearby 30s loop; only modestly decreased activity is observed in this case (Gerber *et al.*, 2000). The conserved 'GP' motif at residues 31–32 in the 30s loop of only the ELR^+ CXC chemokines is also critical for cell binding and receptor activation. Mutation or juxtaposition of these residues causes a reduction of ligand binding and cell activation (Clark-Lewis *et al.*, 1994) similar to that seen in the ELR mutations; it seems that the 30s loop together with the ELR domain may be an important docking site in these chemokines' binding and activation of the neutrophil CXC chemokine receptors.

The non-ELR chemokines active on the CXCR3 receptor, CXCLs 9,10, and 11, have binding determinants that are very similar to those of the ELR^+ class, with the exception that the 30s loop does not seem to be required for activation of the receptor. A detailed structural study of the CXCL11–CXCR3 interaction supports the two-step binding model by showing a receptor docking site along the N-loop of the chemokine as well as the binding of more membrane proximal activating elements of the receptor by a domain in the N-terminal sequence of the ligand (Booth *et al.*, 2004). Although we will not discuss the structural determinants for activity of the remaining CXC chemokines, we will discuss the well-characterized biological signaling and specificities of these chemokines later on (Table I).

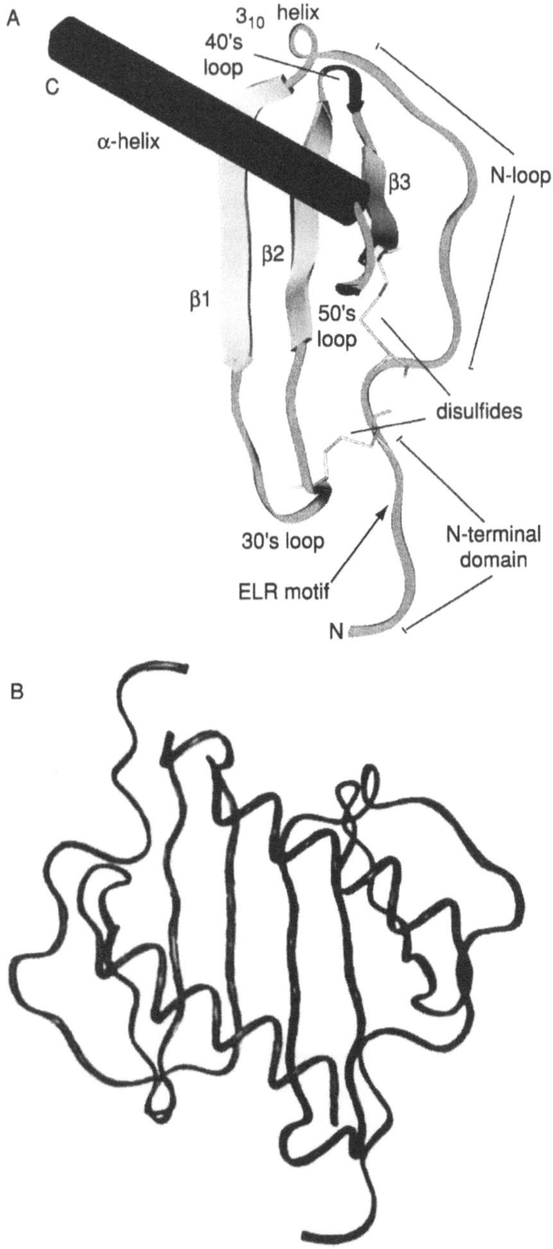

FIGURE 3 *(For legend, see opposite page.)*

Another class of CXC chemokines that will be mentioned only briefly here is those of viral origin. Over millions of years of coevolution, multiple viruses have managed to acquire and mutate chemokine and chemokine-receptor DNA from host genes, eventually leading to the emergence of virus-derived chemokines, which have become important in the infectious cycle of these pathogens. For example, cytomegalovirus encodes and expresses two ELR^+ CXC chemokines, vCXC1 and vCXC2, the first of which can potently ligate CXCR2 and recruit neutrophils into sites of infection acutely (Penfold

TABLE I
The Cellular Sources, Receptor Specificities, and Cellular Targets of the CXC Chemokines

Chemokine source cell type	Ligand	Receptor	Target cell type
Mϕ, Mo, Mc, Eo, Nu, B, T, Ep, En, SM, Ch, Fib, Kc	CXCL8	CXCR1	Neutrophils, macrophages, T-cells
All of the above, as well as Mg, Ac, Pl, Tu	CXCL1–3 CXCL5–8	CXCR2	Neutrophils, macrophages
Mϕ, Mo, Kc, En, Fib, Ac	CXCL9–11	CXCR3	Activated T-cells, B-cells
Stromal cells, bone marrow stromal cells, developing mesenchyme, synoviocytes	CXCL12	CXCR4	B-cells, neurons, splenocytes, germ cells, tumor cells, plasma cells, hematopoietic progenitor cells
Follicular dendritic cells, lymphoid stromal cells	CXCL13	CXCR5	Circulating B-cells, Ag-activated helper T-cells
Fibroblasts	CXCL14	Unknown	Monocytes
Bone marrow stromal cells, Mϕ, B-cells, dendritic cells	CXCL16	CXCR6	Plasma cells, memory T-cells

Mϕ, macrophages; Mo, monocytes; Mc, mast cells; Eo, eosinophils; Nu, neutrophils; B, B-cells; T, T-cells; Ep, epithelial cells; En, endothelial cells; SM, smooth-muscle cells; Ch, chondrocytes; Fib, fibroblasts; Kc, keratinocytes; Mg, microglia; Ac, astrocytes; Pl, platelets; Tu, tumor cells; DCs, dendritic cells.

FIGURE 3 The degenerate structure of CXC chemokines in monomeric form (A) and the dimeric structure of CXCL8 (B). The secondary structural motifs discussed in the text are labeled in A. In B, it can be seen that dimerization occurs along the $\beta 1$ strand of the molecules. Reprinted from Fernandez and Lolis (2002) with permission. (See Color Insert.)

et al., 1999). Because neutrophils are important reservoirs for cytomegalovirus (Revello *et al.*, 1998), this chemokine activity helps enable long-term, disseminated infection of the host, the behavior pattern that is characteristic of cytomegalovirus. Another means by which viruses have assimilated chemokine signaling is through development of chemokine ligand homologs that are potent inhibitors of chemokine receptors; some such homologs are active on multiple receptors and provide broad protection of the virus against host immunity. The best example of this kind of chemokine is the virally encoded macrophage inflammatory protein-2 (vMIP-II), which is actually a CC chemokine capable of blocking signaling through several CCRs as well as through CX3CR1 and CXCR4. Actions of human immunodeficiency virus (HIV) pathology pertinent to CXC receptors will be mentioned in the next section, and more can be learned about the eons-long saga of molecular warfare mediated by these 'virokines' in available reviews on the subject (Murphy, 2001; Liston and McColl, 2003).

B. The CXC Receptors

All chemokine receptors are G-protein-coupled receptors possessing seven transmembrane domains and contacting a heterotrimeric G protein intracellularly, each composed of alpha, beta, and gamma subunits. The CXCRs consist of six family members. As mentioned above, CXCR1 exclusively recognizes CXCL8, though CXCR2 can be ligated by all the ELR+ ligands (CXCLs 1–8, except CXCL4). CXCR3 is bound and activated by CXCLs 9–11, while the other three receptors do not display this kind of ligand promiscuity: CXCRs 4, 5, and 6 are ligated exclusively by CXCLs 12, 13, and 16, respectively (Table I).

CXCR4 is a coreceptor for HIV that may be essential for infectivity, as the gp120 surface molecule of the virus makes a complex with CXCR4 along with CD4 and CCR5 (Littman, 1998). While individuals with a mutation of CCR5 that leads to its underexpression are mostly resistant to HIV infection, allelic variation of CXCR4 has not been documented. Single-nucleotide polymorphism scanning of the human CXCR1 and CXCR2 receptors has revealed an allele in CXCR2 that correlates with systemic sclerosis but has no effect on receptor structure and that may be relevant at the level of receptor mRNA (Renzoni *et al.*, 2000).

CXCRs 1–4 are the best characterized of the CXCRs and will be the subjects of our structural discussion. The ligand-binding determinant regions of the CXCRs are distributed throughout the extracellular portion of the molecule, which consists of the amino terminus and three extracellular loops (ECLs 1–3); Fig. 4 shows the arrangement of a prototypical CXCR in

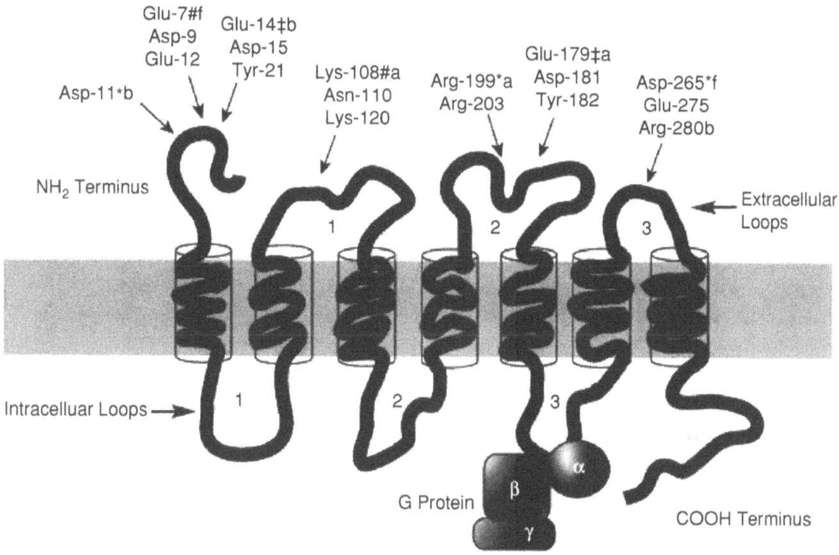

FIGURE 4 Binding determinants identified in the CXCR receptors. Residues on the extracellular side of the molecule important for ligand binding (b), cell activation (a), or unspecified receptor function (f) are displayed for CXCR1 (*), CXCR2 (#), and CXCR4 (‡).

the membrane; determinants of binding and activation for CXCRs 1, 2, and 4 have been identified. The aforementioned two-step model for receptor binding and then activation seems to involve distinct regions of the receptors, and key residues for binding are usually hydrophilic or charged amino acids located in the N-terminus while those implicated in receptor activation tend to be located in the ECL regions (Ahuja *et al.*, 1996). Uniquely, CXCR1 has a key residue in ECL3 (Arg-280) implicated in both ligand binding and receptor activation, though other determinants have also been identified (Leong *et al.*, 1994).

The CXCRs have N-terminal regions with quite different characteristics, which are thought to contribute to their ligand specificities; several hydrophilic or acidic residues in this region of each molecule are implicated in ligand binding, including Asp-11 in the CXCR1 receptor molecule and Glu-7, Asp-9, and Glu-12 in CXCR2. CXCR4 requires the Glu-14, Glu-15, and Tyr-21 residues for ligand binding (Glu-15 is also implicated in the interactions of the receptor with HIV gp120 along with Glu-32, Asp-97, and Asp 187 [Chabot *et al.*, 1999; Fernandez and Lolis, 2002]).

The activation of the CXCRs is dependent on agonist-mediated conformational changes to the receptor molecule's extracellular domain, causing a shift in one or more of the transmembrane domains and a consequent

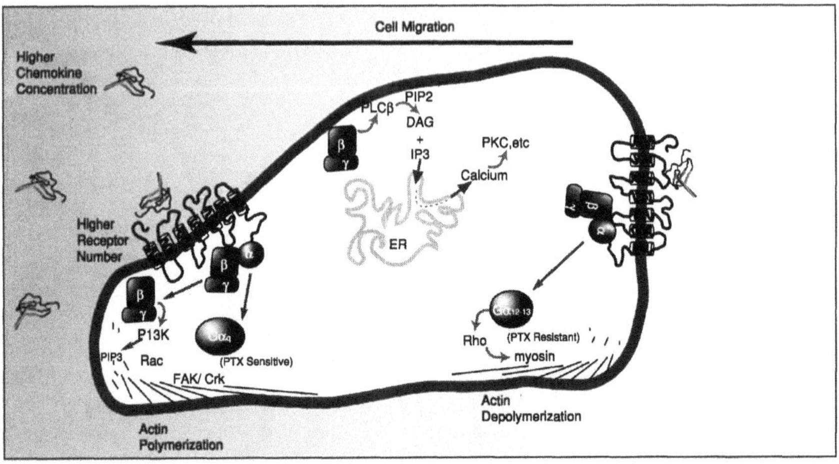

FIGURE 5 Basic cell migration and activation. GTPases Rac and Rho coordinate actin polymerization and depolymerization. They are sequestered in the leading and lagging edges of the cell, respectively. Chemokine receptors are present in higher numbers at the leading edge of the cell, and the G_α proteins active in the leading and lagging edges differ as shown. Endoplasmic reticulum (ER) calcium release is mediated by the active $G_{\beta\gamma}$ proteins' activation of PLCβ and the downstream effects of IP3 on receptors in the ER. FAK, focal adhesion kinase; PI3K, phosphoinsitide 3 kinase; DAG, diacyl glycerol; PTX, persussis toxin; PIP3, phosphotidylinositol-3,4,5-triphosphate; PIP2, phosphoinositol 4,5 bisphosphate; IP3, inositol 1,4,5 triphosphate; PLCβ, phospholipase C beta: PKC, protein kinase C.

change in the intracellular structure of the molecule at its interface with the G protein. Elements identified in the CXCRs for this activity also tend to be hydrophilic residues, and these activation domains seem to be located in the extracellular loops. Examples of such signal switches include a pair of arginine residues in ECL2 and an Asp and a Glu residue in ECL3 of CXCR1, each of which has been mutated to prevent receptor activation despite more-or-less-normal cell-binding properties (Hebert *et al.*, 1993). In CXCR2, three basic residues in ECL1 (Lys-108, Asn-110, and Lys-120) have likewise been identified as key residues for signaling (Katancik *et al.*, 2000); CXCR4 has a highly acidic activation domain in ECL2, 3 of the 4 contiguous residues (179–182) being either Asp or Glu (Doranz *et al.*, 1999).

Based on the characterization of the determinants for ligand-receptor interactions described here, we can visualize that the CXCLs bind to the N-terminus of the receptor based largely on charge associations between the N-terminal portion of the CXCRs and hydrophilic residues on the ligands' N-loops. Once this intimate association between the two molecules has been established, the ligand's orientation to the charged residues of the extracellular loops of the receptor allows the presumptive interaction between these domains on the receptor with specific activating residues of the chemokine. This process initiates the biochemical cascade within the cell, which will eventually cause cell migration and activation through a complex series of biochemical events. We will summarize those events next.

III. THE BIOCHEMICAL SIGNAL

In order for the chemokines to be functional mediators of cell migration, they must use the ligand-receptor interactions just discussed to directionally mobilize the target cells towards higher chemokine concentrations. To accomplish this task, the cellular machinery of the target cell organizes itself into an efficient and sensitive locomotor. In this process, the cell becomes polarized along the direction of the chemokine concentration gradient, placing the essential guanosine triphosphate exchange factors (GTPases) Rac and Rho at each end of the polarized cell (Sanchez-Madrid and del Pozo, 1999; Xu *et al.*, 2003). At the pole with higher chemokine concentration, Rac is present and active as an actin motor and is required to extend a pseudopod outward towards the source of chemokine (Roberts *et al.*, 1999). Meanwhile, at the opposite side of the cell, Rho GTPase–associated activities are retracting the part of the cell in contact with lower chemokine concentrations, thus allowing forward progress of the intact cell (Ridley *et al.*, 1999). These two systems are counter-regulatory: Rac activity on one end of the cell suppresses local Rho activity and vice versa, which allows for consistent, robust cell polarization along the direction of the chemical gradient (Sander *et al.*, 1999). The polarization just described is set up and maintained by a fairly complex system of biochemical signals, notable features of which are that chemokine receptors are more numerous at the leading edge of the cells' movement than at the lagging edge (Nieto *et al.*, 1997) and that the constituents of the G-proteins may differ between the two poles. Figure 5 is a simplified cartoon of this complicated process.

Upon ligand binding and activation of the receptor, transmembrane helices 3 and 6 ratchet themselves in the membrane, causing a shift in the intracellular portions of the receptor, particularly the third intracellular loop of the molecule, which is associated with the G protein heteromer. The Gα subunit then exchanges its guanosine diphosphate (GDP) for GTP, allowing the alpha and beta/gamma subunits that compose the G protein heterotrimer to dissociate (Hamm, 2001). This step frees the $\beta\gamma$ subunit to activate phosphoinositide-3 kinase (PI3K) family members (Sotsios and Ward, 2000), thus generating phosphotidylinositol-3,4,5-triphosphate (PIP_3). PI3K and PIP_3 collocalize with Rac in the leading edge of the cell (Huang *et al.*, 2003), and there PIP_3 can act on signaling proteins that induce both cellular "activation" and migratory responses. In response to CXCR4 activation, PI3Kγ causes activation of focal adhesion kinase and Crk, both of which are involved in signaling for cell migration (Wang *et al.*, 2000; Ridley *et al.*, 2003) and lead to the eventual polymerization of actin in the pseudopod and its forward extension (Sotsios and Ward, 2000).

In the lagging edge of the cell, the Rho GTPases cause activation of myosin, which leads to the generation of actin–myosin complexes and retracts the back side of the cell (Alblas *et al.*, 2001). In order for this to occur, the Rho GTPases must be acted upon by $G_{\alpha 12/13}$ subunits, which are distinct from the $G_{\alpha i}$ proteins active in the pseudopod and associated with Rac activity. Pharmacologically, $G_{\alpha i}$ activity is sensitive to pertussis toxin and the $G_{\alpha 12/13}$ pathway is not. This implies that in addition to modulating the receptor distribution on responsive cells, the chemokine gradient polarizes the character of the G proteins present at each end of the cell (Xu *et al.*, 2003).

Cell migration is not the only effect of CXC chemokine activity. A broad spectrum of cell activation by CXCL8 and related ELR^+ CXC chemokines via CXCRs 1 and 2 has been extremely well characterized in neutrophils (Lee *et al.*, 1992; Jones *et al.*, 1996). PIP_3 generated by isoforms of PI3K in the cell can disinhibit the translocation of the transcription factor nuclear factor kappa-b to the nucleus, leading to *de novo* synthesis of several proinflammatory cytokines, proteases, and oxidant-generating machinery involved in inflammation and host defense. Among the first observations of chemokine activity was an increase in intracellular calcium (Peveri *et al.*, 1988; Samanta *et al.*, 1989; Lee *et al.*, 1992). The mechanism for this activity involves phospholipase C-beta (PLCβ), which is activated by $G_{\beta\gamma}$ or $G_{\alpha q}$ proteins. Active PLCβ cleaves PIP_2 in the membrane, creating diacylglycerol (DAG) and IP3. IP3 activates receptors on intracellular calcium stores, opening store-operated calcium channels and transiently increasing cytosolic Ca^{++}, which, in turn, signals a number of physiological changes in the cell, including activation of protein kinase C (PKC) (Rebecchi and Pentyala, 2000). Another feature of CXCR activation in neutrophils is exocytosis of granules, which are vesicles of preformed enzymes such as proteases, peroxidases, and lysozymes (Thelen *et al.*, 1988). Release of this arsenal of enzymes is a powerful way to kill bacteria in times of infection, though it can also be extremely damaging to host tissue in diseases of inflammation or chronic infection. Granule exocytosis is mediated by the actions of the beta arrestins (Barlic *et al.*, 2000), which are activated in response to CXCR ligation. The beta arrestins are also important in downregulating receptor expression at the membrane in response to chemokine stimulation (Barlic *et al.*, 1999). The 'activation' effects just mentioned are seen only at chemokine concentrations much higher than those required for chemotaxis, indicating that these effects are only mediated once the neutrophil has reached inflammatory sites containing very high concentrations of chemokines.

In studies on human chemokines and receptors, increases in intracellular calcium and granule exocytosis occurred in response to activation of either CXCR1 or CXCR2, while other effects of chemokines have been characterized as exclusive to CXCR1. The genesis of the superoxide molecule by activation of the NADPH oxidase occurs in response to ligation of CXCR1 but not of CXCR2, though the process is dependent on Rac activation, which occurs when either receptor is activated. Along with the induction of superoxide by chemokines, the specific activation of phospholipase D, which is active in cytoskeletal rearrangement and the exocytotic response (Choi *et al.*, 2002), is also transduced by activation of the CXCR1 but not the CXCR2 receptor. In addition to this profile of effectors, other specific biochemical signals have been documented for the chemokines, and the interested reader is encouraged to consult the review literature available on this subject (Sotsios and Ward, 2000; Ridley *et al.*, 2003).

IV. COMMUNICATIONS OF THE CXC CHEMOKINES

The trafficking of leukocytes during innate and adaptive immune responses is a complex and essential process for animals' life maintenance. In the higher vertebrates, the chemokine systems have been specifically developed to effectively direct immune cells of different types to appropriate sites of infection or injury. The CXC chemokines figure prominently in these responses, and the various trafficking roles that they perform help manage innate inflammatory responses, adaptive effector-cell development and signaling, and lymphoid and nonlymphoid organogenesis and development, as well as maintain immune memory. The majority of available literature on the CXC chemokines to date is concerned with their role in inflammation, particularly as it pertains to neutrophil trafficking and activation. However, as the activities of the other CXC signals become clearer, the body of knowledge on their biology is rapidly expanding.

Though neutrophils are the predominant cell type expressing CXCR1 and CXCR2, and have long been considered the lone effector cells responsive to the ELR^+ CXCLs, it is now known that monocytes can also be induced to express these receptors and migrate accordingly in the presence of cytokines like IL-4 and IL-13 (Bonecchi *et al.*, 2000). The ELR^+ chemokines are produced abundantly by macrophages and monocytes in response to a variety of stimuli, including pathogen-associated molecules like endotoxin or flagellin and cytokines like tumor necrosis factor alpha (TNF-α), interleukin-1, and others. While macrophages are probably the most common source of CXCL8 and related chemokines, numerous other cell types have been shown to produce ELR^+ CXCLs in response to a range of stimuli, including mechanical stress, bacterial exposure, or cytokine treatment. These cells include nearly all cells of myeloid lineage (neutrophils, eosinophils, mast cells, kupfer cells, and microglia, among others.), as well as cells not usually associated with the immune system, such as epithelial and endothelial cells, chondrocytes, and smooth muscle cells, to name a few (see Table I). Generally speaking, CXCL8 and/or its counterpart ligands are produced as acute response signals to noxious stimuli, and they serve to recruit CXCR1- and CXCR2-expressing neutrophils into the site of insult. Neutrophils are normally the most abundant leukocytes in circulation, and their most important effector function is to kill bacteria at a site of infection, though their release of toxic mediators is often damaging to surrounding host tissue as well as to the target pathogens.

CXCR3 is expressed exclusively on lymphocytes, and it is upregulated on T-cells in response to activation by interleukin-2. This receptor's ligands are CXCL9 (originally called monokine induced by gamma-interferon or Mig), CXCL10 (also known as interferon-inducible protein-10 or IP10), and CXCL11 (ITAC, or interferon-inducible T-cell alpha chemoattractant), all of which are produced by other cells only upon stimulation by interferon-γ (Loetscher *et al.*, 1996). CXCR3 signaling allows the trafficking of activated memory or effector T-cells into sites of inflammation and infection for immune surveillance (Moser and Willimann, 2004) and stimulates helper T-cells' population expansion and effector cytokine production (Whiting *et al.*, 2004). The CXCR3 ligands have another very interesting feature in common that may prove to be as important as their effector functions: These molecules are endogenous antagonists of the CCR3 chemokine receptor (Loetscher *et al.*, 2001), and as such they suppress trafficking of $CCR3^+$ cells into sites of interferon production. Immunologically, this is important because the so-called ' "Type 1" helper T-cells express CXCR3, while Type-2 cells (including T-cells and eosinophils) express CCR3; thus, the antagonistic actions of CXCLs 9–11 may have counter-regulatory effects that sustain a polarized Type-1 response while suppressing Type-2 immunity (Kim *et al.*, 2001; Moser and Willimann, 2004).

Because CXCR4 is the most ancient of the CXC chemokine receptors, its expression and function are much more extensive than the other CXCRs'. It is required for the proper development of many of the body's organ systems and is expressed on primordial germ cells and developing cells of all components of the cardiovascular and immune systems, including the heart, vasculature, lymphoid organs, and bones. It is also expressed on many of the glial cells of the CNS, where it is essential for proper organization of the nervous system during development. The CXCL12–CXCR4 signal is also active in recruiting hematopoietic cells and both B- and T-lymphocytes in the mature immune system, displaying its highly evolved versatility as a signaling molecule. The only CXCR4 ligand, CXCL12 (originally termed stromal cell–derived factor 1, or SDF-1), is produced by many different cell types throughout development to stimulate proper tissue organization, and it is also expressed in the mature animal for maintenance of homeostasis and proper cellular distributions in lymphoid organs.

CXCR5 and its exclusive ligand CXCL13 are used by the adaptive immune system for B-lymphocyte development and trafficking. CXCL13 was originally characterized as B-Cell Attractant 1 (BCA-1) (Legler *et al.*, 1998). This ligand is constitutively expressed in secondary lymphoid tissue, and its signaling through CXCR5 is essential for proper development of these organs as well as germinal center formation (Forster *et al.*, 1996). CXCL13 is also produced by dendritic cells in the B-cell follicles and can be induced by IL-10 stimulation (Perrier *et al.*, 2004) to attract circulating B-cells, nearly all of which express CXCR5. A subpopulation of helper T-cells (which carry a memory phenotype) also expresses CXCR5 upon antigen stimulation. Thus, CXCL13 and CXCR5 work together to help orchestrate humoral immune responses by attracting circulating B-cells and antigen-primed helper T-cells to germinal centers in the spleen and lymph nodes.

The most recently identified members of the CXC chemokine family are CXCL16 and its receptor CXCR6 (Matloubian *et al.*, 2000; Murphy, 2002). One notable feature of the CXCL16 molecule is that it possesses a single transmembrane domain, and both transmembrane and soluble forms of the ligand are detectable in splenic red pulp. CX3CL1 (fractalkine) is the only other known chemokine that possesses a transmembrane domain. CXCL16 was originally identified in dendritic cells, and it is also expressed on other antigen-presenting cells like B-cells and macrophages. The CXCR6 receptor was originally found on activated helper and memory T-cells (Matloubian *et al.*, 2000; Kim *et al.*, 2001) and was later shown to be expressed on plasma cells (Nakayama *et al.*, 2003), which are terminally differentiated B-lymphocytes that produce antibody. Although the overall function of this ligand receptor pair in immunity is as yet unclear, we can speculate that it plays some role in the recruitment of T-cells to active zones in the spleen for presentation of antigen and that it could be involved in long-term cell contact between T-cells or plasma cells and antigen-presenting cells, since the ligand is sometimes membrane associated.

One of the only "orphan" CXC chemokines documented is CXCL14, which was originally identified as being highly expressed in breast and kidney (and was accordingly named BRAK). This chemokine has no known receptor, but it has been shown to be an active chemoattractant for circulating monocytes possessing the CD14 surface marker. The constitutive expression of CXCL14 by fibroblasts is thought to result in the recruitment of these monocytes out of the bloodstream so that they can later develop into tissue macrophages, whose morphogenesis is dependent on fibroblast-derived factors (Kurth *et al.*, 2001).

A. Implications in Disease and Medicine

The activities of chemokines in normal physiology represent an impressive system of biological plasticity in which the immune system can respond to a lifestyle that encounters offending agents of numerous types. Each type of encounter can elicit a specific chemokine profile, resulting in recruitment of appropriate leukocytes for resolution of the problem. However, this response network can, at times, become overactive, as in the case of exogenous chemokine signaling by viruses or through allergic or autoimmune disorders that inappropriately stimulate cells to produce these chemical signals. This superfluous signaling can begin a self-sustaining process of cell recruitment and activation that leads to chronic inflammatory states and functional decline of the tissue site. Given the reasonably specific signaling modalities of the chemokine superfamily of molecules and the powerful effects they can have on the state of the animal, it is no surprise that our understanding of these signals has also generated great excitement about the therapeutic benefits that may be effected by specific targeting of one or another of these signals (Proudfoot, 2002). Human medicine now stands at the threshold of the implementation of a new and powerful class of chemokine-signaling-targeted drugs that seems to hold great promise for fighting some of the many diseases attributable to misguided immunity.

The most-sought CXC chemokine signal antagonist in current development is one that will disturb CXCR1- and CXCR2-mediated neutrophil recruitment. Neutrophils are implicated in a host of diseases in many systems, including acute respiratory distress syndrome, emphysema, ischemia reperfusion injuries, and inflammatory bowel disease. As such, the selective depletion or the blocking of their recruitment could greatly suppress or alleviate a number of disease states, both chronic and acute. The development of compounds targeting the ELR^+ chemokines has been under way for nearly a decade now, and despite very promising results *in vitro* and in animal studies, only one published trial antagonizing this signal has been published to date (Mahler *et al.*, 2004). In this study, emphysema patients treated with a monoclonal antibody to CXCL8 have less difficulty breathing than untreated patients, despite a lack of differences in other clinical parameters. This early study is the first of perhaps hundreds of clinical interventions targeting chemokine trafficking, and these agents represent the beginning of a new genre of anti-inflammatory therapeutics whose efficacy and specificity will be proven in the years to come.

The CXCR3 chemokine receptor and its ligands may also prove to be important targets for therapeutic interventions. Because of its role in immune surveillance and T-cell function, CXCR3 is implicated as a mediator of acute allograft rejection (Hancock *et al.*, 2000) and complications in allogeneic stem-cell transplantation (Hildebrandt *et al.*, 2004). The use of antibodies targeting CXCLs 9 and 11 relieves the T-cell burden and prevents the onset of transplant-associated pneumonia in one report on stem-cell transplantation (Hildebrandt *et al.*, 2004), while antibody targeting the receptor diminishes the accumulation of T-cells and delayed-type hypersensitivity in another model. However, because of the suppression of CCR3 by the ligands of CXCR3, there may be use for mimetics of these CXCLs to combat Type-2 disorders like asthma, though reports have shown both beneficial and detrimental effects due to these ligands in animal models (Medoff *et al.*, 2002; Fulkerson *et al.*, 2004).

The ELR^+ chemokine signal and the CXCR3-mediated responses may be the total of current development in CXC chemokine signaling targets for therapeutics. Because of the broad spectrum of activity represented by CXCR4 and CXCL12, it is doubtful that therapeutics targeting this pathway will be sufficiently specific for treatment of diseases, especially considering the role of this signal in organogenesis and immune homeostasis. The biology of signaling for CXCR5 and CXCR6 is not yet well enough understood to allow clear associations with pathology to be identified, and disruption of these signaling systems may negatively affect the healthy development of an appropriate adaptive immune response. Early studies of the CXCL14 chemokine showed its expression in normal tissues but not in tumoral cell lines. This association was thought to be pertinent to cancer pathology, and while the idea is interesting, a direct role for a CXCL14 signal in tumor biology remains to be clearly established. As the roles of chemokines in disease are discussed further in explicit detail, the importance of the signaling of all the chemokines and their receptors should become evident. It is tempting to speculate that therapeutics that can effectively modulate the chemokine responses may provide extremely safe and efficacious means by which to treat many of the inflammatory maladies seen in clinics today.

References

Ahuja, S. K., Lee, J. C., and Murphy, P. M. (1996). CXC chemokines bind to unique sets of selectivity determinants that can function independently and are broadly distributed on multiple domains of human interleukin-8 receptor B. Determinants of high affinity binding and receptor activation are distinct. *J. Biol. Chem.* **271,** 225–232.

Ahuja, S. K., and Murphy, P. M. (1996). The CXC chemokines growth-regulated oncogene (GRO) alpha, GRObeta, GROgamma, neutrophil-activating peptide-2, and epithelial cell-derived neutrophil-activating peptide-78 are potent agonists for the type B, but not the type A, human interleukin-8 receptor. *J. Biol. Chem.* **271,** 20545–20550.

Alblas, J., Ulfman, L., Hordijk, P., and Koenderman, L. (2001). Activation of Rhoa and ROCK are essential for detachment of migrating leukocytes. *Mol. Biol. Cell* **12,** 2137–2145.

Baggiolini, M., Dewald, B., and Moser, B. (1997). Human chemokines: An update. *Annu. Rev. Immunol.* **15,** 675–705.

Barlic, J., Andrews, J. D., Kelvin, A. A., Bosinger, S. E., DeVries, M. E., Xu, L., Dobransky, T., Feldman, R. D., Ferguson, S. S., and Kelvin, D. J. (2000). Regulation of tyrosine kinase activation and granule release through beta-arrestin by CXCRI. *Nat. Immunol.* **1,** 227–233.

Barlic, J., Khandaker, M. H., Mahon, E., Andrews, J., DeVries, M. E., Mitchell, G. B., Rahimpour, R., Tan, C. M., Ferguson, S. S., and Kelvin, D. J. (1999). Beta-arrestins regulate interleukin-8-induced CXCR1 internalization. *J. Biol. Chem.* **274,** 16287–16294.

Bonecchi, R., Facchetti, F., Dusi, S., Luini, W., Lissandrini, D., Simmelink, M., Locati, M., Bernasconi, S., Allavena, P., Brandt, E., Rossi, F., Mantovani, A., and Sozzani, S. (2000). Induction of functional IL-8 receptors by IL-4 and IL-13 in human monocytes. *J. Immunol.* **164,** 3862–3869.

Booth, V., Clark-Lewis, I., and Sykes, B. D. (2004). NMR structure of CXCR3 binding chemokine CXCL11 (ITAC). *Protein Sci.* **13,** 2022–2028.

Burrows, S. D., Doyle, M. L., Murphy, K. P., Franklin, S. G., White, J. R., Brooks, I., McNulty, D. E., Scott, M. O., Knutson, J. R., Porter, D., Young, P. R., and Hensley, P. (1994). Determination of the monomer-dimer equilibrium of interleukin-8 reveals it is a monomer at physiological concentrations. *Biochemistry* **33,** 12741–12745.

Chabot, D. J., Zhang, P. F., Quinnan, G. V., and Broder, C. C. (1999). Mutagenesis of CXCR4 identifies important domains for human immunodeficiency virus type 1 X4 isolate envelope-mediated membrane fusion and virus entry and reveals cryptic coreceptor activity for R5 isolates. *J. Virol.* **73,** 6598–6609.

Choi, W. S., Kim, Y. M., Combs, C., Frohman, M. A., and Beaven, M. A. (2002). Phospholipases D1 and D2 regulate different phases of exocytosis in mast cells. *J. Immunol.* **168,** 5682–5689.

Clark-Lewis, I., Dewald, B., Loetscher, M., Moser, B., and Baggiolini, M. (1994). Structural requirements for interleukin-8 function identified by design of analogs and CXC chemokine hybrids. *J. Biol. Chem.* **269,** 16075–16081.

Clore, G. M., Appella, E., Yamada, M., Matsushima, K., and Gronenborn, A. M. (1990). Three-dimensional structure of interleukin-8 in solution. *Biochemistry* **29,** 1689–1696.

Crump, M. P., Gong, J. H., Loetscher, P., Rajarathnam, K., Amara, A., Arenzana-Seisdedos, F., Virelizier, J. L., Baggiolini, M., Sykes, B. D., and Clark-Lewis, I. (1997). Solution structure and basis for functional activity of stromal cell-derived factor-1; dissociation of CXCR4 activation from binding and inhibition of HIV-1. *EMBO J.* **16,** 6996–7007.

Doranz, B. J., Orsini, M. J., Turner, J. D., Hoffman, T. L., Berson, J. F., Hoxie, J. A., Peiper, S. C., Brass, L. F., and Doms, R. W. (1999). Identification of CXCR4 domains that support coreceptor and chemokine receptor functions. *J. Virol.* **73,** 2752–2761.

Fernandez, E. J., and Lolis, E. (2002). Structure, function, and inhibition of chemokines. *Annu. Rev. Pharmacol. Toxicol.* **42,** 469–499.

Forster, R., Mattis, A. E., Kremmer, E., Wolf, E., Brem, G., and Lipp, M. (1996). A putative chemokine receptor, BLR1, directs B cell migration to defined lymphoid organs and specific anatomic compartments of the spleen. *Cell* **87,** 1037–1047.

Fulkerson, P. C., Zimmermann, N., Brandt, E. B., Muntel, E. E., Doepker, M. P., Kavanaugh, J. L., Mishra, A., Witte, D. P., Zhang, H., Farber, J. M., Yang, M., Foster, P. S., and Rothenberg, M. E. (2004). Negative regulation of eosinophil recruitment to the lung by the chemokine monokine induced by IFN-gamma (Mig, CXCL9). *Proc. Natl. Acad. Sci. USA* **101,** 1987–1992.

Gerber, N., Lowman, H., Artis, D. R., and Eigenbrot, C. (2000). Receptor-binding conformation of the "ELR" motif of IL-8: X-ray structure of the L5C/H33C variant at 2.35 A resolution. *Proteins* **38,** 361–367.

Hamm, H. E. (2001). How activated receptors couple to G proteins. *Proc. Natl. Acad. Sci. USA* **98,** 4819–4821.

Hancock, W. W., Lu, B., Gao, W., Csizmadia, V., Faia, K., King, J. A., Smiley, S. T., Ling, M., Gerard, N. P., and Gerard, C. (2000). Requirement of the chemokine receptor CXCR3 for acute allograft rejection. *J. Exp. Med.* **192,** 1515–1520.

Hebert, C. A., Chuntharapai, A., Smith, M., Colby, T., Kim, J., and Horuk, R. (1993). Partial functional mapping of the human interleukin-8 type A receptor. Identification of a major ligand binding domain. *J. Biol. Chem.* **268,** 18549–18553.

Hebert, C. A., Vitangcol, R. V., and Baker, J. B. (1991). Scanning mutagenesis of interleukin-8 identifies a cluster of residues required for receptor binding. *J. Biol. Chem.* **266,** 18989–18994.

Hildebrandt, G. C., Corrion, L. A., Olkiewicz, K. M., Lu, B., Lowler, K., Duffner, U. A., Moore, B. B., Kuziel, W. A., Liu, C., and Cooke, K. R. (2004). Blockade of CXCR3 receptor: Ligand interactions reduces leukocyte recruitment to the lung and the severity of experimental idiopathic pneumonia syndrome. *J. Immunol.* **173,** 2050–2059.

Holmes, W. E., Lee, J., Kuang, W. J., Rice, G. C., and Wood, W. I. (1991). Structure and functional expression of a human interleukin-8 receptor. *Science* **253,** 1278–1280.

Howard, O. M., Oppenheim, J. J., and Wang, J. M. (1999). Chemokines as molecular targets for therapeutic intervention. *J. Clin. Immunol.* **19,** 280–292.

Huang, Y. E., Iijima, M., Parent, C. A., Funamoto, S., Firtel, R. A., and Devreotes, P. (2003). Receptor-mediated regulation of PI3Ks confines PI(3,4,5)P3 to the leading edge of chemotaxing cells. *Mol. Biol. Cell* **14,** 1913–1922.

Huising, M. O., Stet, R. J. M., Kruiswijk, C. P., Savelkoul, H. F. J., and Lidy Verburg-van, Kemenade,B. M. (2003). Molecular evolution of CXC chemokines: Extant CXC chemokines originate from the CNS. *Trends in Immunology* **24,** 306–312.

Jones, S. A., Wolf, M., Qin, S., Mackay, C. R., and Baggiolini, M. (1996). Different functions for the interleukin-8 receptors (IL-8R) of human neutrophil leukocytes: NADPH oxidase and phospholipase D are activated through IL-8R1 but not IL-8R2. *PNAS* **93,** 6682–6686.

Katancik, J. A., Sharma, A., and de Nardin, E. (2000). Interleukin-8, neutrophil-activating peptide-2 and GRO-alpha bind to and elicit cell activation via specific and different amino acid residues of CXCR2. *Cytokine* **12,** 1480–1488.

Kim, C. H., Kunkel, E. J., Boisvert, J., Johnston, B., Campbell, J. J., Genovese, M. C., Greenberg, H. B., and Butcher, E. C. (2001). Bonzo/CXCR6 expression defines type 1-polarized T-cell subsets with extralymphoid tissue homing potential. *J. Clin. Invest.* **107,** 595–601.

Kim, C. H., Rott, L., Kunkel, E. J., Genovese, M. C., Andrew, D. P., Wu, L., and Butcher, E. C. (2001). Rules of chemokine receptor association with T cell polarization *in vivo*. *J. Clin. Invest.* **108,** 1331–1339.

Kurth, I., Willimann, K., Schaerli, P., Hunziker, T., Clark-Lewis, I., and Moser, B. (2001). Monocyte selectivity and tissue localization suggests a role for breast and kidney-expressed chemokine (BRAK) in macrophage development. *J. Exp. Med.* **194,** 855–861.

Lee, J., Horuk, R., Rice, G. C., Bennett, G. L., Camerato, T., and Wood, W. I. (1992). Characterization of two high affinity human interleukin-8 receptors. *J. Biol. Chem.* **267,** 16283–16287.

Legler, D. F., Loetscher, M., Roos, R. S., Clark-Lewis, I., Baggiolini, M., and Moser, B. (1998). B cell-attracting chemokine-1, a human CXC chemokine expressed in lymphoid tissues, selectively attracts B lymphocytes via BLR1/CXCR5. *J. Exp. Med.* **187,** 655–660.

Leong, S. R., Kabakoff, R. C., and Hebert, C. A. (1994). Complete mutagenesis of the extracellular domain of interleukin-8 (IL-8) type A receptor identifies charged residues mediating IL-8 binding and signal transduction. *J. Biol. Chem.* **269,** 19343–19348.

Liston, A., and McColl, S. (2003). Subversion of the chemokine world by microbial pathogens. *Bioessays* **25,** 478–488.

Littman, D. R. (1998). Chemokine receptors: Keys to AIDS pathogenesis? *Cell* **93,** 677–680.

Loetscher, M., Gerber, B., Loetscher, P., Jones, S. A., Piali, L., Clark-Lewis, I., Baggiolini, M., and Moser, B. (1996). Chemokine receptor specific for IP10 and mig: Structure, function, and expression in activated T-lymphocytes. *J. Exp. Med.* **184,** 963–969.

Loetscher, P., Pellegrino, A., Gong, J. H., Mattioli, I., Loetscher, M., Bardi, G., Baggiolini, M., and Clark-Lewis, I. (2001). The ligands of CXC chemokine receptor-3, I-TAC, Mig, and IP10, are natural antagonists for CCR3. *J. Biol. Chem.* **276,** 2986–2991.

Lowman, H. B., Slagle, P. H., De Forge, L. E., Wirth, C. M., Gillece-Castro, B. L., Bourell, J. H., and Fairbrother, W. J. (1996). Exchanging interleukin-8 and melanoma growth-stimulating activity receptor binding specificities. *J. Biol. Chem.* **271,** 14344–14352.

Mahler, D. A., Huang, S., Tabrizi, M., and Bell, G. M. (2004). Efficacy and safety of a monoclonal antibody recognizing interleukin-8 in COPD: A pilot study. *Chest* **126,** 926–934.

Matloubian, M., David, A., Engel, S., Ryan, J. E., and Cyster, J. G. (2000). A transmembrane CXC chemokine is a ligand for HIV-coreceptor Bonzo. *Nat. Immunol.* **1,** 298–304.

Medoff, B. D., Sauty, A., Tager, A. M., Maclean, J. A., Smith, R. N., Mathew, A., Dufour, J. H., and Luster, A. D. (2002). IFN-gamma-inducible protein-10 (CXCL10) contributes to airway hyperreactivity and airway inflammation in a mouse model of asthma. *J. Immunol.* **168,** 5278–5286.

Moser, B., and Willimann, K. (2004). Chemokines: Role in inflammation and immune surveillance. *Ann Rheum Dis* **63**(Suppl 2), ii84–ii89.

Murphy, P. M. (2001). Viral exploitation and subversion of the immune system through chemokine mimicry. *Nat. Immunol.* **2,** 116–122.

Murphy, P. M. (2002). International Union of Pharmacology. XXX. Update on chemokine receptor nomenclature. *Pharmacol. Rev.* **54,** 227–229.

Murphy, P. M., Baggiolini, M., Charo, I. F., Hebert, C. A., Horuk, R., Matsushima, K., Miller, L. H., Oppenheim, J. J., and Power, C. A. (2000). International union of pharmacology. XXII. Nomenclature for chemokine receptors. *Pharmacol. Rev.* **52,** 145–176.

Murphy, P. M., and Tiffany, H. L. (1991). Cloning of complementary DNA encoding a functional human interleukin-8 receptor. *Science* **253,** 1280–1283.

Nakayama, T., Hieshima, K., Izawa, D., Tatsumi, Y., Kanamaru, A., and Yoshie, O. (2003). Cutting edge: Profile of chemokine receptor expression on human plasma cells accounts for their efficient recruitment to target tissues. *J. Immunol.* **170,** 1136–1140.

Nickel, R., Beck, L. A., Stellato, C., and Schleimer, R. P. (1999). Chemokines and allergic disease. *J. Allergy Clin. Immunol.* **104,** 723–742.

Nieto, M., Frade, J. M., Sancho, D., Mellado, M., Martinez, A. C., and Sanchez-Madrid, F. (1997). Polarization of chemokine receptors to the leading edge during lymphocyte chemotaxis. *J. Exp. Med.* **186,** 153–158.

Penfold, M. E., Dairaghi, D. J., Duke, G. M., Saederup, N., Mocarski, E. S., Kemble, G. W., and Schall, T. J. (1999). Cytomegalovirus encodes a potent alpha chemokine. *Proc. Natl. Acad. Sci. USA* **96,** 9839–9844.

Perrier, P., Martinez, F. O., Locati, M., Bianchi, G., Nebuloni, M., Vago, G., Bazzoni, F., Sozzani, S., Allavena, P., and Mantovani, A. (2004). Distinct transcriptional programs activated by interleukin-10 with or without lipopolysaccharide in dendritic cells: Induction of the B cell-activating chemokine, CXC chemokine ligand-13. *J. Immunol.* **172,** 7031–7042.

Peveri, P., Walz, A., Dewald, B., and Baggiolini, M. (1988). A novel neutrophil-activating factor produced by human mononuclear phagocytes. *J. Exp. Med.* **167,** 1547–1559.

Proudfoot, A. E. (2002). Chemokine receptors: Multifaceted therapeutic targets. *Nat. Rev. Immunol.* **2,** 106–115.

Rebecchi, M. J., and Pentyala, S. N. (2000). Structure, function, and control of phosphoinositide-specific phospholipase C. *Physiol. Rev.* **80,** 1291–1335.

Renzoni, E., Lympany, P., Sestini, P., Pantelidis, P., Wells, A., Black, C., Welsh, K., Bunn, C., Knight, C., Foley, P., and du Bois, R. M. (2000). Distribution of novel polymorphisms of the interleukin-8 and CXC receptor 1 and 2 genes in systemic sclerosis and cryptogenic fibrosing alveolitis. *Arthritis. Rheum.* **43,** 1633–1640.

Revello, M. G., Percivalle, E., Arbustini, E., Pardi, R., Sozzani, S., and Gerna, G. (1998). *In Vitro* Generation of Human Cytomegalovirus pp65 Antigenemia, Viremia, and Leuko DNAemia. *J. Clin. Invest.* **101,** 2686–2692.

Ridley, A. J., Allen, W. E., Peppelenbosch, M., and Jones, G. E. (1999). Rho family proteins and cell migration. *Biochem. Soc. Symp.* **65,** 111–123.

Ridley, A. J., Schwartz, M. A., Burridge, K., Firtel, R. A., Ginsberg, M. H., Borisy, G., Parsons, J. T., and Horwitz, A. R. (2003). Cell migration: Integrating signals from front to back. *Science* **302,** 1704–1709.

Roberts, A. W., Kim, C., Zhen, L., Lowe, J. B., Kapur, R., Petryniak, B., Spaetti, A., Pollock, J. D., Borneo, J. B., Bradford, G. B., Atkinson, S. J., Dinauer, M. C., and Williams, D. A. (1999). Deficiency of the hematopoietic cell-specific Rho family GTPase Rac2 is characterized by abnormalities in neutrophil function and host defense. *Immunity* **10,** 183–196.

Rot, A., and von Andrian, U. H. (2004). Chemokines in innate and adaptive host defense: Basic chemokinese grammar for immune cells. *Annu. Rev. Immunol.* **22,** 891–928.

Samanta, A. K., Oppenheim, J. J., and Matsushima, K. (1989). Identification and characterization of specific receptors for monocyte-derived neutrophil chemotactic factor (MDNCF) on human neutrophils. *J. Exp. Med.* **169,** 1185–1189.

Sanchez-Madrid, F., and del Pozo, M. A. (1999). Leukocyte polarization in cell migration and immune interactions. *EMBO J.* **18,** 501–511.

Sander, E. E., ten Klooster, J. P., van Delft, S., van der Kammen, R. A., and Collard, J. G. (1999). Rac downregulates Rho activity: Reciprocal balance between both GTPases determines cellular morphology and migratory behavior. *J. Cell. Biol.* **147,** 1009–1022.

Sotsios, Y., and Ward, S. G. (2000). Phosphoinositide 3-kinase: A key biochemical signal for cell migration in response to chemokines. *Immunol. Rev.* **177,** 217–235.

Thelen, M., Peveri, P., Kernen, P., von Tscharner, V., Walz, A., and Baggiolini, M. (1988). Mechanism of neutrophil activation by NAF, a novel monocyte-derived peptide agonist. *FASEB J.* **2,** 2702–2706.

Walz, A., Peveri, P., Aschauer, H., and Baggiolini, M. (1987). Purification and amino acid sequencing of NAF, a novel neutrophil-activating factor produced by monocytes. *Biochem. Biophys. Res. Commun.* **149,** 755–761.

Wang, J. F., Park, I. W., and Groopman, J. E. (2000). Stromal cell-derived factor-1alpha stimulates tyrosine phosphorylation of multiple focal adhesion proteins and induces migration of hematopoietic progenitor cells: Roles of phosphoinositide-3 kinase and protein kinase C. *Blood* **95,** 2505–2513.

Wang, J. M., Su, S., Gong, W., and Oppenheim, J. J. (1998). Chemokines, receptors, and their role in cardiovascular pathology. *Int. J. Clin. Lab. Res.* **28,** 83–90.

Whiting, D., Hsieh, G., Yun, J. J., Banerji, A., Yao, W., Fishbein, M. C., Belperio, J., Strieter, R. M., Bonavida, B., and Ardehali, A. (2004). Chemokine monokine induced by IFN-gamma/CXC chemokine ligand-9 stimulates T lymphocyte proliferation and effector cytokine production. *J. Immunol.* **172,** 7417–7424.

Xu, J., Wang, F., Van Keymeulen, A., Herzmark, P., Straight, A., Kelly, K., Takuwa, Y., Sugimoto, N., Mitchison, T., and Bourne, H. R. (2003). Divergent signals and cytoskeletal assemblies regulate self-organizing polarity in neutrophils. *Cell* **114,** 201–214.

Yoshimura, T., Matsushima, K., Tanaka, S., Robinson, E. A., Appella, E., Oppenheim, J. J., and Leonard, E. J. (1987). Purification of a human monocyte-derived neutrophil chemotactic factor that has peptide sequence similarity to other host defense cytokines. *Proc. Natl. Acad. Sci. USA* **84,** 9233–9237.

Zou, Y. R., Kottmann, A. H., Kuroda, M., Taniuchi, I., and Littman, D. R. (1998). Function of the chemokine receptor CXCR4 in hematopoiesis and in cerebellar development. *Nature* **393,** 595–599.

CHAPTER 3

The Molecular and Cellular Biology of CC Chemokines and Their Receptors

James M. Fox and James E. Pease
Leukocyte Biology Section, Biomedical Sciences Division,
Faculty of Medicine, Imperial College, London SW7 2AZ,
United Kingdom

I. INTRODUCTION

Chemokines (*chemo*tactic cyto*kines*) are low-molecular-weight proteins, typically basic, that have been shown to activate and mediate the migration of leukocytes (Mackay, 2001). As discussed earlier in this volume, the chemokine family consists of over 40 members, which are divided into two

Current Topics in Membranes, Volume 55

1063-5823/05 $35.00
DOI: 10.1016/S1063-5823(04)55003-2

major and two minor families based on the location of N-terminal cysteine residues. The largest of these families by far is that of the CC chemokines, which together with their receptors have been implicated in the pathogenesis of inflammatory diseases, a variety of autoimmune disorders, and viral infection (Gerard and Rollins, 2001; Murphy, 2001). Consequently, they have attracted much attention, and investigations into their biology have given rise to a number of strategies aimed at producing therapeutic treatments. In this review, we will outline recent developments in the understanding of their molecular and cellular biology.

II. CC CHEMOKINES

Currently, twenty-eight chemokines and ten receptors constitute the CC chemokine and chemokine receptor family (Tables I and II). Like all other classes, CC chemokines have been given a systematic number to distinguish them from other group members; this system updates the anecdotal method of defining a chemokine upon its function (Zlotnik and Yoshie, 2000). Chemokines may be produced either constitutively or in response to inflammatory stimuli. Constitutively produced chemokines have important homeostatic roles in lymphocyte trafficking to and from a variety of organs; for example, CCL19 coordinates the emergence of newly generated T-cells from the thymus to the general circulation (Ueno *et al.*, 2002). In contrast, the production of CCL3 is induced by inflammatory stimuli (Kasama *et al.*, 1993) and the chemokine is therefore implicated in a variety of pathological conditions, including rheumatoid arthritis (Koch *et al.*, 1994) and airway inflammation (Lukacs *et al.*, 1995). These distinctions are not absolute, as some chemokines can have both homeostatic and inflammatory roles. For example, constitutive production of CCL11 by the intestinal epithelial cells recruits eosinophils to the gut (Kitaura *et al.*, 1996; Humbles *et al.*, 2002), yet its production by lung epithelial cells in response to allergens has been implicated in the recruitment of eosinophils to the inflamed lung (Humbles *et al.*, 1997; Rothenberg *et al.*, 1997).

III. CC CHEMOKINE RECEPTORS

The biological effects of chemokines are mediated by their binding to cell surface receptors that belong to Class I of the rhodopsin-like G-protein-coupled receptor (GPCR) superfamily, which is believed to number over 900 members in the human genome (Takeda *et al.*, 2002; Fredriksson *et al.*,

TABLE I
Human CC Chemokines Identified to Date

Systematic name	Colloquial name	Known receptor usage
CCL1	I-309	CCR8
CCL2	MCP-1/MCAF	CCR2
CCL3	MIP-1α/LD78α	CCR1, CCR5
CCL3L1	LD78β	CCR1, CCR5
CCL4	MIP-1β	CCR5
CCL4L1	LAG-1	CCR5
CCL5	RANTES	CCR1, CCR3, CCR5
CCL7	MCP-3	CCR1, CCR2, CCR3
CCL8	MCP-2	CCR3
CCL11	Eotaxin	CCR3
CCL13	MCP-4	CCR2, CCR3
CCL14	HCC-1	CCR1
CCL15	HCC-2/Lkn-1/MIP-1δ/MIP5	CCR1, CCR3
CCL16	HCC-4/LEC	CCR1
CCL17	TARC	CCR4
CCL18	DC-CK-1/AMAC-1/MIP5/PARC/ MIP-4	Unknown
CCL19	MIP 3β/ELC/Exodus-3	CCR7
CCL20	MIP-3α/LARC/Exodus-1	CCR6
CCL21	SLC/Exodus-2/6Ckine	CCR7
CCL22	MDC/STCP-1	CCR4
CCL23	MPIF-1/	CCR1
CCL24	MPIF-2/Eotaxin-2	CCR3
CCL25	TECK	CCR9
CCL26	Eotaxin-3	CCR3
CCL27	CTACK/ALP/ILC/ESkine	CCR10
CCL28	MEC	CCR10, CCR3

Note: Some human chemokines (for example, CCL6) appear to be missing from the list. In such instances, although a chemokine of that name has been identified in the mouse, no human ortholog has been documented.

2003). Eighteen human chemokine receptors have been identified to date (Murphy *et al.*, 2000; Murphy, 2002), ten of which belong to the CC family (Table II). The ligand repertoire of chemokine receptors is typically class restricted, that is to say, CC chemokine receptors are activated only by CC

TABLE II
Human CC Chemokine Receptors Identified to Date

Chemokine receptor	Cellular expression
CCR1	Mo, iDC, Eo, Bs, actT, PMN, NK
CCR2	Mo, iDC, actT, Bs
CCR3	Eo, actT, Bs
CCR4	mDC, actT, Bs, NK
CCR5	Mo, iDC, actT
CCR6	iDC
CCR7	mDC, T, B, NK
CCR8	Mo, actT, NK
CCR9	actT
CCR10	actT

Mo, monocytes; iDC, immature dendritic cell; Eo, eosinophil; Bs, basophil; actT, activated T-lymphocyte; PMN, neutrophil; mDC, mature dendritic cell; B, B-lymphocyte; T, naïve T-lymphocyte; NK, natural killer cell.

chemokines. One exception to this is the Duffy antigen receptor complex (DARC), which binds both CC and CXC chemokines (Lu *et al.*, 1995). Monogamy (i.e., interacting with a single ligand) is quite rare; only one CC chemokine receptor, CCR6, has only a single ligand to date. Rather more common is the incidence of promiscuity—the majority of receptors are able to interact with many different ligands. To complicate matters further, leukocytes typically express several different chemokine receptors at their cell surfaces, giving them the ability to respond to multiple chemokines.

Chemokine receptors are typically 340–380 amino acids in length and possess a common topographical distribution in the plasma membrane, whereby a seven-transmembrane α-helical motif threads its way in and out of the membrane, leaving an extracellular N-terminal region and a cytoplasmic C-terminus, with three intracellular and 3 extracellular domains (Fig. 1). The C-terminus typically contains multiple serine and threonine residues, which are phosphorylated following receptor occupancy, rendering the receptor insensitive. Other common features are single cysteine residues in each of the extracellular domains, which are believed to form disulphide bridges that link the N-terminus and extracellular loops and maintain receptor conformation. Mutation of these residues in the receptor CCR5 leads to a loss of chemokine binding and subsequent function (Blanpain *et al.*, 1999a,b). Also conserved between the majority of CC chemokine receptors is a

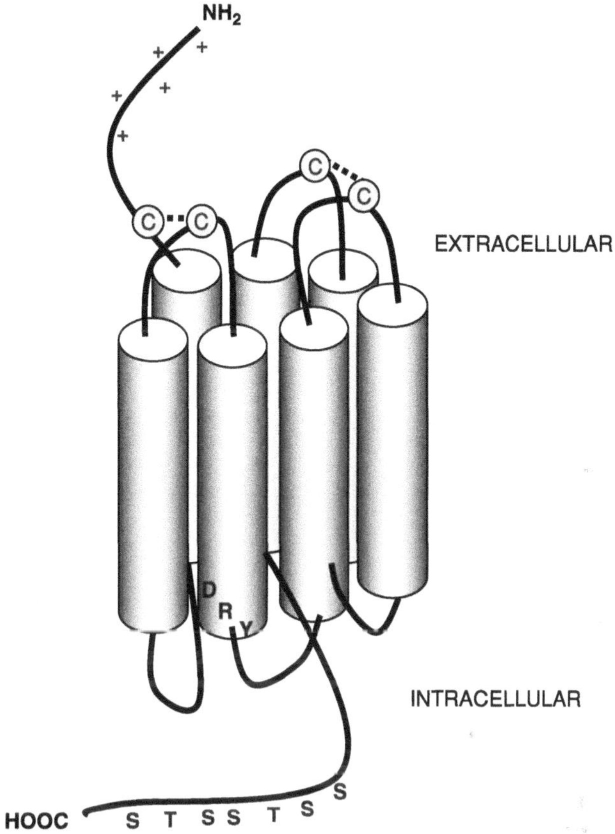

FIGURE 1 This cartoon illustrates the secondary structure typical of a CC chemokine receptor. The seven-transmembrane (TM) helices in the bundle are depicted as cylinders and are held together by disulphide bonding of conserved cysteine residues (yellow). The N-terminus is negatively charged and binds the predominantly basic chemokine, while the intracellular C-terminus is rich in serine and threonine residues, some of which undergo phosphorylation following receptor activation. The DRY motif of TM helix 3 is also illustrated. (See Color Insert.)

"DRY" motif at the cytoplasmic end of the third transmembrane α-helix. This is analogous to the ERY motif of bovine rhodopsin, a common feature of GPCRs. This motif is thought to act as an ionic lock, holding the GPCR in an inactive state, prior to activation by a ligand (Ballesteros *et al.*, 2001). Indeed, nonconservative mutagenesis of this domain in the receptor CCR3 leads to a dramatic loss of function (Auger *et al.*, 2002).

IV. ASPECTS OF CC CHEMOKINE STRUCTURE

Like their CXC counterparts, CC chemokines share a common protein fold, known as a "Greek key" motif, in which three antiparallel β-pleated sheets are overlaid by a C-terminal α-helix (Fig. 2A). Following the first pair of cysteine residues is a ten-residue loop known as the N-loop, and then a succession of three β-strands and a C-terminal α-helix. The three β-strands are positioned antiparallel to each other and form a β-pleated sheet that is overlaid at an angle of approximately 75° by the C-terminal α-helix. As might be expected, the existence of four conserved amino-terminal cysteine residues within CC chemokines has structural implications. Using their

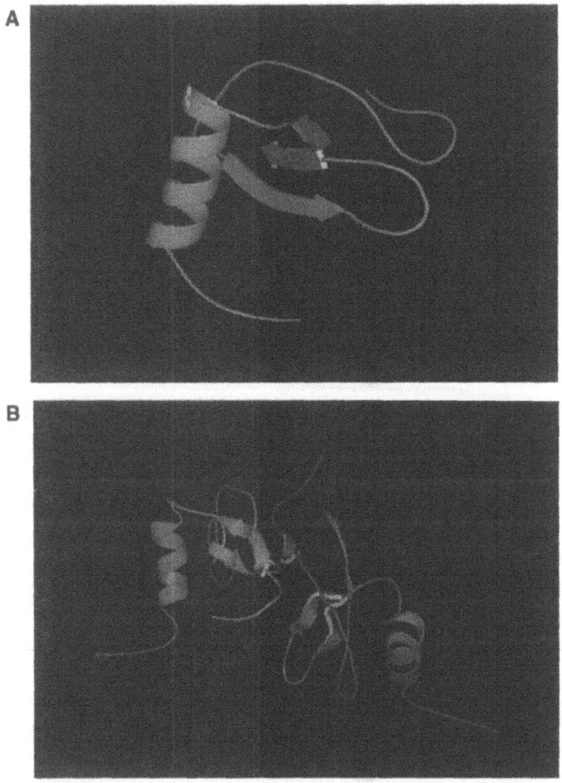

FIGURE 2 Panel A shows the secondary structural "Greek Key" motif of CC chemokines, as typified by CCL2. Three antiparallel β-pleated sheets overlay a C-terminal, α-helical domain. Panel B shows the elongated CCL2 dimer, which is facilitated by interactions of the chemokine N-termini. The CCL2 monomers making up the dimer are depicted in blue and red. (See Color Insert.)

order in the primary sequence for notation, Cys1–Cys3 and Cys2–Cys4 form disulphide bonds, giving stability to the tertiary structure. Interestingly, three CC chemokines, CCL1, CCL15, and CCL23, have two additional cysteine residues that are also disulphide linked (Sticht *et al.*, 1999; Keizer *et al.*, 2000; Rajarathnam *et al.*, 2001). The majority of CC chemokines are known to form dimers and higher-order oligomers, e.g., CCL2 (Handel and Domaille, 1996), and they do so via residues encompassing the CC motif near to the amino terminal region (Fig. 2B). The structures are more elongated than those of CXC dimers due to reduced hydrophobicity around the β-sheet, which means that more of the side-chain residues are exposed. The biological significance of dimerization of chemokines is unknown. Experimental data suggest that chemokines exist and have maximal activity *in vivo* at nanomolar concentrations, and since dissociation constants for oligomeric chemokines are usually in the micromolar range, it is thought that the majority of chemokines are active as monomers (Burrows *et al.*, 1994). In addition, a Pro8Ala mutant of CCL2, which cannot dimerize, has been engineered and retains its biological activity (Paavola *et al.*, 1998).

V. STRUCTURAL DETERMINANTS OF LIGAND BINDING AND RECEPTOR ACTIVATION BY CC CHEMOKINES

The use of chimeric chemokine receptor constructs has successfully dissected the mechanism by which receptors bind chemokines and undergo subsequent activation. This occurs via a two-step model in which a high-affinity interaction with the receptor N-terminus and chemokine tethers the ligand to the receptor. This facilitates a second, lower-affinity interaction in which the chemokine is delivered to the remainder of the receptor, leading to the activation of heterotrimeric G-proteins (Monteclaro and Charo, 1996; Pease *et al.*, 1998). The high-affinity interaction is thought to be dependent upon the typically acidic N-terminus's interacting electrostatically with the basic chemokine, and it may be facilitated further by the sulfation of tyrosine residues within this receptor domain (Farzan *et al.*, 1999). Recent studies of CCR5 have refined the two-step model further with the notion that binding of the chemokine core by extracellular receptor domains facilitates an interaction between the chemokine N-terminus and transmembrane helices of the receptor (Blanpain *et al.*, 2003; Govaerts *et al.*, 2003). Disruption of hydrophobic interactions between the side chains of helix II and helix III is thought to initiate the conformational changes needed for receptor activation (Fig. 3).

The length and composition of the chemokine N-terminus play a crucial role in this process and can determine whether the ligand–receptor

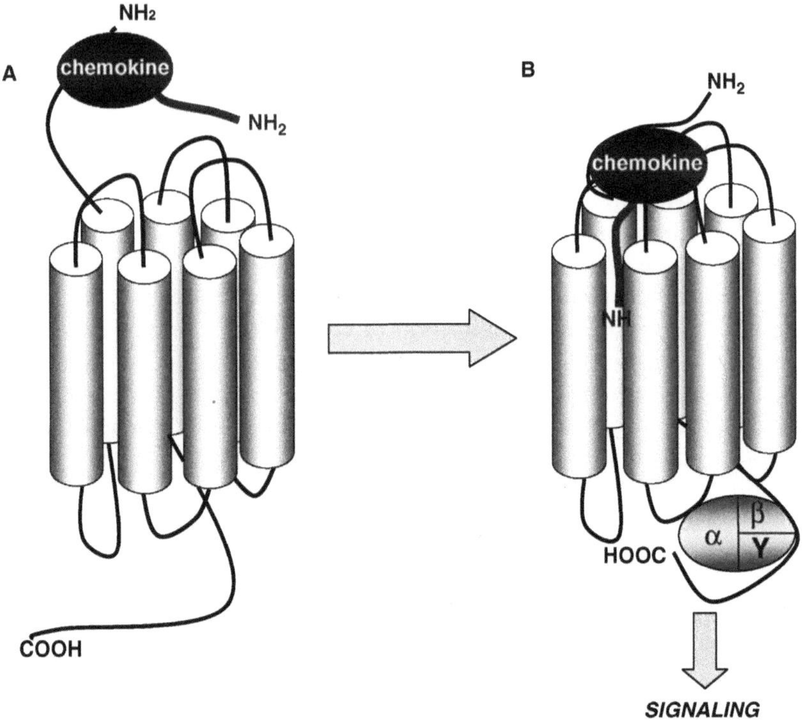

FIGURE 3 The two-step model of receptor activation. The amino terminus is thought to bind the chemokine with high affinity (Panel A) and subsequently deliver the tethered ligand to the remainder of the receptor (Panel B). This second event allows the chemokine N-terminus to activate a ligand-binding pocket composed of the transmembrane helices. The resulting conformational changes result in the activation of heterotrimeric G-proteins and in subsequent downstream signaling cascades.

interaction is agonistic or antagonistic (Clark-Lewis *et al.*, 1995). Changes in chemokine sequence at the N-terminus typically result in altered receptor binding or activation capabilities and are epitomized by studies of CCL5. Truncation of this chemokine to CCL5(9–68) produces a receptor antagonist, as deduced by several assays, blocking responses to CCL2, CCL3, and CCL5 (Gong *et al.*, 1996). Similarly, elongation at the amino terminus of CCL5 with either a single methionine residue (Met-CCL5) or an aminoxypentane moiety (AOP-CCL5) produces potent receptor antagonists (Proudfoot *et al.*, 1996; Simmons *et al.*, 1997).

Understanding of the molecular basis of chemokine receptor activation has been aided considerably by the determination of the crystal structure of the related GPCR bovine rhodopsin (Palczewski *et al.*, 2000), allowing comparative modelling studies to be performed (Govaerts *et al.*, 2003; Onuffer *et al.*, 2003). Activation is thought to involve movement of several receptor helices, producing conformational changes in the intracellular domains of the receptor, which are thought to increase their affinity for heterotrimeric G-proteins (Sakmar *et al.*, 2002). Highly conserved proline residues in helices V, VI, and VII disrupt the hydrogen-bonding network of the helix, introducing a kink in the helices and imparting a degree of flexibility. One particular motif in helix II, the TXP motif, has been the focus of much research and appears to play a key structural role in CCR5 function by conferring the backbone flexibility necessary for the conformational change associated with the activation step induced by ligand binding, while contributing very little to ligand binding (Govaerts *et al.*, 2001; Arias *et al.*, 2003). Since this motif is conserved among many of the CC chemokine receptors, such a mechanism of receptor activation may exist in other members of the family. The introduction of metal ion binding sites into chemokine receptor sequences has also shed light on the dynamics of other helices; movement of helix VI is also implicated during the activation process (Gerlach *et al.*, 2003; Rosenkilde *et al.*, 2004).

VI. G-PROTEIN COUPLING OF CC CHEMOKINE RECEPTORS AND DOWNSTREAM SIGNALING EVENTS

Chemokine receptors, along with other members of the GPCR family, transduce signals via heterotrimeric G-proteins, consisting of α, β, and γ subunits. About twenty α subunits, six β subunits, and twelve γ subunits have been identified in mammals (Maghazachi, 2000). The α subunits belong to four subfamilies: α_s (α_s and α_{olf}); α_q (α_q, α_{11}, α_{14}, α_{15}, and α_{16}); α_i (α_{i1}, α_{i2}, α_{i3}, α_{o1}, α_{o2}, α_z, α_t, α_{gus}, α_{con}, and α_{rod}); and α_{12} (α_{12} and α_{13}). Each of these α subunits has a unique biological function. Gα_s activates adenylyl cyclase, leading to the accumulation of cyclic adenosine monophosphate (cAMP), while Gα_i inhibits adenylyl cyclase, leading to depletion of cellular cAMP. Gα_q activates phospholipase Cβ, resulting in phosphatidylinositol bisphosphate hydrolysis and generation of diacylglycerol (DAG) and inositol 1,4,5-trisphosphate (IP_3). These activate protein kinase C (PKC) and the release of intracellular Ca^{2+}, respectively. Gα_o is linked to the activation of ion channels such as Ca^{2+}, Cl^-, and K^+, while Gα_{12} and Gα_{13} provide links between GPCRs and activation of the small GTPase Rho.

Intracellular signalling by chemokine receptors is thought to occur mainly through Gα_i proteins, as physiological responses are readily inhibited by preincubation of cells with pertussis toxin, which ADP-ribosylates a cysteine residue in the C-terminus of the Gα_i subunit, rendering it non-functional.

The α subunits of G-proteins are located in close proximity to the plasma membrane by a series of posttranslational modifications, such as palmitoylation and myristoylation. These posttranslational modifications greatly increase the ability of the α subunits to interact with other protein effectors; for instance, myristoylation greatly increases the affinity of $G\alpha_i$ for $\beta\gamma$, and myristoylation of α_i is necessary for the inhibition of adenylyl cyclase (Gallego *et al.*, 1992). Furthermore, the $\beta\gamma$ subunit is also modified; prenylation of γ by geranylgeranyl is necessary for normal $\beta\gamma$ dimer function (Higgins and Casey, 1994). Solution of the crystal structure of the G-protein heterotrimer $G\alpha_i$ $\beta\gamma$ reveals that the carboxy terminal helix of the α subunit of the G-protein binds the third transmembrane domain of the GPCR, though binding is also capable of occurring between this region and the serine/threonine-rich carboxy terminal (Wall *et al.*, 1995).

The $\beta\gamma$ subunit interacts with the amino-terminal region of the α subunit at a site called the switch II region, the area at which guanosine diphosphate (GDP) binds when the receptor is in an inactive state. Upon ligand activation of the GPCR, conformational changes occur within the receptor that allow the exchange of guanosine trisphosphate (GTP), which is abundant in the transmembrane, for GDP. This results in the displacement of the $\beta\gamma$ from the α subunit and the release of both the α-GTP complex and the $\beta\gamma$ subunit, leaving them to activate various effectors and multiple downstream signaling events. Activation is terminated by the intrinsic GTPase activity of the α subunit and results in the reassociation of the α subunit with the $\beta\gamma$ dimer. Activation of G-proteins is thought to occur catalytically, where a single activated GPCR is able to activate several G-proteins (Janetopoulos *et al.*, 2001).

A model of GPCR activation termed the "extended ternary complex model" was established in 1980 and has been applied to chemokine receptor activation (De Lean *et al.*, 1980). This model states that the receptor exists in an equilibrium between an inactive conformation (R) and an active conformation (R*). In the absence of any agonist, the inactive R state dominates. However, a certain proportion of the receptors can spontaneously assume the R* state because the energy barrier between the R and R* states is sufficiently low. Agonists are predicted to bind with highest affinity to the R* conformation and, in this way, shift the equilibrium and increase the proportion of receptor in the activated R* state. Some receptors can spontaneously adopt an active conformation or R* state and couple to the G-protein in the absence of agonist, a phenomenon known as constitutive activation (Samama *et al.*, 1993). Moreover, inverse agonists, able to bind to the receptor and switch off constitutive signaling, have been described; for example, the small molecule antagonist Banyu (I) is an inverse agonist of constitutive signaling in transfectants expressing the chemokine receptor CCR3 (Wan *et al.*, 2002).

The subsequent signaling events downstream of chemokine-mediated G-protein activation are complicated. It has already been mentioned that signaling events from chemokine receptors are commonly inhibited by pertussis toxin, leading to the assumption that chemokine receptors are $G\alpha_i$ linked. One of the most rapid occurrences following chemokine receptor activation is a rise in the intracellular calcium concentration ($[Ca^{2+}]_i$), and this has commonly been used as a measurement for the functional response of receptor activation. Increases in $[Ca^{2+}]_i$ are mediated via the $\beta\gamma$ subunit of the G-protein that activates phospholipase $C\beta$, leading to 1) hydrolysis of phosphatidylinositol to IP_3 and 2) DAG formation, which respectively induce the release of Ca^{2+} from intracellular stores and subsequent activation of PKC. PKC activation is necessary for certain leukocyte responses, such as neutrophil respiratory burst (Li *et al.*, 2000), and it also plays a role in receptor phosphorylation, leading to desensitization and internalization, which will be discussed later.

The $\beta\gamma$ subunit is also responsible for the activation of phosphatidylinositide 3-kinase (PI3K), which, via a dephosphorylation cascade, forms phosphatidylinositol (3,4) bisphosphate [PtdIns(3,4)P_2]. One of the hallmark responses of leukocytes is their chemotaxis (directed migration) towards a source of chemokine. The use of PI3K inhibitors *in vitro* has demonstrated a significant role for PI3K in this process (Turner *et al.*, 1995), which has been corroborated by studies using mice in which the γ isoform of PI3K has been deleted. These mice show impaired neutrophil recruitment in models of *Escherichia coli*–induced peritonitis and have leukocytes defective in chemotaxis to CCL3 (Li *et al.*, 2000). PI3K also activates Akt/protein kinase B and the mitogen activated protein kinase (MAPK) pathway, also known as extracellular signal–regulated kinase (ERK). Collectively, these are responsible for the regulation of a variety of cellular processes, such as stress-induced cell-cycle arrest, activation of transcription, and apoptosis. Signaling events are not only initiated via the $\beta\gamma$ subunit: The α subunit, in addition to activating adenylyl cyclase, is also capable of activating the Src family of kinases, which are known to be able to activate Ras and thus Rho via Shc, Grb2, and SOS (Ma *et al.*, 2000).

More recent evidence has emerged that implies that, at least in T-lymphocytes, a more complicated signaling mechanism may be at work and that PI3K is dispensable for CCR4-stimulated chemotaxis of both Th2 cells and a T-lymphocyte cell line (Cronshaw *et al.*, 2004). These studies identified a pathway involving small GTPases, and more specifically Rho-associated coiled-coil-forming protein kinase (ROCK), as the critical PI3K independent pathway for T-lymphocyte migration. ROCK is an effector of Rho, a small GTPase that is also activated by the $\beta\gamma$ G-protein subunit via the GTPases Ras/Rac, which have previously been shown to be associated with cell cytoskeletal rearrangements and to facilitate such cellular responses as shape change, firm adhesion, and chemotaxis (Bokoch, 1995).

VII. CHEMOKINE RECEPTOR DIMERIZATION

While the dimerization of several GPCRs has been well documented (reviewed by Milligan *et al.* [2003]), the ability of chemokine receptors to dimerize has only recently undergone close scrutiny (Rodriguez-Frade *et al.*, 2001). Both homodimerization and heterodimerization have been demonstrated for the CC chemokine receptors CCR2 and CCR5. Homodimerization following ligand activation has been shown to occur for CCR2 (Rodriguez-Frade *et al.*, 1999a), CCR5 (Rodriguez-Frade *et al.*, 1999b), and the CXC chemokine receptor CXCR4 (Vila-Coro *et al.*, 1999). Heterodimerization of CCR2 and CCR5 induces distinct signaling pathways that involve the recruitment of Janus kinases (JAKs), which in turn transphosphorylate the C-terminus of the dimer, allowing the subsequent activation of signal transducers and activators of transcription (STATs), which can then induce G-protein activation and downstream signaling events (Mellado *et al.*, 2001).

Many in the field considered the evidence for receptor dimerization to be unconvincing because dimerization had been demonstrated by SDS-PAGE analysis, and these researchers suggested that more compelling evidence could be obtained via the implementation of more sensitive techniques that could detect real-time receptor redistribution and oligomerization (Thelen and Baggiolini, 2001). Results utilizing fluorescence resonance energy transfer (FRET) support the earlier studies and have shown that both CCR2 and CCR5 form dimers following stimulation with CCL5 and CCL2 via interactions between residues in transmembrane helices 1 and 4 (Hernanz-Falcon *et al.*, 2004). If cells are able to form chemokine receptor heterodimers, the possibility is raised that, in doing so, they may increase their responsiveness to a wider repertoire of ligands and also activate distinct signaling pathways, adding yet another level of complexity to receptor signaling.

VIII. RECEPTOR DESENSITIZATION, INTERNALIZATION, AND RECYCLING

Receptor activation and downstream signaling events are typically followed by receptor desensitization, the process by which the receptor becomes refractory to continued stimulation within seconds to minutes of agonist exposure. This is followed in turn by internalization, a process whereby the receptor is sequestered from the cell surface to intracellular compartments. Desensitization can be divided into homologous desensitization and heterologous desensitization. Homologous desensitization is carried out by specific G-protein-coupled receptor kinases (GRKs) following occupation of the receptor by the ligand. GRKs are serine/threonine kinases that are capable of phosphorylating the serine/threonine residues typically found in abundance in the C-termini of GPCRs. Following GPCR activation and then dissociation of the $\beta\gamma$ subunit from the α subunit, the $\beta\gamma$ complex and membrane-bound phosphatidylinositol bisphosphate appear to bind to the pleckstrin homology domain in the C-terminus of the GRKs. This induces translocation of the GRK to the membrane and into close proximity with the activated receptor, allowing phosphorylation to take place (reviewed by Pitcher *et al.*, 1998).

In contrast to homologous desensitization, heterologous desensitization does not require activation of the receptor by the ligand and is affected by PKA (activated by G_s-coupled receptors) and PKC (activated by G_q) following the activation of downstream signaling pathways by other receptors (Bohm *et al.*, 1997; Ali *et al.*, 1999). The accompanying phosphorylation is carried out in a region of the GPCR that contains an appropriate PKA and/or PKC phosphorylation site, typically the C-terminus. In both homologous and heterologous desensitization, the phosphorylation events serve to uncouple the receptor from G-protein activation, thereby terminating the relay of extracellular signals. Chemokine receptors have also been reported to be uncoupled from the intracellular signaling machinery of dendritic cells by the actions of the cytokine IL-10, in conjunction with LPS stimulation (D'Amico *et al.*, 2000). The mechanisms by which this is achieved are yet to be fully elucidated, but initial studies on gene-expression profiling show that PI3Kγ is downregulated, while suppressor of cytokine signaling (SOCS)-3 expression is upregulated (Perrier *et al.*, 2004).

Receptor desensitization is commonly caused by receptor internalization, whereby the receptor is removed from its location on the membrane into the cytoplasm, and then either degraded or recycled back to the surface by two main mechanisms. The level of cell-surface expression of a chemokine receptor is a balance between the rate of internalization and the rate of recovery, which can occur via *de novo* receptor production or receptor recycling, and which in some cases is a constitutive process (Pelchen-Matthews *et al.*, 1999). Internalization is dramatically increased in the presence of agonist, and it is an active process, as internalization is almost entirely prevented at temperatures below 16°C (von Zastrow and Kobilka, 1994). There appear to be two main mechanisms of chemokine receptor internalization: 1) the classical recruitment of arrestin to the receptor, resulting in the formation of a clathrin-coated vesicle around the receptor and subsequent internalization, and 2) internalization via clathrin-independent pathways.

Internalization via clathrin-coated pits is by far the best-understood and most common mechanism of internalization, and in the case of CC chemokine receptors, CCR5 has undergone the most thorough examination. Following the process of desensitization just described, arrestins are recruited to the receptor, and they act as adapter proteins, bringing together phosphorylated receptors, clathrin, and the adaptin-2 adaptor complex. Arrestins 2 and 3 are able to bind directly and with high affinity to clathrin, which is followed by sequestration of the receptor–arrestin complex to clathrin-coated pits (Goodman *et al.*, 1996). At this point, dynamin is involved in the process of "pinching off" the clathrin-coated pits to form vesicles. There, the receptor is dissociated from the ligand and undergoes dephosphorylation and subsequent trafficking from early to late endosomes, at which time the fate of the receptor may be degradation or recycling back to the cell surface (Mueller *et al.*, 2002). The use of arrestin mutants or inhibitors of internalization has further characterized these mechanisms. Dominant negative arrestins, which are unable to bind either receptor or clathrin, impair the receptor internalization process (Goodman *et al.*, 1996). Hypertonic sucrose (Heuser and Anderson, 1989) and chlorpromazine treatment (Wang *et al.*, 1993) are known to inhibit receptor internalization via clathrin-coated pits 1) by inhibiting clathrin pit formation and 2) via inhibition of the clathrin subunit AP2, which holds clathrin onto the plasma membrane, respectively. Both treatments have been shown to inhibit internalization of CCR5 (Mueller *et al.*, 2002) and further demonstrate the importance of this pathway in the cessation of signaling via chemokine receptors.

A second, less rapid pathway of internalization that is independent of arrestin and clathrin involves caveolae, microdomains of 50–100nm that appear as flask-like invaginations in the plasma membrane. Caveolae have a lipid composition similar to that of lipid rafts (indeed, their function is maintained by cholesterol) but can be distinguished by the localization of caveolin proteins such as caveolin-1. It is thought that after agonist binding GPCRs can move into lipid rafts/caveolae, as is the case for the somatostatin receptor sst2 (Krisch *et al.*, 1998). Caveolae have also been shown to be involved in the internalization of CCR5 (Mueller *et al.*, 2002). Both filipin (a polyene antibiotic and sterol-binding agent) and nystatin (a cholesterol-sequestering agent) can be used to disrupt lipid rafts and caveolae (Schnitzer *et al.*, 1994). Alongside standard immunostaining and confocal microscopy techniques, their use has demonstrated that CCR4 and CCR5 are internalized both by the classical clathrin-coated vesicle pathway and via caveolae (Mueller *et al.*, 2002; Mariani *et al.*, 2004). In contrast, CCR3 internalization occurs solely via clathrin-coated pits (Zimmermann and Rothenberg, 2003). The importance of each mechanism in the internalization of other CC chemokine receptors and the existence of additional, as yet undefined pathways remains to be established.

Following internalization, chemokine receptors may be recycled to the membrane or undergo photolytic degradation. For instance, CCR5 is recycled immediately (Signoret *et al.*, 2000), whereas the CXC chemokine receptor CXCR4 undergoes degradation with little receptor recycling, with cell-surface receptor numbers maintained by *de novo* protein synthesis (Tarasova *et al.*, 1998). Ligand dissociation does not appear to be required for receptor internalization, as agonist-occupied CCR5 can undergo multiple cycles of receptor internalization and recycling (Signoret *et al.*, 2000). Such a process has been used to explain the efficacy of a chemokine receptor antagonist. Aminooxypentane (AOP)-CCL5 remains bound to CCR5 and, through the continued stimulation of receptor internalization, affects receptor expression and subsequent responses. Receptor fate after internalization appears to be dependent upon its location following intracellular trafficking, which involves several steps in the cell machinery. It is common for recycling of GPCR to be dependent upon endocytic pH, but this has been shown not to be the case for CCR5 (Signoret *et al.*, 2004). This receptor appears to be trafficked after internalization to early sorting endosomes, which accumulate around the Golgi apparatus before transfer of the receptor to recycling endosomes, where the receptor is dephosphorylated and recycled to the surface (Signoret *et al.*, 2000). Until recently, the paradigm of receptor recycling was that receptors are transported to early and then late endosomes prior to recycling, but independent studies of CCR5 dispute this conceptualization. Nocodazole is known to inhibit transport of vesicles from early to late endosomes, and its use was shown to have no effect on the recycling of CCR5 (Mueller *et al.*, 2002). Furthermore, staining for late endosomes showed that this mechanism does not involve late endosomes (Signoret *et al.*, 2000). Moreover, pharmacological tools, such as monensin (which inhibits the acidification of intracellular compartments) and brefeldin A (which blocks the translocation of proteins from the endoplasmic reticulum to the Golgi) have been used to demonstrate that CCR5 is recycled without passing through the Golgi. Cycloheximide, an inhibitor of protein synthesis, has also been utilized to demonstrate that *de novo* protein synthesis is not a factor in the recovery of CCR5 (Mueller *et al.*, 2002), nor indeed in that of CCR4 (Mariani *et al.*, 2004).

IX. PROTEOLYTIC PROCESSING OF CC CHEMOKINES

In addition to receptor desensitization and internalization, the modulation of chemokine receptor signaling can be achieved by proteolytic processing of CC chemokines themselves (reviewed by Struyf *et al.*, 2003). The majority of information known about chemokine processing has been obtained from research into the lymphocyte membrane–expressed peptidase CD26/dipeptidyl-peptidase IV (DPP IV), a serine-type peptidase that cleaves some, but not all, CC chemokines (Lambeir *et al.*, 2001).

CD26/DPP IV exists as a type II membrane protein with a short intracellular N-terminal region, a single transmembrane region, and an extracellular catalytic domain that can be cleaved in an active form from the cell surface (Mentlein, 1999). CD26/DPP IV has a unique aminopeptidase activity and is known to selectively cleave dipeptides from the amino terminus of proteins with a proline or alanine residue at the second position of their sequence. While agonistic responses by CD26/DPP IV truncated CXCL6 are unaffected, a variety of CC chemokine ligands have vastly impaired responses at their receptors following cleavage (Proost *et al.*, 1998). In the majority of cases, CD26/DPP IV processing reduces agonistic properties, the most prominent exception being the CCL3 isoform CCL3L1 (LD78β). CD26/DPP IV processed CCL3L1 (CCL3L1(3–70)) becomes the most potent CCR1 and CCR5 agonist and the most efficient monocyte chemoattractant described to date (Proost *et al.*, 2000). This is in stark contrast to CD26/DPP IV processed CCL5(3–68), which has less than one-tenth the chemotactic potency for monocytes and eosinophils via CCR1 and CCR3, while being fully active on CCR5 (Oravecz *et al.*, 1997). The same conditions result in a double truncation of the chemokine CCL22 in which not only Gly1–Pro2 but Tyr3–Gly4 is lost. Compared with the intact chemokine, CCL22(5–69) has a reduced chemotactic activity for lymphocytes and monocyte-derived dendritic cells, but it remains as chemotactic as intact CCL22 for monocytes (Proost *et al.*, 1999). CCL11, a potent eosinophil attractant signaling via CCR3, has its chemotactic potency for blood eosinophils reduced 30 times upon truncation by CD26/DPP IV (Struyf *et al.*, 1999). In summary, these data confirm the importance of the N-terminal residues of the chemokine in activating the chemokine receptor.

Several matrix metalloproteinases (MMPs) have been implicated in the degradation of CC chemokines. CCL7 is processed by gelatinase A/MMP-2, resulting in the removal of 4 N-terminal residues. This removal results in the creation of a receptor antagonist, which is able to bind to the receptor but not signal (McQuibban *et al.*, 2000). This antagonist form of CCL7 is able to reduce mononuclear cell infiltration *in vivo* and attenuate cellular infiltration in a mouse model of inflammation (McQuibban *et al.*, 2000). In addition, it is now also known that additional MMPs are able to process CCL2 (MMP-1, -2, -3, and -13), CCL8 (MMP-1 and -3), and CCL13 (MMP-1 and -3) (McQuibban *et al.*, 2002; Overall *et al.*, 2002). Such a process has not been lost on the hookworm *Necator americanus*, which secretes as yet unidentified metalloproteases to degrade CCL11 and thereby evade host eosinophils (Culley *et al.*, 2000).

X. CHEMOKINE BINDING TO GLYCOSAMINOGLYCANS

In addition to binding to their receptors, chemokines are also capable of binding to proteoglycans, endothelial cell surface–expressed molecules that consist of a core protein and glycosaminoglycan (GAG) side chains. These GAG chains are highly acidic and facilitate electrostatic interactions of micromolar affinity with the basic chemokines (Witt and Lander, 1994; Hoogewerf *et al.*, 1997; Kuschert *et al.*, 1999). The interaction of chemokines with endothelially expressed proteoglycans has been theorized to immobilize a locally high concentration of chemokine at the endothelium, an interaction that is maintained during conditions of shear flow observed in blood vessels. The resulting haptotactic gradient of chemokine can induce the firm adhesion of leukocytes to the endothelium via the modulation of integrin affinity and ultimately coordinate the migration of leukocytes from the circulation to the site of chemokine production in the surrounding tissue (Constantin *et al.*, 2000). GAG binding may also induce the dimerization and oligomerization of chemokines (McCornack *et al.*, 2003). Studies of CCL2 oligomers have shown that the residues that contribute most to GAG binding also contribute to binding to the cell-surface receptor CCR2, so it is highly unlikely that monomeric CCL2 binds both to GAGs and to its receptor in the same instance. One possibility is the "handoff" model, in which a single subunit of the chemokine tetramer is delivered to a receptor, inducing dissociation of the chemokine oligomer and activation of the receptor (Lau *et al.*, 2004). Soluble GAGs have been used *in vitro* to inhibit the actions of chemokines on eosinophils by sequestering the chemokine such that it can no longer bind to the specific receptor (Burns *et al.*, 1999; Culley *et al.*, 2003). Heparin, produced *in vivo* exclusively by mast cells, has been shown to inhibit eosinophil recruitment in a guinea-pig model of allergic lung inflammation and to attenuate the late asthmatic response to allergen challenge in patients with atopic asthma (Seeds *et al.*, 1995; Diamant *et al.*, 1996). This suggests that it may represent a natural anti-inflammatory agent able to decrease allergic inflammation in the lung following mast-cell degranulation.

The GAG binding domains of some CC chemokines have been extensively characterized by mutagenesis. In CCL5, a BBXB motif in the 40s loop (where B is a basic amino acid) has been identified as a major determinant of heparin binding, and a mutant CCL5 (^{44}AANA^{47}) lacking all three basic residues retains its *in vitro* chemotactic activity although it is unable to bind to GAGs (Proudfoot *et al.*, 2001). Similarly, studies of chemokine receptors expressed in CHO cells defective in GAG synthesis have shown that GAG expression is unnecessary for the biological activity of some CC chemokines (Ali *et al.*, 2000). Interestingly, the ^{44}AANA^{47} CCL5 mutant is unable to recruit cells when administered to mice *in vivo*, suggesting that GAG binding is essential for the *in vivo* activity of some chemokines (Proudfoot *et al.*, 2003). This work raises the possibility that antagonism of the chemokine: GAG interaction may be a useful therapeutic angle.

XI. SCAVENGING OF CC CHEMOKINES BY D6 AND DARC

Several inflammatory CC chemokines, including CCL2, CCL3L1, CCL4, CCL5, CCL7, CCL11, and CCL16, can be readily bound with high affinity by the seven-transmembrane receptor D6 (Nibbs *et al.*, 1997). This receptor is expressed at high levels on the surface of lymphatic endothelial cells and on synctial trophoblasts of the placenta (Nibbs *et al.*, 2001). Despite much effort, no cell signals have been documented to be transduced following chemokine binding, which suggests that D6 may act as a scavenger of inflammatory chemokines (Nibbs *et al.*, 2003). D6 undergoes rapid constitutive internalization, enabling it to rapidly remove chemokines from the endothelial cell surface (Fra *et al.*, 2003; Bonecchi *et al.*, 2004; Galliera *et al.*, 2004). The fate of the internalized chemokine appears to be proteolytic degradation, while the receptor is recycled back to the cell surface (Weber *et al.*, 2004). In this way, D6 is seen as a "gatekeeper," preserving the integrity of lymphoid tissue (Fra *et al.*, 2003).

Likewise, the DARC acts as a scavenger of both CC and CXC chemokines. DARC was originally defined serologically in the 1950s as a minor red-blood-cell antigen (Young *et al.*, 1955) and later at the molecular level as a receptor for the chemokine CXCL8, with considerable homology to chemokine receptors (Chaudhuri *et al.*, 1993; Neote *et al.*, 1994). In addition to its expression on erythrocytes, it has also been identified on the post-capillary venule endothelial cells of several organs (Peiper *et al.*, 1995) and on subsets of neurons in the central nervous system (Horuk *et al.*, 1997). Studies of mice in which the DARC gene has been deleted suggest that it functions as a biological 'sink' for chemokines, with both an anti-inflammatory role and an antiangiogenic role (Dawson *et al.*, 2000). As with D6, chemokine binding by DARC does not appear to result in signal transduction (Neote *et al.*, 1994). Both receptors lack a DRY motif in the putative third-transmembrane helix, present in the majority of signaling chemokine receptors and thought to play a critical role in maintaining GPCR conformation (Ballesteros *et al.*, 2001).

XII. REGULATION OF CC CHEMOKINE RECEPTOR EXPRESSION

Chemokine receptors can be broadly divided into those that are expressed exclusively by a particular subset of leukocytes and those that are more widely expressed (see Table II). For example, CCR6 expression is largely restricted to immature dendritic cells (Greaves *et al.*, 1997; Power *et al.*, 1997) and memory T-lymphocytes (Liao *et al.*, 1999), their migration to lymphoid organs being facilitated by production of the specific ligand CCL20 (Dieu *et al.*, 1998). In contrast, CCR3 is expressed on a variety of cells involved in allergic responses, such as eosinophils (Daugherty *et al.*, 1996; Kitaura *et al.*, 1996; Ponath *et al.*, 1996), basophils (Uguccioni *et al.*, 1997), and mast cells (De Paulis *et al.*, 2001), and the ligand CCL11 selectively recruits all three cell types. T-helper lymphocytes of either the Th1 or the Th2 subset also exhibit differential chemokine receptor expression profiles. Th1 lymphocytes selectively express CCR1, CCR5, and CXCR3, whereas CCR3, CCR4, and CCR8 are found on Th2 lymphocytes (Sallusto *et al.*, 1997, 1998; Bonecchi *et al.*, 1998). The effect of fine-tuning the surface-expression receptor profiles of leukocytes is to enable the cells to respond to a variety of chemokine gradients, allowing them to leave one tissue compartment and migrate to another. Such flexibility is undoubtedly important for a focused adaptive immune response.

XIII. NATURAL ANTAGONISM OF CC CHEMOKINE RECEPTORS

It is not uncommon for chemokines to be able to interact with more than one receptor; in fact, the majority of chemokines that activate only a single receptor, for example CXCL12, CCL19, and CCL21, are responsible for coordinating lymphocyte homing to tissues. However, chemokines are usually promiscuous (with respect to receptor interaction) within the same subfamily of chemokines. For example, CC chemokines bind CC chemokine receptors, and the interactions are usually agonistic. Nevertheless, it has been shown more recently that there are examples of CC chemokines that act in an antagonistic manner at CC chemokine receptors; additionally, both CC and CXC chemokines have been shown to be capable of interacting with CXC and CC chemokine receptors. CCL7, an agonist of CCR1, CCR2, and CCR3, has been reported to be a natural antagonist at CCR5, inhibiting the functional responses of CCR5 to other ligands (Blanpain *et al.*, 1999a,b). Other examples of natural antagonists include CCL18, which is antagonistic at CCR3 (Nibbs *et al.*, 2000); CCL11, which is an antagonist of CCR2 (Martinelli *et al.*, 2001; Ogilvie *et al.*, 2001); and CCL26, which is an antagonist of CCR1, CCR2, and CCR5 (Ogilvie *et al.*, 2003; Petkovic *et al.*, 2004). The CXCR3 ligands CXCL9, CXCL10, and CXCL11 have also been described as natural antagonists for CCR3 (Loetscher *et al.*, 2001; Xanthou *et al.*, 2003), and CXCL11 is a natural antagonist for CCR5 (Petkovic *et al.*, 2004). The functional significance of natural antagonism is not fully understood, but it may well represent an internal mechanism by which adaptive immune responses are self-regulated by the lymphocytes involved. For example, the expression of CXCR3 ligands (induced by IFN-γ) may be perceived to further polarize the immune response towards a Th1 response, by both activating a Th1-associated receptor (CXCR3) and inhibiting a Th2-associated receptor (CCR3).

XIV. FUTURE DIRECTIONS AND UNANSWERED QUESTIONS

Since the discovery of the first chemokine in 1977, over 40 chemokines have been described in the human, of which the vast majority belong to the CC family. The application of molecular biological approaches has done much to further our understanding of how these molecules exert their functions at the cellular level. Indeed, with the exception of CCL18, the CC chemokines are no longer orphans; they have had their agonist activity defined at specific receptors. The recent resolution of the rhodopsin crystal structure (Palczewski *et al.*, 2000), coupled with the development of powerful modeling techniques, is facilitating the development of models of chemokine receptor tertiary structure, which can be tested in the laboratory by mutagenesis strategies. While it is relatively straightforward to model the helical regions of CC chemokine receptors on those of rhodopsin, the modeling of the extracellular and intracellular loops, which facilitate ligand binding and G-protein coupling, respectively, is difficult, to say the least. *Ab initio* modeling of chemokine receptors, as has been carried out with some success for other GPCRs (Vaidehi *et al.*, 2002; Freddolino *et al.*, 2004; Trabanino *et al.*, 2004), might be helpful here, as would the determination of the crystal structure of an actual chemokine receptor.

Seen from the exterior of the cell, the signaling pathways employed by chemokines appear at first glance to resemble an inverted pyramid, with 28 CC chemokines activating 10 CC receptors, which couple directly to only a handful of G-proteins (predominantly $G\alpha_i$ in the leukocyte). In turn, these signals appear to be directly regulated by only five GRKs and two arrestin molecules. Such a simplistic view does not explain the diversity of the downstream signals observed at both the temporal and the spatial level, that are now being described following GPCR activation. The discovery that the arrestins function not only as inhibitors of β-adrenergic receptor signaling but as adaptor molecules for the recruitment of additional signaling molecules such as Src kinase may shed some light in this direction (Luttrell *et al.*, 1999). The arrestin molecule binds to the phosphorylated GPCR and serves as a scaffold, activating kinases such as ERK and also directing the GPCR into cytoplasmic vesicles, thereby circumventing its nuclear action (Luttrell *et al.*, 2001). Early experiments in the chemokine field have shown arrestin-2 to be essential for CXCR4 and CCR7 signaling *in vitro* (Sun *et al.*, 2002;

Kohout *et al.*, 2004), and leukocytes isolated from mice deficient in arrestin-2 exhibit defective chemotaxis (Fong *et al.*, 2002).

It is observable that chemokine receptors couple differentially to either alternative intracellular G-proteins (of which there are several), or additional scaffolding molecules (perhaps enriched in discrete signaling domains such as lipid rafts or caveolae). This would allow the leukocyte to switch from an early signal response, such as migration to a site of chemokine generation, to a later response, for example, the transcription of appropriate genes at the inflammatory site. Such a system can be likened to an hourglass, with chemokine receptors expertly funneling the responses of several different chemokine ligands on the extracellular face to a similarly diverse array of intracellular signaling molecules, and, in doing so, activating a multitude of downstream signaling cascades. The application of a proteomic approach, coupled with the knockdown of specific signaling molecules, should complement well-established methodologies and greatly increase our understanding of how these fascinating molecules exert their effects on cellular processes.

Acknowledgments

We are grateful to Philip Murphy, Massimo Locati, and William Glass for helpful discussions and to the Medical Research Council, the Wellcome Trust, the British Heart Foundation, and the Arthritis Research Council for their funding of our research.

References

Ali, H., Richardson, R. M., Haribabu, B., and Snyderman, R. (1999). Chemoattractant receptor cross-desensitization. *J. Biol. Chem.* **274,** 6027–6030.

Ali, S., Palmer, A. C., Banerjee, B., Fritchley, S. J., and Kirby, J. A. (2000). Examination of the function of RANTES, MIP-1alpha, and MIP-1beta following interaction with heparin-like glycosaminoglycans. *J. Biol. Chem.* **275,** 11721–11727.

Arias, D. A., Navenot, J. M., Zhang, W. B., Broach, J., and Peiper, S. C. (2003). Constitutive activation of CCR5 and CCR2 induced by conformational changes in the conserved TXP motif in transmembrane helix 2. *J. Biol. Chem.* **278,** 36513–36521.

Auger, G. A., Pease, J. E., Shen, X., Xanthou, G., and Barker, M. D. (2002). Alanine scanning mutagenesis of CCR3 reveals that the three intracellular loops are essential for functional receptor expression. *Eur. J. Immunol.* **32,** 1052–1058.

Ballesteros, J. A., Jensen, A. D., Liapakis, G., Rasmussen, S. G., Shi, L., Gether, U., and Javitch, J. A. (2001). Activation of the beta 2–adrenergic receptor involves disruption of an ionic lock between the cytoplasmic ends of transmembrane segments 3 and 6. *J. Biol. Chem.* **276,** 29171–29177.

Blanpain, C., Doranz, B. J., Bondue, A., Govaerts, C., De Leener, A., Vassart, G., Doms, R. W., Proudfoot, A., and Parmentier, M. (2003). The core domain of chemokines binds CCR5 extracellular domains while their amino terminus interacts with the transmembrane helix bundle. *J. Biol. Chem.* **278,** 5179–5187.

Blanpain, C., Lee, B., Vakili, J., Doranz, B. J., Govaerts, C., Migeotte, I., Sharron, M., Dupriez, V., Vassart, G., Doms, R. W., and Parmentier, M. (1999a). Extracellular cysteines of CCR5 are required for chemokine binding, but dispensable for HIV-1 coreceptor activity. *J. Biol. Chem.* **274,** 18902–18908.

Blanpain, C., Migeotte, I., Lee, B., Vakili, J., Doranz, B. J., Govaerts, C., Vassart, G., Doms, R. W., and Parmentier, M. (1999b). CCR5 binds multiple CC-chemokines: MCP-3 acts as a natural antagonist. *Blood* **94,** 1899–1905.

Bohm, S. K., Grady, E. F., and Bunnett, N. W. (1997). Regulatory mechanisms that modulate signalling by G-protein-coupled receptors. *Biochem J.* **322**(Pt 1), 1–18.

Bokoch, G. M. (1995). Chemoattractant signaling and leukocyte activation. *Blood* **86,** 1649–1660.

Bonecchi, R., Bianchi, G., Bordignon, P. P., D'Ambrosio, D., Lang, R., Borsatti, A., Sozzani, S., Allavena, P., Gray, P. A., Mantovani, A., and Sinigaglia, F. (1998). Differential expression of chemokine receptors and chemotactic responsiveness of type 1 T helper cells (Th1s) and Th2s. *J. Exp. Med.* **187,** 129–134.

Bonecchi, R., Locati, M., Galliera, E., Vulcano, M., Sironi, M., Fra, A. M., Gobbi, M., Vecchi, A., Sozzani, S., Haribabu, B., Van Damme, J., and Mantovani, A. (2004). Differential recognition and scavenging of native and truncated macrophage-derived chemokine (macrophage-derived chemokine/CC chemokine ligand 22) by the D6 decoy receptor. *J. Immunol.* **172,** 4972–4976.

Burns, J. M., Lewis, G. K., and DeVico, A. L. (1999). Soluble complexes of regulated upon activation, normal T cells expressed and secreted (RANTES) and glycosaminoglycans suppress HIV-1 infection but do not induce Ca(2+) signaling. *Proc. Natl. Acad. Sci. USA* **96,** 14499–14504.

Burrows, S. D., Doyle, M. L., Murphy, K. P., Franklin, S. G., White, J. R., Brooks, I., McNulty, D. E., Scott, M. O., Knutson, J. R., Porter, D., Young, PR, Hensley, and P (1994). Determination of the monomer-dimer equilibrium of interleukin-8 reveals it is a monomer at physiological concentrations. *Biochemistry* **33,** 12741–12745.

Chaudhuri, A., Polyakova, J., Zbrzezna, V., Williams, K., Gulati, S., and Pogo, A. O. (1993). Cloning of glycoprotein D cDNA, which encodes the major subunit of the Duffy blood group system and the receptor for the Plasmodium vivax malaria parasite. *Proc. Natl. Acad. Sci. USA* **90,** 10793–10797.

Clark-Lewis, I., Kim, K. S., Rajarathnam, K., Gong, J. H., Dewald, B., Moser, B., Baggiolini, M., and Sykes, B. D. (1995). Structure-activity relationships of chemokines. *J. Leukoc. Biol.* **57,** 703–711.

Constantin, G., Majeed, M., Giagulli, C., Piccio, L., Kim, J. Y., Butcher, E. C., and Laudanna, C. (2000). Chemokines trigger immediate beta2 integrin affinity and mobility changes: Differential regulation and roles in lymphocyte arrest under flow. *Immunity* **13,** 759–769.

Cronshaw, D. G., Owen, C., Brown, Z., and Ward, S. G. (2004). Activation of phosphoinositide 3–kinases by the CCR4 ligand macrophage-derived chemokine is a dispensable signal for T lymphocyte chemotaxis. *J. Immunol.* **172,** 7761–7770.

Culley, F. J., Brown, A., Conroy, D. M., Sabroe, I., Pritchard, D. I., and Williams, T. J. (2000). Eotaxin is specifically cleaved by hookworm metalloproteases preventing its action *in vitro* and *in vivo*. *J. Immunol.* **165,** 6447–6453.

Culley, F. J., Fadlon, E. J., Kirchem, A., Williams, T. J., Jose, P. J., and Pease, J. E. (2003). Proteoglycans are potent modulators of the biological responses of eosinophils to chemokines. *Eur. J. Immunol.* **33,** 1302–1310.

D'Amico, G., Frascaroli, G., Bianchi, G., Transidico, P., Doni, A., Vecchi, A., Sozzani, S., Allavena, P., and Mantovani, A. (2000). Uncoupling of inflammatory chemokine receptors by IL-10: Generation of functional decoys. *Nat. Immunol.* **1,** 387–391.

Daugherty, B. L., Siciliano, S. J., DeMartino, J., Malkowitz, L., Sirontino, A., and Springer, M. S. (1996). Cloning, Expression and Characterization of the Human Eosinophil Eotaxin Receptor. *J. Exp. Med.* **183,** 2349–2354.

Dawson, T. C., Lentsch, A. B., Wang, Z., Cowhig, J. E., Rot, A., Maeda, N., and Peiper, S. C. (2000). Exaggerated response to endotoxin in mice lacking the Duffy antigen/receptor for chemokines (DARC). *Blood.* **96,** 1681–1684.

De Lean, A., Stadel, J. M., and Lefkowitz, R. J. (1980). A ternary complex model explains the agonist-specific binding properties of the adenylate cyclase-coupled beta-adrenergic receptor. *J. Biol. Chem.* **255,** 7108–7117.

De Paulis, A., Annunziato, F., Di Gioia, L., Romagnani, S., Carfora, M., Beltrame, C., Marone, G., and Romagnani, P. (2001). Expression of the chemokine receptor CCR3 on human mast cells. *Int. Arch. Allergy Immunol.* **124,** 146–150.

Diamant, Z., Timmers, M. C., van, d,V., Page, C. P., van der Meer, F. J., and Sterk, P. J. (1996). Effect of inhaled heparin on allergen-induced early and late asthmatic responses in patients with atopic asthma. *Am. J. Respir. Crit. Care. Med.* **153,** 1790–1795.

Dieu, M. C., Vanbervliet, B., Vicari, A., Bridon, J. M., Oldham, E., Ait-Yahia, S., Briere, F., Zlotnik, A., Lebecque, S., and Caux, C. (1998). Selective recruitment of immature and mature dendritic cells by distinct chemokines expressed in different anatomic sites. *J. Exp. Med.* **188,** 373–386.

Farzan, M., Mirzabekov, T., Kolchinsky, P., Wyatt, R., Cayabyab, M., Gerard, N. P., Gerard, C., Sodroski, J., and Choe, H. (1999). Tyrosine sulfation of the amino terminus of CCR5 facilitates HIV-1 entry. *Cell* **96,** 667–676.

Fong, A. M., Premont, R. T., Richardson, R. M., Yu, Y. R., Lefkowitz, R. J., and Patel, D. D. (2002). Defective lymphocyte chemotaxis in beta-arrestin2- and GRK6-deficient mice. *Proc. Natl. Acad. Sci. USA* **99,** 7478–7483.

Fra, A. M., Locati, M., Otero, K., Sironi, M., Signorelli, P., Massardi, M. L., Gobbi, M., Vecchi, A., Sozzani, S., and Mantovani, A. (2003). Cutting edge: Scavenging of inflammatory CC chemokines by the promiscuous putatively silent chemokine receptor D6. *J. Immunol.* **170,** 2279–2282.

Freddolino, P. L., Kalani, M. Y., Vaidehi, N., Floriano, W. B., Hall, S. E., Trabanino, R. J., Kam, V. W., and Goddard, W. A., 3rd. (2004). Predicted 3D structure for the human beta 2 adrenergic receptor and its binding site for agonists and antagonists. *Proc. Natl. Acad. Sci. USA* **101,** 2736–2741.

Fredriksson, R., Lagerstrom, M. C., Lundin, L. G., and Schioth, H. B. (2003). The G-protein-coupled receptors in the human genome form five main families. Phylogenetic analysis, paralogon groups, and fingerprints. *Mol. Pharmacol.* **63,** 1256–1272.

Gallego, C., Gupta, S. K., Winitz, S., Eisfelder, B. J., and Johnson, G. L. (1992). Myristoylation of the G alpha i2 polypeptide, a G-protein alpha subunit, is required for its signaling and transformation functions. *Proc. Natl. Acad. Sci. USA* **89,** 9695–9699.

Galliera, E., Jala, V. R., Trent, J. O., Bonecchi, R., Signorelli, P., Lefkowitz, R. J., Mantovani, A., Locati, M., and Haribabu, B. (2004). beta-Arrestin-dependent constitutive internalization of the human chemokine decoy receptor D6. *J. Biol. Chem.* **279,** 25590–25597.

Gerard, C., and Rollins, B. J. (2001). Chemokines and disease. *Nat. Immunol.* **2,** 108–115.

Gerlach, L. O., Jakobsen, J. S., Jensen, K. P., Rosenkilde, M. R., Skerlj, R. T., Ryde, U., Bridger, G. J., and Schwartz, T. W. (2003). Metal ion enhanced binding of AMD3100 to Asp262 in the CXCR4 receptor. *Biochemistry* **42,** 710–717.

Gong, J. H., Uguccioni, M., Dewald, B., Baggiolini, M., and Clark-Lewis, I. (1996). RANTES and MCP-3 antagonists bind multiple chemokine receptors. *J. Biol. Chem.* **271,** 10521–10527.

Goodman, O. B., Jr., Krupnick, J. G., Santini, F., Gurevich, V. V., Penn, R. B., Gagnon, A. W., Keen, J. H., and Benovic, J. L. (1996). Beta-arrestin acts as a clathrin adaptor in endocytosis of the beta2-adrenergic receptor. *Nature* **383,** 447–450.

Govaerts, C., Blanpain, C., Deupi, X., Ballet, S., Ballesteros, J. A., Wodak, S. J., Vassart, G., Pardo, L., and Parmentier, M. (2001). The TXP motif in the second transmembrane helix of CCR5. A structural determinant of chemokine-induced activation. *J. Biol. Chem.* **276,** 13217–13225.

Govaerts, C., Bondue, A., Springael, J. Y., *et al.* (2003). Activation of CCR5 by chemokines involves an aromatic cluster between transmembrane helices 2 and 3. *J. Biol. Chem.* **278,** 1892–1903.

Greaves, D. R., Wang, W., Dairaghi, D. J., Dieu, M. C., Saint-Vis, B., Franz-Bacon, K., Rossi, D., Caux, C., McClanahan, T., Gordon, S., Zlotnik, A., and Schall, T. J. (1997). CCR6, a CC chemokine receptor that interacts with macrophage inflammatory protein 3alpha and is highly expressed in human dendritic cells. *J. Exp. Med.* **186,** 837–844.

Handel, T. M., and Domaille, P. J. (1996). Heteronuclear (1H, 13C, 15N) NMR assignments and solution structure of the monocyte chemoattractant protein-1 (MCP-1) dimer. *Biochemistry* **35,** 6569–6584.

Hernanz-Falcon, P., Rodriguez-Frade, J. M., Serrano, A., Juan, D., del Sol, A., Soriano, S. F., Roncal, F., Gomez, L., Valencia, A., Martinez, A. C., and Mellado, M. (2004). Identification of amino acid residues crucial for chemokine receptor dimerization. *Nat. Immunol.* **5,** 216–223.

Heuser, J. E., and Anderson, R. G. (1989). Hypertonic media inhibit receptor-mediated endocytosis by blocking clathrin-coated pit formation. *J. Cell. Biol.* **108,** 389–400.

Higgins, J. B., and Casey, P. J. (1994). *In vitro* processing of recombinant G-protein gamma subunits. Requirements for assembly of an active beta gamma complex. *J. Biol. Chem.* **269,** 9067–9073.

Hoogewerf, A. J., Kuschert, G. S., Proudfoot, A. E., Borlat, F., Clark-Lewis, I., Power, C. A., and Wells, T. N. (1997). Glycosaminoglycans mediate cell surface oligomerization of chemokines. *Biochemistry* **36,** 13570–13578.

Horuk, R., Martin, A. W., Wang, Z., Schweitzer, L., Gerassimides, A., Guo, H., Lu, Z., Hesselgesser, J., Perez, H. D., Kim, J., Parker, J., Hadley, T. J., and Peiper, S. C. (1997). Expression of chemokine receptors by subsets of neurons in the central nervous system. *J. Immunol.* **158,** 2882–2890.

Humbles, A. A., Conroy, D. M., Marleau, S., Rankin, S. M., Palframan, R. T., Proudfoot, A. E. I., Wells, T. N. C., Li, D., Jeffery, P. K., Griffiths-Johnson, D. A., Williams, T. J., and Jose, P. J. (1997). Kinetics of eotaxin generation and its relationship to eosinophil accumulation in allergic airways disease: Analysis in a guinea pig model *in vivo*. *J. Exp. Med.* **186,** 601–612.

Humbles, A. A., Lu, B., Friend, D. S., Okinaga, S., Lora, J., Al-Garawi, A., Martin, T. R., Gerard, N. P., and Gerard, C. (2002). The murine CCR3 receptor regulates both the role of eosinophils and mast cells in allergen-induced airway inflammation and hyperresponsiveness. *Proc. Natl. Acad. Sci. USA* **99,** 1479–1484.

Janetopoulos, C., Jin, T., and Devreotes, P. (2001). Receptor-mediated activation of heterotrimeric G-proteins in living cells. *Science* **291,** 2408–2411.

Kasama, T., Strieter, R. M., Standiford, T. J., Burdick, M. D., and Kunkel, S. L. (1993). Expression and regulation of human neutrophil-derived macrophage inflammatory protein 1a. *J. Exp. Med.* **178,** 63–72.

Keizer, D. W., Crump, M. P., Lee, T. W., Slupsky, C. M., Clark-Lewis, I., and Sykes, B. D. (2000). Human CC chemokine I-309, structural consequences of the additional disulfide bond. *Biochemistry* **39,** 6053–6059.

Kitaura, M., Nakajima, T., Imai, T., Harada, S., Combadiere, C., Tiffany, H. L., Murphy, P. M., and Yoshie, O. (1996). Molecular cloning of human eotaxin, an eosinophil-selective CC chemokine, and identification of a specific eosinophil eotaxin receptor, CC chemokine receptor 3. *J. Biol. Chem.* **271,** 7725–7730.

Koch, A. E., Kunkel, S. L., Harlow, L. A., Mazarakis, D. D., Haines, G. K., Burdick, M. D., Pope, R. M., and Strieter, R. M. (1994). Macrophage inflammatory protein-1 alpha. A novel chemotactic cytokine for macrophages in rheumatoid arthritis. *J. Clin. Invest.* **93,** 921–928.

Kohout, T. A., Nicholas, S. L., Perry, S. J., Reinhart, G., Junger, S., and Struthers, R. S. (2004). Differential desensitization, receptor phosphorylation, beta-arrestin recruitment, and ERK1/2 activation by the two endogenous ligands for the CC chemokine receptor 7. *J. Biol. Chem.* **279,** 23214–23222.

Krisch, B., Feindt, J., and Mentlein, R. (1998). Immunoelectronmicroscopic analysis of the ligand-induced internalization of the somatostatin receptor subtype 2 in cultured human glioma cells. *J. Histochem. Cytochem.* **46,** 1233–1242.

Kuschert, G. S., Coulin, F., Power, C. A., Proudfoot, A. E., Hubbard, R. E., Hoogewerf, A. J., and Wells, T. N. (1999). Glycosaminoglycans interact selectively with chemokines and modulate receptor binding and cellular responses. *Biochemistry* **38,** 12959–12968.

Lambeir, A. M., Proost, P., Durinx, C., Bal, G., Senten, K., Augustyns, K., Scharpe, S., Van Damme, J., and De Meester, I. (2001). Kinetic investigation of chemokine truncation by CD26/dipeptidyl peptidase IV reveals a striking selectivity within the chemokine family. *J. Biol. Chem.* **276,** 29839–29845.

Lau, E. K., Paavola, C. D., Johnson, Z., Gaudry, J. P., Geretti, E., Borlat, F., Kungl, A. J., Proudfoot, A. E., and Handel, T. M. (2004). Identification of the glycosaminoglycan binding site of the CC chemokine, MCP-1: implications for structure and function in vivo. *J. Biol. Chem.* **279,** 22294–22305.

Li, Z., Jiang, H., Xie, W., Zhang, Z., Smrcka, A. V., and Wu, D. (2000). Roles of PLC-beta2 and -beta3 and PI3Kgamma in chemoattractant-mediated signal transduction. *Science* **287,** 1046–1049.

Liao, F., Rabin, R. L., Smith, C. S., Sharma, G., Nutman, T. B., and Farber, J. M. (1999). CC-chemokine receptor 6 is expressed on diverse memory subsets of T cells and determines responsiveness to macrophage inflammatory protein-3 alpha. *J. Immunol.* **162,** 186–194.

Loetscher, P., Pellegrino, A., Gong, J. H., Mattioli, I., Loetscher, M., Bardi, G., Baggiolini, M., and Clark Lewis, I. (2001). The ligands of CXC chemokine receptor 3, I-TAC, Mig and IP10, are natural antagonists for CCR3. *J. Biol. Chem.* **276,** 2986–2991.

Lu, Z. H., Wang, Z. X., Horuk, R., Hesselgesser, J., Lou, Y. C., Hadley, T. J., and Peiper, S. C. (1995). The promiscuous chemokine binding profile of the Duffy antigen/receptor for chemokines is primarily localized to sequences in the amino-terminal domain. *J. Biol. Chem.* **270,** 26239–26245.

Lukacs, N. W., Strieter, R. M., Shaklee, C. L., Chensue, S. W., and Kunkel, S. L. (1995). Macrophage inflammatory protein-1α influences eosinophil recruitment in antigen-specific airway inflammation. *Eur. J. Immunol.* **25,** 245–251.

Luttrell, L. M., Ferguson, S. S., Daaka, Y., Miller, W. E., Maudsley, S., Della Rocca, G. J., Lin, F., Kawakatsu, H., Owada, K., Luttrell, D. K., Caron, M. G., and Lefkowitz, R. J. (1999). Beta-arrestin-dependent formation of beta2 adrenergic receptor-Src protein kinase complexes. *Science* **283,** 655–661.

Luttrell, L. M., Roudabush, F. L., Choy, E. W., Miller, W. E., Field, M. E., Pierce, K. L., and Lefkowitz, R. J. (2001). Activation and targeting of extracellular signal-regulated kinases by beta-arrestin scaffolds. *Proc. Natl. Acad. Sci. USA* **98,** 2449–2454.

Ma, Y. C., Huang, J., Ali, S., Lowry, W., and Huang, X. Y. (2000). Src tyrosine kinase is a novel direct effector of G-proteins. *Cell* **102,** 635–646.

Mackay, C. R. (2001). Chemokines: Immunology's high impact factors. *Nat. Immunol.* **2,** 95–101.

Maghazachi, A. A. (2000). Intracellular signaling events at the leading edge of migrating cells. *Int. J. Biochem. Cell. Biol.* **32,** 931–943.

Mariani, M., Lang, R., Binda, E., Panina-Bordignon, P., and D'Ambrosio, D. (2004). Dominance of CCL22 over CCL17 in induction of chemokine receptor CCR4 desensitization and internalization on human Th2 cells. *Eur. J. Immunol.* **34,** 231–240.

Martinelli, R., Sabroe, I., LaRosa, G., Williams, T. J., and Pease, J. E. (2001). The CC chemokine eotaxin (CCL11) is a partial agonist of the chemokine receptor CCR2b. *J. Biol. Chem.* **276,** 42957–42964.

McCornack, M. A., Cassidy, C. K., and LiWang, P. J. (2003). The binding surface and affinity of monomeric and dimeric chemokine macrophage inflammatory protein-1 beta for various glycosaminoglycan disaccharides. *J. Biol. Chem.* **278,** 1946–1956.

McQuibban, G. A., Gong, J. H., Tam, E. M., McCulloch, C. A., Clark-Lewis, I., and Overall, C. M. (2000). Inflammation dampened by gelatinase A cleavage of monocyte chemoattractant protein-3. *Science* **289,** 1202–1206.

McQuibban, G. A., Gong, J. H., Wong, J. P., Wallace, J. L., Clark-Lewis, I., and Overall, C. M. (2002). Matrix metalloproteinase processing of monocyte chemoattractant proteins generates CC chemokine receptor antagonists with anti-inflammatory properties *in vivo*. *Blood* **100,** 1160–1167.

Mellado, M., Rodriguez-Frade, J. M., Manes, S., and Martinez, A. C. (2001). Chemokine signaling and functional responses: the role of receptor dimerization and TK pathway activation. *Annu. Rev. Immunol.* **19,** 397–421.

Mentlein, R. (1999). Dipeptidyl-peptidase IV (CD26)–role in the inactivation of regulatory peptides. *Regul. Pept.* **85,** 9–24.

Milligan, G., Ramsay, D., Pascal, G., and Carrillo, J. J. (2003). GPCR dimerisation. *Life Sci.* **74,** 181–188.

Monteclaro, F. S., and Charo, I. F. (1996). The amino-terminal extracellular domain of the MCP-1 receptor, but not the RANTES/MIP-1alpha receptor, confers chemokine selectivity. Evidence for a two-step mechanism for MCP-1 receptor activation. *J. Biol. Chem.* **271,** 19084–19092.

Mueller, A., Kelly, E., and Strange, P. G. (2002). Pathways for internalization and recycling of the chemokine receptor CCR5. *Blood* **99,** 785–791.

Murphy, P. M. (2001). Viral exploitation and subversion of the immune system through chemokine mimicry. *Nat. Immunol.* **2,** 116–122.

Murphy, P. M. (2002). International Union of Pharmacology. XXX. Update on chemokine receptor nomenclature. *Pharmacol. Rev.* **54,** 227–229.

Murphy, P. M., Baggiolini, M., Charo, I. F., *et al.* (2000). International union of pharmacology. XXII. Nomenclature for chemokine receptors. *Pharmacol. Rev.* **52,** 145–176.

Neote, K., Mak, J. Y., Kolakowski, L. F., Jr., and Schall, T. J. (1994). Functional and biochemical analysis of the cloned Duffy antigen: Identity with the red blood cell chemokine receptor. *Blood* **84,** 44–52.

Nibbs, R. J., Kriehuber, E., Ponath, P. D., Parent, D., Qin, S., Campbell, J. D., Henderson, A., Kerjaschki, D., Maurer, D., Graham, G. J., and Rot, A. (2001). The beta-chemokine receptor D6 is expressed by lymphatic endothelium and a subset of vascular tumors. *Am. J. Pathol.* **158,** 867–877.

Nibbs, R. J., Salcedo, T. W., Campbell, J. D., Yao, X. T., Li, Y., Nardelli, B., Olsen, H. S., Morris, T. S., Proudfoot, A. E., Patel, V. P., and Graham, G. J. (2000). C-C chemokine receptor 3 antagonism by the beta-chemokine macrophage inflammatory protein-4, a property strongly enhanced by an amino-terminal alanine-methionine swap. *J. Immunol.* **164,** 1488–1497.

Nibbs, R. J., Wylie, S. M., Yang, J., Landau, N. R., and Graham, G. J. (1997). Cloning and characterization of a novel promiscuous human beta-chemokine receptor D6. *J. Biol. Chem.* **272,** 32078–32083.

Nibbs, R., Graham, G., and Rot, A. (2003). Chemokines on the move: Control by the chemokine "interceptors" Duffy blood group antigen and D6. *Semin. Immunol.* **15,** 287–294.

Ogilvie, P., Bardi, G., Clark-Lewis, I., Baggiolini, M., and Uguccioni, M. (2001). Eotaxin is a natural antagonist for CCR2 and an agonist for CCR5. *Blood* **97,** 1920–1924.

Ogilvie, P., Paoletti, S., Clark-Lewis, I., and Uguccioni, M. (2003). Eotaxin-3 is a natural antagonist for CCR2 and exerts a repulsive effect on human monocytes. *Blood* **102,** 789–794.

Onuffer, J., McCarrick, M. A., Dunning, L., Liang, M., Rosser, M., Wei, G. P., Ng, H., and Horuk, R. (2003). Structure function differences in nonpeptide CCR1 antagonists for human and mouse CCR1. *J. Immunol.* **170,** 1910–1916.

Oravecz, T., Pall, M., Roderiquez, G., Gorrell, M. D., Ditto, M., Nguyen, N. Y., Boykins, R., Unsworth, E., and Norcross, M. A. (1997). Regulation of the receptor specificity and function of the chemokine RANTES (regulated on activation, normal T cell expressed and secreted) by dipeptidyl peptidase IV (CD26)-mediated cleavage. *J. Exp. Med.* **186,** 1865–1872.

Overall, C. M., McQuibban, G. A., and Clark-Lewis, I. (2002). Discovery of chemokine substrates for matrix metalloproteinases by exosite scanning: A new tool for degradomics. *Biol. Chem.* **383,** 1059–1066.

Paavola, C. D., Hemmerich, S., Grunberger, D., Polsky, I., Bloom, A., Freedman, R., Mulkins, M., Bhakta, S., McCarley, D., Wiesent, L., Wong, B., Jarnagin, K., and Handel, T. M. (1998). Monomeric monocyte chemoattractant protein-1 (MCP-1) binds and activates the MCP-1 receptor CCR2B. *J. Biol. Chem.* **273,** 33157–33165.

Palczewski, K., Kumasaka, T., Hori, T., Behnke, C. A., Motoshima, H., Fox, B. A., Le, T. I., Teller, D. C., Okada, T., Stenkamp, R. E., Yamamoto, M., and Miyano, M. (2000). Crystal structure of rhodopsin: A G protein-coupled receptor. *Science* **289,** 739–745.

Pease, J. E., Wang, J., Ponath, P. D., and Murphy, P. M. (1998). The N terminal extracellular segments of the chemokine receptors CCR1 and CCR3 are determinants for MIP-1a and eotaxin binding, respectively, but a second domain is essential for receptor activation. *J. Biol. Chem.* **273,** 19972–19976.

Peiper, S. C., Wang, Z. X., Neote, K., Martin, A. W., Showell, H. J., Conklyn, M. J., Ogborne, K., Hadley, T. J., Lu, Z. H., Hesselgesser, J., and Horuk, R. (1995). The Duffy antigen/receptor for chemokines (DARC) is expressed in endothelial cells of Duffy negative individuals who lack the erythrocyte receptor. *J. Exp. Med.* **181,** 1311–1317.

Pelchen-Matthews, A., Signoret, N., Klasse, P. J., Fraile-Ramos, A., and Marsh, M. (1999). Chemokine receptor trafficking and viral replication. *Immunol. Rev.* **168,** 33–49.

Perrier, P., Martinez, F. O., Locati, M., Bianchi, G., Nebuloni, M., Vago, G., Bazzoni, F., Sozzani, S., Allavena, P., and Mantovani, A. (2004). Distinct transcriptional programs activated by interleukin-10 with or without lipopolysaccharide in dendritic cells: Induction of the B cell-activating chemokine, CXC chemokine ligand 13. *J. Immunol.* **172,** 7031–7042.

Petkovic, V., Moghini, C., Paoletti, S., Uguccioni, M., and Gerber, B. (2004). Eotaxin-3/CCL26 is a natural antagonist for CC chemokine receptors 1 and 5. A human chemokine with a regulatory role. *J. Biol. Chem.* **279,** 23357–23363.

Petkovic, V., Moghini, C., Paoletti, S., Uguccioni, M., and Gerber, B. (2004). I-TAC/CXCL11 is a natural antagonist for CCR5. *J. Leukoc. Biol.* **76,** 701–708.

Pitcher, J. A., Freedman, N. J., and Lefkowitz, R. J. (1998). G-protein-coupled receptor kinases. *Annu. Rev. Biochem.* **67,** 653–692.

Ponath, P. D., Qin, S., Post, T. W., Wang, J., Wu, L., Gerard, N. P., Newman, W., Gerard, C., and Mackay, C. R. (1996). Molecular cloning and characterization of a human eotaxin receptor expressed selectively on eosinophils. *J. Exp. Med.* **183,** 2437–2448.

Power, C. A., Church, D. J., Meyer, A., Alouani, S., Proudfoot, A. E., Clark-Lewis, I., Sozzani, S., Mantovani, A., and Wells, T. N. (1997). Cloning and characterization of a specific receptor for the novel CC chemokine MIP-3alpha from lung dendritic cells. *J. Exp. Med.* **186,** 825–835.

Proost, P., De, M. I., Schols, D., Struyf, S., Lambeir, A. M., Wuyts, A., Opdenakker, G., De Clercq, E., Scharpe, S., and Van Damme, J. (1998). Amino-terminal truncation of chemokines by CD26/dipeptidyl-peptidase IV. Conversion of RANTES into a potent inhibitor of monocyte chemotaxis and HIV-1-infection. *J. Biol. Chem.* **273,** 7222–7227.

Proost, P., Menten, P., Struyf, S., Schutyser, E., De Meester, I., and Van Damme, J. (2000). Cleavage by CD26/dipeptidyl peptidase IV converts the chemokine LD78beta into a most efficient monocyte attractant and CCR1 agonist. *Blood* **96,** 1674–1680.

Proost, P., Struyf, S., Schols, D., Opdenakker, G., Sozzani, S., Allavena, P., Mantovani, A., Augustyns, K., Bal, G., Haemers, A., Lambeir, A. M., Scharpe, S., Van Damme, J., and De, M. I. (1999). Truncation of macrophage-derived chemokine by CD26/dipeptidyl-peptidase IV beyond its predicted cleavage site affects chemotactic activity and CC chemokine receptor 4 interaction. *J. Biol. Chem.* **274,** 3988–3993.

Proudfoot, A. E. I., Power, C. A., Hoogerwerf, A. J., Montjovent, M.-C., Borlat, F., Offord, R. E., and Wells, T. N. C. (1996). Extension of recombinant human RANTES by the retention of the initiating methionine produces a potent antagonist. *J. Biol. Chem.* **271,** 2599–2603.

Proudfoot, A. E., Fritchley, S., Borlat, F., Shaw, J. P., Vilbois, F., Zwahlen, C., Trkola, A., Marchant, D., Clapham, P. R., and Wells, T. N. (2001). The BBXB motif of RANTES is the principal site for heparin binding and controls receptor selectivity. *J. Biol. Chem.* **276,** 10620–10626.

Proudfoot, A. E., Handel, T. M., Johnson, Z., Lau, E. K., LiWang, P., Clark-Lewis, I., Borlat, F., Wells, T. N., and Kosco-Vilbois, M. H. (2003). Glycosaminoglycan binding and oligomerization are essential for the *in vivo* activity of certain chemokines. *Proc. Natl. Acad. Sci. USA* **100,** 1885–1890.

Rajarathnam, K., Li, Y., Rohrer, T., and Gentz, R. (2001). Solution structure and dynamics of myeloid progenitor inhibitory factor-1 (MPIF-1), a novel monomeric CC chemokine. *J. Biol. Chem.* **276,** 4909–4916.

Rodriguez-Frade, J. M., Mellado, M., and Martinez, A. C. (2001). Chemokine receptor dimerization: Two are better than one. *Trends Immunol.* **22,** 612–617.

Rodriguez-Frade, J. M., Vila-Coro, A. J., de Ana, A. M., Albar, J. P., Martinez, A., and Mellado, M. (1999a). The chemokine monocyte chemoattractant protein-1 induces functional responses through dimerization of its receptor CCR2. *Proc. Natl. Acad. Sci. USA* **96,** 3628–3633.

Rodriguez-Frade, J. M., Vila-Coro, A. J., Martin, A., Nieto, M., Sanchez-Madrid, F., Proudfoot, A. E., Wells, T. N., Martinez, A., and Mellado, M. (1999b). Similarities and differences in RANTES- and (AOP)-RANTES-triggered signals: Implications for chemotaxis. *J. Cell. Biol.* **144,** 755–765.

Rosenkilde, M. M., Gerlach, L. O., Jakobsen, J. S., Skerlj, R. T., Bridger, G. J., and Schwartz, T. W. (2004). Molecular mechanism of AMD3100 antagonism in the CXCR4 receptor: Transfer of binding site to the CXCR3 receptor. *J. Biol. Chem.* **279,** 3033–3041.

Rothenberg, M. E., MacLean, J. A., Pearlman, E., Luster, A. D., and Leder, P. (1997). Targeted disruption of the chemokine eotaxin partially reduces antigen-induced tissue eosinophilia. *J. Exp. Med.* **185,** 785–790.

Sakmar, T. P., Menon, S. T., Marin, E. P., and Awad, E. S. (2002). Rhodopsin: Insights from recent structural studies. *Annu. Rev. Biophys. Biomol. Struct.* **31,** 443–484.

Sallusto, F., Lenig, D., Mackay, C. R., and Lanzavecchia, A. (1998). Flexible Programs of Chemokine Receptor Expression on Human Polarised T Helper 1 and 2 Lymphocytes. *J. Exp. Med.* **187,** 875–883.

Sallusto, F., Mackay, C. R., and Lanzavecchia, A. (1997). Selective expression of the eotaxin receptor CCR3 by human T helper 2 cells. *Science* **277,** 2005–2007.

Samama, P., Cotecchia, S., Costa, T., and Lefkowitz, R. J. (1993). A mutation-induced activated state of the beta 2–adrenergic receptor. Extending the ternary complex model. *J. Biol. Chem.* **268,** 4625–4636.

Schnitzer, J. E., Oh, P., Pinney, E., and Allard, J. (1994). Filipin-sensitive caveolae-mediated transport in endothelium: Reduced transcytosis, scavenger endocytosis, and capillary permeability of select macromolecules. *J. Cell. Biol.* **127,** 1217–1232.

Seeds, E. A., Horne, A. P., Tyrrell, D. J., and Page, C. P. (1995). The effect of inhaled heparin and related glycosaminoglycans on allergen-induced eosinophil infiltration in guinea-pigs. *Pulm. Pharmacol.* **8,** 97–105.

Signoret, N., Christophe, T., Oppermann, M., and Marsh, M. (2004). pH-independent endocytic cycling of the chemokine receptor CCR5. *Traffic* **5,** 529–543.

Signoret, N., Pelchen-Matthews, A., Mack, M., Proudfoot, A. E., and Marsh, M. (2000). Endocytosis and recycling of the HIV coreceptor CCR5. *J. Cell. Biol.* **151,** 1281–1294.

Simmons, G., Clapham, P. R., Picard, L., Offord, R. E., Rosenkilde, M. M., Schwartz, T. W., Buser, R., Wells, T. N. C., and Proudfoot, A. E. I. (1997). Potent inhibition of HIV-1 infectivity in macrophage and lymphocytes by a novel CCR5 antagonist. *Science* **276,** 276–279.

Sticht, H., Escher, S. E., Schweimer, K., Forssmann, W. G., Rosch, P., and Adermann, K. (1999). Solution structure of the human CC chemokine 2: A monomeric representative of the CC chemokine subtype. *Biochemistry* **38,** 5995–6002.

Struyf, S., Proost, P., Schols, D., De Clercq, E., Opdenakker, G., Lenaerts, J. P., Detheux, M., Parmentier, M., De, M. I., Scharpe, S., and Van Damme, J. (1999). CD26/dipeptidyl-peptidase IV down-regulates the eosinophil chemotactic potency, but not the anti-HIV activity of human eotaxin by affecting its interaction with CC chemokine receptor 3. *J. Immunol.* **162,** 4903–4909.

Struyf, S., Proost, P., and Van Damme, J. (2003). Regulation of the immune response by the interaction of chemokines and proteases. *Adv. Immunol.* **81,** 1–44.

Sun, Y., Cheng, Z., Ma, L., and Pei, G. (2002). Beta-arrestin2 is critically involved in CXCR4–mediated chemotaxis, and this is mediated by its enhancement of p38 MAPK activation. *J. Biol. Chem.* **277,** 49212–49219.

Takeda, S., Kadowaki, S., Haga, T., Takaesu, H., and Mitaku, S. (2002). Identification of G-protein-coupled receptor genes from the human genome sequence. *FEBS Lett.* **520,** 97–101.

Tarasova, N. I., Stauber, R. H., and Michejda, C. J. (1998). Spontaneous and ligand-induced trafficking of CXC-chemokine receptor 4. *J. Biol. Chem.* **273,** 15883–15886.

Thelen, M., and Baggiolini, M. (2001). Is dimerization of chemokine receptors functionally relevant? *Sci. STKE.* **2001,** PE34.

Trabanino, R. J., Hall, S. E., Vaidehi, N., Floriano, W. B., Kam, V. W., and Goddard, W. A., 3rd. (2004). First principles predictions of the structure and function of g-protein-coupled receptors: Validation for bovine rhodopsin. *Biophys. J.* **86,** 1904–1921.

Turner, L., Ward, S. G., and Westwick, J. (1995). A role for phosphoinositide 3-kinase in RANTES induced chemotaxis of T lymphocytes. *Bio. Chem. Soc. Trans.* **23,** 283S.

Ueno, T., Hara, K., Willis, M. S., *et al.* (2002). Role for CCR7 ligands in the emigration of newly generated T lymphocytes from the neonatal thymus. *Immunity* **16,** 205–218.

Uguccioni, M., Mackay, C. R., Ochensberger, B., Loetscher, P., Rhis, S., LaRosa, G. J., Rao, P., Ponath, P. D., Baggiolini, M., and Dahinden, C. A. (1997). High expression of the chemokine receptor CCR3 in human blood basophils. Role in activation by eotaxin, MCP-4, and other chemokines. *J. Clin. Invest.* **100,** 1137–1143.

Vaidehi, N., Floriano, W. B., Trabanino, R., Hall, S. E., Freddolino, P., Choi, E. J., Zamanakos, G., and Goddard, W. A., 3rd. (2002). Prediction of structure and function of G-protein-coupled receptors. *Proc. Natl. Acad. Sci. USA* **99,** 12622–12627.

Vila-Coro, A. J., Rodriguez-Frade, J. M., Martin, D. A., Moreno-Ortiz, M. C., Martinez, A., and Mellado, M. (1999). The chemokine SDF-1alpha triggers CXCR4 receptor dimerization and activates the JAK/STAT pathway. *FASEB J.* **13,** 1699–1710.

von Zastrow, M., and Kobilka, B. K. (1994). Antagonist-dependent and -independent steps in the mechanism of adrenergic receptor internalization. *J. Biol. Chem.* **269,** 18448–18452.

Wall, M. A., Coleman, D. E., Lee, E., Iniguez-Lluhi, J. A., Posner, B. A., Gilman, A. G., and Sprang, S. R. (1995). The structure of the G-protein heterotrimer Gi alpha 1 beta 1 gamma 2. *Cell* **83,** 1047–1058.

Wan, Y., Jakway, J. P., Qiu, H., Shah, H., Garlisi, C. G., Tian, F., Ting, P., Hesk, D., Egan, R. W., Billah, M. M., and Umland, S. P. (2002). Identification of full, partial and inverse CC chemokine receptor 3 agonists using [35S]GTPgammaS binding. *Eur. J. Pharmacol.* **456,** 1–10.

Wang, L. H., Rothberg, K. G., and Anderson, R. G. (1993). Mis-assembly of clathrin lattices on endosomes reveals a regulatory switch for coated pit formation. *J. Cell Biol.* **123,** 1107–1117.

Weber, M., Blair, E., Simpson, C. V., O'Hara, M., Blackburn, P. E., Rot, A., Graham, G. J., and Nibbs, R. J. (2004). The chemokine receptor D6 constitutively traffics to and from the cell surface to internalize and degrade chemokines. *Mol. Biol. Cell.* **15,** 2492–2508.

Witt, D. P., and Lander, A. D. (1994). Differential binding of chemokines to glycosaminoglycan subpopulation. *Curr. Biol.* **4,** 394–400.

Xanthou, G., Duchesnes, C. E., Williams, T. J., and Pease, J. E. (2003). CCR3 functional responses are regulated by both CXCR3 and its ligands CXCL9, CXCL10 and CXCL11. *Eur. J. Immunol.* **33,** 2241–2250.

Young, M. D., Eyles, D. E., Burgess, R. W., and Jeffery, G. M. (1955). Experimental testing of the immunity of Negroes to *Plasmodium vivax. J. Parasitol.* **41,** 315–318.

Zimmermann, N., and Rothenberg, M. E. (2003). Receptor internalization is required for eotaxin-induced responses in human eosinophils. *J. Allergy Clin. Immunol.* **111,** 97–105.

Zlotnik, A., and Yoshie, O. (2000). Chemokines: A new classification system and their role in immunity. *Immunity* **12,** 121–127.

CHAPTER 4

The Molecular and Cellular Biology of C and CX3C Chemokines and Their Receptors

Tadaatsu Imaizumi, Hidemi Yoshida, and Kei Satoh
Department of Vascular Biology, Hirosaki University School of Medicine,
Hirosaki 036-8562, Japan

I. C AND CX3C CHEMOKINES

The family of cytokines that recruits specific types of leukocytes is called the chemokines. Chemokines have conserved cysteine residues and are classified into C, CC, CXC, and CX3C subfamilies. We will focus here on the discussions of C and CX3C chemokines.

Only a single member is identified in each of the C and CX3C subfamilies of chemokines. Lymphotactin, which is also called "single C motif-1" (SCM-1) or "activation-induced T-cell-derived and chemokine-related molecule" (ATAC), is the only member of the C chemokine family. This unique chemokine was first identified in activated pro-T cells (Kelner *et al.*, 1994) and was designated as XCL1. The only CX3C chemokine is fractalkine, which was first identified in endothelial cells with a unique CX3C motif (Bazan *et al.*, 1997) and was designated as CX3CL1.

Current Topics in Membranes, Volume 55 1063-5823/05 $35.00
DOI: 10.1016/S1063-5823(05)55004-X

Lymphotactin is produced and secreted by activated T-cells, thymocytes, natural killer (NK) cells, and mast cells (Hedrick *et al.*, 1997). Fractalkine mRNA is expressed in cells such as endothelial cells, astrocytes, and dendritic cells (Imaizumi *et al.*, 2004) and in most human organs, including the heart, brain, lung, kidney, intestines, and skeletal muscles (Bazan *et al.*, 1997). Most generated lymphotactin is secreted from the cells; however, fractalkine is synthesized as a membrane protein, and the cell surface–bound molecule serves as a potent adhesion molecule along with other factors such as vascular cell adhesion molecule-1 (VCAM-1). The N-terminal portion of the fractalkine molecule, which contains the CX3C motif, can be cleaved by a protease called tumor necrosis factor-α (TNF-α) converting enzyme (TACE, also known as ADAM17) (Garton *et al.*, 2001) and released from the membrane-associated molecules. TACE is expressed in human endothelial cells (Imaizumi *et al.*, 2000b). The secreted form of fractalkine has potent chemotactic activity; therefore, fractalkine may function as both an adhesion molecule and a chemokine for target leukocytes. This is a unique feature of fractalkine that is similar to other adhesion molecules or chemokines. For instance, the N-terminal domain of VCAM-1 can also be released from the membrane-bound molecules, and the secreted form of VCAM-1 binds its ligand on leukocytes. However, the soluble form does not induce chemotaxis of leukocytes but is regarded to function as an antagonist against leukocyte adhesion.

II. RECEPTORS FOR LYMPHOTACTIN AND FRACTALKINE

Chemokines stimulate leukocyte chemotaxis by activating specific seven-transmembrane G-protein-coupled receptors expressed on their target cell population. The receptor for lymphotactin is designated as XCR1, and XCR1 is expressed selectively in T-cells and NK cells. High levels of mRNA for XCR1 were detected in tissues such as the placenta, spleen, and thymus (Yoshida *et al.*, 1998). The cell-surface receptor for fractalkine, CX3CR1, is expressed in T-cells and monocytes (Imai *et al.*, 1997) and microglia (Nishiyori *et al.*, 1998). Neither XCR1 nor CX3CR1 is expressed on neutrophils, and this may be regarded as a common feature of lymphotactin and fractalkine since most other types of chemokines are active on neutrophils. T-cells and NK cells express both lymphotactin and its receptor XCR1, and this suggests the possibility that lymphotactin may function in an autocrine manner. On the other hand, the target cells of fractalkine are different from its sources (Fig. 1), except that cultured neurons express both fractalkine and CX3CR1 (Hatori *et al.*, 2002).

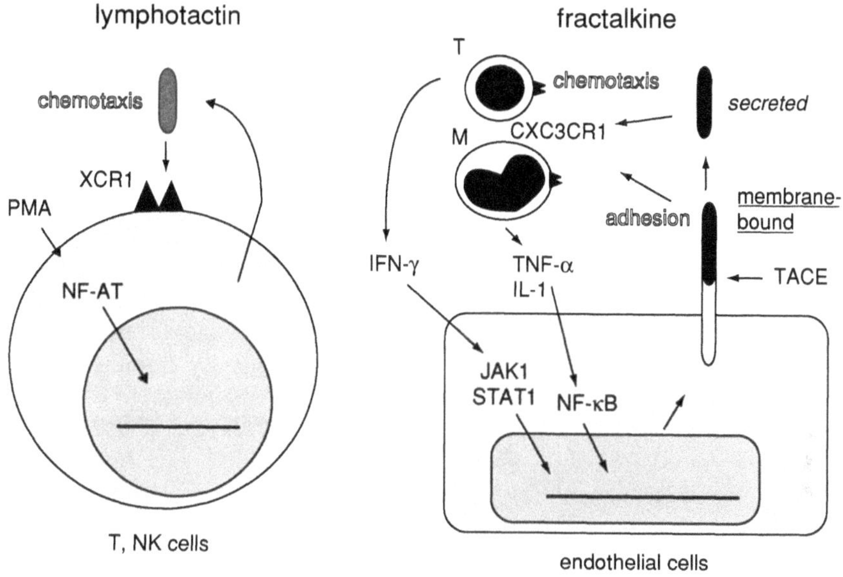

FIGURE 1 Production and function of lymphotactin and fractalkine. Phorbol ester (PMA) induces the expression of lymphotactin in T-cells and NK cells, and lymphotactin is secreted and activates T-cells and NK cells that express the receptor CRX1. Proinflammatory cytokines induce the expression of fractalkine in endothelial cells. Fractalkine is produced as a membrane-bound form, which works as an adhesion molecule for T-cells (T) and monocytes (M) expressing the receptor CX3CR1. The secreted form is released by processing by TNF–α converting enzyme (TACE) and functions as a chemotactic factor.

III. REGULATION OF LYMPHOTACTIN AND FRACTALKINE EXPRESSION

Both lymphotactin and fractalkine are expressed in an activation-dependent manner. The expression of lymphotactin is upregulated by phorbol ester and concanavalin 1 in T-cells isolated from peripheral blood (Middel *et al.*, 2001). It can be inhibited by anti-inflammatory steroid dexamethasone or immunosuppressive drugs such as cyclosporine A and FK506 (Middel *et al.*, 2001). These agents also regulate the expression of lymphotactin receptor XCR1 on T-cells in a manner similar to the regulation of lymphotactin expression. The "nuclear factor of activated T-cells" (NF–AT) family's transcription factors are likely to be involved in the activation-induced expression and cyclosporine A–dependent inhibition of lymphotactin (Yoshida *et al.*, 1999).

Fractalkine was first documented as a factor induced by TNF-α, interleukin-1 (IL-1), and bacterial lipopolysaccharide (LPS) (Bazan *et al.*, 1997). TNF-α, IL-1, and LPS stimulate a similar spectrum of signal-transduction systems and activate both nuclear factor-κB (NF-κB) and mitogen-activated protein kinases (MAPKs). In rat aortic endothelial cells, the induction of fractalkine by these agonists is mediated through the NF-κB system (Garcia *et al.*, 2000).

The two subtypes of helper T-cells, Th1 and Th2, produce distinct profiles of cytokines. Interferon-γ (IFN-γ), one of the potent proinflammatory and immunomodulatory cytokines, is thought to be produced by Th1-type cells. Th2-type cells produce cytokines such as IL-4 and IL-5 and are involved in allergic reactions. IFN-γ exerts its biological activity by binding to the specific receptor, and it sequentially activates (1) Janus kinase 1 (JAK1), a signal transducer and activator of transcription 1 (STAT1), and (2) STAT1-dependent transcription of genes. Vascular endothelial cells express the receptor for IFN-γ, and IFN-γ induces various factors, including a CXC chemokine interferon γ-inducible protein-10 (IP-10) (Luster and Ravetech, 1987). IFN-γ also induces the expression of fractalkine in rat aortic endothelial cells (Garcia *et al.*, 2000) and in human umbilical vein endothelial cells (HUVEC) (Imaizumi *et al.*, 2000a). Human fractalkine gene is located on chromosome 16 (Nomiyama *et al.*, 1998), and there are potential binding sites for NF-κB and STAT in the promoter region of the gene.

Fractalkine is constitutively expressed in dendritic cells and upregulated upon dendritic cell maturation (Kanazawa *et al.*, 1999). This suggests that fractalkine plays an important role in the intercellular communications during immune reactions.

In the nervous system, fractalkine is constitutively expressed in the olfactory bulb, cerebral cortex, and hippocampus, among other places (Nishiyori *et al.*, 1998), but its expression is not affected by experimentally induced inflammation of central nervous tissue (Schwaeble *et al.*, 1998). TNF-α and IFN-γ synergistically enhance the expression of fractalkine in cultured human astrocytes (Yoshida *et al.*, 2001). IFN-γ also enhances fractalkine expression in human bronchial epithelial cells, and the chemokine mediates adherence of macrophages (Fujimoto *et al.*, 2001). Fractalkine is induced by TNF-α, IFN-γ, and IL-4 in fibroblasts (Fahy *et al.*, 2003; Yoshikawa *et al.*, 2004).

It has been known for some time that the soluble form of the IL-6 receptor possesses an agonistic activity towards vascular endothelial cells (Modur *et al.*, 1997). IL-6 is a multifunctional cytokine and mediates a variety of inflammatory and cell-growth responses. The cell-surface receptor for IL-6 consists of a cytokine-binding α subunit (IL-6Rα) and a transmembrane-signaling gp130 subunit. Vascular endothelial cells lack IL-6Rα, but gp130 signaling of endothelial cells can be activated by the soluble form of IL-6Rα (sIL-6Rα), which is secreted from activated leukocytes. sIL-6Rα stimulates the endothelial cells to express various factors, including CXC chemokines IL-8 (Modur *et al.*, 1997) and growth-related oncogene-α (GRO-α) (Matsumiya *et al.*, 2002). The expression of fractalkine induced by IL-1 or IFN-γ in endothelial cells is inhibited by sIL-6Rα (Matsumiya *et al.*, 2001). Thus, sIL-6Rα inhibits not only the secretion of the mononuclear cell-chemotactic activity from IFN-γ-treated endothelial cells, but the adhesion of mononuclear cells to endothelial monolayers stimulated with IFN-γ.

A fatty acid, arachidonic acid, can be released from membrane phospholipids when the cells are activated. The arachidonic acid is metabolized to prostaglandins via the arachidonic acid cascade, and cyclooxygenase-2 is a key enzyme for the synthesis of prostaglandins (Jones *et al.*, 1993). 15-Deoxy-$\Delta^{12,14}$-prostaglandin J_2 (15d-PGJ_2) is a metabolite of PGD_2. This prostaglandin was detected in the foamy macrophages in human atherosclerotic plaques (Shibata *et al.*, 2002) and thought to be involved in the pathogenesis of atherosclerosis. 15d-PGJ_2 is an agonist for peroxisome proliferator-activated receptor-γ (PPAR-γ), a member of the nucleic hormone receptor superfamily. PPAR-γ was originally identified in adipose tissue as a factor that induces adipocyte differentiation and triglyceride synthesis. It is known that 15d-PGJ_2 regulates the expression of various genes in a PPAR-γ-dependent or PPAR-γ-independent manner. 15d-PGJ_2 induces the expression of IL-8 (Jozkowicz *et al.*, 2001), but it does not affect the expression of GRO-α (Imaizumi *et al.*, 2003a). However, 15d-PGJ_2 inhibits the expression of IP-10 (Marx *et al.*, 2000) and another CXC chemokine, epithelial neutrophil activating peptide-78 (ENA-78) (Imaizumi *et al.*, 2003a). We found that 15d-PGJ_2 also inhibits the induction of fractalkine by IFN-γ or IL-1β in endothelial cells (Imaizumi *et al.*, 2002). However, ciglitazone, a specific agonist for PPAR-γ, did not inhibit the cytokine-induced fractalkine expression, and the effect of 15d-PGJ_2 may be independent on PPAR-γ. 15d-PGJ_2 is shown to inhibit the signaling through NF-κB, activating protein-1 (AP-1), and STAT1 (Rossi *et al.*, 2000; Imaizumi *et al.*, 2003b), and these transcriptional factors are tentatively regarded as a cellular mechanism for the downregulation of fractalkine by 15d-PGJ_2.

Hypoxia is one of the constituent pathophysiological states in various conditions, including ischemic diseases, inflammatory diseases, and cancers. Hypoxia modulates the expression of various genes that may represent part of the mechanisms for adaptation of the cells to hypoxic stress. A transcriptional factor, hypoxia-inducible factor-1α (HIF-1α), is thought to be involved in the activation of most hypoxia-responsive genes, such as erythropoietin and vascular endothelial growth factor (Namiki *et al.*, 1995). Hypoxia also impairs the functioning of many cell types by inhibiting the expression of several genes: argininosuccinate synthase in lung endothelial cells (Su and Block, 1997) and heme oxygenase 1 (HO-1) in human endothelial cells and astrocytes (Nakayama *et al.*, 2000). We found that hypoxia inhibits the IFN-γ-induced expression of fractalkine in HUVEC (Yamashita *et al.*, 2003). Desferrioxamine (DFX) is an iron chelator that mimics hypoxia *in vitro*, pression of fractalkine in HUVEC stimulated with IFN-γ (Yamashita *et al.*, 2003). The inhibition of fractalkine expression by hypoxia was reversed when the HUVEC cultures were returned to the reoxygenation condition (Yamashita *et al.*, 2003). Thus, the inhibition of fractalkine expression by hypoxia is reversible, and this suggests the existence of a specific intracellular regulatory mechanism for the hypoxia-mediated inhibition of fractalkine expression. HIF-1α is unlikely to be involved in this regulation, and the mechanism for hypoxia-induced fractalkine downregulation is still unknown.

The surfaces of various cells are covered by a dense layer of proteoglycans. The main functional component of this layer is polyanionic glycosaminoglycans (GAGs), and heparan sulfate is one of the most abundant GAGs. Heparan sulfate is structurally and biosynthetically related to heparin, which is commonly used as an anticoagulant. Interactions between chemokines and GAGs may result in the confinement of chemokines in inflammatory foci, and various cytokines are known to bind to heparin or heparin-related molecules (Salmivirta *et al.*, 1996). Lymphotactin binds to heparin with high affinity, and lymphotactin–GAG interactions are important for the regulation of the chemokine function (Peterson *et al.*, 2004). Heparin and heparin-related molecules are also known to affect leukocyte–endothelial interactions by binding chemokines (Miller and Krangel, 1992). IFN-γ also binds to certain GAGs, and exogenous heparin may regulate activities of IFN-γ by competing with cell-surface GAGs or proteoglycans for IFN-γ binding. In fact, heparin inhibited the fractalkine expression in HUVEC stimulated with IFN-γ but not with IL-1β (Hatakeyama *et al.*, 2004). Heparin and related GAGs may affect the inflammatory and immune responses by regulating the expression and function of chemokines.

IV. ROLE OF LYMPHOTACTIN AND FRACTALKINE IN DISEASES

Expressions of chemokines are tightly regulated under physiological conditions; however, dysregulation of these chemokines may lead to inflammatory diseases.

Crohn's disease is a chronic inflammatory bowel disease. Recruitment of lymphocytes is a prominent feature of the inflammatory process in Crohn's disease, and the expressions of lymphotactin (Middel *et al.*, 2001) and fractalkine (Muehlhoefer *et al.*, 2000) have been detected at the inflammatory foci.

Rheumatoid arthritis is an autoimmune inflammatory disease, and various proinflammatory cytokines are thought to be involved in the pathogenesis of this disease. Possible roles of lymphotactin (Blaschke *et al.*, 2003a) and fractalkine (Blaschke *et al.*, 2003b) in rheumatoid arthritis have also been suggested.

Fractalkine is reported to be involved in the pathogenesis of renal diseases like glomerulonephritis (Cockwell *et al.*, 2002) and nephropathy in Streptozotocin-induced diabetic rats (Kikuchi *et al.*, 2004). Involvement of fractalkine is also shown in cardiac allograft rejection (Robinson *et al.*, 2000), and the inhibition of lymphotactin expression by cyclosporine A and FK506 suggests that this chemokine contributes to allograft rejection.

There is a known relationship between viral infection and chemokines or their receptors. Infection by HIV-1 induces immunodeficiency and related pathological disorders, including neurological diseases. A polymorphism of the CX3CR1 gene that impairs fractalkine binding is reported to be associated with an increased risk of HIV-1 disease (Faure *et al.*, 2000). HIV-1 tat protein, which is secreted from HIV-1-infected cells, is proposed to be a key agent in HIV-related neurological disease, and lymphotactin expression in astrocytes is significantly increased in HIV-1 tat transgenic mice (Kim *et al.*, 2004). These facts suggest that lymphotactin and the receptor for fractalkine may contribute to the development of HIV-1-associated diseases.

The interactions of vascular endothelial cells with leukocytes are important in inflammatory responses. However, they may trigger cellular processes that eventually may lead to atherosclerotic vascular changes (Zimmerman *et al.*, 1996). ApoE knockout mice are known to develop atherosclerosis by a high-fat diet, and knockout of the gene encoding CX3CR1 in apoE$^{-/-}$ mice resulted in a significant reduction of atherosclerotic lesions (Lesnik *et al.*, 2003). In humans, polymorphism of the CX3CR1 gene is associated with an increased risk of atherosclerotic cardiovascular diseases (McDermott *et al.*, 2003) and internal carotid artery occlusive diseases (Ghilardi *et al.*, 2004). These observations suggest that fractalkine and its receptor are important players in atherogenesis.

V. FUTURE DIRECTIONS AND UNANSWERED QUESTIONS

Lymphotactin and fractalkine are involved in various diseases, and regulatory measures on the functions of these chemokines may lead to the development of new strategies for therapeutic approaches to human diseases. Both lymphotactin and fractalkine are chemotactic for T-cells and NK cells, both of which play important roles in immunological responses against tumor and infection. Recruitment of these cells will be valuable in cancer vaccines, whereas antagonists or chemokine-neutralizing agents may be useful for treating certain autoimmune or inflammatory diseases. In fact, the antitumor effect of the combined transgene of lymphotactin and T-cell-activating cytokine IL-2 has been shown in murine breast cancer models (Emtage *et al.*, 1999) and in human clinical trials for the treatment of neuroblastoma (Rousseau *et al.*, 2003). However, the biological and physiological roles of lymphotactin are not fully clarified. Accumulating evidence has shown the contribution of fractalkine to various diseases to be related to inflammatory reactions. However, these observations have not lead to the development of new therapeutic measures. The details of the roles of these chemokines in the immune system and inflammation should be investigated by future studies, and this may lead to the development of new therapeutic strategies for various diseases.

Acknowledgments

The authors thank Kumiko Munakata for her help. Part of this study was supported by Karoji Memorial Fund and a grant from the Aomori Bank for Medical Research at Hirosaki University.

References

Bazan, I. F., Bacon, K. B., Hardiman, G., Wang, W., Soo, K., Rossi, D., Greaves, D. A., Zlotnik, A., and Schall, T. J. (1997). A new class of membrane-bound chemokine with a CX3C motif. *Nature* **385,** 640–644.

Blaschke, S., Middel, P., Dorner, B. G., Blaschke, V., Hummel, K. M., Kroczek, R. A., Reich, K., Benoehr, P., Koziolek, M., and Muller, G. A. (2003a). Expression of activation-induced, T cell-derived, and chemokine-related cytokine/lymphotactin and its functional role in rheumatoid arthritis. *Arthritis Rheum.* **48,** 1858–1872.

Blaschke, S., Koziolek, M., Schwarz, A., Benohr, P., Middel, P., Schwarz, G., Hummel, K. M., and Muller, G. A. (2003b). Proinflammatory role of fractalkine (CX3CL1) in rheumatoid arthritis. *J. Rheumatol.* **30,** 1918–1927.

Cockwell, P., Chakravorty, S. J., Girdlestone, J., and Savage, C. O. (2002). Fractalkine expression in human renal inflammation. *J. Pathol.* **196,** 85–90.

Emtage, P. C., Wan, Y., Hitt, M., Graham, F. L., Muller, W. J., Zlotnik, A., and Gauldie, J. (1999). Adenoviral vectors expressing lymphotactin and interleukin 2 or lymphotactin and interleukin 12 synergize to facilitate tumor regression in murine breast cancer models. *Hum. Gene. Ther.* **10,** 697–709.

Fahy, O. L., Coates, N. J., and McColl, S. R. (2003). Inhibition of cytokine-induced fractalkine production by bacterial invasion of human-dermal fibroblasts. *Lab. Invest.* **83,** 721–730.

Faure, S., Meyer, L., Costagliola, D., Vaneensberghe, C., Genin, E., Autran, B., Delfraissy, J. F., McDermott, D. H., Murphy, P. M., Debre, P., Theodorou, I., and Combadiere, C.

(2000). Rapid progression to AIDS in HIV+ individuals with a structural variant of the chemokine receptor CX3CR1. *Science* **287,** 2274–2277.

Fujimoto, K., Imaizumi, T., Yoshida, H., Takanashi, S., Okumura, K., and Satoh, K. (2001). Interferon-γ stimulates fractalkine expression in human bronchial epithelial cells and regulates mononuclear cell adherence. *Am. J. Respir. Cell Mol. Biol.* **25,** 233–238.

Garcia, G. E., Xia, Y., Chen, S., Wang, Y., Ye, R. D., Harrison, J. K., Bacon, K. B., Zerwes, H. G., and Feng, L. (2000). NF-κB-dependent fractalkine induction in rat aortic endothelial cells stimulated by IL-1β, TNF-α, and LPS. *J. Leukoc. Biol.* **67,** 577–584.

Garton, K. J., Gough, P. J., Blobel, C. P., Murphy, G., Greaves, D. R., Dempsey, P. J., and Raines, E. W. (2001). Tumor necrosis factor-α-converting enzyme (ADAM17) mediates the cleavage and shedding of fractalkine (CX3CL1). *J. Biol. Chem.* **276,** 37993–38001.

Ghilardi, G., Biondi, M. L., Turri, O., Guagnellini, E., and Scorza, R. (2004). Internal carotid artery occlusive disease and polymorphisms of fractalkine receptor CX3CR1: A genetic risk factor. *Stroke* **35,** 1276–1279.

Hatakeyama, M., Imaizumi, T., Tamo, W., Yamashita, K., Yoshida, H., Fukuda, I., and Satoh, K. (2004). Heparin inhibits IFN-γ-induced fractalkine/CX3CL1 expression in human endothelial cells. *Inflammation* **28,** 7–13.

Hatori, K., Nagai, A., Heisel, R., Ryu, J. K., and Kim, S. U. (2002). Fractalkine and fractalkine receptors in human neurons and glial cells. *J. Neurosci. Res.* **69,** 418–426.

Hedrick, J. A., Saylor, V., Figueroa, D., Mizoue, L., Xu, Y., Menon, S., Abrams, J., Handel, T., and Zlotnik, A. (1997). Lymphotactin is produced by NK cells and attracts both NK cells and T cells *in vivo. J. Immunol.* **158,** 1533–1540.

Imai, T., Hieshima, K., Haskell, C., Baba, M., Nagira, M., Nishimura, M., Kakizaki, M., Takagi, S., Nomiyama, H., Schall, T. J., and Yoshie, O. (1997). Identification and molecular characterization of fractalkine receptor CX3CR1, which mediates both leukocyte migration and adhesion. *Cell* **91,** 521–530.

Imaizumi, T., Matsumiya, T., Fujimoto, K., Okamoto, K., Cui, X. F., Ohtaki, U., Yoshida, H., and Satoh, K. (2000a). Interferon-γ stimulates the expression of CX3CL1/fractalkine in cultured human endothelial cells. *Tohoku J. Exp. Med.* **192,** 127–139.

Imaizumi, T., Itaya, H., Fujita, K., Kudoh, D., Kudoh, S., Mori, K., Fujimoto, K., Matsumiya, T., and Satoh, K. (2000b). Expression of tumor-necrosis factor-α in cultured human endothelial cells stimulated with lipopolysaccharide or interleukin-1α. *Arterioscler. Thromb. Vasc. Biol.* **20,** 410–415.

Imaizumi, T., Matsumiya, T., Tamo, W., Shibata, T., Fujimoto, K., Kumagai, M., and Satoh, K. (2002). 15-Deoxy-$\Delta^{12,14}$ prostaglandin J_2 inhibits CX3CL1/fractalkine expression in human endothelial cells. *Immunol. Cell Biol.* **80,** 531–536.

Imaizumi, T., Kumagai, M., Hatakeyama, M., Tamo, W., Yamashita, K., Yoshida, H., Munakata, H., and Satoh, K. (2003a). Effect of 15-deoxy-$\Delta^{12,14}$-prostaglandin J_2 on IL-1β-induced expression of epithelial neutrophil-activating protein-78 in human endothelial cells. *Prostaglandins Leukot. Essent. Fatty Acids* **69,** 323–327.

Imaizumi, T., Kumagai, M., Nishi, N., Hirashima, M., Hatakeyama, M., Tamo, W., Yoshida, H., Nakamura, T., Okumura, K., and Satoh, K. (2003b). 15-Deoxy-$\Delta^{12,14}$-prostaglandin J_2 inhibits IFN-γ-induced galectin-9 expression in human umbilical vein endothelial cells. *Int. Arch. Allergy Immunol.* **131,** 57–61.

Imaizumi, T., Yoshida, H., and Satoh, K. (2004). Regulation of CX3CL1/fractalkine expression on endothelial cells. *J. Atheroscler. Thromb.* **11,** 15–21.

Jones, D. A., Carlton, D. P., McIntyre, T. M., Zimmerman, G. A., and Prescott, S. M. (1993). Molecular cloning of human prostaglandin endoperoxide synthase type II and demonstration of expression in response to cytokines. *J. Biol. Chem.* **268,** 9049–9054.

Jozkowicz, A., Dulak, J., Prager, M., Nanobashvili, J., Nigisch, A., Winter, B., Weigel, G., and Huk, I. (2001). Prostaglandin-J_2 induces synthesis of interleukin-8 by endothelial cells in a PPAR-γ-independent manner. *Prostaglandins Other Lipid Mediat.* **66,** 165–177.

Kanazawa, N., Nakamura, T., Tashiro, K., Muramatsu, M., Morita, K., Yoneda, K., Inaba, K., Imura, S., and Honjo, T. (1999). Fractalkine and macrophage-derived chemokine: T cell-attracting chemokines expressed in T-cell area dendritic cells. *Eur. J. Immunol.* **29,** 1925–1932.

Kelner, G. S., Kennedy, J., Bacon, K. B., Kleyensteuber, S., Largaespada, D. A., Jenkins, N. A., Copeland, N. G., Bazan, J. F., Moore, K. W., Schall, T. J., and Zlotnik, A. (1994). Lymphotactin: A cytokine that represents a new class of chemokine. *Science* **266,** 1395–1399.

Kikuchi, Y., Ikee, R., Hemmi, N., Hyodo, N., Saigusa, T., Namikoshi, T., Yamada, M., Suzuki, S., and Miura, S. (2004). Fractalkine and its receptor, CX3CR1, upregulation in streptozotocin-induced diabetic kidneys. *Nephron Exp. Nephrol.* **97,** e17–e25.

Kim, B. O., Liu, Y., Zhou, B. Y., and He, J. J. (2004). Induction of C chemokine XCL1 (lymphotactin/single C motif-1 alpha/activation-induced, T-cell-derived and chemokine-related cytokine) expression by HIV-1 Tat protein. *J. Immunol.* **172,** 1888–1895.

Lesnik, P., Hakell, C. A., and Charo, I. F. (2003). Decreased atherosclerosis in $CX3CR^{-/-}$ mice reveals a role for fractalkine in atherogenesis. *J. Clin. Invest.* **111,** 333–340.

Luster, A. D., and Ravetech, J. V. (1987). Biochemical characterization of a gamma interferon-inducible cytokine (IP-10). *J. Exp. Med.* **166,** 1084–1097.

Marx, N., Mach, F., Sauty, A., Leung, J. H., Sarafi, M. N., Ransohoff, R. M., Libby, P., Plutzky, J., and Luster, A. D. (2000). Peroxisome proliferator-activated receptor-γ activators inhibit IFN-γ-induced expression of the T-cell-active CXC chemokines IP-10, Mig, and I-TAC in human endothelial cells. *J. Immunol.* **164,** 6503–6508.

Matsumiya, T., Imaizumi, T., Fujimoto, K., Cui, X., Shibata, T., Tamo, W., Kumagai, M., Tanji, K., Yoshida, H., Kimura, H., and Satoh, K. (2001). Soluble interleukin-6 receptor α inhibits the cytokine-Induced fractalkine/CX3CL1 expression in human vascular endothelial cells in culture. *Exp. Cell Res.* **269,** 35–41.

Matsumiya, T., Imaizumi, T., Yoshida, H., Cui, X.-F., Kimura, H., and Satoh, K. (2002). Production of growth related oncogene protein α in human umbilical vein endothelial cells stimulated with soluble interleukin-6 receptor α: Role of signal transducers, janus kinase 2 and mitogen-activated protein kinase kinase. *Life Sci.* **70,** 3179–3190.

McDermott, D. H., Fong, A. M., Yang, Q., Sechler, J. M., Cupples, L. A., Merrell, M. N., Wilson, P. W. F., D'Agostino, R. B., O'Donnell, C. J., Patel, D. D., and Murphy, P. M. (2003). Chemokine receptor mutant CX3CR1-M280 has impaired adhesive fuction and correlates with protection from cardiovascular disease in humans. *J. Clin. Invest.* **111,** 1241–1250.

Middel, P., Thelen, P., Blaschke, S., Polzien, F., Reich, K., Blaschke, V., Wrede, A., Hummel, K. M., Gunawan, B., and Radzun, H. J. (2001). Expression of the T-cell chemoattractant chemokine lymphotactin in Crohn's disease. *Am. J. Pathol.* **159,** 1751–1761.

Miller, M. D., and Krangel, M. S. (1992). Biology and biochemistry of the chemokines: A family of chemotactic and inflammatory cytokines. *Crit. Rev. Immunol.* **12,** 17–46.

Modur, V., Li, Y., Zimmerman, G. A., Prescott, S. M., and McIntyre, T. M. (1997). Retrograde inflammatory signaling from neutrophil to endothelial cells by soluble interleukin-6 receptor α. *J. Clin. Invest.* **100,** 2752–2756.

Muehlhoefer, A., Saubermann, L. J., Gu, X., Luedtke-Heckenkamp, K., Xavier, R., Blumberg, R. S., Podolsky, D. K., MacDermott, R. P., and Reinecker, H. C. (2000). Fractalkine is an epithelial and endothelial cell-derived chemoattractant for intraepithelial lymphocytes in the small intestinal mucosa. *J. Immunol.* **164,** 3368–3376.

Nakayama, M., Takahashi, K., Kitamuro, T., Yasumoto, K., Katayose, D., Shirato, K., Fujii-Kuriyama, Y., and Shibahara, S. (2000). Repression of heme oxygenase-1 by hypoxia in vascular endothelial cells. *Biochem. Biophys. Res. Commun.* **271,** 665–671.

Namiki, A., Brogi, E., Kearney, M., Kim, E. A., Wu, T., Couffinhal, T., Varticovski, L., and Isner, J. M. (1995). Hypoxia induces vascular endothelial growth factor in cultured human endothelial cells. *J. Biol. Chem.* **270,** 31189–31195.

Nishiyori, A., Minami, M., Ohtani, Y., Takami, S., Yamamoto, J., Kawaguchi, N., Kume, T., Akaike, A., and Satoh, M. (1998). Localization of fractalkine and CX3CR1 mRNAs in rat brain: Does fractalkine play a role in signaling from neuron to microglia? *FEBS Lett.* **429,** 167–172.

Nomiyama, H., Imai, T., Kusuda, J., Miura, R., Callen, D. F., and Yoshie, O. (1998). Human chemokines fractalkine (SCYD1), MDC (SCYA22) and TARC (SCYA17) are clustered on chromosome 16q13. *Cytogenet. Cell Genet.* **81,** 10–11.

Peterson, F. C., Elgin, E. S., Nelson, T. J., Zhang, F., Hoeger, T. J., Linhardt, R. J., and Volkman, B. F. (2004). Identification and characterization of a glycosaminoglycan recognition element of the C chemokine lymphotactin. *J. Biol. Chem.* **279,** 12598–12604.

Robinson, L. A., Nataraj, C., Thomas, D. W., Howell, D. N., Griffiths, R., Bautch, V., Patel, D. D., Feng, L., and Coffman, T. M. (2000). A role for fractalkine and its receptor (CX3CR1) in cardiac allograft rejection. *J. Immunol.* **165,** 6067–6072.

Rossi, A., Kapahi, P., Natoli, G., Takahashi, T., Chen, Y., Karin, M., and Santoro, M. G. (2000). Anti-inflammatory cyclopentenone prostaglandins are direct inhibitors of IκB kinase. *Nature* **403,** 103–108.

Rousseau, R. F., Haight, A. E., Hirschmann-Jax, C., Yvon, E. S., Rill, D. R., Mel, Z., Smith, S. C., Inman, S., Cooper, K., Alcoser, P., Grilley, B., Gee, Popek, E., Davidoff, A., Bowman, L. C., Brenner, M. K., and Strother, D. (2003). Local and systematic effects of an allogeneic tumor cell vaccine combining transgenic human lymphotactin with interleukin-2 in patients with advanced or refractory neuroblastoma. *Blood* **101,** 1718–1726.

Salmivirta, M., Lidholt, K., and Lindahl, U. (1996). Heparan sulfate: A piece of information. *FASEB J.* **10,** 1270–1279.

Schwaeble, W. J., Stover, C. M., Schall, T. J., Dairaghi, D. J., Trinder, P. K., Linington, C., Iglesias, A., Schubart, A., Lynch, N. J., Weihe, E., and Schafer, M. K. (1998). Neuronal expression of fractalkine in the presence and absence of inflammation. *FEBS Lett.* **429,** 203–207.

Shibata, T., Kondo, M., Osawa, T., Shibata, N., Kobayashi, M., and Uchida, K. (2002). 15-Deoxy-$\Delta^{12,14}$-prostaglandin J_2. A prostaglandin D_2 metabolite generated during inflammatory processes. *J. Biol. Chem.* **277,** 10459–10466.

Su, Y., and Block, E. R. (1997). Hypoxia inhibits the induction of argininosuccinate synthetase by endotoxin in lung endothelial cells. *Am. J. Physiol.* **272,** 934–938.

Yamashita, K., Imaizumi, T., Hatakeyama, M., Tamo, Y., Kimura, D., Kumagai, M., Yoshida, H., and Satoh, K. (2003). Effect of hypoxia on the expression of fractalkine in human endothelial cells. *Tohoku J. Exp. Med.* **200,** 187–194.

Yoshida, T., Imai, T., Kakizaki, M., Nishimura, M., Takagi, S., and Yoshie, O. (1998). Identification of single C motif-1/lymphotactin receptor XCR1. *J. Biol. Chem.* **273,** 16551–16554.

Yoshida, T., Ishikawa, I., Ono, Y., Imai, T., Suzuki, R., and Yoshie, O. (1999). An activation-responsive element in single C motif-1/lymphotactin promoter is a site of constitutive and inducible DNA-protein interactions involving nuclear factor of activated T-cell. *J. Immunol.* **163,** 3295–3303.

Yoshida, H., Imaizumi, T., Fujimoto, K., Matsuo, N., Kimura, K., Cui, X.-F., Matsumiya, T., Tanji, K., Shibata, T., Tamo, W., Kumagai, M., and Satoh, K. (2001). Synergistic stimulation, by tumor necrosis factor-α and interferon-γ, of fractalkine expression in human strocytes. *Neurosci. Lett.* **303,** 132–136.

Yoshikawa, M., Nakajima, T., Matsumoto, K., Okada, N., Tsukidate, T., Iida, M., Otori, N., Haruna, S., Moriyama, H., Imai, T., and Saito, H. (2004). TNF-α and IL-4 regulate expression of fractalkine (CX3CL1) as a membrane-anchored proadhesive protein and soluble chemotactic peptide on human fibroblasts. *FEBS Lett.* **561,** 105–110.

Zimmerman, G. A., McIntyre, T. M., and Prescott, S. M. (1996). Adhesion and signaling in vascular cell-cell interactions. *J. Clin. Invest.* **98,** 1699–1702.

CHAPTER 5

Chemokines and Their Receptors in Hematopoietic Cell Development and Functioning

Chang H. Kim

Laboratory of Immunology and Hematopoiesis, Department of Pathobiology; Purdue Cancer Center; Bindley Bioscience Center, Purdue University, West Lafayette, Indiana 47907

I. INTRODUCTION

Chemokines play indispensable roles in the hematopoietic system. Most of all, chemokines regulate the migration of hematopoietic cells into various tissue sites and organs by inducing chemotaxis and activating integrins. The

Current Topics in Membranes, Volume 55 1063-5823/05 $35.00

DOI: 10.1016/S1063-5823(04)55005-6

positioning of hematopoietic cells in appropriate tissue environments is critical for normal hematopoiesis, which is the process by which immune cells, red blood cells, and platelets are generated. Chemokines regulate the migration of primitive hematopoietic stem and progenitor cells to the bone marrow and the migration of T-cell progenitors to the thymus. They also play important roles in the recruitment and retention of B-cell progenitors in the bone marrow. Ordered migration of T-cells and B-cells to secondary lymphoid tissues, upon completion of their development in the thymus and bone marrow, respectively, is also regulated, in part, by chemokines. Many chemokines have been discovered, because they are induced in inflammatory conditions during infection or injury. After chemokines are induced, they recruit appropriate immune cells to clear pathogens and repair tissue damages. Chemokines also have other functions, such as regulating cell proliferation/cycling, survival/apoptosis, and cell differentiation. In this chapter, I will review the functions of chemokines in the biology of hematopoietic stem/progenitor cells, T-cells, and B-cells.

II. CHEMOKINES FOR HEMATOPOIETIC STEM AND PROGENITOR CELLS

A. *CXCL12: A Chemokine Critical for the Migration of Hematopoietic Stem and Progenitor Cells and B-Lymphopoiesis in the Bone Marrow*

Hematopoietic stem cells are found in the yolk sac, aorta–gonad–mesonephros (AGM), and liver during embryonic development. After birth and during adult life, small numbers of stem cells circulate the periphery while most stem cells reside in the bone marrow. Recently, hematopoietic stem cells have become a valuable tool with which to treat or cure a number of diseases. Transplantations of hematopoietic stem cells into cancer patients and patients with congenital immunodeficiency or autoimmune diseases are good examples of their capabilities. Cord blood, bone marrow aspirates, and mobilized peripheral blood stem cells are good sources of hematopoietic stem cells for transplantation. In transplantation, it is important for the hematopoietic stem cells to migrate to appropriate tissue sites, most ideally the bone marrow, for repopulation. In the last decade, we have learned that chemokines regulate the homing and mobilization of stem and progenitor cells in and out of the bone marrow.

The first evidence that hinted at the link between chemokines and hematopoiesis was obtained from a gene knockout study where the gene for stromal cell–derived factor-1 (SDF-1, also called PBSF-1/CXCL12) was deleted (Nagasawa *et al.*, 1996; Ma *et al.*, 1998; Zou *et al.*, 1998). CXCL12 is now the standard term for SDF-1, and it will be used hereafter in this chapter. CXCL12 was first cloned from the cDNAs of a bone-marrow stromal-cell line by a cloning method called "signal sequence trap" (Tashiro *et al.*, 1993). The CXCL12-deficient mice had multiple defects in hematopoiesis and organ development (Nagasawa *et al.*, 1996). These mice died perinatally, and they were defective in vascular development. Although the numbers of B-cell progenitors in $CXCL12^{-/-}$ embryos were severely reduced in the fetal liver and bone marrow, the numbers of myeloid progenitors were reduced only in the bone marrow, not in the fetal liver. This implies that CXCL12 may be important for general B-lymphopoiesis and bone marrow myelopoiesis. While many chemokines have multiple receptors, CXCL12 uses only one receptor, CXCR4. Not surprisingly, the mice, deficient in CXCR4, had the identical phenotype to that of CXCL12-/- mice (Ma *et al.*, 1998; Tachibana *et al.*, 1998; Zou *et al.*, 1998).

The defective B-lymphopoiesis in CXCL12-deficient mice can be explained by the fact that CXCL12 acts as a hematopoietic factor for B-cell progenitors. Nagasawa *et al.* (1994) found that a stromal cell line, PA6, produces a soluble mediator that supports the proliferation of a stromal cell–dependent pre-B-cell clone. Using an expression cloning method, they isolated a cDNA clone encoding a B-cell growth-stimulating factor. They designated this protein "pre-B-cell growth-stimulating factor" (PBSF), and it turned out to be CXCL12. CXCL12 synergistically augments the growth of bone marrow B-cell progenitors with IL-7 (Nagasawa *et al.*, 1994). The same group found that $CXCR4^{+/+}$, but not $CXCR4^{-/-}$, fetal liver cells can repopulate long-term lymphoid and myeloid compartments (Kawabata *et al.*, 1999). In line with the hematopoietic activity of CXCL12, $CXCR4^{-/-}$ fetal liver cells had more severely reduced numbers of B-cells than did other lineages in the bone marrow, while $CXCL12^{-/-}$ fetal cells had no obvious problems in reconstituting all the lineages of hematopoietic cells. The defect was more severe when the mice were reconstituted with c-kit^{+} Sca-1 lin$^{low/-}$ cells than with more primitive c-kit^{+} Sca-1^{+} lin$^{low/-}$ cells (Kawabata *et al.*, 1999). The Nagasawa group further demonstrated that, immediately after the commitment to B lineage, precursors become dependent on the chemokine CXCL12 and its receptor CXCR4 for proliferation and differentiation (Egawa *et al.*, 2001).

The functions of CXCL12 for hematopoietic stem and progenitor cells (HSPCs) can largely be classified into two groups: regulation of cell growth and regulation of cell motility. The hematopoietic effect of CXCL12 on B-cell progenitors, as discussed above, involves the effect of CXCL12 on cell growth. However, the regulatory functions of CXCL12 on cell motility are very important for the hematopoietic stem cells and all lineages of progenitors. In humans, CD34^{+} lineage^{-} cells represent a hematopoietic population enriched with stem and progenitor cells. CXCL12 is an efficient chemoattractant for bone marrow CD34$^+$ cells (Aiuti *et al.*, 1997; Kim and Broxmeyer, 1998). While nonchemokine chemoattractants such as steel factor (SLF, also called stem cell factor) and thrombopoietin (TPO) have both chemotactic and chemokinetic activities in cell migration, CXCL12 has only the chemotactic activity (Kim and Broxmeyer, 1998, 1999b). SLF and CXCL12 are synergistic in inducing chemotaxis of HSPCs. A negative concentration gradient of CXCL12 is inhibitory for the cell migration induced by CXCL12 in a positive gradient. Based upon this fact, it has been proposed that CXCL12 recruits HSPCs to the bone marrow and prevents spontaneous HSPC mobilization out of the bone marrow (Kim and Broxmeyer, 1998).

In an adoptive transfer study of CXCR4-deficient HSPCs into irradiated mice, it has been actually demonstrated *in vivo* that CXCL12 and CXCR4 are required for the retention of B lineage and granulocytic precursors within the fetal liver and bone marrow (Ma *et al.*, 1999). In the reconstituted mice, the numbers of progenitor cells were reduced in the bone marrow but the numbers were abnormally high in the circulation. CXCL12 and CXCR4 are required for human CD34$^+$ cells to engraft the bone marrow of mice with severe combined immunodeficiency (SCID) (Peled *et al.*, 1999b). Neutralizing antibodies to CXCR4 prevented the engraftment of CD34$^+$ cells into the bone marrow (Peled *et al.*, 1999b). CXCL12 stimulates human bone marrow CD34$^+$ cells so that they adhere to vascular endothelium via integrin activation under shear flow (Peled *et al.*, 1999a) and in a static condition (Hidalgo *et al.*, 2001). In this regard, CXCL12 activates integrins such as LFA-1, VLA-4, and VLA-5 on human CD34$^+$ cells (Peled *et al.*, 2000).

B. *Chemokines in the Mobilization of Bone Marrow Stem and Progenitor Cells into the Periphery*

The fact that mobilization of human and murine stem cells out of the bone marrow is inhibited by antibodies that neutralize CXCR4 or CXCL12 suggests that the CXCL12–CXCR4 axis regulates not only the homing but the mobilization of these cells (Ponomaryov *et al.*, 2000). Granulocyte colony-stimulating factor (G-CSF) is a well-known mobilizer of HSPCs. It appears that G-CSF mobilizes HSPCs by modulating the CXCL12–CXCR4 axis. G-CSF induces the release of neutrophil elastase and cathepsin G (Levesque *et al.*, 2002), which can degrade CXCL12 (Ponomaryov *et al.*, 2000; Valenzuela-Fernandez *et al.*, 2002; van Os *et al.*, 2002; Levesque *et al.*, 2003). Another protease, this one implicated in the mobilization of hematopoietic stem and progenitor cells, is CD26/DPPIV, which is a membrane-bound extracellular peptidase that cleaves dipeptides from the N-terminus of polypeptide chains after a proline or an alanine (Christopherson *et al.*, 2002). Interestingly, CD26 can selectively cleave dipeptides from the CXCL12 and CCL22 polypeptides but does not degrade many other chemokines (Lambeir *et al.*, 2001). A subset of human $CD34^+$ cord blood cells expresses CD26. N-terminal-truncated CXCL12, generated by CD26, not only failed to induce the migration of $CD34^+$ cells, but also inhibited the cell migration to intact CXCL12 (Christopherson *et al.*, 2002). In mice, CD26 is expressed on a subpopulation of Sca-1^+ c-kit^+ lin^- and Sca-1^+c-kit^- lin^- hematopoietic cells, isolated from the bone marrow (Christopherson *et al.*, 2003a). Treatment of mice with CD26 inhibitors during G-CSF-induced mobilization resulted in a reduction in mobilization of bone-marrow cells to the periphery (Christopherson *et al.*, 2003a). HSPC mobilization was also significantly reduced in G-CSF-treated $CD26^{-/-}$ mice versus wild-type mice (Christopherson *et al.*, 2003b), suggesting that CD26 plays an important role in G-CSF-induced mobilization of HSPCs. Matrix metalloproteinase-9 (MMP-9) is another protease implicated in the mobilization of HSPCs (Heissig *et al.*, 2002). CXCL12 is induced by DNA-damaging agents, such as ionizing irradiation, cyclophosphamide, and 5-fluorouracil, in immature osteoblasts and endothelial cells in the bone marrow (Ponomaryov *et al.*, 2000). Bone-marrow ablation by these agents induces CXCL12, and the induced CXCL12 upregulates MMP-9 expression. MMP-9, in turn, cleaves membrane-bound SLF and mobilizes stem/progenitor cells. It is known that other chemokines, such as IL-8, Gro-β, and MIP1-α, also induce HSPC mobilization. These chemokines release several proteases in a manner similar to that of G-CSF, leading to degradation of SDF-1/CXCR4, extracellular matrix molecules, and adhesion molecules (Pruijt *et al.*, 1999; Fibbe *et al.*, 2000; Lapidot and Petit, 2002; Pruijt *et al.*, 2002; Gazitt, 2004). The roles of CXCL12 and proteases in the homing and mobilization of HSPCs are summarized in Fig. 1.

III. CHEMOKINES IN B-CELL MIGRATION

A. *B-Cells Change Their Chemotactic Responsiveness During Maturation in the Bone Marrow and Secondary Lymphoid Tissues*

It is interesting that B-cells are made in the bone marrow and migrate to the periphery for functional maturation but that many migrate back to the bone marrow as plasma B-cells. Chemokines regulate the growth and trafficking of B-cells at various developmental stages. The role of CXCL12 in B-lymphopoiesis was discussed early in this article, and in this section we

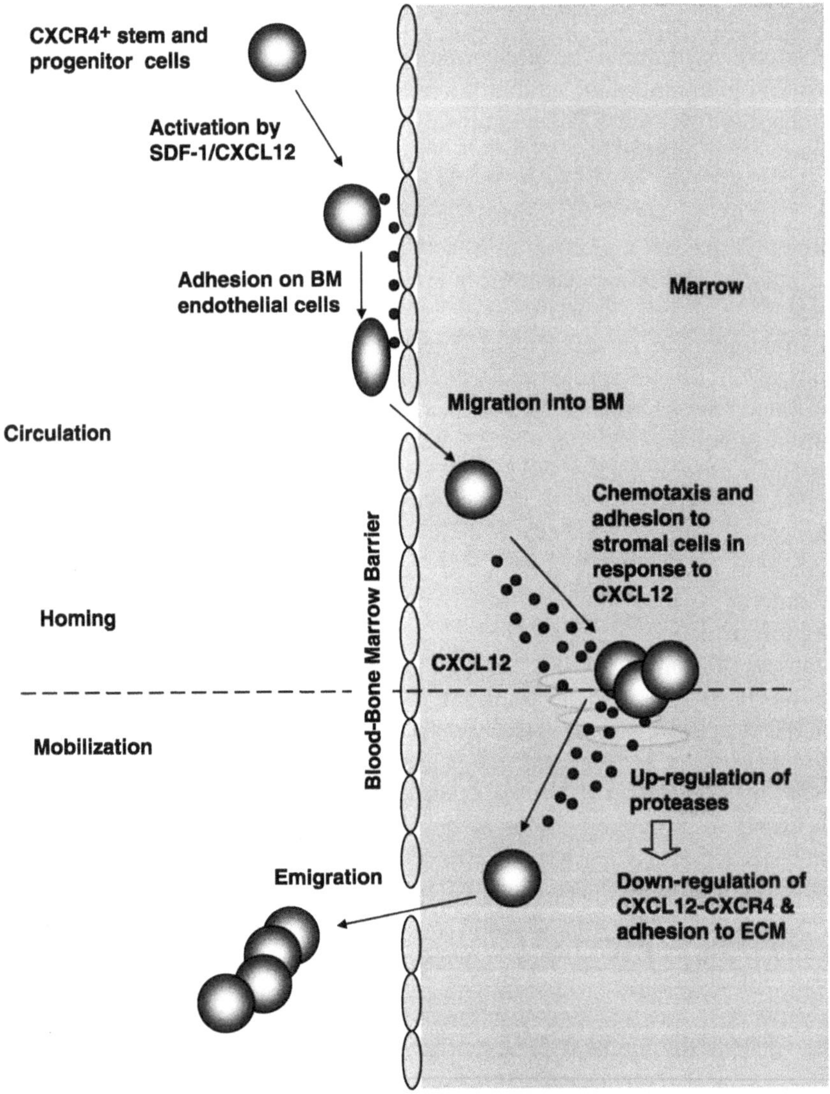

FIGURE 1 CXCL12, produced by bone-marrow stromal cells and endothelial cells, attracts hematopoietic stem and progenitor cells into the bone marrow and regulates the homing and mobilization of hematopoietic stem cells. By activating integrins, CXCL12 promotes the interaction between stem/progenitor cells and bone-marrow stromal cells. Mobilizing agents, such as G-CSF and IL-8, release stem/progenitor cells from the bone marrow by releasing or activating proteases (elastases, cathepsins, and CD26) from neutrophils or stem and progenitor cells. These proteases inactivate the CXCL12–CXCR4 axis and weaken the retention capacity of bone marrow. In addition, CXCL12, unlike many other chemokines, enhances the survival of stem and progenitor cells.

will focus our discussion on B-cell trafficking. B-cell precursors undergo phenotypically identifiable stages of differentiation in the bone marrow (pre-pro → pro → pre → immature B-cells) (Hardy and Hayakawa, 2001). D'Apuzzo *et al.* (1997) demonstrated that CXCL12 is a potent chemoattractant for pro- and pre-B-cells in the bone marrow but is inactive on more mature B-cells at later stages of development. Unexpectedly, mature B-cells express higher levels of CXCR4 than do uncommitted progenitors and pro-B-cells (Fedyk *et al.*, 1999; Honczarenko *et al.*, 1999), suggesting that the maturation-dependent unresponsiveness to CXCL12 is regulated by a mechanism or mechanisms other than the expression levels of CXCR4. In contrast, the chemotactic responsiveness to CCL19 and CCL21, the chemokines expressed in the T zone of secondary lymphoid tissues, increases as B-cell progenitors become more mature in the bone marrow (Kim and Broxmeyer, 1999a; Kim *et al.*, 1999). Bowman *et al.* (2000) made extensive investigation into the changes in chemotactic responsiveness during B-cell development in the mouse bone marrow and secondary lymphoid tissues. Bone-marrow pre-pro-B-cells migrate to CCL25/TECK. The migration response to CXCL13 is first displayed by a pro-B-cell subset, but it is lost at the pre-B-cell stage. As in humans, the responses to the CC chemokine receptor 7 (CCR7) ligands, CCL19 and CCL21, are upregulated before B-cells exit from the mouse marrow. Unlike in humans, chemotaxis to CXCL12 is observed at all stages of B-cell differentiation (Bowman *et al.*, 2000).

Naïve B-cells, after emigration out of the bone marrow, circulate through the blood system and secondary lymphoid tissues. Migrated into the secondary lymphoid tissues, they form primary follicles. Upon activation with antigens and activated T-cells at the T–B area border, naïve B-cells undergo activation and differentiation processes to become plasma B-cells. Some of the activated B-cells form germinal centers and further undergo somatic hypermutation and class switch recombination to become memory and plasma B-cells that are capable of producing class-switched and high-affinity antibodies. Germinal center B-cells do not chemotax to most chemokines and do not express key adhesion receptors such as L-selectin, $\alpha4\beta7$, and cutaneous lymphocyte antigen (CLA) (Butcher *et al.*, 1982; Reichert *et al.*, 1983; Bleul *et al.*, 1998; Corcione *et al.*, 2002). In contrast, memory B-cells respond to various chemotactic signals and express L-selectin, $\alpha4\beta7$ and CLA. As GC B-cells become memory B-cells, chemotactic responses to CXCL12, CXCL13, and CCL19 and adhesion receptors such as L-selectin, $\alpha4\beta7$ and CLA are upregulated (Butcher *et al.*, 1982; Reichert *et al.*, 1983). Stimulation of peripheral blood B-lymphocytes with lipopolysaccharide or anti-CD40 plus IL-4 resulted in marked enhancement of the migration response to key lymphoid tissue chemokines: CXCL12, CXCL13, CCL19, and CCL21 (Brandes *et al.*, 2000). CCR6 is expressed by naive and memory B-cells but is absent from germinal center B-cells. CCR6 expression is downregulated on activated B-cells, but it is re-expressed on memory B-cells (Krzysiek *et al.*, 2000). The process of naïve B-cells' migration to secondary lymphoid tissues to become plasma B-cells is summarized in Fig. 2.

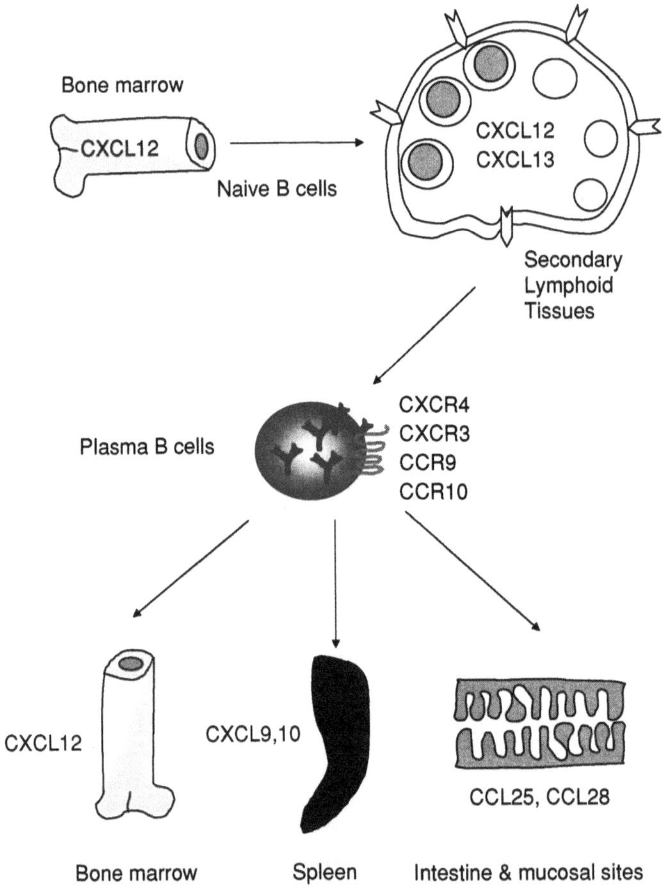

FIGURE 2 Naïve B-cells are generated in the bone marrow, where CXCL12 is highly expressed. CXCL12 acts as a pre-B-cell growth factor and keeps immature B-cells within the bone marrow. Naïve B-cells migrate to secondary lymphoid tissues in response to the chemokine ligands that bind CXCR4 and CXCR5. After migrating into secondary lymphoid tissues, B-cells migrate to and form follicles. After activation in response to antigens and T-cells in the T-cell-rich area and GC, B-cells become plasma cells. Plasma cells migrate mainly to bone marrow, spleen, intestine, and/or other mucosal tissues. Chemokines, such as CXCL12, CXCL9, CXCL10, CCL25, and CCL28, regulate the migration of plasma cells to these organs in a tissue-specific manner.

B. Chemokines That Regulate Plasma-Cell Migration

Plasma blasts, formed during memory immune responses, emigrate from the secondary lymphoid tissues to the bone marrow, intestine, and chronically inflamed tissues. Hargreaves *et al.* (2001) demonstrated that CXCL12 is expressed within splenic red pulp and lymph-node medullary cords as well as in the bone marrow. Plasma cells exhibit increased chemotactic sensitivity to CXCL12 but downregulate the expression of CXCR5 and CCR7. Furthermore, CXCR4-deficient plasma cells fail to accumulate normally in the bone marrow. Plasma blasts also gain both the expression of CXCR3 and the responsiveness to its chemokine ligands, CXCL9 and CXCL10 (Hauser *et al.*, 2002). Nakayama *et al.* (2003) comprehensively examined the expression of chemokine receptors by human plasma B-cells. Bone-marrow plasma cells express CXCR4, CXCR6, CCR10, and CCR3, and they respond to their respective ligands. In this regard, CXCL12 (the CXCR4 ligand), CXCL16 (the CXCR6 ligand), and CC chemokine ligand 28 (a CCR10 and CCR3 ligand) are expressed in the bone marrow. Plasma cells, depending on their status with regard to immunoglobulin class switch, can produce IgM, IgG, IgA, or IgE. CCL25, expressed in the thymus and small-intestinal crypts, is a potent and selective chemoattractant for a small subset of IgA-secreting B-cells in spleen, Peyer's patches, and mesenteric lymph node. Also, CCL28, expressed by the epithelia in diverse mucosal tissues, attracts IgA- but not IgG- or IgM-producing B-cells in the intestine, lungs, and lymph nodes (Lazarus *et al.*, 2003). In humans, CCR10, a receptor for CCL28, is selectively expressed by IgA-secreting cells in the salivary gland, small intestine, large intestine, appendix, and tonsils (Kunkel *et al.*, 2003). Migration of plasma B-cells is summarized in Fig. 2.

IV. CHEMOKINES AND T-CELLS

A. Chemokines and T-Cell Development in the Thymus

T-cell precursors, originating from the bone marrow, continuously repopulate the thymus in adult mice (Foss *et al.*, 2001). T-cell precursors undergo maturation and selection processes to become functional T-cells that do not respond to self-antigens. Initially, thymocytes do not express CD4 or CD8 and are, thus, double negative (DN). DN T-cells upregulate both CD4 and CD8 on the cell surface to become double positive (DP) cells. DP cells become either $CD4^+$ or $CD8^+$ single-positive (SP) T-cells. At the double-positive stage, thymocytes undergo massive proliferation and selection processes (Sebzda *et al.*, 1999). Positive selection refers to the process of rescuing major histocompatibility complex (MHC)-restricted thymocytes from apoptosis, and it allows the T-cells that recognize self-MHC molecules to survive. Negative selection refers to the deletion of autoreactive thymocytes that strongly bind self-peptides and MHC molecules. After the selection processes, only 3% of DP thymocytes become SP cells (Shortman *et al.*, 1990).

Thymocytes are located in different places in the thymus depending on their stage of differentiation (Prockop and Petrie, 2000; Lind *et al.*, 2001). It is thought that early T-cell precursors seed the thymus through the postcapillary venules located deep in the tissue, near the corticomedullary junction. DN cells are found in the subcapsular region. Differentiation of DN cells into DP cells correlates with their migration inward to the thymic medulla. SP cells are located in the medulla and migrate out to the blood system (a journey called emigration) through the same vessels in the corticomedullary junction as their progenitors used to migrate into the thymus.

In 1998, we reported for the first time that DN, DP, and SP thymocytes selectively migrate to different chemokines (Kim *et al.*, 1998b). CXCL12 showed chemotactic preference for immature DN and DP thymocytes over mature SP thymocytes. In contrast, CCL19 weakly attracts DN and DP thymocytes but strongly attracts SP thymocytes. The chemotactic responsiveness of thymocytes to CXCL12 and CCL19 increases after the positive selection process at the DP stage. It has been reported that CXCL12 is important for the development of early thymocytes (Hernandez-Lopez *et al.*, 2002). In this regard, CXCR4 is expressed by many thymocytes (Berkowitz *et al.*, 1998; Zaitseva *et al.*, 1998; Zamarchi *et al.*, 2002). Using chimeric human–mouse fetal thymus organ culture (FTOC), seeded with purified $CD34^+$ thymic progenitors and treated with neutralizing antibodies against CXCL12 or CXCR4, Hernandez-Lopez *et al.* found that the thymocyte differentiation from $CD34^+$ precursor cells to $CD4^+$ SP thymocytes was arrested. CCL19 and its receptor CCR7 play an important role in the emigration of mature SP thymocytes. Premature expression of CCR7 in transgenic mice repositioned DP cells into the medulla (Kwan and Killeen, 2004). Emigration of thymocytes in newborn mice was defective in CCL19-neutralized mice as well as in CCR7-deficient mice (Ueno *et al.*, 2002). In this regard, CCL19 is expressed in the medullary areas but not in the cortex (Annunziato *et al.*, 2000).

Chemokines such as CCL22 and CCL25 and their receptors CCR4 and CCR9 appear to play important roles in positioning T-cells in the thymus. Campbell *et al.* reported that CCL25, a CCR9 ligand, attracts DP and L-selectin^{low} SP thymocytes but not $CD4^+$ L-selectin^{h} emigrating SP cells (Campbell *et al.*, 1999). It has also been reported that CCL25 is expressed by thymic dendritic cells (Vicari *et al.*, 1997). CCR9 is expressed by all thymocyte subsets, but it is more upregulated following pre-TCR signaling at the transition of DN to DP cells (Norment *et al.*, 2000). A subset of $CD8^+$ SP thymocytes continues to express CCR9 after emigrating to the thymus, while $CD4^+$ SP thymocytes stop expressing CCR9 (Carramolino *et al.*, 2001; Uehara *et al.*, 2002). Interestingly, one-half of $\gamma\delta$ TCR^+ thymocytes express CCR9 on their surfaces (Uehara *et al.*, 2002). In the human, CCL25 efficiently attracts all the DN, DP, and SP thymocytes (Youn *et al.*, 1999). CCL22, another chemokine expressed in the medullary area, attracts DP cells that exhibit reduced levels of CD8 or memory $CD4^+$ cells infiltrating the thymus (Annunziato *et al.*, 2000). In this regard, it has also been reported that CCL22 attracts thymocytes during a brief period of development between the late cortical and early medullary stages (Campbell *et al.*, 1999). Figure 3 displays T-cell migration in the thymus as well as the chemokines important for this process.

B. Chemokines and T-Cell Migration in Secondary Lymphoid Tissues

Naïve T-cells, once they are generated in the thymus, migrate to secondary lymphoid tissues. Although many chemokines are expressed in the secondary lymphoid tissues, the three chemokines CCL19, CCL21, and CXCL13 are especially important in compartmentalizing these tissues. CCL21 is a chemokine highly expressed in secondary lymphoid tissues (Hedrick and Zlotnik, 1997; Nagira *et al.*, 1997, 1998; Tanabe *et al.*, 1997; Yoshie *et al.*, 1997; Gunn *et al.*, 1998b; Yoshida *et al.*, 1998b). It is specifically expressed in the high endothelial venules of lymph nodes and Peyer's patches and in the T-cell areas of spleen tissue, lymph nodes, and Peyer's patches (Gunn *et al.*, 1998b). CCL21 has the ability to arrest rolling $CCR7^+$ lymphocytes on endothelium by activating integrins (Campbell *et al.*, 1998; Pachynski *et al.*, 1998; Tangemann *et al.*, 1998; Willimann *et al.*, 1998). CCL19 is another chemokine highly expressed in the T-cell area of secondary lymphoid tissues (Rossi *et al.*, 1997; Yoshida *et al.*, 1997). CCL19 is specifically expressed by dendritic cells within the T-cell zone of secondary lymphoid tissues (Ngo *et al.*, 1998), and is regulated by lymphotoxins (Ngo *et al.*, 1999). While CCL19 and CCL21 are the chemokines marking the T-cell territory, CXCL13 is specifically expressed in the B-cell area (Gunn *et al.*, 1998a; Legler *et al.*, 1998). CXCL13 is expressed in the follicles of Peyer's patches, the spleen, and lymph nodes. It has been reported that follicular stromal cells (Ngo *et al.*, 1999), dendritic cells (Ishikawa *et al.*, 2001; Vissers *et al.*, 2001), and germinal-center T-cells (Kim *et al.*, 2004) are the cell types that can produce CXCL13. At the RNA level, CXCL13 is expressed in the

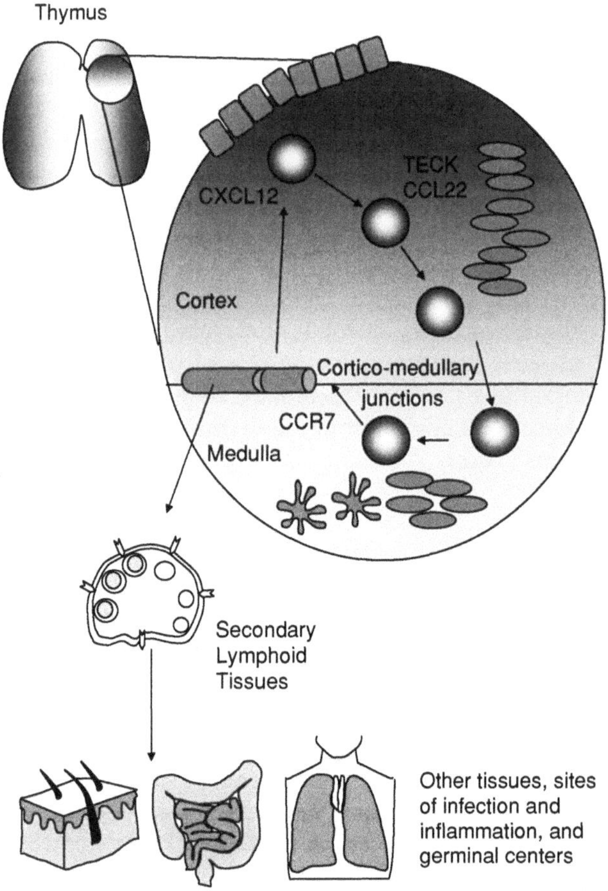

FIGURE 3 T-cell progenitors migrate into the thymus through postcapillary venules in the corticomedullary junctions. Primitive T-cell progenitors seed the cortex and migrate inward as they differentiate. Mature single-positive T-cells are located in the medulla and migrate out of the thymus through the same route (postcapillary venules in the corticomedullary junctions). CXCL12 may play a role in the thymus seeding by T-cell progenitors, and CCL25 and CCL22 may position developing T-cell progenitors in the thymic cortex and corticomedullary area. CCL19 and CCL21 play important roles in the emigration of mature naïve T-cells.

mantle zone and germinal centers (Ansel *et al.*, 1999). CXCL13 is a potent lymphoid tissue-inducing chemokine. Luther *et al.* (2000) made transgenic mice expressing CXCL13 in the pancreatic islets. As a result of this expression, the pancreas was transformed into lymph node–like structures that contained B- and T-cell zones, high endothelial venules, and stromal cells.

CCL19 and CCL21 are potent chemoattractants for naïve T-cells and many memory T-cells (Kim *et al.*, 1998a; Ngo *et al.*, 1998; Yoshida *et al.*, 1998a). In this regard, all naïve T-cells and the majority of memory T-cells express their receptor CCR7 (Sallusto *et al.*, 1999). In human peripheral blood, ~80% of memory T-cells express CCR7 (Kim *et al.*, 2001b). On the assumption that $CCR7^+$ T-cells migrate to secondary lymphoid tissues while $CCR7^-$ T-cells migrate to nonlymphoid tissues, Sallusto *et al.* (1999) hypothesized the presence of $CCR7^+$ central-memory T-cells and $CCR7^-$ effector-memory T-cells (Sallusto *et al.*, 1999). Some T-cells in the circulation and secondary lymphoid tissues express CXCR5. $CXCR5^+$ T-cells are generated rapidly upon priming—within a few days—and are located in the areas surrounding and within B-cell follicles (Breitfeld *et al.*, 2000; Schaerli *et al.*, 2000; Kim *et al.*, 2001c; Schaerli *et al.*, 2001; Kim *et al.*, 2003). Most $CXCR5^+$ T-cells co-express CCR7. However, some $CXCR5^+$ T-cells lose the expression of CCR7, and these are located in germinal centers (Kim *et al.*, 2001c). Interestingly, these germinal-center T-cells produce CXCL13 themselves (Kim *et al.*, 2004), which is important for the recruitment of $CXCR5^+$ B-cells, dendritic cells, and T-cells into germinal centers. Other chemokines, such as CXCL9, CXCL10, and CCL17, may play important roles in the interactions among T-cells, B-cells, and dendritic cells. In this regard, $CXCL9^{-/-}$ mice were defective in production of antibodies after immunization (Foti *et al.*, 1999; Lieberam and Forster, 1999; Schaniel *et al.*, 2000; Park *et al.*, 2002). Naïve T-cells switch their expression pattern of chemokine receptors from the "$CCR7^+$ naïve type" to diverse "memory/effector types" during their differentiation in the T-cell zone. Some downregulate CCR7, while many upregulate CXCR3, CCR4, CRTH2, CCR5, CXCR6, CCR2, CCR6, CCR9, CCR10, CXCR5, CX3CR1, and/or CR1, receptors implicated in the migration to effector sites. Figure 4 shows T-cell migration in secondary lymphoid tissues as well as the chemokines that regulate this process.

C. Chemokines and T-Cell Polarization

Many chemokines, such as CCL2-5, CXCL9, and CXCL10, have been identified as inducible genes, expressed in response to cell activation or inflammatory signals. Since 1996, many chemokines have been discovered by a different method: bioinformatics, which finds new chemokine-like genes deposited in DNA sequence databases rather than using classical cloning studies. It is now clear that some chemokines are constitutively expressed at many tissue sites (as described earlier in this chapter). These chemokines are expressed at certain tissue sites, especially primary (generative) and

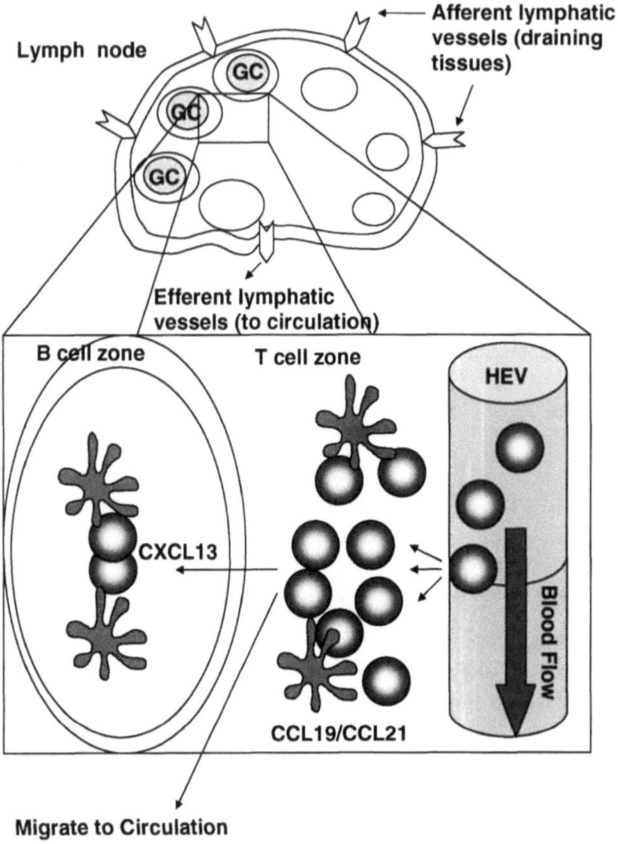

FIGURE 4 In secondary lymphoid tissues, $CCR7^+$ T-cells are located in the T-cell-rich area after they enter through high endothelial venules. CCL19 and CCL21 are highly expressed in the T-cell-rich area and recruit naïve T-cells and many memory T-cells. During antigen priming by dendritic cells, T-cells switch their chemokine receptors so that GC-homing $CXCR5^+$ $CCR7^-$ T-cells and nonlymphoid tissue–homing $CXCR5^-$ $CXCR3^{+/-}$ $CCR5^{+/-}$ $CCR4^{+/-}$ T-cells are generated. $CXCR5^+CCR7^-$ T-cells migrate to GC within the secondary lymphoid tissues, while $CXCR5^-$ $CXCR3^{+/-}$ $CCR5^{+/-}$ $CCR4^{+/-}$ T-cells migrate out of secondary lymphoid tissues and migrate to nonlymphoid tissue sites. In this regard, CXCL13 is highly expressed in GC, while CXCL9, CXCL10, CCL4, CCL5, CCL17, and CCL22 are expressed at sites of inflammation.

secondary lymphoid tissues such as thymus, bone marrow, lymph nodes, and intestinal sites. During infection or inflammation, many other chemokines are additionally induced to regulate leukocyte recruitment and immune responses.

It has been established that some of these inflammatory chemokines are induced by Th1 cytokines, while others are induced by Th2 cytokines.

CXCL10 (a CXCR3 ligand) was cloned from an IFN-γ-treated U937 cell line with monocytic characteristics (Luster *et al.*, 1985). Another ligand of CXCR3, CXCL9, is also induced by IFN-γ but not by IFN-α or β in a mouse macrophage cell line, RAW 264.7 (Farber, 1990). Another CXCR3 ligand was cloned from IFN-γ-activated primary human astrocytes and was named interferon-inducible T-cell α-chemoattractant (I-TAC) (now called CXCL11) (Cole *et al.*, 1998). On the other hand, chemokines such as CCL17 and CCL22 are induced in a variety of tissue sites in response to Th2 cytokines (Andrew *et al.*, 1998; Kumagai *et al.*, 2000; Sekiya *et al.*, 2000). Both CCL17 and CCR22 use CCR4 as their receptor (Imai *et al.*, 1997, 1998). Importantly, CXCR3 (the receptor for CXCL9, CXCL10, and CXCL11) is primarily expressed by Th1 cells. On the other hand, CCR4 is primarily expressed by Th2 cells and nonpolarized T-cells and is expressed by few Th1 cells (Annunziato *et al.*, 1998; Bonecchi *et al.*, 1998; Sallusto *et al.*, 1998; Kim *et al.*, 2001b). In addition, CCR5 and CXCR6 are preferentially expressed by Th1 cells (Sallusto *et al.*, 1997; Annunziato *et al.*, 1998; Bonecchi *et al.*, 1998; Sallusto *et al.*, 1998; Kim *et al.*, 2001a,b). Chemokine (or chemoattractant) receptors such as CCR3, CCR8, and CRTH2 are expressed by some Th2 cells, but they are expressed by few Th1 cells (Sallusto *et al.*, 1997; D'Ambrosio *et al.*, 1998; Zingoni *et al.*, 1998; Nagata *et al.*, 1999; Cosmi *et al.*, 2000). These chemokines and their receptors are thought to play important roles in the amplification of Th1 or Th2 inflammation by recruiting polarized T-cells. Recruited Th1 or Th2 cells produce cytokines, which further induce the production of the chemokines and then the recruitment of many more polarized T-cells (Fig. 5). The expression of CCR6, CCR7, CCR9, and CCR10 is not associated with T-cell polarization.

D. Shared Chemokine Receptor Expression by Immune Cells to Maximize Their Interaction at Effector Sites

Many chemokine receptors are shared by specialized T-cells and their downstream effector cells. The chemokine receptors, expressed by Th1 cells such as CCR1, CCR2, CCR5, CXCR3, and CXCR6, are often expressed by other immune cells, such as monocytes, macrophages, B-cells, and NK cells, which act as downstream effector cells for Th1 cells (Inngjerdingen *et al.*, 2001; Katschke *et al.*, 2001). On the other hand, Th2 chemokine receptors, such as CCR3, CCR4, CCR8, and CRTH2, are shared by Th2 cells, eosinophils, basophils, and/or mast cells. Chemokines, expressed at tissue sites of inflammation, recruit and collocalize Th1 cells, macrophages, and/or NK cells. Th1 cells produce IFN-γ, which in turn activates the collocalized macrophages and NK cells, a necessary step for clearance of pathogens.

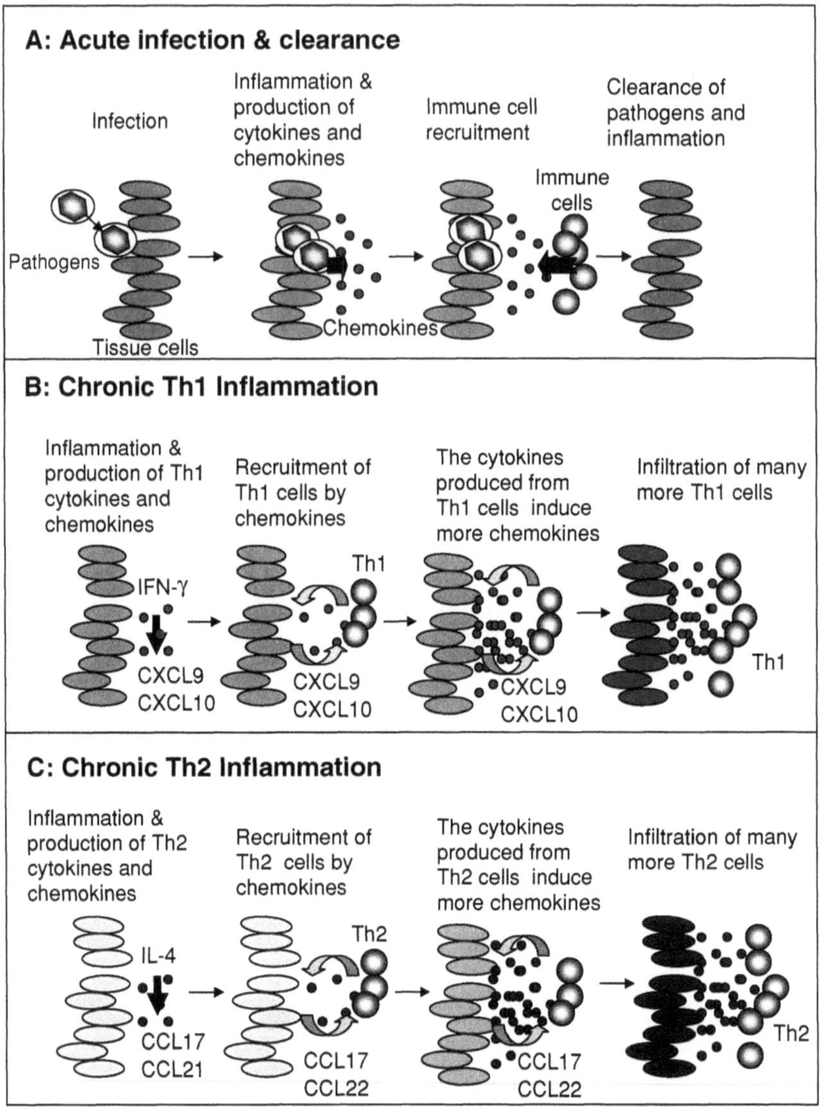

FIGURE 5 In most infections, chemokines are induced to attract effector cells, a process that is important for the clearance of pathogens. Once pathogens are cleared, inflammation is quickly resolved at tissue sites (A). In cases of autoimmune diseases and allergy, chronic inflammation responses are commonly polarized to Th1 or Th2. Chemokines play important roles in this polarization. Some chemokines, such as CXCL9 and CXCL10, are induced by IFN-γ, while CCL17 and CCL22 are induced by IL-4. Th1 cells express CXCR3, CCR5, and CXCR6, while Th2 cells express CCR4, CCR8, and CRTH2. Therefore, CXCL9 and CXCL10 (along with many other Th1 type chemokines) can mediate the recruitment of Th1 cells and the amplification of Th1 inflammation (B), while CCL17 and CCL22 (along with other Th2 chemokines) may mediate the recruitment of Th2 cells that lead to chronic Th2 inflammation (C).

Chemokines such as CCL1, CCL11, CCL17, and CCL22 attract and collocalize Th2 cells, eosinophils, and/or mast cells. Upon activation by antigens, Th2 cells produce cytokines that can activate the collocalized eosinophils and/or mast cells.

V. NONCHEMOTACTIC FUNCTIONS OF CHEMOKINES

By activating a number of signaling pathways, ones such as phosphoinositide 3-kinase (PI3K), janus kinase (JAK)–signal transducers and activators of transcription (STAT), focal adhesion kinases (FAKs) and mitogen-activated protein kinase (MAPK) (Wong and Fish, 2003), chemokines and their receptors modulate a number of biological responses in addition to cell trafficking. In an effort to search the factors that regulate the proliferation of hematopoietic stem and progenitor cells, Graham *et al.* (1990) discovered a reversible inhibitor of hematopoietic stem cell proliferation. This molecule was found to be a chemokine, CCL3. Broxmeyer *et al.* (1990) found that, when stimulated together by granulocyte–macrophage colony-stimulating factor (GM-CSF) or macrophage colony-stimulating factor (M-CSF), CCL3 can act as a growth factor for more mature granulocyte and macrophage progenitors. Overall, the suppressive effect of CCL3 on primitive stem and progenitor cells is dominant over the enhancing effect on more mature cells. *In vivo* administration of CCL3 was inhibitory to hematopoiesis, as evidenced by decreased cycling rates and absolute numbers of myeloid progenitor cells in the marrow and spleen (Dunlop *et al.*, 1992; Maze *et al.*, 1992). This inhibitory activity seems to be mediated in part by inducement of apoptosis in immature progenitor cells (Haneline *et al.*, 1998). Later it was found that many chemokines in addition to CCL3 have the suppressive activity (Broxmeyer and Kim, 1999). These are CCL1–3, CCL6, CCL9, CCL13, CCL15–16, CCL18, CCL20–21, CCL23–25, CXCL2, CXCL4–6, CXCL9–10, and CL1. Chemokines that do not have the suppressive activity include CCL4–5, CCL7–8, CCL11, CCL14, CCL17, CCL22, CXCL12, CXCL7, CXCL1, CXCL3, and CX3CL1. Consistent with the suppressive activity of the chemokines, mice deficient in CCR2 had greatly enhanced myelopoiesis (Boring *et al.*, 1997; Gao *et al.*, 1997). However, the myelopoiesis in CCR1-deficient mice was suppressed, suggesting that the effect of the suppressive activity of chemokines may not be mediated through CCR1, or, unexpectedly, that CCR1 may play a positive role in hematopoiesis (Gao *et al.*, 1997; Broxmeyer *et al.*, 1999).

Interestingly, CXCL12, a chemokine highly produced in the bone marrow, has no suppressive activity on hematopoietic stem and progenitor cells. Actually, CXCL12 enhances the survival of hematopoietic stem and progenitor cells. We generated transgenic mice expressing CXCL12 under the regulation of the Rous sarcoma virus long terminal repeat (RSV-LTR) promoter, and we found that transgenic myeloid progenitors expressing CXCL12 survive much better than their wild-type counterparts (Broxmeyer *et al.*, 2003a). Hematopoietic progenitors, isolated from the CXCL12-transgenic mice, displayed greatly increased cell cycling. In this regard, the progenitor cell numbers were more enhanced in the transgenic mice than the wild-type mice. When added exogenously, CXCL12 was effective in preventing the apoptosis of stem and progenitor cells *in vitro*. Pretreatment of stem and progenitor cells with CXCL12 greatly enhanced the engrafting capability of mouse long-term, marrow-competitive, and repopulating stem cells (Broxmeyer *et al.*, 2003b). Other chemokines have similar survival functions for various cell types. For example, CCL25 makes T-cells survive better (Youn *et al.*, 2001). In this case, CCL25 was protective against the apoptosis signals induced by FAS ligand or cycloheximide (a translation inhibitor).

Taub *et al.* reported that CCL2, CCL3, and CCL5 can costimulate T-cell activation (Wong and Fish, 2003). Since the receptors for CCL2, CCL3, and CCL5 are not expressed by naïve T-cells, these chemokines would not co-stimulate naïve T-cells in secondary lymphoid tissues. However, CXCL12, CCL19, and CCL21 may do so (Nanki and Lipsky, 2000), as their receptors CXCR4 and CCR7 are expressed by all naïve T-cells. In addition to the costimulation function, chemokines are known to regulate the T-cell polarization process to Th1 or Th2 cells. For example, it has been shown that CCL2/MCP-1 promotes IL-4 production and Th2 cell development when it was used with polyclonal T-cell activators to activate T-cells, while CCL3/MIP-1α promotes Th1 cell development (Lukacs *et al.*, 1997; Karpus *et al.*, 1998). Oral administration of antigens induced tolerance where CCL2 and IL-4 were induced in oral mucosa while IL-12 expression was reduced (Karpus *et al.*, 1998). Although the major receptor for CCL2 is CCR2, CCR2 promotes Th1 rather than Th2 responses. When challenged with Mycobacterium bovis antigens, CCR2$^{-/-}$ mice had a decrease in granuloma size and IFN-γ production in the draining lymph nodes (Boring *et al.*, 1997). The mechanism by which CCL2 regulates T-cell polarization is still unclear. It is possible that CCL2 may work on T-cells during T-cell priming in secondary lymphoid tissues. In this case, CCL2 may need to work through other unknown receptors, not CCR2, as naïve T-cells do not express CCR2. It is also possible that the polarizing effect of CCL2 may be indirectly mediated through other cell types, such as monocytes, macrophages, and/or dendritic cells. In this case, CCL2 may modulate these antigen-presenting cells so that they promote the Th2 response over that of Th1 response. It is also highly likely that the CCL2–CCR2 axis regulates the migration of antigen-presenting cells to sites of antigen priming and effector function.

VI. CONCLUDING REMARKS

The functions of chemokines for hematopoietic cells appear to be comprehensive, complicated, and redundant. Largely, chemokines regulate trafficking and activation/suppression of hematopoietic cells. The functions of chemokines in cell trafficking are mediated by inducement of chemotaxis and activation of integrins. Although there are over 40 chemokines signaling through ~18 different receptors, many chemokines perform unique functions in the positioning of specialized cell types in certain microenvironments. During their development and activation processes, hematopoietic cells switch their patterns of expressing chemokine receptors so that they can migrate out of one organ and into another. In the future, selective control of hematopoiesis, migration, and immune responses in certain microenvironments may be made possible, in part, by manipulation of the chemokine–chemokine receptor network.

Acknowledgments

This work has been supported by grants from the Leukemia and Lymphoma Society, the American Cancer Society (#IRG-58-006-44), the United States Department of Agriculture (IND73077AH), and the Leukemia Research Foundation.

References

Aiuti, A., Webb, I. J., Bleul, C., Springer, T., and Gutierrez-Ramos, J. C. (1997). The chemokine SDF-1 is a chemoattractant for human CD34+ hematopoietic progenitor cells and provides a new mechanism to explain the mobilization of CD34+ progenitors to peripheral blood. *J. Exp. Med.* **185,** 111–120.

Andrew, D. P., Chang, M. S., McNinch, J., Wathen, S. T., Rihanek, M., Tseng, J., Spellberg, J. P., and Elias, C. G., III (1998). STCP-1 (MDC) CC chemokine acts specifically on chronically activated Th2 lymphocytes and is produced by monocytes on stimulation with Th2 cytokines IL-4 and IL-13. *J. Immunol.* **161,** 5027–5038.

Annunziato, F., Galli, G., Cosmi, L., Romagnani, P., Manetti, R., Maggi, E., and Romagnani, S. (1998). Molecules associated with human Th1 or Th2 cells. *Eur. Cytokine Netw.* **9,** 12–16.

Annunziato, F., Romagnani, P., Cosmi, L., Beltrame, C., Steiner, B. H., Lazzeri, E., Raport, C. J., Galli, G., Manetti, R., Mavilia, C., Vanini, V., Chantry, D., Maggi, E., and Romagnani, S. (2000). Macrophage-derived chemokine and EBI1-ligand chemokine attract human thymocytes in different stage of development and are produced by distinct subsets of medullary epithelial cells: possible implications for negative selection. *J. Immunol.* **165,** 238–246.

Ansel, K. M., McHeyzer-Williams, L. J., Ngo, V. N., McHeyzer-Williams, M. G., and Cyster, J. G. (1999). *In vivo*-activated CD4 T cells upregulate CXC chemokine receptor 5 and reprogram their response to lymphoid chemokines. *J. Exp. Med.* **190,** 1123–1134.

Berkowitz, R. D., Beckerman, K. P., Schall, T. J., and McCune, J. M. (1998). CXCR4 and CCR5 expression delineates targets for HIV-1 disruption of T-cell differentiation. *J. Immunol.* **161,** 3702–3710.

Bleul, C. C., Schultze, J. L., and Springer, T. A. (1998). B lymphocyte chemotaxis regulated in association with microanatomic localization, differentiation state, and B-cell receptor engagement. *J. Exp. Med.* **187,** 753–762.

Bonecchi, R., Bianchi, G., Bordignon, P. P., D'Ambrosio, D., Lang, R., Borsatti, A., Sozzani, S., Allavena, P., Gray, P. A., Mantovani, A., and Sinigaglia, F. (1998). Differential expression of chemokine receptors and chemotactic responsiveness of type 1 T helper cells (Th1s) and Th2s. *J. Exp. Med.* **187,** 129–134.

Boring, L., Gosling, J., Chensue, S. W., Kunkel, S. L., Farese, R. V., Jr., Broxmeyer, H. E., and Charo, I. F. (1997). Impaired monocyte migration and reduced type 1 (Th1) cytokine responses in C-C chemokine receptor 2 knockout mice. *J. Clin. Invest.* **100,** 2552–2561.

Bowman, E. P., Campbell, J. J., Soler, D., Dong, Z., Manlongat, N., Picarella, D., Hardy, R. R., and Butcher, E. C. (2000). Developmental switches in chemokine response profiles during B cell differentiation and maturation. *J. Exp. Med.* **191,** 1303–1318.

Brandes, M., Legler, D. F., Spoerri, B., Schaerli, P., and Moser, B. (2000). Activation-dependent modulation of B lymphocyte migration to chemokines. *Int. Immunol.* **12,** 1285–1292.

Breitfeld, D., Ohl, L., Kremmer, E., Ellwart, J., Sallusto, F., Lipp, M., and Forster, R. (2000). Follicular B helper T-cells express CXC chemokine receptor 5, localize to B-cell follicles, and support immunoglobulin production. *J. Exp. Med.* **192,** 1545–1552.

Broxmeyer, H. E., and Kim, C. H. (1999). Regulation of hematopoiesis in a sea of chemokine family members with a plethora of redundant activities. *Exp. Hematol.* **27,** 1113–1123.

Broxmeyer, H. E., Cooper, S., Hangoc, G., Gao, J. L., and Murphy, P. M. (1999). Dominant myelopoietic effector functions mediated by chemokine receptor CCR1. *J. Exp. Med.* **189,** 1987–1992.

Broxmeyer, H. E., Cooper, S., Kohli, L., Hangoc, G., Lee, Y., Mantel, C., Clapp, D. W., and Kim, C. H. (2003a). Transgenic expression of stromal cell-derived factor-1/CXC chemokine ligand 12 enhances myeloid progenitor cell survival/antiapoptosis *in vitro* in response to growth factor withdrawal and enhances myelopoiesis *in vivo*. *J. Immunol.* **170,** 421–429.

Broxmeyer, H. E., Kohli, L., Kim, C. H., Lee, Y., Mantel, C., Cooper, S., Hangoc, G., Shaheen, M., Li, X., and Clapp, D. W. (2003b). Stromal cell-derived factor-1/CXCL12 directly enhances survival/antiapoptosis of myeloid progenitor cells through CXCR4 and G(alpha)i proteins and enhances engraftment of competitive, repopulating stem cells. *J. Leukoc. Biol.* **73,** 630–638.

Broxmeyer, H. E., Sherry, B., Lu, L., Cooper, S., Oh, K. O., Tekamp-Olson, P., Kwon, B. S., and Cerami, A. (1990). Enhancing and suppressing effects of recombinant murine macrophage inflammatory proteins on colony formation *in vitro* by bone marrow myeloid progenitor cells. *Blood* **76,** 1110–1116.

Butcher, E. C., Rouse, R. V., Coffman, R. L., Nottenburg, C. N., Hardy, R. R., and Weissman, I. L. (1982). Surface phenotype of Peyer's patch germinal center cells: Implications for the role of germinal centers in B-cell differentiation. *J. Immunol.* **129,** 2698–2707.

Campbell, J. J., Bowman, E. P., Murphy, K., Youngman, K. R., Siani, M. A., Thompson, D. A., Wu, L., Zlotnik, A., and Butcher, E. C. (1998). 6-C-kine (SLC), a lymphocyte adhesion-triggering chemokine expressed by high endothelium, is an agonist for the MIP-3beta receptor CCR7. *J. Cell Biol.* **141,** 1053–1059.

Campbell, J. J., Pan, J., and Butcher, E. C. (1999). Cutting edge: Developmental switches in chemokine responses during T-cell maturation. *J. Immunol.* **163,** 2353–2357.

Carramolino, L., Zaballos, A., Kremer, L., Villares, R., Martin, P., Ardavin, C., Martinez, A. C., and Marquez, G. (2001). Expression of CCR9 beta-chemokine receptor is modulated in thymocyte differentiation and is selectively maintained in CD8(+) T-cells from secondary lymphoid organs. *Blood* **97,** 850–857.

Christopherson, K. W., 2nd, Hangoc, G., and Broxmeyer, H. E. (2002). Cell surface peptidase CD26/dipeptidylpeptidase IV regulates CXCL12/stromal cell-derived factor-1 alpha-mediated chemotaxis of human cord blood CD34+ progenitor cells. *J. Immunol.* **169,** 7000–7008.

Christopherson, K. W., 2nd, Cooper, S., and Broxmeyer, H. E. (2003a). Cell surface peptidase CD26/DPPIV mediates G-CSF mobilization of mouse progenitor cells. *Blood* **101,** 4680–4686.

Christopherson, K. W., Cooper, S., Hangoc, G., and Broxmeyer, H. E. (2003b). CD26 is essential for normal G-CSF-induced progenitor cell mobilization as determined by CD26-/- mice. *Exp. Hematol.* **31,** 1126–1134.

Cole, K. E., Strick, C. A., Paradis, T. J., Ogborne, K. T., Loetscher, M., Gladue, R. P., Lin, W., Boyd, J. G., Moser, B., Wood, D. E., Sahagan, B. G., and Neote, K. (1998). Interferon-inducible T cell alpha chemoattractant (I-TAC): a novel non-ELR CXC chemokine with potent activity on activated T-cells through selective high affinity binding to CXCR3. *J. Exp. Med.* **187,** 2009–2021.

Corcione, A., Tortolina, G., Bonecchi, R., Battilana, N., Taborelli, G., Malavasi, F., Sozzani, S., Ottonello, L., Dallegri, F., and Pistoia, V. (2002). Chemotaxis of human tonsil B lymphocytes to CC chemokine receptor (CCR) 1, CCR2 and CCR4 ligands is restricted to non-germinal center cells. *Int. Immunol.* **14,** 883–892.

Cosmi, L., Annunziato, F., Galli, M. I. G., Maggi, R. M. E., Nagata, K., and Romagnani, S. (2000). CRTH2 is the most reliable marker for the detection of circulating human type 2 Th and type 2 T cytotoxic cells in health and disease. *Eur. J. Immunol.* **30,** 2972–2979.

D'Ambrosio, D., Iellem, A., Bonecchi, R., Mazzeo, D., Sozzani, S., Mantovani, A., and Sinigaglia, F. (1998). Selective up-regulation of chemokine receptors CCR4 and CCR8 upon activation of polarized human type 2 Th cells. *J. Immunol.* **161,** 5111–5115.

D'Apuzzo, M., Rolink, A., Loetscher, M., Hoxie, J. A., Clark-Lewis, I., Melchers, F., Baggiolini, M., and Moser, B. (1997). The chemokine SDF-1, stromal cell-derived factor 1, attracts early stage B-cell precursors via the chemokine receptor CXCR4. *Eur. J. Immunol.* **27,** 1788–1793.

Dunlop, D. J., Wright, E. G., Lorimore, S., Graham, G. J., Holyoake, T., Kerr, D. J., Wolpe, S. D., and Pragnell, I. B. (1992). Demonstration of stem cell inhibition and myeloprotective effects of SCI/rhMIP1 alpha *in vivo*. *Blood* **79,** 2221–2225.

Egawa, T., Kawabata, K., Kawamoto, H., Amada, K., Okamoto, R., Fujii, N., Kishimoto, T., Katsura, Y., and Nagasawa, T. (2001). The earliest stages of B-cell development require a chemokine stromal cell-derived factor/pre-B-cell growth-stimulating factor. *Immunity* **15,** 323–334.

Farber, J. M. (1990). A macrophage mRNA selectively induced by gamma-interferon encodes a member of the platelet factor 4 family of cytokines. *Proc. Natl. Acad. Sci. USA* **87,** 5238–5242.

Fedyk, E. R., Ryyan, D. H., Ritterman, I., and Springer, T. A. (1999). Maturation decreases responsiveness of human bone marrow B lineage cells to stromal-derived factor 1 (SDF-1). *J. Leukoc. Biol.* **66,** 667–673.

Fibbe, W. E., Pruijt, J. F., van Kooyk, Y., Figdor, C. G., Opdenakker, G., and Willemze, R. (2000). The role of metalloproteinases and adhesion molecules in interleukin-8-induced stem-cell mobilization. *Semin. Hematol.* **37,** 19–24.

Foss, D. L., Donskoy, E., and Goldschneider, I. (2001). The importation of hematogenous precursors by the thymus is a gated phenomenon in normal adult mice. *J. Exp. Med.* **193,** 365–374.

Foti, M., Granucci, F., Aggujaro, D., Liboi, E., Luini, W., Minardi, S., Mantovani, A., Sozzani, S., and Ricciardi-Castagnoli, P. (1999). Upon dendritic cell (DC) activation chemokines and chemokine receptor expression are rapidly regulated for recruitment and maintenance of DC at the inflammatory site. *Int. Immunol.* **11,** 979–986.

Gao, J. L., Wynn, T. A., Chang, Y., Lee, E. J., Broxmeyer, H. E., Cooper, S., Tiffany, H. L., Westphal, H., Kwon-Chung, J., and Murphy, P. M. (1997). Impaired host defense, hematopoiesis, granulomatous inflammation and type 1-type 2 cytokine balance in mice lacking CC chemokine receptor 1. *J. Exp. Med.* **185,** 1959–1968.

Gazitt, Y. (2004). Homing and mobilization of hematopoietic stem cells and hematopoietic cancer cells are mirror image processes, utilizing similar signaling pathways and occurring concurrently: Circulating cancer cells constitute an ideal target for concurrent treatment with chemotherapy and antilineage-specific antibodies. *Leukemia* **18,** 1–10.

Graham, G. J., Wright, E. G., Hewick, R., Wolpe, S. D., Wilkie, N. M., Donaldson, D., Lorimore, S., and Pragnell, I. B. (1990). Identification and characterization of an inhibitor of haemopoietic stem cell proliferation. *Nature* **344,** 442–444.

Gunn, M. D., Ngo, V. N., Ansel, K. M., Ekland, E. H., Cyster, J. G., and Williams, L. T. (1998a). A B-cell-homing chemokine made in lymphoid follicles activates Burkitt's lymphoma receptor-1. *Nature* **391,** 799–803.

Gunn, M. D., Tangemann, K., Tam, C., Cyster, J. G., Rosen, S. D., and Williams, L. T. (1998b). A chemokine expressed in lymphoid high endothelial venules promotes the adhesion and chemotaxis of naive T lymphocytes. *Proc. Natl. Acad. Sci. USA* **95,** 258–263.

Haneline, L. S., Broxmeyer, H. E., Cooper, S., Hangoc, G., Carreau, M., Buchwald, M., and Clapp, D. W. (1998). Multiple inhibitory cytokines induce deregulated progenitor growth and apoptosis in hematopoietic cells from Fac-/- mice. *Blood* **91,** 4092–4098.

Hardy, R. R., and Hayakawa, K. (2001). B-cell development pathways. *Annu. Rev. Immunol.* **19,** 595–621.

Hargreaves, D. C., Hyman, P. L., Lu, T. T., Ngo, V. N., Bidgol, A., Suzuki, G., Zou, Y. R., Littman, D. R., and Cyster, J. G. (2001). A coordinated change in chemokine responsiveness guides plasma cell movements. *J. Exp. Med.* **194,** 45–56.

Hauser, A. E., Debes, G. F., Arce, S., Cassese, G., Hamann, A., Radbruch, A., and Manz, R. A. (2002). Chemotactic responsiveness toward ligands for CXCR3 and CXCR4 is regulated on plasma blasts during the time course of a memory immune response. *J. Immunol.* **169,** 1277–1282.

Hedrick, J. A., and Zlotnik, A. (1997). Identification and characterization of a novel beta chemokine containing six conserved cysteines. *J. Immunol.* **159,** 1589–1593.

Heissig, B., Hattori, K., Dias, S., Friedrich, M., Ferris, B., Hackett, N. R., Crystal, R. G., Besmer, P., Lyden, D., and Moore, M. A. (2002). Recruitment of stem and progenitor cells from the bone marrow niche requires MMP-9 mediated release of kit-ligand. *Cell* **109,** 625–637.

Hernandez-Lopez, C., Varas, A., Sacedon, R., Jimenez, E., Munoz, J. J., Zapata, A. G., and Vicente, A. (2002). Stromal cell–derived factor 1/CXCR4 signaling is critical for early human T-cell development. *Blood* **99,** 546–554.

Hidalgo, A., Sanz-Rodriguez, F., Rodriguez-Fernandez, J. L., Albella, B., Blaya, C., Wright, N., Cabanas, C., Prosper, F., Gutierrez-Ramos, J. C., and Teixido, J. (2001). Chemokine stromal cell-derived factor-1alpha modulates VLA-4 integrin-dependent adhesion to fibronectin and VCAM-1 on bone marrow hematopoietic progenitor cells. *Exp. Hematol.* **29,** 345–355.

Honczarenko, M., Douglas, R. S., Mathias, C., Lee, B., Ratajczak, M. Z., and Silberstein, L. E. (1999). SDF-1 responsiveness does not correlate with CXCR4 expression levels of developing human bone marrow B cells. *Blood* **94,** 2990–2998.

Imai, T., Baba, M., Nishimura, M., Kakizaki, M., Takagi, S., and Yoshie, O. (1997). The T cell-directed CC chemokine TARC is a highly specific biological ligand for CC chemokine receptor 4. *J. Biol. Chem.* **272,** 15036–15042.

Imai, T., Chantry, D., Raport, C. J., Wood, C. L., Nishimura, M., Godiska, R., Yoshie, O., and Gray, P. W. (1998). Macrophage-derived chemokine is a functional ligand for the CC chemokine receptor 4. *J. Biol. Chem.* **273,** 1764–1768.

Inngjerdingen, M., Damaj, B., and Maghazachi, A. A. (2001). Expression and regulation of chemokine receptors in human natural killer cells. *Blood* **97,** 367–375.

Ishikawa, S., Sato, T., Abe, M., Nagai, S., Onai, N., Yoneyama, H., Zhang, Y., Suzuki, T., Hashimoto, S., and Shirai, T. (2001). Aberrant high expression of B lymphocyte chemokine (BLC/CXCL13) by C11b+CD11c+ dendritic cells in murine lupus and preferential chemotaxis of B1 cells towards BLC. *J. Exp. Med.* **193,** 1393v1402.

Karpus, W. J., Kennedy, K. J., Kunkel, S. L., and Lukacs, N. W. (1998). Monocyte chemotactic protein 1 regulates oral tolerance induction by inhibition of T helper cell 1-related cytokines. *J. Exp. Med.* **187,** 733–741.

Katschke, K. J., Jr., Rottman, J. B., Ruth, J. H., Qin, S., Wu, L., LaRosa, G., Ponath, P., Park, C. C., Pope, R. M., and Koch, A. E. (2001). Differential expression of chemokine receptors on peripheral blood, synovial fluid, and synovial tissue monocytes/macrophages in rheumatoid arthritis. *Arthritis. Rheum.* **44,** 1022–1032.

Kawabata, K., Ujikawa, M., Egawa, T., Kawamoto, H., Tachibana, K., Iizasa, H., Katsura, Y., Kishimoto, T., and Nagasawa, T. (1999). A cell-autonomous requirement for CXCR4 in long-term lymphoid and myeloid reconstitution. *Proc. Natl. Acad. Sci. USA* **96,** 5663–5667.

Kim, C. H., and Broxmeyer, H. E. (1998). *In vitro* behavior of hematopoietic progenitor cells under the influence of chemoattractants: Stromal cell-derived factor-1, steel factor, and the bone marrow environment. *Blood* **91,** 100–110.

Kim, C. H., and Broxmeyer, H. E. (1999a). Chemokines: Signal lamps for trafficking of T and B cells for development and effector function. *J. Leukoc. Biol.* **65,** 6–15.

Kim, C. H., and Broxmeyer, H. E. (1999b). Thrombopoietin and interleukin-3 are chemotactic and chemokinetic chemoattractants for a factor-dependent hematopoietic progenitor cell line. *Ann. New York Acad. Sci.* **872,** 395–398.

Kim, C. H., Pelus, L. M., White, J. R., Applebaum, E., Johanson, K., and Broxmeyer, H. E. (1998a). CK beta-11/macrophage inflammatory protein-3 beta/EBI1-ligand chemokine is an efficacious chemoattractant for T and B cells. *J. Immunol.* **160,** 2418–2424.

Kim, C. H., Pelus, L. M., White, J. R., and Broxmeyer, H. E. (1998b). Differential chemotactic behavior of developing T cells in response to thymic chemokines. *Blood* **91,** 4434–4443.

Kim, C. H., Pelus, L. M., Appelbaum, E., Johanson, K., Anzai, N., and Broxmeyer, H. E. (1999). CCR7 ligands, SLC/6Ckine/Exodus2/TCA4 and CKbeta-11/MIP-3beta/ELC, are chemoattractants for CD56(+)CD16(−) NK cells and late stage lymphoid progenitors. *Cell Immunol.* **193,** 226–235.

Kim, C. H., Kunkel, E. J., Boisvert, J., Johnston, B., Campbell, J. J., Genovese, M. C., Greenberg, H. B., and Butcher, E. C. (2001a). Bonzo/CXCR6 expression defines type 1-polarized T-cell subsets with extralymphoid tissue homing potential. *J. Clin. Invest.* **107,** 595–601.

Kim, C. H., Rott, L., Kunkel, E. J., Genovese, M. C., Andrew, D. P., Wu, L., and Butcher, E. C. (2001b). Rules of chemokine receptor association with T cell polarization *in vivo*. *J. Clin. Invest.* **108,** 1331–1339.

Kim, C. H., Rott, L. S., Clark-Lewis, I., Campbell, D. J., Wu, L., and Butcher, E. C. (2001c). Subspecialization of CXCR5+ T cells: B helper activity is focused in a germinal center-localized subset of CXCR5+ T cells. *J. Exp. Med.* **193,** 1373–1381.

Kim, C. H., Nagata, K., and Butcher, E. C. (2003). Dendritic cells support sequential reprogramming of chemoattractant receptor profiles during naive to effector T cell differentiation. *J. Immunol.* **171,** 152–158.

Kim, C. H., Lim, H. W., Kim, J. R., Rott, L., Hillsamer, P., and Butcher, E. C. (2004). Unique gene expression program of human germinal center T helper cells. *Blood* **104,** 1952–1960.

Krzysiek, R., Lefevre, E. A., Bernard, J., Foussat, A., Galanaud, P., Louache, F., and Richard, Y. (2000). Regulation of CCR6 chemokine receptor expression and responsiveness to macrophage inflammatory protein-3alpha/CCL20 in human B cells. *Blood* **96,** 2338–2345.

Kumagai, N., Fukuda, K., and Nishida, T. (2000). Synergistic effect of TNF-alpha and IL-4 on the expression of thymus- and activation-regulated chemokine in human corneal fibroblasts. *Biochem. Biophys. Res. Commun.* **279,** 1–5.

Kunkel, E. J., Kim, C. H., Lazarus, N. H., Vierra, M. A., Soler, D., Bowman, E. P., and Butcher, E. C. (2003). CCR10 expression is a common feature of circulating and mucosal epithelial tissue IgA Ab-secreting cells. *J. Clin. Invest.* **111,** 1001–1010.

Kwan, J., and Killeen, N. (2004). CCR7 directs the migration of thymocytes into the thymic medulla. *J. Immunol.* **172,** 3999–4007.

Lambeir, A. M., Proost, P., Durinx, C., Bal, G., Senten, K., Augustyns, K., Scharpe, S., Van Damme, J., and De Meester, I. (2001). Kinetic investigation of chemokine truncation by CD26/dipeptidyl peptidase IV reveals a striking selectivity within the chemokine family. *J. Biol. Chem.* **276,** 29839–29845.

Lapidot, T., and Petit, I. (2002). Current understanding of stem cell mobilization: The roles of chemokines, proteolytic enzymes, adhesion molecules, cytokines, and stromal cells. *Exp. Hematol.* **30,** 973–981.

Lazarus, N. H., Kunkel, E. J., Johnston, B., Wilson, E., Youngman, K. R., and Butcher, E. C. (2003). A common mucosal chemokine (mucosae-associated epithelial chemokine/CCL28) selectively attracts IgA plasmablasts. *J. Immunol.* **170,** 3799–3805.

Legler, D. F., Loetscher, M., Roos, R. S., Clark-Lewis, I., Baggiolini, M., and Moser, B. (1998). B cell-attracting chemokine 1, a human CXC chemokine expressed in lymphoid tissues, selectively attracts B lymphocytes via BLR1/CXCR5. *J. Exp. Med.* **187,** 655–660.

Levesque, J. P., Hendy, J., Takamatsu, Y., Simmons, P. J., and Bendall, L. J. (2003). Disruption of the CXCR4/CXCL12 chemotactic interaction during hematopoietic stem cell mobilization induced by GCSF or cyclophosphamide. *J. Clin. Invest.* **111,** 187–196.

Levesque, J. P., Hendy, J., Takamatsu, Y., Williams, B., Winkler, I. G., and Simmons, P. J. (2002). Mobilization by either cyclophosphamide or granulocyte colony-stimulating factor transforms the bone marrow into a highly proteolytic environment. *Exp. Hematol.* **30,** 440–449.

Lieberam, I., and Forster, I. (1999). The murine beta-chemokine TARC is expressed by subsets of dendritic cells and attracts primed CD4+ T cells. *Eur. J. Immunol.* **29,** 2684–2694.

Lind, E. F., Prockop, S. E., Porritt, H. E., and Petrie, H. T. (2001). Mapping precursor movement through the postnatal thymus reveals specific microenvironments supporting defined stages of early lymphoid development. *J. Exp. Med.* **194,** 127–134.

Lukacs, N. W., Chensue, S. W., Karpus, W. J., Lincoln, P., Keefer, C., Strieter, R. M., and Kunkel, S. L. (1997). C-C chemokines differentially alter interleukin-4 production from lymphocytes. *Am. J. Pathol.* **150,** 1861–1868.

Luster, A. D., Unkeless, J. C., and Ravetch, J. V. (1985). Gamma-interferon transcriptionally regulates an early-response gene containing homology to platelet proteins. *Nature* **315,** 672–676.

Luther, S. A., Lopez, T., Bai, W., Hanahan, D., and Cyster, J. G. (2000). BLC expression in pancreatic islets causes B cell recruitment and lymphotoxin-dependent lymphoid neogenesis. *Immunity* **12,** 471–481.

Ma, Q., Jones, D., Borghesani, P. R., Segal, R. A., Nagasawa, T., Kishimoto, T., Bronson, R. T., and Springer, T. A. (1998). Impaired B-lymphopoiesis, myelopoiesis, and derailed cerebellar neuron migration in CXCR4- and SDF-1-deficient mice. *Proc. Natl. Acad. Sci. USA* **95,** 9448–9453.

Ma, Q., Jones, D., and Springer, T. A. (1999). The chemokine receptor CXCR4 is required for the retention of B lineage and granulocytic precursors within the bone marrow microenvironment. *Immunity* **10,** 463–471.

Maze, R., Sherry, B., Kwon, B. S., Cerami, A., and Broxmeyer, H. E. (1992). Myelosuppressive effects *in vivo* of purified recombinant murine macrophage inflammatory protein-1 alpha. *J. Immunol.* **149,** 1004–1009.

Nagasawa, T., Kikutani, H., and Kishimoto, T. (1994). Molecular cloning and structure of a pre-B-cell growth-stimulating factor. *Proc. Natl. Acad. Sci. USA* **91,** 2305–2309.

Nagasawa, T., Hirota, S., Tachibana, K., Takakura, N., Nishikawa, S., Kitamura, Y., Yoshida, N., Kikutani, H., and Kishimoto, T. (1996). Defects of B-cell lymphopoiesis and bone-marrow myelopoiesis in mice lacking the CXC chemokine PBSF/SDF-1. *Nature* **382,** 635–638.

Nagata, K., Tanaka, K., Ogawa, K., Kemmotsu, K., Imai, T., Yoshie, O., Abe, H., Tada, K., Nakamura, M., Sugamura, K., and Takano, S. (1999). Selective expression of a novel surface molecule by human Th2 cells *in vivo*. *J. Immunol.* **162,** 1278–1286.

Nagira, M., Imai, T., Yoshida, R., Takagi, S., Iwasaki, M., Baba, M., Tabira, Y., Akagi, J., Nomiyama, H., and Yoshie, O. (1998). A lymphocyte-specific CC chemokine, secondary lymphoid tissue chemokine (SLC), is a highly efficient chemoattractant for B cells and activated T cells. *Eur. J. Immunol.* **28,** 1516–1523.

Nagira, M., Imai, T., Hieshima, K., Kusuda, J., Ridanpaa, M., Takagi, S., Nishimura, M., Kakizaki, M., Nomiyama, H., and Yoshie, O. (1997). Molecular cloning of a novel human CC chemokine secondary lymphoid-tissue chemokine that is a potent chemoattractant for lymphocytes and mapped to chromosome 9p13. *J. Biol. Chem.* **272,** 19518–19524.

Nakayama, T., Hieshima, K., Izawa, D., Tatsumi, Y., Kanamaru, A., and Yoshie, O. (2003). Cutting edge: Profile of chemokine receptor expression on human plasma cells accounts for their efficient recruitment to target tissues. *J. Immunol.* **170,** 1136–1140.

Nanki, T., and Lipsky, P. E. (2000). Cutting edge: Stromal cell–derived factor-1 is a costimulator for CD4+ T cell activation. *J. Immunol.* **164,** 5010–5014.

Ngo, V. N., Tang, H. L., and Cyster, J. G. (1998). Epstein–Barr virus–induced molecule 1 ligand chemokine is expressed by dendritic cells in lymphoid tissues and strongly attracts naive T cells and activated B cells. *J. Exp. Med.* **188,** 181–191.

Ngo, V. N., Korner, H., Gunn, M. D., Schmidt, K. N., Riminton, D. S., Cooper, M. D., Browning, J. L., Sedgwick, J. D., and Cyster, J. G. (1999). Lymphotoxin alpha/beta and tumor necrosis factor are required for stromal cell expression of homing chemokines in B and T cell areas of the spleen. *J. Exp. Med.* **189,** 403–412.

Norment, A. M., Bogatzki, L. Y., Gantner, B. N., and Bevan, M. J. (2000). Murine CCR9, a chemokine receptor for thymus-expressed chemokine that is up-regulated following pre-TCR signaling. *J. Immunol.* **164,** 639–648.

Pachynski, R. K., Wu, S. W., Gunn, M. D., and Erle, D. J. (1998). Secondary lymphoid-tissue chemokine (SLC) stimulates integrin alpha 4 beta 7-mediated adhesion of lymphocytes to mucosal addressin cell adhesion molecule-1 (MAdCAM-1) under flow. *J. Immunol.* **161,** 952–956.

Park, M. K., Amichay, D., Love, P., Wick, E., Liao, F., Grinberg, A., Rabin, R. L., Zhang, H. H., Gebeyehu, S., Wright, T. M., Iwasaki, A., Weng, Y., DeMartino, J. A., Elkins, K. L., and Farber, J. M. (2002). The CXC chemokine murine monokine induced by IFN-gamma (CXC chemokine ligand 9) is made by APCs, targets lymphocytes including activated B cells, and supports antibody responses to a bacterial pathogen *in vivo. J. Immunol.* **169,** 1433–1443.

Peled, A., Grabovsky, V., Habler, L., Sandbank, J., Arenzana-Seisdedos, F., Petit, I., Ben-Hur, H., Lapidot, T., and Alon, R. (1999a). The chemokine SDF-1 stimulates integrin-mediated arrest of CD34(+) cells on vascular endothelium under shear flow. *J. Clin. Invest.* **104,** 1199–1211.

Peled, A., Petit, I., Kollet, O., Magid, M., Ponomaryov, T., Byk, T., Nagler, A., Ben-Hur, H., Many, A., Shultz, L., *et al.* (1999b). Dependence of human stem cell engraftment and repopulation of NOD/SCID mice on CXCR4. *Science* **283,** 845–848.

Peled, A., Kollet, O., Ponomaryov, T., Petit, I., Franitza, S., Grabovsky, V., Slav, M. M., Nagler, A., Lider, O., Alon, R., *et al.* (2000). The chemokine SDF-1 activates the integrins LFA-1, VLA-4, and VLA-5 on immature human CD34(+) cells: Role in transendothelial/stromal migration and engraftment of NOD/SCID mice. *Blood* **95,** 3289–3296.

Ponomaryov, T., Peled, A., Petit, I., Taichman, R. S., Habler, L., Sandbank, J., Arenzana-Seisdedos, F., Magerus, A., Caruz, A., Fujii, N., *et al.* (2000). Induction of the chemokine stromal-derived factor-1 following DNA damage improves human stem cell function. *J. Clin. Invest.* **106,** 1331–1339.

Prockop, S., and Petrie, H. T. (2000). Cell migration and the anatomic control of thymocyte precursor differentiation. *Semin. Immunol.* **12,** 435–444.

Pruijt, J. F., Fibbe, W. E., Laterveer, L., Pieters, R. A., Lindley, I. J., Paemen, L., Masure, S., Willemze, R., and Opdenakker, G. (1999). Prevention of interleukin-8-induced mobilization of hematopoietic progenitor cells in rhesus monkeys by inhibitory antibodies against the metalloproteinase gelatinase B (MMP-9). *Proc. Natl. Acad. Sci. USA* **96,** 10863–10868.

Pruijt, J. F., Verzaal, P., van Os, R., de Kruijf, E. J., van Schie, M. L., Mantovani, A., Vecchi, A., Lindley, I. J., Willemze, R., Starckx, S., Opdenakker, G., and Fibbe, W. E. (2002). Neutrophils are indispensable for hematopoietic stem cell mobilization induced by interleukin-8 in mice. *Proc. Natl. Acad. Sci. USA* **99,** 6228–6233.

Reichert, R. A., Gallatin, W. M., Weissman, I. L., and Butcher, E. C. (1983). Germinal center B cells lack homing receptors necessary for normal lymphocyte recirculation. *J. Exp. Med.* **157,** 813–827.

Rossi, D. L., Vicari, A. P., Franz-Bacon, K., McClanahan, T. K., and Zlotnik, A. (1997). Identification through bioinformatics of two new macrophage proinflammatory human chemokines: MIP-3alpha and MIP-3beta. *J. Immunol.* **158,** 1033–1036.

Sallusto, F., Mackay, C. R., and Lanzavecchia, A. (1997). Selective expression of the eotaxin receptor CCR3 by human T helper 2 cells. *Science* **277,** 2005–2007.

Sallusto, F., Lenig, D., Mackay, C. R., and Lanzavecchia, A. (1998). Flexible programs of chemokine receptor expression on human polarized T helper 1 and 2 lymphocytes. *J. Exp. Med.* **187,** 875–883.

Sallusto, F., Palermo, B., Lenig, D., Miettinen, M., Matikainen, S., Julkunen, I., Forster, R., Burgstahler, R., Lipp, M., and Lanzavecchia, A. (1999). Distinct patterns and kinetics of chemokine production regulate dendritic cell function. *Eur. J. Immunol.* **29,** 1617–1625.

Schaerli, P., Willimann, K., Lang, A. B., Lipp, M., Loetscher, P., and Moser, B. (2000). CXC chemokine receptor 5 expression defines follicular homing T cells with B cell helper function. *J. Exp. Med.* **192,** 1553–1562.

Schaerli, P., Loetscher, P., and Moser, B. (2001). Cutting edge: Induction of follicular homing precedes effector Th cell development. *J. Immunol.* **167,** 6082–6086.

Schaniel, C., Melchers, F., and Rolink, A. G. (2000). The cluster of ABCD chemokines which organizes T cell-dependent B cell responses. *Curr. Top Microbiol. Immunol.* **251,** 181–189.

Sebzda, E., Mariathasan, S., Ohteki, T., Jones, R., Bachmann, M. F., and Ohashi, P. S. (1999). Selection of the T cell repertoire. *Annu. Rev. Immunol.* **17,** 829–874.

Sekiya, T., Miyamasu, M., Imanishi, M., Yamada, H., Nakajima, T., Yamaguchi, M., Fujisawa, T., Pawankar, R., Sano, Y., Ohta, K., Ishii, A., Morita, Y., Yamamoto, K., Matsushima, K., Yoshie, O., and Hirai, K. (2000). Inducible expression of a Th2-type CC chemokine thymus- and activation-regulated chemokine by human bronchial epithelial cells. *J. Immunol.* **165,** 2205–2213.

Shortman, K., Egerton, M., Spangrude, G. J., and Scollay, R. (1990). The generation and fate of thymocytes. *Semin. Immunol.* **2,** 3–12.

Tachibana, K., Hirota, S., Iizasa, H., Yoshida, H., Kawabata, K., Kataoka, Y., Kitamura, Y., Matsushima, K., Yoshida, N., Nishikawa, S., Kishimoto, T., and Nagasawa, T. (1998). The chemokine receptor CXCR4 is essential for vascularization of the gastrointestinal tract. *Nature* **393,** 591–594.

Tanabe, S., Lu, Z., Luo, Y., Quackenbush, E. J., Berman, M. A., Collins-Racie, L. A., Mi, S., Reilly, C., Lo, D., Jacobs, K. A., and Dorf, M. E. (1997). Identification of a new mouse beta-chemokine, thymus-derived chemotactic agent 4, with activity on T lymphocytes and mesangial cells. *J. Immunol.* **159,** 5671–5679.

Tangemann, K., Gunn, M. D., Giblin, P., and Rosen, S. D. (1998). A high endothelial cell-derived chemokine induces rapid, efficient, and subset-selective arrest of rolling T lymphocytes on a reconstituted endothelial substrate. *J. Immunol.* **161,** 6330–6337.

Tashiro, K., Tada, H., Heilker, R., Shirozu, M., Nakano, T., and Honjo, T. (1993). Signal sequence trap: A cloning strategy for secreted proteins and type I membrane proteins. *Science* **261,** 600–603.

Uehara, S., Song, K., Farber, J. M., and Love, P. E. (2002). Characterization of CCR9 expression and CCL25/thymus expressed chemokine responsiveness during T-cell development: CD3(high)CD69+ thymocytes and gammadeltaTCR+ thymocytes preferentially respond to CCL25. *J. Immunol.* **168,** 134–142.

Ueno, T., Hara, K., Willis, M. S., Malin, M. A., Hopken, U. E., Gray, D. H., Matsushima, K., Lipp, M., Springer, T. A., Boyd, R. L., Yoshie, O., and Takahama, Y. (2002). Role for CCR7 ligands in the emigration of newly generated T lymphocytes from the neonatal thymus. *Immunity* **16,** 205–218.

Valenzuela-Fernandez, A., Planchenault, T., Baleux, F., Staropoli, I., Le-Barillec, K., Leduc, D., Delaunay, T., Lazarini, F., Virelizier, J. L., Chignard, M., Pidard, D., and Arenzana-Seisdedos, F. (2002). Leukocyte elastase negatively regulates Stromal cell-derived factor-1 (SDF-1)/CXCR4 binding and functions by amino-terminal processing of SDF-1 and CXCR4. *J. Biol. Chem.* **277,** 15677–15689.

van Os, R., van Schie, M. L., Willemze, R., and Fibbe, W. E. (2002). Proteolytic enzyme levels are increased during granulocyte colony-stimulating factor-induced hematopoietic stem cell mobilization in human donors but do not predict the number of mobilized stem cells. *J. Hematother. Stem. Cell Res.* **11,** 513–521.

Vicari, A. P., Figueroa, D. J., Hedrick, J. A., Foster, J. S., Singh, K. P., Menon, S., Copeland, N. G., Gilbert, D. J., Jenkins, N. A., Bacon, K. B., and Zlotnik, A. (1997). TECK: A novel CC chemokine specifically expressed by thymic dendritic cells and potentially involved in T-cell development. *Immunity* **7,** 291–301.

Vissers, J. L., Hartgers, F. C., Lindhout, E., Figdor, C. G., and Adema, G. J. (2001). BLC (CXCL13) is expressed by different dendritic cell subsets *in vitro* and *in vivo*. *Eur. J. Immunol.* **31,** 1544–1549.

Willimann, K., Legler, D. F., Loetscher, M., Roos, R. S., Delgado, M. B., Clark-Lewis, I., Baggiolini, M., and Moser, B. (1998). The chemokine SLC is expressed in T-cell areas of lymph nodes and mucosal lymphoid tissues and attracts activated T cells via CCR7. *Eur. J. Immunol.* **28,** 2025–2034.

Wong, M. M., and Fish, E. N. (2003). Chemokines: Attractive mediators of the immune response. *Semin. Immunol.* **15,** 5–14.

Yoshida, R., Imai, T., Hieshima, K., Kusuda, J., Baba, M., Kitaura, M., Nishimura, M., Kakizaki, M., Nomiyama, H., and Yoshie, O. (1997). Molecular cloning of a novel human CC chemokine EBI1-ligand chemokine that is a specific functional ligand for EBI1, CCR7. *J. Biol. Chem.* **272,** 13803–13809.

Yoshida, R., Nagira, M., Imai, T., Baba, M., Takagi, S., Tabira, Y., Akagi, J., Nomiyama, H., and Yoshie, O. (1998a). EBI1-ligand chemokine (ELC) attracts a broad spectrum of lymphocytes: Activated T cells strongly up-regulate CCR7 and efficiently migrate toward ELC. *Int. Immunol.* **10,** 901–910.

Yoshida, R., Nagira, M., Kitaura, M., Imagawa, N., Imai, T., and Yoshie, O. (1998b). Secondary lymphoid-tissue chemokine is a functional ligand for the CC chemokine receptor CCR7. *J. Biol. Chem.* **273,** 7118–7122.

Yoshie, O., Imai, T., and Nomiyama, H. (1997). Novel lymphocyte-specific CC chemokines and their receptors. *J. Leukoc. Biol.* **62,** 634–644.

Youn, B. S., Kim, C. H., Smith, F. O., and Broxmeyer, H. E. (1999). TECK, an efficacious chemoattractant for human thymocytes, uses GPR-9-6/CCR9 as a specific receptor. *Blood* **94,** 2533–2536.

Youn, B. S., Kim, Y. J., Mantel, C., Yu, K. Y., and Broxmeyer, H. E. (2001). Blocking of c-FLIP(L)–independent cycloheximide-induced apoptosis or Fas-mediated apoptosis by the CC chemokine receptor 9/TECK interaction. *Blood* **98,** 925–933.

Zaitseva, M. B., Lee, S., Rabin, R. L., Tiffany, H. L., Farber, J. M., Peden, K. W., Murphy, P. M., and Golding, H. (1998). CXCR4 and CCR5 on human thymocytes: Biological function and role in HIV-1 infection. *J. Immunol.* **161,** 3103–3113.

Zamarchi, R., Allavena, P., Borsetti, A., Stievano, L., Tosello, V., Marcato, N., Esposito, G., Roni, V., Paganin, C., Bianchi, G., Titti, F., Verani, P., Gerosa, G., and Amadori, A. (2002). Expression and functional activity of CXCR-4 and CCR-5 chemokine receptors in human thymocytes. *Clin. Exp. Immunol.* **127,** 321–330.

Zingoni, A., Soto, H., Hedrick, J. A., Stoppacciaro, A., Storlazzi, C. T., Sinigaglia, F., D'Ambrosio, D., O'Garra, A., Robinson, D., Rocchi, M., Santoni, A., Zlotnik, A., and Napolitano, M. (1998). The chemokine receptor CCR8 is preferentially expressed in Th2 but not Th1 cells. *J. Immunol.* **161,** 547–551.

Zou, Y. R., Kottmann, A. H., Kuroda, M., Taniuchi, I., and Littman, D. R. (1998). Function of the chemokine receptor CXCR4 in haematopoiesis and in cerebellar development. *Nature* **393,** 595–599.

CHAPTER 6

Chemokines in Immune Surveillance of the Intestine

Jan Hendrik Niess and Hans-Christian Reinecker
Gastrointestinal Unit and Center for the Study of Inflammatory Bowel Disease, Massachusetts General Hospital and Harvard Medical School, Boston, Massachusetts 02141

I. INTRODUCTION

The gastrointestinal mucosa is in constant interaction with the luminal microenvironment, which contains commensal microbiota as well as a variety of bacterial, viral, and protozoan pathogens. A single layer of intestinal epithelial cells (IECs) as big as a tennis court (260–300m^2) covers this large inner surface preventing uncontrolled dissemination of pathogens within host tissues (Nagler-Anderson, 2001). The intestine has a highly adaptable host defense system in which the intestinal epithelium and the mucosal immune system tightly cooperate through chemokines to facilitate barrier function while allowing the development of immunity. Dendritic cells (DCs) and macrophages play an important role in the immune defense of the gut, particularly in sampling pathogens and bacteria (Neutra *et al.*, 1996; Kelsall *et al.*, 2002; Uhlig and Powrie, 2003). Interactions of antigen-presenting DCs with

Current Topics in Membranes, Volume 55

1063-5823/05 $35.00
DOI: 10.1016/S1063-5823(04)55006-8

antigen-specific T-lymphocytes trigger the induction of adaptive immune responses and are also critical for the control of tissue inflammation and the maintenance of immune tolerance (Kelsall *et al.*, 2002; Steinman *et al.*, 2003). Evidence from mouse models of inflammatory bowel disease (IBD) indicates that dysregulation of antigen recognition and processing of intestinal microbiota are common disease mechanisms (Podolsky, 2002). Furthermore, mutations in proteins involved in pathogen pattern recognition have been associated with Crohn's disease (CD) (Hugot *et al.*, 2001; Ogura *et al.*, 2001).

Chemokines and their receptors orchestrate the intercellular communication required to maintain the complex architecture of immune compartments in the intestine at multiple levels. Homeostatic chemokines are responsible for a steady-state turnover of lymphocytes and DCs in the lamina propria and Peyer's patches (PPs) and their migration to and from mesenteric lymph nodes (MLNs). A particular role of epithelial cell-derived chemokines has been established for the direction of DCs, T-cells, and B-cells within these immune-cell compartments of the intestine. While a number of trafficking signals are ubiquitously shared among leukocyte populations that respond to inflammation throughout the body, intestine-specific chemokine signals begin to emerge that are linked to the intrinsic functional programs and subspecifications of leukocytes in the mucosal immune system.

II. CHEMOKINES REGULATE THE FUNCTION OF INTESTINAL DENDRITIC CELLS

The intestinal mucosa and immune system coexist in intimate association with the intestinal microbiota, and the outcome of this interaction is determined by the functional properties of DCs; furthermore, a seemingly key role is played by immune responses because of the facilitation of antigen sampling, pathogen recognition, and innate host defenses. Besides acquiring antigens in the lamina propria upon intestinal epithelial barrier breakdown, intestinal DCs located in intraepithelial pockets below the M-cells can detect incoming pathogens and phagocytose them constitutively (Iwasaki and Kelsall, 1999). Intestinal villous M-cells may provide an additional system for passing intestinal microbiota to intestinal DCs. Furthermore, intestinal DCs appear to transport captured proteins from the gut lumen and fragments of epithelial cells themselves (Huang *et al.*, 2000). Recently, evidence was presented that intestinal DCs can also monitor the contents of the intestinal lumen directly by entering or extending dendrites through the intestinal epithelium, although the molecular basis of this mechanism remains to be elucidated (Maric *et al.*, 1996; Rescigno *et al.*, 2001).

Most studies of migration of intestinal DCs have focused on the analysis of PPs and MLNs. CCR6 has been demonstrated to be involved in the migration of CD11CD11b$^+$ DCs in response to CCL20/LARC, expressed by IECs, into the dome regions of PPs, a mechanism that is linked to the regulation of antigen recognition in the intestine (Iwasaki and Kelsall, 2000; Kobayashi *et al.*, 2004). However, CD11c$^+$CD11b$^+$ DCs are still present in the PPs of four different CCR6-deficient mice, which indicates a previously unappreciated heterogeneity among intestinal DCs (Zhao *et al.*, 2003). The same study also demonstrated CCL9/MIP 1-γ/CCF18/MRP-2 to be a regulator of the migration of CD11c$^+$CD11b$^+$ DC into PPs (Zhao *et al.*, 2003).

Myeloid-derived intestinal DCs throughout the lamina propria are characterized by the expression of CX3CR1, the receptor for the membrane-bound chemokine CX3CL1/fractalkine (Niess *et al.*, 2005). CX3CL1, the only member of the CX3C chemokine family, is a membrane-bound chemokine that combines properties of both adhesion molecules and chemokines (Bazan *et al.*, 1997). Lamina propria CX3CR1$^+$ CD11c$^+$ DCs directly sample antigens from the intestinal lumen in the terminal ileum and the basolateral surfaces of IEC in the jejunum, proximal ileum, and colon. These intestinal DCs extend transepithelial dendrites in the terminal ileum in a process that requires CX3CR1. The formation of transepithelial dendrites may be regulated by direct DC-IEC interactions, since IECs constitutively express CX3CL1 (Muehlhoefer *et al.*, 2000). Thus, intestinal epithelium-derived chemokines may regulate intestinal antigen sampling and presentation. Furthermore, fractalkine induced the activation of IECs, which express the fractalkine receptor CX3CR1 in an autocrine fashion, resulting in increased proliferation rate, NF-κB activation, and expression of proinflammatory mediators such as CXCL8/IL-8 (Brand *et al.*, 2002). In response to fractalkine, polymononuclear neutrophils (PMNs) accumulate within the intestinal epithelium, indicating that CX3CL1 has homeostatic as well as inducible inflammatory functions in the recruitment of leukocytes into the intestine (Brand *et al.*, 2002).

Expression of CX3CR1 has been shown to distinguish two distinct monocyte subpopulations, a short-lived CX3CR1^{lo}Gr1$^+$ subset that is actively recruited to inflamed tissues and a CX3CR1^{hi}Gr1$^-$ subset characterized by CX3CR1-dependent recruitment to noninflamed tissues (Geissmann *et al.*, 2003). All subsets have the potential to differentiate into DCs *in vivo* and could be the source of DCs in the intestine. The level of CX3CR1 expression also defines the two major human monocyte subsets, the CD14$^+$CD16$^-$ and CD14^{lo}CD16$^+$ monocytes, which share phenotype and homing potential with the mouse subsets (Geissmann *et al.*, 2003).

Additional chemokine receptors implicated in the migration of myeloid-derived macrophages and immature DCs into the intestine include CCR2, CCR5 (Andres *et al.*, 2000), and CXCR4 (Ancuta *et al.*, 2003). Consequently, DCs respond to a vast array of chemokines, such as CCL25/TECK, CXCL12/SDF-1, CCL2/MCP-1, CCL8/MCP-2, and CCL7/MCP-3 (D'Amico *et al.*, 1998; Delgado *et al.*, 1998; Dieu *et al.*, 1998). The contact of DCs with antigens results in activation and maturation of DCs, which together lead to the downregulation of the chemokine receptors CCR1, CCR5, and CCR6 (Sozzani *et al.*, 1997). At the same time, DCs upregulate CCR7 expression, allowing them to respond to the chemokines CCL19/ELC and CCL21/SLC, which are selectively expressed in lymph nodes (Sozzani *et al.*, 1998). Consequently, DCs leave the lamina propria and migrate to the MLNs (Dieu *et al.*, 1998; Iwasaki and Kelsall, 2000).

III. CHEMOKINES RECRUIT T-CELLS TO GUT-ASSOCIATED LYMPHOID TISSUE

Naïve $CD4^+$ and $CD8^+$ T-cells, which have not encountered antigen, leave the thymus, display tropism to secondary lymphoid tissues, such as gut-associated lymphoid tissue (GALT), reside for 12–24 hours in lymphoid tissues, and then recirculate back to the bloodstream. The homeostatic chemokines CCL19/ELC and CCL21/SLC are expressed by high-endothelial venules (HEV) of secondary lymphoid tissues, including MLNs and PPs that interact with CCR7 expressed by naïve T-cells (Gunn *et al.*, 1998; Breitfeld *et al.*, 2000). Homing of naïve T-cells to secondary lymphoid tissues was impaired in CCL19/ELC and CCL21/SLC mutant mice, as well as in mice that lack CCR7 expression (Moser and Loetscher, 2001). Antigen contact results in not only T-cell activation but differentiation in memory/effector T-cells, which express specific adhesion and chemokine subsets that direct them to intestinal tissues. IECs in small intestinal but not in colonic crypts express the chemokine CCL25/TECK, which interacts with CCR9 expressed by small intestinal $CD8\alpha\beta^+$ T-lymphocytes (Fig. 1) (Kunkel *et al.*, 2000; Papadakis *et al.*, 2000). Blocking of CCL25 by antibody application prevents the migration of $\alpha_E\beta_7$ integrin$^+$$CD8\alpha\beta^+$ T-lymphocytes into the small intestine *in vivo* (Svensson *et al.*, 2002). In contrast, T-cells isolated from the large intestine do not express CCR9, indicating that naïve T-cell activation by antigen contact in inductive sites may determine the migratory properties of $CD8\alpha\beta^+$ effector cells to nonlymphoid tissues (Kunkel *et al.*, 2000).

However, the majority of lamina propria lymphocytes (LPLs) and intraepithelial lymphocytes (IELs) are unconventional T-cells, such as T-cell subsets expressing the $CD8\alpha\alpha$ homodimer or the $\gamma\delta$ T-cell receptor (TCRs) with

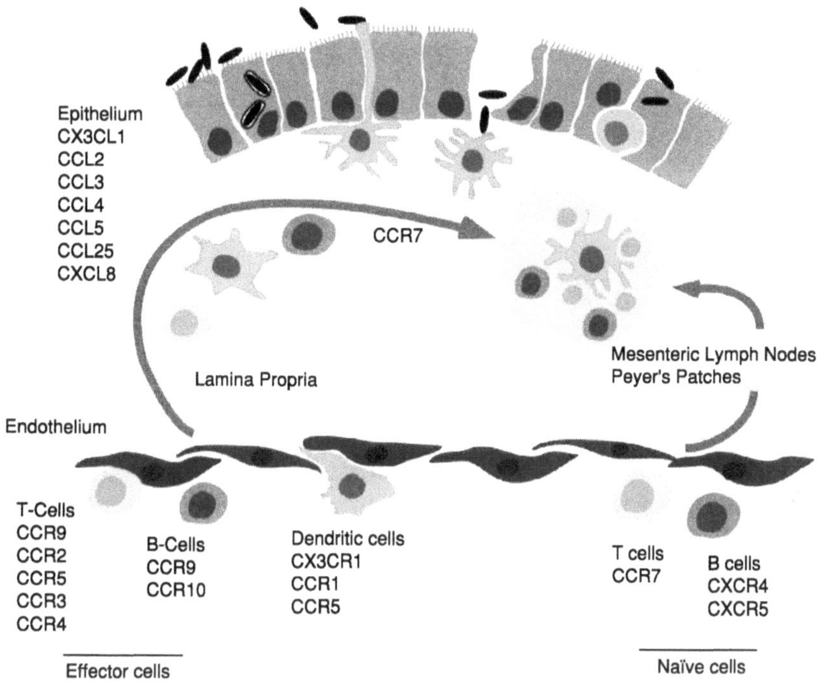

FIGURE 1 Binding of chemokines to their receptors expressed by T-cells, B-cells, and dendritic cells directs them to the intestinal lamina propria or to mesenteric lymph nodes and Peyer's patches. (See Color Insert.)

limited diversity for antigens that appear not to recirculate and to remain fixed in intestinal tissues where they may preserve tissue integrity and serve a regulatory function. Most LPLs and IELs express high levels of CCR2, CCR5, CXCR3, and CXCR4; furthermore, their ligands, such as CCL2/MCP-1, CCL7/MCP-3, CCL5/RANTES, CXCL10/IP-10, and CXCL-12/SDF-1, are all constitutively expressed in the intestinal mucosa (Reinecker *et al.*, 1995; MacDermott *et al.*, 1998; Agace *et al.*, 2000). Subsets of CD8$\alpha\alpha^+$ IELs, NK cells, and CD16$^+$ monocytes express CX3CR1 that interacts with CX3CL1/fractalkine, which is also expressed by IECs (Muehlhoefer *et al.*, 2000). In mice, CD3$^+$ $\gamma\delta$ T-cells can be detected by day 14 of gestation and are seen to peak between days 17 and 18 and then decline until birth, when they account for <5% of thymocytes. In fetal mice, before exogenous antigen and the commensal flora are fully developed, which could influence the homing potential of these cells, $\gamma\delta$ T-cells localize to specific intestinal tissues (Asarnow *et al.*, 1989). In CCR9 mutant mice, the accumulation of $\gamma\delta$ T-cells in the intestinal epithelium is decreased, which may be secondary to developmental defects of $\gamma\delta$ T-cells, because in these mice thymic as well as extrathymic $\gamma\delta$ T-cell development is impaired (Campbell *et al.*, 2003).

IV. CHEMOKINES REGULATE B-CELLS IN THE INTESTINE

Contact of bone marrow-released naïve mature B-cells with thymus-dependent antigens induce B-cell activation and clonal selection that lead to generation of plasma and memory cells. Interaction of CXCR4 expressed by naïve mature B-cells with CXCL12/SDF-1 expressed by HEV of MLN mediates entry of mature B-cells into lymph nodes (Okada *et al.*, 2002). Within MLNs, CXCR5, which is also expressed by specialized subsets of T-cells (the so-called follicular T helper cells), guides B-cells to B-cell follicular regions (Forster *et al.*, 1996). In contrast, entry of mature B-cells into PPs depends on the interactions of CXCR5 with CXCL13/BCA-1, which is expressed by follicular PPs' HEVs (Okada *et al.*, 2002). Binding of CXCL13 to CXCR5 induces production of lymphotoxin $\alpha 1\beta 1$ by B-cells, which is required for regular PP development (Ansel *et al.*, 2000). Despite the organized lymphoid structures of the GALT, single B-cells are dispersed throughout the lamina propria and the epithelium. A large percentage of IgA-secreting plasma cells in the lamina propria express CCR9 and CCR10 that migrate towards CCL25/TECK expressed by IECs in small intestinal crypts (Fig. 1) (Bowman *et al.*, 2002; Kunkel *et al.*, 2003).

V. IEC-DERIVED CHEMOKINES SHAPE THE MUCOSAL IMMUNE SYSTEM

Within the intestine, chemokines are secreted by IECs, endothelial cells, myofibroblasts, and leukocytes (MacDermott *et al.*, 1998). Among these distinct cell types, the ability of IECs to secrete a diverse array of chemokines is well documented (Table I). In response to intestinal pathogens or inflammatory stimuli, IECs synthesize chemokines that facilitate first-line defense against intestinal pathogens (Eckmann *et al.*, 1993; Reinecker *et al.*, 1995). Using *in vitro* model systems, McCormick *et al.* (1993, 1995) demonstrated that the application of *Salmonella typhimurium* to the surface of T84 epithelial cell monolayers led to the basolateral secretion of chemotactic factors, such as CXCL8/IL-8, that caused the translocation of neutrophils across the epithelial monolayer. IECs sense attached noninvasive enteropathogenic *Escherichia coli* strains, which lead to the secretion of chemotactic factors and subsequently to the transmigration of neutrophils (Savkovic *et al.*, 1996). The importance of these observations was highlighted by recent studies identifying the presence of adherent-invasive *E. coli* strains in patients with CD but not in healthy

TABLE I
Chemokines Expressed by Intestinal Epithelial Cells

IEC-derived chemokine	Reference
CXCL1/GRO-α	(Yang *et al.*, 1997; Song *et al.*, 1999)
CXCL5/ENA-78	(Z'Graggen *et al.*, 1997)
CXCL8/IL-8	(Eckmann *et al.*, 1993; Casola *et al.*, 1998)
CXCL9/MIG	(Dwinell *et al.*, 2001)
CXCL10/IP-10	(Uguccioni *et al.*, 1999)
CXCL11/I-TAC	(Dwinell *et al.*, 2001)
CCL2/MCP-1	(Reinecker *et al.*, 1995)
CCL3/MIP-1α	(Yang *et al.*, 1997)
CCL4/MIP-1β	(Yang *et al.*, 1997)
CCL5/RANTES	(Yang *et al.*, 1997)
CCL7/MCP-3	(Uguccioni *et al.*, 1999)
CCL20/MIP-3α	(Izadpanah *et al.*, 2001)
CCL22/MDC	(Berin *et al.*, 2001)
CCL25/TECK	(Kunkel *et al.*, 2000)
CCL26/eotaxin-3	(Garcia-Zepeda *et al.*, 1996)
CCL28/MEC	(Pan *et al.*, 2000)
CX3CL1/fractalkine	(Muehlhoefer *et al.*, 2000)

control patients, which may contribute to the pathophysiology of IBD (Darfeuille-Michaud *et al.*, 1998; Swidsinski *et al.*, 2002).

Further studies demonstrated that IECs not only produce chemokines in response to bacterial invasion but also respond to products of the bacterial metabolism, such as sodium butyrate, or to proinflammatory stimuli, such as the cytokines IL-1β, TNF-α, and IFN-γ (Fusunyan *et al.*, 1998). Secretion of CXCL8/IL-8, CXCL1/Gro-α, and CXCL2/Gro-β occurs within 2 to 3 hours after infection with *Salmonella dublin*, whereas CXCL5/ENA-78 peaks after 6 to 10 hours (Yang *et al.*, 1997), suggesting that danger signals regulate the chemokine secretion via IECs in a time-dependent manner that likely shapes the inflammatory immune response.

IECs do not only secrete chemokines; they also express chemokine receptors, including the CC receptors CCR1-8 (Dwinell *et al.*, 1999), CXCR4 (Delezay *et al.*, 1997), and CX3CR1 (Brand *et al.*, 2002) (Table II). CXCR4 and CCR5 expression by human IECs is mainly apically localized, as shown by immunostaining of specimens obtained from patients undergoing partial colectomy for diverticular disease (Dwinell *et al.*, 1999), whereas

TABLE II

Chemokine Receptors and Their Ligands Expressed by Intestinal Epithelial Cells

Chemokine receptors expressed by IECs	Chemokine ligand(s)
CXCR4 (Delezay *et al.*, 1997)	CXCL12/SDF-1
CCR1 (Dwinell *et al.*, 1999)	CCL2/MCP-1, CCL3/MIP-1α, CCL4/MIP-1β, CCL5/RANTES, CCL7/MCP-3, CCL8/MCP-2, CCL13/MCP-4, CCL14/HCC-1, CCL15/Lkn-1, CCL23/MPIF-1
CCR2 (Dwinell *et al.*, 1999)	CCL2/MCP-1, CCL7/MCP-3, CCL8/MCP-2, CCL13/MCP-4
CCR3 (Dwinell *et al.*, 1999)	CCL5/RANTES, CCL7/MCP-3, CCL8/MCP-2, CCL11/eotaxin, CCL13/MCP-4, CCL15/Lkn-1, CCL24/MPIF-2, CCL26/eotaxin-3
CCR4 (Dwinell *et al.*, 1999)	CCL17/TARC, CCL22/MDC
CCR5 (Dwinell *et al.*, 1999)	CCL3/MIP-1α, CCL4/MIP-1β, CCL5/RANTES, CCL7/MCP-3, CCL8/MCP-2, CCL11/eotaxin, CCL13/MCP-4
CCR6 (Dwinell *et al.*, 1999)	CCL20/LARC/MIP-3α
CCR7 (Dwinell *et al.*, 1999)	CCL19/ELC, CCL21/SLC
CCR8 (Dwinell *et al.*, 1999)	CXCL14/BRAK, CCL4/MIP-1β, CCL17/TARC
CX3CR1 (Brand *et al.*, 2002)	CX3CL1/fractalkine

CX3CR1 is basolaterally expressed, suggesting that localization of receptor expression on IECs assists chemokines in their specific functions. The stimulation of HT-29 epithelial cells with CCL3/MIP-1α, CCL4/MIP-1β, and CXCL12/SDF-1 increased the secretion of CXCL8/IL-8 and CXCL1/Gro-α (Dwinell *et al.*, 1999). Signaling via CX3CR1 results in activation of the $G_i\alpha$-dependent MAPK, P53, and NF-κB, which induces IL-8 and CX3CL1/fractalkine expression. Together, these findings demonstrate that signaling via chemokine receptors in IECs facilitates chemokine and interleukin secretion in order to shape immune responses required for the maintenance of gut homeostasis.

VI. CHEMOKINES HAVE DIRECT ANTIMICROBIAL EFFECTS

The defensins are a family of small antimicrobial peptides with three to four intramolecular cysteine disulfide bonds. They are found in mammals, insects, and plants and are divided on the basis of six conserved cysteine residues into two categories designated α- and β-defensins (Ganz, 2003).

β-defensins (HBDs), which were shown to play a role in the pathogenesis of IBD (Wehkamp *et al.*, 2003), not only exhibit direct antimicrobial activities but chemotactically mobilize immature DCs. Migration of immature DCs in response to HBDs is inhibited by treatment of cells with pertussis toxin, suggesting that HBDs utilize seven transmembrane domain receptors or $G_i\alpha$-protein-coupled receptors for their action on DCs (Yang *et al.*, 1999). Additional analysis showed that HBDs unfold their effects on DCs via the chemokine receptor CCR6 to a lesser extent than CCL20/LARC/MIP-3α (Yang *et al.*, 1999). Surprisingly, further studies demonstrated that CCL20 has greater antimicrobial potential than does HBD in response to *Escherichia coli* and *Staphylococcus aureus* (Hoover *et al.*, 2002). CXCL4/PF-4, CXCL7/NAP-2, CXCL9/Mig, CXCL10/γ-IP-10, CXCL11/eosin, CCL5/RANTES, and CCL29 also function as antimicrobial agents (Yang *et al.*, 2002).

VII. CHEMOKINES IN INFLAMMATORY BOWEL DISEASE

IBD encompasses two chronic diseases, CD and ulcerative colitis (UC), which are both relapsing-remitting diseases often characterized by striking swings between intestinal inflammation and quiescence. The current concepts suggest that genetic, environmental, and immunological factors interact in the pathophysiology of IBD (Niess *et al.*, 2002; Podolsky, 2002). Distinct cell populations, such as neutrophils, monocytes/macrophages, lymphocytes, eosinophils, and mast cells, contribute to the inflammatory infiltration in CD and UC tissues. The recruitment of leukocytes to intestinal tissues is mediated by the upregulation of adhesion molecules on the intestinal vascular endothelium as well as the production of chemokines by endothelial and epithelial cells, as demonstrated by various studies.

CXCL8/IL-8 expression was found to be significantly elevated in colonic biopsies from patients with CD as well as UC. In patients with active CD, CXCL8/IL-8 expression showed a focal distribution in active lesions, whereas patients with active UC demonstrated a diffuse distribution of CXCL8/IL-8 expression throughout the intestinal mucosa (Mazzucchelli *et al.*, 1994). CXCL8/IL-8 is expressed by neutrophils and macrophages, which shape cryptitis and crypt abscesses, histopathological hallmarks of IBD (Daig *et al.*, 1996). IECs also express CXCL8/IL-8, but to a lesser extent than do leukocytes (Mazzucchelli *et al.*, 1994; Mitsuyama *et al.*, 1994). CXCL5/ENA-78 expressed by IECs is preferentially located in intestinal crypts, as demonstrated by *in situ* hybridization of surgical colonic specimens from patients with mild to moderate stages of CD or UC. Interestingly, CXCL5/ENA-78 expression was lower in specimens with high histological disease features, suggesting that CXCL5/ENA-78 is involved in the onset of disease when the destruction of the epithelium is still limited and when the recruitment of neutrophils is mediated by binding of CXCL5/ENA-78 to CXCR2 (Z'Graggen *et al.*, 1997).

Increased CCL2/MCP-1 (Reinecker *et al.*, 1995), CCL3/MIP-1α (Banks *et al.*, 2003), CCL4/MIP-1β (Mazzucchelli *et al.*, 1996; Banks *et al.*, 2003), CCL5/RANTES (Berrebi *et al.*, 1997; Vainer *et al.*, 1998), CCL7/MCP-3 (Banks *et al.*, 2003), CCL8/MCP-2 (Banks *et al.*, 2003), CXCL10/IP-10 (Uguccioni *et al.*, 1999), and CCL11/eotaxinin (Garcia-Zepeda *et al.*, 1996) expression was also found in patients with active IBD. CCL2/MCP-1 unfolds chemotactic effects on monocytes, immature DCs, NK cells, and T-cells, which may contribute to the development of polymono nuclear cell (PMN) infiltration in patients with IBD. Most T-cells isolated from noninflamed large- or small-intestinal tissues express CXCR3, but increased CXCL10/IP-10 expression in patients with IBD further attracts T-cells to inflamed tissues (Uguccioni *et al.*, 1999). Activated $CCR5^{+}$ Th1 effector cells are attracted by increased expression of CCL3/MIP-1α, CCL4/MIP-1β, and CCL5/RANTES, which are also involved in differentiation of naïve T-cells into Th1 effector cells characterized by the expression of the chemokine receptors CXCR3, CXCR6, and CCR5 and the secretion of the interleukins IFN-γ, IL-12, and IL-18 (Romagnani, 1999). In UC patients, the presence of $CCR3^{+}$ Th2 effector cells characterized by the chemokine receptors CCR3, CCR4, and CCR8 and the secretion of the interleukins IL-4, IL-5, and IL-13 was also observed, suggesting that increased CCL11/eosin expression is involved in the pathophysiology of UC (Romagnani, 1999). Blocking of the CCR5 ligand CXCL10-protected mice from dextran sodium sulphate (DSS)-induced colitis, which is an acute model of intestinal inflammation (Sasaki *et al.*, 2002). CCR2- and CCR5-deficient mice are also protected from DSS-induced colitis. CCR5-deficient mice showed an increased infiltration of $CD4^{+}$ and $NK1.1^{+}$ lymphocytes and Th1 shift in mucosal cytokine profile compared to wild-type and CCR2-deficient mice emphasizing the role that CCR5 plays in promoting mucosal Th1 immune responses (Andres *et al.*, 2000).

VIII. CONCLUSION

Chemokines play a major role in regulating the complex interactions between the gut flora, the intestinal epithelium, immune mediators, and effector cells. The interplay between nonimmune cells (particularly epithelial cells) and classical immune competent cell subsets is probably best documented and understood in the intestine. However, the task of unraveling the complexity of the emerging chemokine-mediated cell interactions and functions in the healthy and the diseased intestine remains monumental. Nevertheless, the investigation of chemokine biology in the intestine has yielded important contributions to the understanding of innate and adaptive immune defenses. Genetic mouse model systems allowing *in vivo* analysis of lymphocyte and DC migration are becoming increasingly available to help researchers understand the role of specific chemokines, their receptors, and the leukocyte subsets they control in the intestine. The development of new chemokine-based strategies for the prevention and treatment of IBD will require the ability to differentiate homeostatic chemokine signals responsible for constitutive immune surveillance from those that balance the function of inflammatory leukocyte subsets.

Acknowledgments

This work was supported by grants DK54427 and DK33506 to HCR. JHN is supported by a Research Fellowship Award from the Crohn's and Colitis Foundation of America (CCFA).

References

Agace, W. W., Amara, A., Roberts, A. I., Pablos, J. L., Thelen, S., Uguccioni, M., Li, X. Y., Marsal, J., Arenzana-Seisdedos, F., Delaunay, T., Ebert, E. C., Moser, B., and Parker, C. M. (2000). Constitutive expression of stromal derived factor-1 by mucosal epithelia and its role in HIV transmission and propagation. *Curr. Biol.* **10**, 325–328.

Ancuta, P., Rao, R., Moses, A., Mehle, A., Shaw, S. K., Luscinskas, F. W., and Gabuzda, D. (2003). Fractalkine preferentially mediates arrest and migration of CD16+ monocytes. *J. Exp. Med.* **197**, 1701–1707.

Andres, P. G., Beck, P. L., Mizoguchi, E., Mizoguchi, A., Bhan, A. K., Dawson, T., Kuziel, W. A., Maeda, N., MacDermott, R. P., Podolsky, D. K., and Reinecker, H. C. (2000). Mice with a selective deletion of the CC chemokine receptors 5 or 2 are protected from dextran sodium sulfate-mediated colitis: Lack of CC chemokine receptor 5 expression results in a NK1.1+ lymphocyte-associated Th2-type immune response in the intestine. *J. Immunol.* **164**, 6303–6312.

Ansel, K. M., Ngo, V. N., Hyman, P. L., Luther, S. A., Forster, R., Sedgwick, J. D., Browning, J. L., Lipp, M., and Cyster, J. G. (2000). A chemokine-driven positive feedback loop organizes lymphoid follicles. *Nature* **406**, 309–314.

Asarnow, D. M., Goodman, T., LeFrancois, L., and Allison, J. P. (1989). Distinct antigen receptor repertoires of two classes of murine epithelium-associated T cells. *Nature* **341**, 60–62.

Banks, C., Bateman, A., Payne, R., Johnson, P., and Sheron, N. (2003). Chemokine expression in IBD. Mucosal chemokine expression is unselectively increased in both ulcerative colitis and Crohn's disease. *J. Pathol.* **199**, 28–35.

Bazan, J. F., Bacon, K. B., Hardiman, G., Wang, W., Soo, K., Rossi, D., Greaves, D. R., Zlotnik, A., and Schall, T. J. (1997). A new class of membrane-bound chemokine with a CX3C motif. *Nature* **385**, 640–644.

Berin, M. C., Dwinell, M. B., Eckmann, L., and Kagnoff, M. F. (2001). Production of MDC/CCL22 by human intestinal epithelial cells. *Am. J. Physiol. Gastrointest. Liver Physiol.* **280**, G1217–G1226.

Berrebi, D., Banerjee, A., Paris, R., Potet, F., Aigrain, Y., Emilie, D., Cezard, J. P., Hugot, J. P., Navarro, J., and Peuchmaur, M. (1997). *In situ* Rantes and interferon-gamma gene expression in pediatric small bowel Crohn's disease. *J. Pediatr. Gastroenterol. Nutr.* **25,** 371–376.

Bowman, E. P., Kuklin, N. A., Youngman, K. R., Lazarus, N. H., Kunkel, E. J., Pan, J., Greenberg, H. B., and Butcher, E. C. (2002). The intestinal chemokine thymus-expressed chemokine (CCL25) attracts IgA antibody-secreting cells. *J. Exp. Med.* **195,** 269–275.

Brand, S., Sakaguchi, T., Gu, X., Colgan, S. P., and Reinecker, H. C. (2002). Fractalkine-mediated signals regulate cell-survival and immune-modulatory responses in intestinal epithelial cells. *Gastroenterology* **122,** 166–177.

Breitfeld, D., Ohl, L., Kremmer, E., Ellwart, J., Sallusto, F., Lipp, M., and Forster, R. (2000). Follicular B helper T cells express CXC chemokine receptor 5, localize to B cell follicles, and support immunoglobulin production. *J. Exp. Med.* **192,** 1545–1552.

Campbell, D. J., Debes, G. F., Johnston, B., Wilson, E., and Butcher, E. C. (2003). Targeting T cell responses by selective chemokine receptor expression. *Semin. Immunol.* **15,** 277–286.

Casola, A., Estes, M. K., Crawford, S. E., Ogra, P. L., Ernst, P. B., Garofalo, R. P., and Crowe, S. E. (1998). Rotavirus infection of cultured intestinal epithelial cells induces secretion of CXC and CC chemokines. *Gastroenterology* **114,** 947–955.

D'Amico, G., Bianchi, G., Bernasconi, S., Bersani, L., Piemonti, L., Sozzani, S., Mantovani, A., and Allavena, P. (1998). Adhesion, transendothelial migration, and reverse transmigration of *in vitro* cultured dendritic cells. *Blood* **92,** 207–214.

Daig, R., Andus, T., Aschenbrenner, E., Falk, W., Scholmerich, J., and Gross, V. (1996). Increased interleukin-8 expression in the colon mucosa of patients with inflammatory bowel disease. *Gut* **38,** 216–222.

Darfeuille-Michaud, A., Neut, C., Barnich, N., Lederman, E., Di Martino, P., Desreumaux, P., Gambiez, L., Joly, B., Cortot, A., and Colombel, J. F. (1998). Presence of adherent *Escherichia coli* strains in ileal mucosa of patients with Crohn's disease. *Gastroenterology* **115,** 1405–1413.

Delezay, O., Koch, N., Yahi, N., Hammache, D., Tourres, C., Tamalet, C., and Fantini, J. (1997). Co-expression of CXCR4/fusin and galactosylceramide in the human intestinal epithelial cell line HT-29. *AIDS* **11,** 1311–1318.

Delgado, E., Finkel, V., Baggiolini, M., Mackay, C. R., Steinman, R. M., and Granelli-Piperno, A. (1998). Mature dendritic cells respond to SDF-1, but not to several beta-chemokines. *Immunobiology* **198,** 490–500.

Dieu, M. C., Vanbervliet, B., Vicari, A., Bridon, J. M., Oldham, E., Ait-Yahia, S., Briere, F., Zlotnik, A., Lebecque, S., and Caux, C. (1998). Selective recruitment of immature and mature dendritic cells by distinct chemokines expressed in different anatomic sites. *J. Exp. Med.* **188,** 373–386.

Dwinell, M. B., Eckmann, L., Leopard, J. D., Varki, N. M., and Kagnoff, M. F. (1999). Chemokine receptor expression by human intestinal epithelial cells. *Gastroenterology* **117,** 359–367.

Dwinell, M. B., Lugering, N., Eckmann, L., and Kagnoff, M. F. (2001). Regulated production of interferon-inducible T-cell chemoattractants by human intestinal epithelial cells. *Gastroenterology* **120,** 49–59.

Eckmann, L., Kagnoff, M. F., and Fierer, J. (1993). Epithelial cells secrete the chemokine interleukin-8 in response to bacterial entry. *Infect. Immun.* **61,** 4569–4574.

Forster, R., Mattis, A. E., Kremmer, E., Wolf, E., Brem, G., and Lipp, M. (1996). A putative chemokine receptor, BLR1, directs B cell migration to defined lymphoid organs and specific anatomic compartments of the spleen. *Cell* **87,** 1037–1047.

Fusunyan, R. D., Quinn, J. J., Ohno, Y., MacDermott, R. P., and Sanderson, I. R. (1998). Butyrate enhances interleukin (IL)-8 secretion by intestinal epithelial cells in response to IL-1beta and lipopolysaccharide. *Pediatr. Res.* **43,** 84–90.

Ganz, T. (2003). Defensins: Antimicrobial peptides of innate immunity. *Nat. Rev. Immunol.* **3,** 710–720.

Garcia-Zepeda, E. A., Rothenberg, M. E., Ownbey, R. T., Celestin, J., Leder, P., and Luster, A. D. (1996). Human eotaxin is a specific chemoattractant for eosinophil cells and provides a new mechanism to explain tissue eosinophilia. *Nat. Med.* **2,** 449–456.

Geissmann, F., Jung, S., and Littman, D. R. (2003). Blood monocytes consist of two principal subsets with distinct migratory properties. *Immunity* **19,** 71–82.

Gunn, M. D., Tangemann, K., Tam, C., Cyster, J. G., Rosen, S. D., and Williams, L. T. (1998). A chemokine expressed in lymphoid high endothelial venules promotes the adhesion and chemotaxis of naive T lymphocytes. *Proc. Natl. Acad. Sci. USA* **95,** 258–263.

Hoover, D. M., Boulegue, C., Yang, D., Oppenheim, J. J., Tucker, K., Lu, W., and Lubkowski, J. (2002). The structure of human macrophage inflammatory protein-3alpha/CCL20: Linking antimicrobial and CC chemokine receptor-6-binding activities with human beta-defensins. *J. Biol. Chem.* **277,** 37647–37654.

Huang, F. P., Platt, N., Wykes, M., Major, J. R., Powell, T. J., Jenkins, C. D., and MacPherson, G. G. (2000). A discrete subpopulation of dendritic cells transports apoptotic intestinal epithelial cells to T cell areas of mesenteric lymph nodes. *J. Exp. Med.* **191,** 435–444.

Hugot, J. P., Chamaillard, M., Zouali, H., Lesage, S., Cezard, J. P., Belaiche, J., Almer, S., Tysk, C., O'Morain, C. A., Gassull, M., Binder, V., Finkel, Y., Cortot, A., Modigliani, R., Laurent-Puig, P., Gower-Rousseau, C., Macry, J., Colombel, J. F., Sahbatou, M., and Thomas, G. (2001). Association of NOD2 leucine-rich repeat variants with susceptibility to Crohn's disease. *Nature* **411,** 599–603.

Iwasaki, A., and Kelsall, B. L. (1999). Freshly isolated Peyer's patch, but not spleen, dendritic cells produce interleukin-10 and induce the differentiation of T helper type 2 cells. *J. Exp. Med.* **190,** 229–239.

Iwasaki, A., and Kelsall, B. L. (2000). Localization of distinct Peyer's patch dendritic cell subsets and their recruitment by chemokines macrophage inflammatory protein (MIP)-3alpha, MIP-3beta, and secondary lymphoid organ chemokine. *J. Exp. Med.* **191,** 1381–1394.

Izadpanah, A., Dwinell, M. B., Eckmann, L., Varki, N. M., and Kagnoff, M. F. (2001). Regulated MIP-3alpha/CCL20 production by human intestinal epithelium: Mechanism for modulating mucosal immunity. *Am. J. Physiol. Gastrointest. Liver. Physiol.* **280,** G710–719.

Kelsall, B. L., Biron, C. A., Sharma, O., and Kaye, P. M. (2002). Dendritic cells at the host-pathogen interface. *Nat. Immunol.* **3,** 699–702.

Kobayashi, H., Miura, S., Nagata, H., Tsuzuki, Y., Hokari, R., Ogino, T., Watanabe, C., Azuma, T., and Ishii, H. (2004). *In situ* demonstration of dendritic cell migration from rat intestine to mesenteric lymph nodes: Relationships to maturation and role of chemokines. *J. Leukoc. Biol.* **75,** 434–442.

Kunkel, E. J., Campbell, J. J., Haraldsen, G., Pan, J., Boisvert, J., Roberts, A. I., Ebert, E. C., Vierra, M. A., Goodman, S. B., Genovese, M. C., Wardlaw, A. J., Greenberg, H. B., Parker, C. M., Butcher, E. C., Andrew, D. P., and Agace, W. W. (2000). Lymphocyte CC chemokine receptor 9 and epithelial thymus-expressed chemokine (TECK) expression distinguish the small intestinal immune compartment: Epithelial expression of tissue-specific chemokines as an organizing principle in regional immunity. *J. Exp. Med.* **192,** 761–768.

Kunkel, E. J., Kim, C. H., Lazarus, N. H., Vierra, M. A., Soler, D., Bowman, E. P., and Butcher, E. C. (2003). CCR10 expression is a common feature of circulating and mucosal epithelial tissue IgA Ab-secreting cells. *J. Clin. Invest.* **111,** 1001–1010.

MacDermott, R. P., Sanderson, I. R., and Reinecker, H. C. (1998). The central role of chemokines (chemotactic cytokines) in the immunopathogenesis of ulcerative colitis and Crohn's disease. *Inflamm. Bowel Dis.* **4,** 54–67.

Maric, I., Holt, P. G., Perdue, M. H., and Bienenstock, J. (1996). Class II MHC antigen (Ia)-bearing dendritic cells in the epithelium of the rat intestine. *J. Immunol.* **156,** 1408–1414.

Mazzucchelli, L., Hauser, C., Zgraggen, K., Wagner, H. E., Hess, M. W., Laissue, J. A., and Mueller, C. (1996). Differential *in situ* expression of the genes encoding the chemokines MCP-1 and RANTES in human inflammatory bowel disease. *J. Pathol.* **178,** 201–206.

Mazzucchelli, L., Hauser, C., Zgraggen, K., Wagner, H., Hess, M., Laissue, J. A., and Mueller, C. (1994). Expression of interleukin-8 gene in inflammatory bowel disease is related to the histological grade of active inflammation. *Am. J. Pathol.* **144,** 997–1007.

McCormick, B. A., Colgan, S. P., Delp-Archer, C., Miller, S. I., and Madara, J. L. (1993). Salmonella typhimurium attachment to human intestinal epithelial monolayers: Transcellular signalling to subepithelial neutrophils. *J. Cell. Biol.* **123,** 895–907.

McCormick, B. A., Miller, S. I., Carnes, D., and Madara, J. L. (1995). Transepithelial signaling to neutrophils by salmonellae: A novel virulence mechanism for gastroenteritis. *Infect. Immun.* **63,** 2302–2309.

Mitsuyama, K., Toyonaga, A., Sasaki, E., Watanabe, K., Tateishi, H., Nishiyama, T., Saiki, T., Ikeda, H., Tsuruta, O., and Tanikawa, K. (1994). IL-8 as an important chemoattractant for neutrophils in ulcerative colitis and Crohn's disease. *Clin. Exp. Immunol.* **96,** 432–436.

Moser, B., and Loetscher, P. (2001). Lymphocyte traffic control by chemokines. *Nat. Immunol.* **2,** 123–128.

Muehlhoefer, A., Saubermann, L. J., Gu, X., Luedtke-Heckenkamp, K., Xavier, R., Blumberg, R. S., Podolsky, D. K., MacDermott, R. P., and Reinecker, H. C. (2000). Fractalkine is an epithelial and endothelial cell-derived chemoattractant for intraepithelial lymphocytes in the small intestinal mucosa. *J. Immunol.* **164,** 3368–3376.

Nagler-Anderson, C. (2001). Man the barrier! Strategic defenses in the intestinal mucosa. *Nat. Rev. Immunol.* **1,** 59–67.

Neutra, M. R., Frey, A., and Kraehenbuhl, J. P. (1996). Epithelial M cells: Gateways for mucosal infection and immunization. *Cell* **86,** 345–348.

Niess, J. H., Brand, S., Gu, X., Landsmann, L., Jung, S., McCormick, B. A., Vyas, J. M., Boes, M., Ploegh, P. L., Fox, J. G., Littman, D. R., and Reinecker, H. C. (2005). CX3CRI – mediated dendritic cell access to the intestinal lumen and bacterial clearance. *Science* **307,** 254–258.

Niess, J. H., Monnikes, H., Dignass, A. U., Klapp, B. F., and Arck, P. C. (2002). Review on the influence of stress on immune mediators, neuropeptides and hormones with relevance for inflammatory bowel disease. *Digestion* **65,** 131–140.

Ogura, Y., Bonen, D. K., Inohara, N., Nicolae, D. L., Chen, F. F., Ramos, R., Britton, H., Moran, T., Karaliuskas, R., Duerr, R. H., Achkar, J. P., Brant, S. R., Bayless, T. M., Kirschner, B. S., Hanauer, S. B., Nunez, G., and Cho, J. H. (2001). A frameshift mutation in NOD2 associated with susceptibility to Crohn's disease. *Nature* **411,** 603–606.

Okada, T., Ngo, V. N., Ekland, E. H., Forster, R., Lipp, M., Littman, D. R., and Cyster, J. G. (2002). Chemokine requirements for B cell entry to lymph nodes and Peyer's patches. *J. Exp. Med.* **196,** 65–75.

Pan, J., Kunkel, E. J., Gosslar, U., Lazarus, N., Langdon, P., Broadwell, K., Vierra, M. A., Genovese, M. C., Butcher, E. C., and Soler, D. (2000). A novel chemokine ligand for CCR10 and CCR3 expressed by epithelial cells in mucosal tissues. *J. Immunol.* **165,** 2943–2949.

Papadakis, K. A., Prehn, J., Nelson, V., Cheng, L., Binder, S. W., Ponath, P. D., Andrew, D. P., and Targan, S. R. (2000). The role of thymus-expressed chemokine and its receptor CCR9 on lymphocytes in the regional specialization of the mucosal immune system. *J. Immunol.* **165,** 5069–5076.

Podolsky, D. K. (2002). Inflammatory bowel disease. *N. Engl. J. Med.* **347,** 417–429.

Reinecker, H. C., Loh, E. Y., Ringler, D. J., Mehta, A., Rombeau, J. L., and MacDermott, R. P. (1995). Monocyte-chemoattractant protein 1 gene expression in intestinal epithelial cells and inflammatory bowel disease mucosa. *Gastroenterology* **108,** 40–50.

Rescigno, M., Urbano, M., Valzasina, B., Francolini, M., Rotta, G., Bonasio, R., Granucci, F., Kraehenbuhl, J. P., and Ricciardi-Castagnoli, P. (2001). Dendritic cells express tight junction proteins and penetrate gut epithelial monolayers to sample bacteria. *Nat. Immunol.* **2,** 361–367.

Romagnani, S. (1999). Th1/Th2 cells. *Inflamm. Bowel. Dis.* **5,** 285–294.

Sasaki, S., Yoneyama, H., Suzuki, K., Suriki, H., Aiba, T., Watanabe, S., Kawauchi, Y., Kawachi, H., Shimizu, F., Matsushima, K., Asakura, H., and Narumi, S. (2002). Blockade of CXCL10 protects mice from acute colitis and enhances crypt cell survival. *Eur. J. Immunol.* **32,** 3197–3205.

Savkovic, S. D., Koutsouris, A., and Hecht, G. (1996). Attachment of a noninvasive enteric pathogen, enteropathogenic *Escherichia coli,* to cultured human intestinal epithelial monolayers induces transmigration of neutrophils. *Infect. Immun.* **64,** 4480–4487.

Song, F., Ito, K., Denning, T. L., Kuninger, D., Papaconstantinou, J., Gourley, W., Klimpel, G., Balish, E., Hokanson, J., and Ernst, P. B. (1999). Expression of the neutrophil chemokine KC in the colon of mice with enterocolitis and by intestinal epithelial cell lines: Effects of flora and proinflammatory cytokines. *J. Immunol.* **162,** 2275–2280.

Sozzani, S., Allavena, P., D'Amico, G., Luini, W., Bianchi, G., Kataura, M., Imai, T., Yoshie, O., Bonecchi, R., and Mantovani, A. (1998). Differential regulation of chemokine receptors during dendritic cell maturation: A model for their trafficking properties. *J. Immunol.* **161,** 1083–1086.

Sozzani, S., Luini, W., Borsatti, A., Polentarutti, N., Zhou, D., Piemonti, L., D'Amico, G., Power, C. A., Wells, T. N., Gobbi, M., Allavena, P., and Mantovani, A. (1997). Receptor expression and responsiveness of human dendritic cells to a defined set of CC and CXC chemokines. *J. Immunol.* **159,** 1993–2000.

Steinman, R. M., Hawiger, D., Liu, K., Bonifaz, L., Bonnyay, D., Mahnke, K., Iyoda, T., Ravetch, J., Dhodapkar, M., Inaba, K., and Nussenzweig, M. (2003). Dendritic cell function *in vivo* during the steady state: A role in peripheral tolerance. *Ann. New York Acad. Sci.* **987,** 15–25.

Svensson, M., Marsal, J., Ericsson, A., Carramolino, L., Broden, T., Marquez, G., and Agace, W. W. (2002). CCL25 mediates the localization of recently activated CD8alphabeta(+) lymphocytes to the small-intestinal mucosa. *J. Clin. Invest.* **110,** 1113–1121.

Swidsinski, A., Ladhoff, A., Pernthaler, A., Swidsinski, S., Loening-Baucke, V., Ortner, M., Weber, J., Hoffmann, U., Schreiber, S., Dietel, M., and Lochs, H. (2002). Mucosal flora in inflammatory bowel disease. *Gastroenterology* **122,** 44–54.

Uguccioni, M., Gionchetti, P., Robbiani, D. F., Rizzello, F., Peruzzo, S., Campieri, M., and Baggiolini, M. (1999). Increased expression of IP-10, IL-8, MCP-1, and MCP-3 in ulcerative colitis. *Am. J. Pathol.* **155,** 331–336.

Uhlig, H. H., and Powrie, F. (2003). Dendritic cells and the intestinal bacterial flora: A role for localized mucosal immune responses. *J. Clin. Invest.* **112,** 648–651.

Vainer, B., Nielsen, O. H., and Horn, T. (1998). Expression of E-selectin, sialyl Lewis X, and macrophage inflammatory protein-1alpha by colonic epithelial cells in ulcerative colitis. *Dig. Dis. Sci.* **43,** 596–608.

Wehkamp, J., Harder, J., Weichenthal, M., Mueller, O., Herrlinger, K. R., Fellermann, K., Schroeder, J. M., and Stange, E. F. (2003). Inducible and constitutive beta-defensins are differentially expressed in Crohn's disease and ulcerative colitis. *Inflamm. Bowel Dis.* **9,** 215–223.

Yang, D., Biragyn, A., Kwak, L. W., and Oppenheim, J. J. (2002). Mammalian defensins in immunity: More than just microbicidal. *Trends. Immunol.* **23,** 291–296.

Yang, D., Chertov, O., Bykovskaia, S. N., Chen, Q., Buffo, M. J., Shogan, J., Anderson, M., Schroder, J. M., Wang, J. M., Howard, O. M., and Oppenheim, J. J. (1999). Beta-defensins: Linking innate and adaptive immunity through dendritic and T cell CCR6. *Science* **286,** 525–528.

Yang, S. K., Eckmann, L., Panja, A., and Kagnoff, M. F. (1997). Differential and regulated expression of C-X-C, C-C, and C-chemokines by human colon epithelial cells. *Gastroenterology* **113,** 1214–1223.

Z'Graggen, K., Walz, A., Mazzucchelli, L., Strieter, R. M., and Mueller, C. (1997). The C-X-C chemokine ENA-78 is preferentially expressed in intestinal epithelium in inflammatory bowel disease. *Gastroenterology* **113,** 808–816.

Zhao, X., Sato, A., Dela Cruz, C. S., Linehan, M., Luegering, A., Kucharzik, T., Shirakawa, A. K., Marquez, G., Farber, J. M., Williams, I., and Iwasaki, A. (2003). CCL9 is secreted by the follicle-associated epithelium and recruits dome region Peyer's patch CD11b+ dendritic cells. *J. Immunol.* **171,** 2797–2803.

CHAPTER 7

Chemokines and Central Nervous System Physiology

Robyn S. Klein,* Joshua B. Rubin,† and Andrew D. Luster‡
*Division of Infectious Diseases and Departments of Molecular Microbiology, Pathology and Immunology, and Anatomy and Neurobiology, Washington University School of Medicine, St. Louis, Missouri 63110
†Division of Pediatric Oncology, Departments of Pediatrics and Anatomy and Neurobiology, Washington University School of Medicine, St. Louis, Missouri 63110
‡Division of Rheumatology, Allergy, and Immunology, Massachusetts General Hospital, Boston, Massachusetts 02114

I. OVERVIEW

Until recently, chemokines and their receptors were believed to have evolved solely for the purpose of recruiting and activating immune cells during inflammation. As the known roles for chemokines increased through studies of their effects on hematopoiesis and lymphopoiesis, it became clear that they are important in the development and homeostasis of almost all tissues. In the central nervous system (CNS), investigations of the chemokines responsible for leukocyte recruitment into the CNS during various neuroinflammatory diseases lead to additional discoveries regarding their

Current Topics in Membranes, Volume 55

1063-5823/05 $35.00
DOI: 10.1016/S1063-5823(04)55007-X

roles in development, normal neurophysiology, and diseases ranging from neurodegeneration to neoplasia. In some regards, these newly identified functions recapitulate their effects on leukocytes. However, many of the ways that chemokines influence CNS cell biology represent novel adaptations to the special requirements involved in neural signaling.

This chapter will discuss some of the highlights of these recent discoveries. In keeping with their effects on leukocytes, chemokines have been shown to effect the proliferation, survival, and migration of neural precursors. In the adult CNS, which has more fully differentiated neurons, chemokines exert trophic effects and participate in synaptogenesis, while chemokine receptors exert modulatory influences on neuronal responses to classical neurotransmitters. During inflammatory diseases, chemokines not only perform their typical functions in leukocyte trafficking but can act as neurotoxins, activating pro-apoptic pathways. Finally, overexpression of chemokine receptors in neural cell types is associated with the development of CNS neoplastic diseases. The involvement of these molecules in multiple aspects of various pathologic processes makes them attractive therapeutic targets. The challenge will be to design treatments that target specific chemokine effects while preserving the chemokine actions required for normal brain function.

II. CHEMOKINES AND CNS DEVELOPMENT

The various functions of the CNS are dependent upon a vast network of precise connections between diverse neural cell types. Developmentally, this is accomplished in several steps in which populations of progenitor cells expand in proliferative zones, migrate to their final positions, and extend processes toward target cells. In the case of neurons, this may mean extending axons to synapse with other neurons many centimeters away. All of these steps require that neural cells have molecular mechanisms for detecting and responding to spatially and temporally restricted proliferative, survival, migrational, and path-finding cues from their microenvironment. Recent work has established that the chemokine receptor CXCR4 and its sole ligand, CXCL12, play an important part in each phase of development in multiple brain regions as well as in axonal path-finding within the spinal cord. Similarly, the chemokine receptor CXCR2 and one of its ligands, CXCL1, have been shown to affect the proliferation and migration of oligodendrocyte precursor cells in the developing spinal cord. In this section, we will review these interesting findings as well as additional work suggesting that other chemokines and their receptors participate in CNS development.

While CXCR4 was one of the first chemokine receptors detected on normal neurons (Lavi *et al.*, 1997), clues to its function in normal CNS physiology were discovered only after attempts to evaluate its role in hematopoiesis via generation of mice with targeted deletion of CXCL12 or CXCR4 (Ma *et al.*, 1998; Zou *et al.*, 1998). The deletions turned out to be embryonically lethal due to abnormalities in the development of multiple organ systems; however, examination of CNS tissues in both CXCR4 and CXCL12 knockout mice revealed startling abnormalities in multiple regions of the brain, including the developing cerebellum, hippocampus, neocortex, spinal cord, and retina, due primarily to deleterious effects on various aspects of neuronal development. Further studies of normal embryonic and neonatal mice revealed complimentary patterns of developmentally regulated expression of CXCL12 and CXCR4 in all of these CNS regions, supporting the notion that neural cell expression of CXCR4 is critical to normal neuronal development.

The basis for the CNS phenotype of the CXCL12- and CXCR4-deficient mice was first evaluated in the cerebellum (Ma *et al.*, 1998). Several features make the cerebellum a popular model system for the study of CNS development. Normally, cerebellar development includes the massive expansion of cerebellar granule precursor cells (GPCs) during a brief postnatal period (ten Donkelaar *et al.*, 2003). The near-simultaneous proliferation and subsequent migration of this relatively homogeneous population of CNS precursor cells make the cerebellar system ideal for the study of those factors that regulate proliferation and migration of neural precursors. Granule-cell proliferation occurs in a specialized proliferative zone known as the external granule-cell layer (EGL). This transient cerebellar structure forms immediately below the layer of meningeal cells (the pia mater) that cover the cerebellar surface. Postmitotic granule cells differentiate and migrate inward along Bergmann glial processes past Purkinje cell bodies to form the internal granule cell layer.

The critical role that CXCL12 and CXCR4 play in this orderly process of proliferation and migration was revealed when the targeted deletion of CXCL12 or CXCR4 resulted in premature migration of GPCs out of the EGL and two subsequent results: a decrease in granule-cell number and aberrant localization of those remaining granule cells within or under the Purkinje cell layer (Ma *et al.*, 1998; Zou *et al.*, 1998). Previous studies had shown that destruction of the meninges overlying the EGL resulted in disruption of internal granule-cell layer (IGL) development and that a diffusible factor was responsible for attracting granule cell precursors towards the pial surface (Sievers *et al.*, 1994). The phenotype of the CXCR4 and CXCL12 knockout mice as well as the demonstrations that CXCL12 is expressed in the pia mater during embryonic development and that CXCR4 is expressed in the EGL during the same period (McGrath *et al.*, 1999), suggested that CXCL12 might be the pial-derived factor whose function it is to retain granule-cell precursors within the EGL. The importance of maintaining the position of granule-precursor cells in the EGL lies, in part, in regulating their exposure to the proliferative factors expressed there. Various factors have been shown to promote granule cell proliferation, including Sonic hedgehog (SHH) (Wechsler-Reya and Scott, 1999), fibroblast growth factor (FGF) (Tao *et al.*, 1996), epidermal growth factor (EGF) (Carrasco *et al.*, 2003), and insulin-like growth factor (IGF) (Ye *et al.*, 1996). SHH, which is produced by Purkinje neurons, is the single most potent mitogen for granule cells. The hypothesis that CXCL12 functions to maintain granule-cell position in the EGL was supported by *in vitro* studies that not only demonstrated that granule-cell precursors will migrate toward a source of CXCL12, such as the pia mater, during development but that further suggested that CXCL12 also directly enhances granule-cell proliferative responses to SHH (Klein *et al.*, 2001; Zhu *et al.*, 2002). This effect is similar to that of CXCL12 on lymphocyte localization in secondary lymphoid organs. At these sites, subsets of B-lymphocytes that respond to CXCL12 localize to specific microenvironments where they proliferate and differentiate (Bleul *et al.*, 1998). How CXCL12 acts with SHH to promote granule-cell proliferation is unknown. Among the potential mechanisms is CXCR4's inhibition of adenylyl cyclase and the resultant reduction in the SHH antagonist cAMP (Dwinell *et al.*, 2003). This is supported by the demonstration that activation of another G-protein-coupled receptor that increases cAMP, the pituitary adenylate cyclase activating protein (PACAP) receptor, opposes the effects of CXCR4-mediated enhancement of SHH-induced proliferation (Nicot *et al.*, 2002). CXCL12 is also likely to regulate cell numbers in the nervous system by acting as a survival factor. Additional studies showed that CXCL12 promotes the survival of retinal ganglion cells' neurons via a mitogen-activated protein kinase (MAPK)-dependent pathway (Chalasani *et al.*, 2003a).

While CXCL12 expression is maintained during development, CXCR4 expression peaks during the first postnatal week and declines thereafter, coinciding with the exit of GPCs from the cell cycle and their subsequent migration to the IGL (Klein *et al.*, 2001). This downregulation of CXCR4 expression and loss of localizing effect to the surface of the cerebellum may be important for the radial migration of granule-cell precursors into the IGL. Other developmentally regulated proteins participate in the formation of the IGL, including one that appears to function as a CXCR4 antagonist. Signaling responses to CXCL12 are inhibited by the Eph-B/ephrin-B bidirectional signaling pathway via activation of a regulator of G-protein signaling (RGS) protein via a PDZ domain (PDZ-RGS) (Lu *et al.*, 2001). Migration to IGL has been also shown to occur in response to BDNF, which is expressed by maturing granule cells in the developing IGL (Borghesani

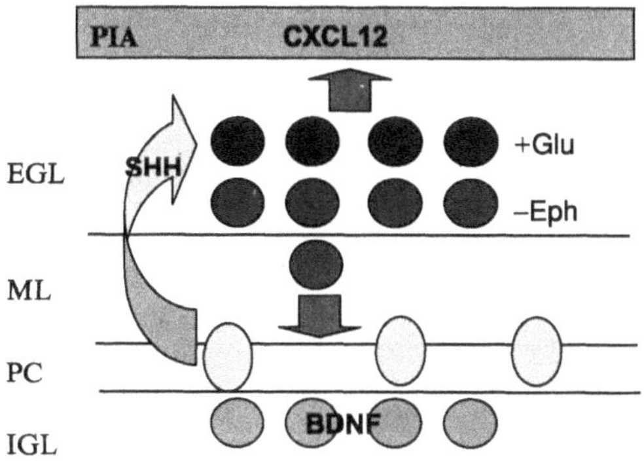

FIGURE 1 CXCL12 is critical for the normal formation of the cerebellum. CXCR4-expressing granule precursor cells (GPCs; dark purple circles) migrate from the rhombic lip to form the external granule cell layer (EGL) in proximity to CXCL12-expressing pia mater. CXCR4 signaling responses are enhanced by glutamate receptor activation (+Glu) (Klein *et al.*, 2001). Purkinje cell (PC)–derived Sonic hedgehog (SHH) induces the proliferation of GPCs in the EGL, which is enhanced by CXCL12. Migration of GPCs out of the EGL to form the internal granule cell layer (IGL) is believed to occur through alterations in CXCL12 chemoattraction achieved, in part, by decreased expression of CXCR4 by maturing GPCs (lighter purple circles), increased expression of ephrin ligands and receptors (-Eph), which antagonize CXCR4 responses to CXCL12, and by the increased expression of brain-derived neurotrophic factor (BDNF) by mature granule cell neurons (lavender ovals), which enhance the movement of GPCs into the IGL. ML, molecular layer containing Purkinje cell and granule cell neuronal axons and dendrites. (See Color Insert.)

et al., 2002). Thus, the movement of maturing granule cells from the EGL to the IGL is accomplished through the interactions of a variety of factors (Fig. 1).

The role of CXCR4 in cerebellar development is not limited to the control of granule cells' number and positions. CXCL12 also functions to control granule-cell axon elongation. This occurs through a fascinating balance of activity in two distinct Rho-dependent pathways (Arakawa *et al.*, 2003). High concentrations of CXCL12, such as may occur directly below the pia mater in the EGL, repress axon formation via Rho-induced ROCK activity. This contrasts with the effect of lower concentrations of CXCL12, which facilitate axon elongation via activation of a Rho/mDia pathway. High doses of CXCL12 and the downstream activation of protein kinase C have also been reported to induce growth cone repulsion (Xiang *et al.*, 2002). This effect could be converted to attraction with cyclic nucleotide agonists, supporting the notion that it is through the regulation of adenylyl cyclase and the level of cyclic neucleotides that CXCR4 exerts its effects. Recently, analyses of CXCR4 knockout mice have revealed some additional neuropathological phenotypes characterized by perturbations in multiple axonal trajectories within the spinal cord and decreased numbers of retinal ganglion neurons (Chalasani *et al.*, 2003b). *In vitro* studies examining axonal pathfinding using dorsal-root and retinal-ganglion cells and sympathetic neurons demonstrated that CXCL12 reduces the activity of multiple axonal repellents, which also appeared to involve Rho-dependent pathways. Together, these data suggest that CXCL12 is a critical element promoting the normal patterning of the cerebellum. This reflects a role for CXCR4 at every stage of GPC development, including proliferation, migration, growth cone movements, and axon elongation.

Expression of a variety of other chemokines and their receptors has been observed in the developing cerebellum and other CNS regions *in situ* as well as in primary cultures of neurons and glia (see Table I). Similarly to CXCR4, some of these receptors and their ligands undergo developmentally regulated changes in the patterns of their expression, suggesting that there are distinct roles in different stages of development. In the case of CCR1/CCL3, for example, the expression of CCR1 by multiple cell types has been temporally correlated with CCL3 expression in Purkinje cells (Cowell and Silverstein, 2003). The precise roles of CCL3 and these other chemokines in specific phases of neuronal and/or glial development in the cerebellum are presently unknown.

Examination of brains from mice with targeted deletion of CXCR4 also revealed failure in the normal formation of the hippocampal dentate gyrus (DG), an essential structure for learning and memory (Lu *et al.*, 2002). Interestingly, the DG is comprised of neurons that, based on their morphology, are referred to as granule cells and that, like cerebellar granule cells, undergo successive cycles of proliferation and migration (Gould and Cameron, 1996). Similar to the establishment of the EGL in the cerebellum, hippocampal GPCs establish a nonperiventricular proliferative zone comprised of precursor neurons that have migrated away from a more standard periventricular germinal matrix. Analysis of normal CXCL12 and CXCR4 expression during hippocampal development suggests that CXCR4-expressing granule-cell precursors that form the migratory stream migrate in proximity to CXCL12-expressing cells (Fig. 2) (Bagri *et al.*, 2002). This is analogous to the migration of cerebellar GPCs from the rhombic lip across the cerebellar primordium beneath the CXCL12-expressing meningeal layer. Loss of CXCR4 in the hippocampus, however, results in a failure of the migratory stream, suggesting that (in contrast to what occurs in the cerebellum) migration from the periventricular neuroepithelium to the DG is directed by CXCL12. In the cerebellum, GPCs successfully migrate across the entire distance of the EGL, but their position is not maintained, suggesting that this initial migration is not directed by CXCL12. The hippocampus and the cerebellum in CXCR4-deficient animals are similar only in the aberrant localization and failed proliferation of granule-cell progenitors.

TABLE I

Developmental Expression of Chemokines and Their Receptors by CNS Cell Types

Cells	Chemokine/Receptor	Species/Source	References
Neurons			
Granule	CXCL12, CXCR4	mouse/*in situ*	Ma *et al.*, 1998; Zou *et al.*, 1998
		rat/*in vitro*	Bonavia *et al.*, 2003
	CXCL12	rat/*in situ*	Tham *et al.*, 2001
	CCR1	rat/*in situ*	Cowell and Silverstein, 2003
	CXCR4, CCR5	macaque/*in situ*	Westmoreland *et al.*, 2002
Cortical	CXCL12, CXCR4	rat/*in vitro, situ*	Bajetto *et al.*, 1999; Stumm *et al.*, 2003
	CXCR1, CCR2	rat/*ex vivo*	Danik *et al.*, 2003
Purkinje	CCR1, CCR3	rat/*in situ*	Cowell and Silverstein, 2003
	CCR5	macaque/*in situ*	Westmoreland *et al.*, 2002
	CCL2	human/*in situ*	Meng *et al.*, 1999
Pyramidal	CXCR4, CCR5	macaque/*in situ*	Westmoreland *et al.*, 2002
Glia			
Astrocyte	CXCL12	rat/*in vitro*	Bajetto *al.*, 1999
		rat/*in vitro*	Bonavia *et al.*, 2003
	CXCL1	rat/*in vitro*	Robinson and Franic, 2001
	CCR1	rat/*in situ*	Cowell and Silverstein, 2003
	CCL2, CCL5, CCR2	human/*in situ*	Rezaie *et al.*, 2002
	CX3CL1	human/*in situ*	Hatori *et al.*, 2002
	CXCR3	human/*in situ*	Van Der Meer *et al.*, 2001
	CXCR4	human/*in vitro*	Rezaie *et al.*, 2002
	CCL3	human/*in vitro*	Miyamoto and Kim, 1999
	CXCL8, CXCL9, CXCL10, CXCL12	human/*in vitro*	Croitoru-Lamoury *et al.*, 2003
	CCL2, CCL3, CCL4		
	CCR3, CCR5, CCR8		
	CXCR3, CXCR5, CXCR6		
	CX3CR1		

(*Continued*)

TABLE I *(Continued)*

Cells	Chemokine/Receptor	Species/Source	References
Microglia	CCR1	rat/*in situ*	Cowell and Silverstein, 2003
	CXCR4, CCL3	human/*in vitro*	Rezaie *et al.*, 2002; Tsai *et al.*, 2002
Oligodendrocyte	CXCR2	rat/*in situ*	Tsai *et al.*, 2002
	CXCL1	rat/*in situ*	Robinson and Franic, 2001
Ependyma	CCR5	macaque/*in situ*	Westmoreland *et al.*, 2002
Vasculature			
Choroids plexus/	CCR5	macaque/*in situ*	Tsai *et al.*, 2002
Endothelium	CXCR4, CCR5	macaque/*ex vivo*	MacLean *et al.*, 2001

It is intriguing to speculate that CXCL12 and CXCR4 are critically involved in the development of these neurons because of the alternating cycles of proliferation and migration and the necessity to maintain both capabilities.

CXCR4 also plays a role in the development of other areas of the neocortex (Stumm *et al.*, 2003). CXCR4 is expressed by interneuron precursors migrating from the basal forebrain to the neocortex. Loss of CXCR4 results in underrepresentation of interneurons in the superficial cortical layers and ectopic placement in the deep layers of the neocortex. Thus, CXCL12 and CXCR4 regulate the layering of the cerebellum, the hippocampus, and other cortical structures.

Development of the posterior spinal cord also requires chemokine activity. Similarly to granule-cell precursors, oligodendrocyte precursors undergo a period of proliferation prior to migrating to their final destinations during the first postnatal week in the mouse (Dubois-Dalcq and Murray, 2000). These cells are generated in the ventral ventricular zone of the embryonic spinal cord, where they have been observed to express the chemokine receptor CXCR2 while astrocytes within the same region express the CXCR2 ligand CXCL1 (Robinson and Franic, 2001; Tsai *et al.*, 2002). Interestingly, in contrast to its usual role as a chemoattractant, here CXCL1 acts to inhibit the migration of oligodendrocyte precursors from this proliferative zone by enhancing adhesive interactions of these cells with their extracellular matrix. However, similarly to the role of CXCL12 in the cerebellum and hippocampus, CXCL1 also enhances the proliferation of oligodendrocyte precursors to PDGF, a mitogen secreted by astrocytes and neurons in the developing spinal cord. CXCL1 expression is downregulated by postnatal day 5, when

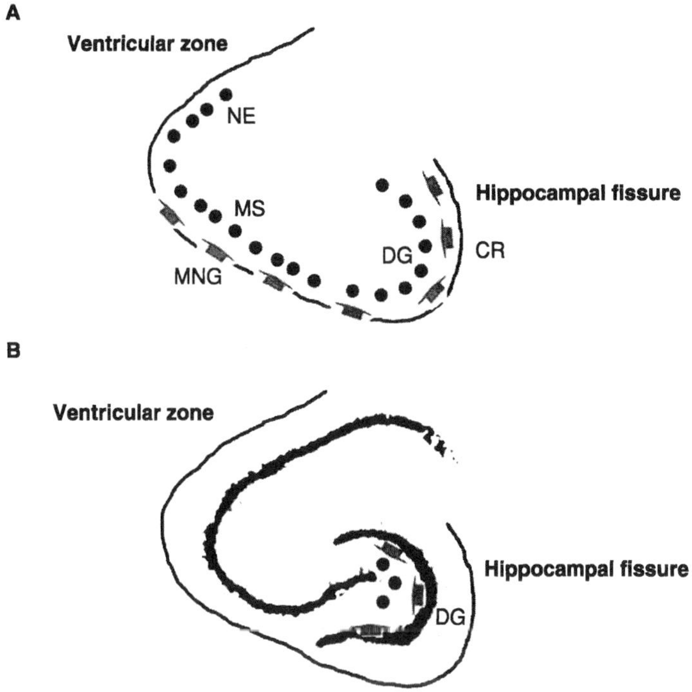

FIGURE 2 CXCL12 and CXCR4 are critical for the normal formation of the dentate gyrus (DG). (A) Dentate granule cells (blue circles) first arise in the dentate neuroepithelium (NE), a typical subventricular primary neural germinal zone. The dentate migratory stream (MS) arises from this and constitutes a secondary proliferative zone in which granule precursor cells can be seen to migrate and divide. Finally, a tertiary proliferative zone forms in the hilus of the DG. During embryonic and early postnatal development, granule cells are expressed by the meninges (MNG) as well as by Cajal–Retzius cells (CR) lining the hippocampal fissure. Targeted deletion of CXCR4 results in the mislocalization of dentate granule cells and a reduction in their numbers. During early hippocampal development, CXCL12 may also modulate neurotransmission in developing hippocampal circuits, further affecting hippocampal patterning. (B) The pattern of CXCL12 and CXCR4 expression is different beyond the postnatal period. Now CXCL12 is highly expressed by cells populating the blades of the DG (green arrows), and proliferating cells within the dentate hilus can be seen to express CXCR4 (blue circles). (See Color Insert.)

oligodendrocyte precursors are beginning to migrate into the dorsal portion of the spinal cord. Astrocytes in the cord begin to express CXCL1 at this time, leading to another wave of proliferation of oligodendrocyte precursors, which then proceed in their myelination of this portion of the spinal cord.

In all of these processes, chemokines and their receptors perform many of the same functions as described for immune cells, namely, enhancement of proliferative responses, recruitment to new locations, promotion of survival, and alterations of the cytoskeletal framework to facilitate the performance of more highly differentiated cellular functions. Indeed, many of the cells that express chemokines and/or their receptors during CNS development cease producing these molecules once the nervous system has properly formed. Thus, once neural cell types have attained appropriate cell number and placement, chemokine expression is dramatically downregulated, with the exception of several homeostatic chemokines and their receptors whose role in normal CNS physiology has become an area of intense investigation.

III. CHEMOKINES IN ADULT CNS FUNCTION

There are two patterns of chemokine expression, constitutive and regulated (DeVries *et al.*, 1999). Lymphoid and parenchymal tissues, which orchestrate the recirculation of leukocytes during normal immune surveillance, may constitutively express many chemokines. During inflammation, all chemokine expression may potentially be upregulated. In the CNS, constitutive expression of the chemokines CXCL12, CX3CL1, CCL19, and CCL21 has been observed (Nishiyori *et al.*, 1998; Rossi *et al.*, 1998; Alt *et al.*, 2002; Stumm *et al.*, 2002). CX3CL1 is expressed by neurons, CCL19 and CCL21 are expressed by endothelial cells, and CXCL12 is expressed by both cell types. Though it is unclear whether any or all of these chemokines participate in the recruitment of leukocytes into the CNS for the purpose of immune surveillance, the neuronal expression of chemokines has been linked to a novel neuromodulatory function for these molecules in the normal CNS. In this section, we will detail studies of the constitutive expression of chemokines and chemokine receptors in adult CNS and discuss the potential roles for these molecules in CNS function.

CXCL12 is expressed by many normal tissues, including spleen, liver, lung, kidney, and brain, and it is thought to be important for the trafficking of leukocytes in the absence of inflammation (Nagasawa *et al.*, 1994; Kim and Broxmeyer, 1999). Indeed, the level of expression of the receptor for CXCL12, CXCR4, on lymphoblastic leukemic cells correlates with their infiltration into organs, and the CNS is the most common site of extramedullary leukemia (Crazzolara *et al.*, 2001). Recent studies examining CNS expression of CXCL12 isoforms as well as CXCR4 have demonstrated cellular and regional specificity in the distribution patterns of these proteins. CXCL12 and CXCR4 demonstrate overlapping expression patterns found exclusively in neurons in the cerebral cortex, hippocampal formation, and amygdala (Stumm *et al.*, 2002). CXCL12 is also expressed by endothelial cells of cerebral microvessels throughout the CNS. Inflammations induced by injection of lipopolysaccharide or focal ischemia was observed to differentially regulate both cellular and regional patterns of expression of CXCL12 isoforms, suggesting that this chemokine has a role in both neurotransmission and leukocyte trafficking during inflammatory states (Stumm *et al.*, 2002). However, a role for CXCL12 in normal neurophysiology and in the baseline trafficking patterns of leukocytes during noninflammatory states remains to be established.

Early studies demonstrated that CXCL12 signaling through neuronal CXCR4 induced increases in intracellular calcium through a $G_{\alpha i}$-coupled pathway, as in leukocytes, leading to various downstream effects (Meucci *et al.*, 1998; Klein *et al.*, 1999). One effect was neuroprotection; CXCL12 treatment of fully differentiated neurons was observed to increase the survival of neurons deprived of trophic support or treated with virotoxins (Meucci *et al.*, 1998; Gillard *et al.*, 2002). The anti-apoptic effects of CXCL12 have been linked to activation of MAPK signal-transduction pathways and the regulation of cell-cycle proteins in neurons (Khan *et al.*, 2003; Chalasani *et al.*, 2003a). Another profound and novel effect of CXCL12 is neuromodulation. Activation of neuronal CXCR4 was shown to affect both preexcitatory and postexcitatory currents through actions on ion channels and glutamate release (Limatola *et al.*, 2000b; Bezzi *et al.*, 2001; Oh *et al.*, 2002). The expression of CXCL12 and CXCR4 by neurons in several regions of the hippocampus and in the cerebellum further supports the notion that this chemokine, through effects on glutaminergic synapses, might play a significant role in synaptic plasticity. In support of this, CXCR4 activation produces presynaptic inhibition of glutamate release at parallel fibers in cerebellar slices and decreases excitatory transmissions between hippocampal neurons in culture (Meucci *et al.*, 1998; Ragozzino *et al.*, 2002; Liu *et al.*, 2003). In addition, in hippocampal neurons, CXCL12 modulates the frequency of calcium spikes, whose synchronization has been linked to the plasticity of neuronal networks (Liu *et al.*, 2003). CXCL12 may also play a role in the modulation of synaptic transmissions due to other classical neurotransmitters. Studies of CXCR4 expression on neuron subtypes have demonstrated its co-localization with enzymes involved in the synthesis of acetylcholine and dopamine in neurons of the basal ganglia and substantia nigra, respectively (Banisadr *et al.*, 2002a). The possible role of CXCL12 as a novel neurotransmitter in its own right has also been suggested. Peptide-sequence analysis of CXCL12 reveals that it contains proteolytic cleavage sites that could generate novel neuropeptides (Gleichmann *et al.*, 2000). These peptides could be generated extracellularly, adding an additional means of regulating neurotransmission directly within the synaptic cleft.

CX3CL1 is a unique chemokine distinguished structurally by the presence of a CX3C motif as well as transmembrane spanning and mucin-like domains; it shows atypical constitutive expression in a number of nonhematopoietic tissues, including brain tissue (Rossi *et al.*, 1998). In rodent tissues, CX3CL1 expression has been localized principally to neurons in the olfactory bulb, cortex, and hippocampus, while CX3CR1, its receptor, is expressed by resident microglia and by hippocampal neurons (Combadiere *et al.*, 1998; Nishiyori *et al.*, 1998). In human tissues, CX3CL1 was also found to be constitutively expressed by astrocytes (Hulshof *et al.*, 2003). In contrast to the research on the roles of CXCL12 and CXCR4 in CNS development, targeted deletion of CX3CL1 and CX3CR1 did not produce any neurological abnormalities (Cook *et al.*, 2001; Haskell *et al.*, 2001). Thus, the physiologic roles of this chemokine and its receptor in normal CNS function are unclear. The expression of both CX3CL1 and CX3CR1 by healthy hippocampal neurons suggests an autocrine function for this chemokine in these neurons (Meucci *et al.*, 2000). In addition, the findings that CX3CL1 can act directly on neurons to activate survival pathways and the upregulation of neuronal CX3CL1 levels during certain types of CNS inflammation suggests that this chemokine has a neuroprotective role (Tong *et al.*, 2000). Neurons expressing CX3CL1 have also been observed to be associated with CX3CR1-expressing microglia in normal CNS, and neuronal–microglial interactions via CX3CL1 have been observed *in vitro* (Harrison *et al.*, 1998; Nishiyori *et al.*, 1998). Additional studies will be required to determine the role of these molecules in normal CNS function.

CCL19 and CCL21 are ligands for chemokine receptor CCR7 that are implicated in the migration of lymphocytes into secondary lymphoid organs (Mori *et al.*, 2001). Interest in the role of these chemokines in CNS function was generated when lymphocytes obtained from the cerebrospinal fluid (CSF) of multiple sclerosis (MS) patients were observed to express CCR7 (Giunti *et al.*, 2003). CCL19 is expressed by a subpopulation of venules in noninflamed CNS tissues, suggesting that it may play a role in the trafficking of lymphocytes during CNS immune surveillance (Chen *et al.*, 2002). Transgenic mice expressing CCL21, but not CCL19, in oligodendrocytes were observed to develop CNS inflammatory disease with granulocyte infiltration (Chen *et al.*, 2002), although both chemokines are upregulated in CNS inflammatory models of mononuclear cell infiltration. Although additional studies have demonstrated that both chemokines can induce adhesion of myelin-specific T-cells to inflamed CNS microvessels (Alt *et al.*, 2002), it is unclear whether either of these chemokines plays a role in the trafficking of leukocytes into noninflamed CNS tissues.

A variety of other chemokine receptors have been detected on neurons *in vitro* and *in situ* (see Table II). Treatment of cultured neurons with chemokine ligands for these receptors generally induces increases in intracellular calcium. Activation of these receptors appears to induce survival benefits and modulation of neurotransmission along the same lines as CXCR4.

TABLE II
Neuronal Chemokine Receptor Expression in the Adult Central Nervous System

Neurons	Chemokine/Receptor	Species	References
Purkinje	Duffy Ag	human	Horuk *et al.*, 1997
Cortical, spinal cord	CXCR2	human	Horuk *et al.*, 1997; Xia, 1997
Granule, Purkinje	CXCR1	mouse	Giovannelli *et al.*, 1998
Septal	CXCR2	rat	Danik *et al.*, 2003
Cortical	CXCR4	human	Lavi *et al.*, 1998
		rat	Stumm *et al.*, 2002
Midbrain		rat	Banisadr *et al.*, 2003
Cortical	CCR5	macaque	Rottman *et al.*, 1997
Pyramidal	CCR3, -5, CXCR4	macaque	Westmoreland *et al.*, 1998
Cortical, subcortical	CXCR3	human	Xia *et al.*, 2000
Cortex, midbrain	CCR2	rat	Banisadr *et al.*, 2002b

For example, CXCR2 activation in cerebellar slices protects granule cells from apoptosis through a mechanism that requires AMPA receptor signaling (Limatola *et al.*, 2000a) and leads to enhanced neurotransmitter release from both granule and Purkinje neurons (Giovannelli *et al.*, 1998; Ragozzino *et al.*, 1998). A common theme in the neuronal chemokine receptor literature is the balance between cell death and survival through the activation of multiple MAPK pathways linked to proapoptotic and antiapoptotic effects (e.g., Akt versus p38 activation) (Limatola *et al.*, 2002; Chalasani *et al.*, 2003a). There is speculation that chemokines are released during stressful states to relay information about the need to increase antiapoptotic machinery. Each neural network may have a threshold above which proapoptotic machinery kicks in, but the proapoptotic machinery may kick in only after a neuron has relayed information about the inevitability of its death to other neurons in its network. This could function to strengthen some synapses in compensation for loss of others. For example, CX3CL1 has been shown to inhibit neurotoxicity induced by the HIV-1 viral envelope protein gp120 via Akt-phosphorylation in a PI3-kinase-dependent pathway, but it can also potentiate the toxic effects of PI3-kinase inhibitors (Meucci *et al.*, 2000). This suggests that in the absence of activation of antiapoptotic pathways, CX3CL1 instead activates proapoptotic pathways such as ERK1/2 and spurs the mobilization of calcium.

The normal separation of neural and immune cell types has allowed both systems to utilize chemokines and their receptors for various aspects of development and normal physiology. During CNS inflammation, however, the two systems become superimposed, occupying the same space within a single tissue, as glia upregulate both homeostatic and inflammatory chemokines in order to recruit leukocytes. The presence of elevated levels of chemokines in the brain not only serves to recruit immune cells but also could potentially disrupt normal brain function through direct actions on neurons.

IV. CHEMOKINES AND CNS INFLAMMATORY DISEASES

Leukocyte infiltration into brain and spinal-cord parenchyma accompanies many types of CNS diseases, including viral encephalitis, autoimmune demyelinating diseases, and neurodegenerative disorders. The importance of chemokines in leukocyte trafficking into sites of inflammation has been appreciated in various inflammatory conditions. Thus, examination of CSF and CNS tissues for chemokine expression was a logical approach to the identification of novel therapeutic targets for the treatment of CNS inflammatory diseases for which there are currently no effective treatments. Elevated levels of many mononuclear cell chemoattractants have been observed in the CSF of patients with aseptic and bacterial meningitis and with encephalitides caused by herpes simplex virus (HSV)-1, human immunodeficiency virus (HIV)-1, and cytomegalovirus (CMV), as well as patients with MS (see Table III). Along these lines, mononuclear cells expressing many of the chemokine receptors to which these ligands bind have been observed in the CSF and serum of patients with many of these conditions (Sorensen *et al.*, 1999; Nansen *et al.*, 2000; Misu *et al.*, 2001; Pashenkov *et al.*, 2002; Giunti *et al.*, 2003). The sources of these inflammatory chemokines include astrocytes, microglia, and neurons. However, the regional and temporal patterns of expression and how they relate to the wide range of pathologies observed in CNS inflammatory diseases remain unknown.

In the case of MS, as well as its animal model, experimental autoimmune encephalomyelitis (EAE), the expression of chemokines within CNS lesions suggests that these molecules play an important role in the propagation of injury and maintenance of the disease cycle through the recruitment of autoreactive T-cells. Upon encountering antigen, these cells become activated and begin to secrete chemokines that recruit more leukocytes but also express inflammatory cytokines such as TNF-α, which upregulate the expression of chemokines from resident glia, amplifying the initial inflammatory response (Huang *et al.*, 2000). As additional leukocytes are recruited, the cascade of inflammation continues; activated astrocytes and microglia continue to

TABLE III

Chemokines in the Cerebrospinal Fluid of Patients with Neuroinflammatory Diseases

Bacterial Meningitis	
CXCL8	Handa, 1992
CXCL1, CCL2–4	Spanaus *et al.*, 1997
CXCL10	Lahrtz *et al.*, 1997
CX3CL1	Kastenbauer *et al.*, 2003
CXCL5	Zwijnenburg *et al.*, 2003
Viral Encephalitis	
HSV-1	
CXCL8	Rosler *et al.*, 1998
CCL2	
CCL3	
CCL5	
HIV-1	
CCL2	Cinque *et al.*, 1998
CXC3L1	Cotter *et al.*, 2002
CMV	
CCL2	Bernasconi *et al.*, 1996
JEV	
CXCL8	Singh *et al.*, 2000
Multiple Sclerosis	
CCL2	Franciotta *et al.*, 2001
CCL3	Miyagishi *et al.*, 1995
CXCL9, CXCL10, CCL5	Sorensen *et al.*, 1999
CCL19, CCL21	Pashenkov *et al.*, 2003
Lupus	
CXCL8	Trysberg *et al.*, 2000

express inflammatory cytokines and chemokines, and autoreactive T-cells and macrophages attack oligodendrocytes within the white matter.

While the chemokines that target T-cells for homing into the CNS have yet to be described, recent studies using fluorescence videomicroscopic analysis of white-matter microcirculation have observed that the α4-integrin-mediated adhesion of T-cells to normal CNS endothelial cells and their subsequent transmigration into normal CNS involves PTX-sensitive G-proteins (Vajkoczy *et al.*, 2001), suggesting that this process is indeed mediated by unknown chemoattractants expressed constitutively by the blood–brain barrier. In a recent placebo-controlled trial, inhibition of α4-integrin led to fewer new inflammatory brain lesions and fewer relapses over a six-month period in patients with relapsing MS (Miller *et al.*, 2003). Thus, the process of T-cell extravasation into the CNS is likely to require interactions between various molecules, including those that mediate adhesion to endothelium and antigen recognition as well as chemoattraction.

The expression of a variety of chemokines within CNS tissues of MS patients and mice with EAE (see Table III) correlates with the appearance of pathogenic T-cells bearing chemokine receptors that bind them. CD4+ T-cells isolated from the CNS of encephalitic animals express CCR1, CCR2, CCR5, CCR7, and CXCR3 (Fife *et al.*, 2001b; Alt *et al.*, 2002). Use of antibodies and targeted deletion to elucidate the chemokines and chemokine receptors responsible for the trafficking of T-cells has yielded conflicting results. Antibody treatments that neutralize the effects of CCL3 and CXCL10, which bind CCR5 and CXCR3, respectively, have resulted in the suppression of acute EAE (Karpus and Kennedy, 1997; Fife *et al.*, 2001a), while targeted deletion of these chemokines or their receptors has not (Tran *et al.*, 2000; Klein *et al.*, 2004). In contrast, use of antibodies that neutralize the effect of CCL2, which recruits monocytes, and targeted gene deletion of CCL2, or its receptor CCR2, all resulted in altered susceptibility to EAE (Karpus and Kennedy, 1997; Fife *et al.*, 2000; Izikson *et al.*, 2000; Huang *et al.*, 2001). Thus, the question of signals responsible for T-lymphocyte recruitment in this model remains to be fully investigated.

Infection with mouse hepatitis model induces an acute inflammatory encephalomyelitis followed by a chronic demyelinating disease similar to MS (Glass *et al.*, 2002). As in EAE, infiltrating T-cells appear to accelerate the inflammatory phase of the disease, expressing macrophage chemoattractants within the CNS and upregulating expression of multiple chemokines by astrocytes (see Table III). Antibody treatment of mice against CCL5, which binds CCR1 and CCR5, halted the macrophage-mediated demyelination, suggesting that this chemokine plays a role in acute demyelination, and diminished the initial recruitment of T-cells, resulting in higher viral titers (Lane *et al.*, 2000). This effect was not sustained, however; compensatory mechanisms resulted in control of the virus by day 12 postinfection. Administration of anti-CXCL10 antibodies during acute murine hepatitis virus (MHV) encephalomyelitis leads to decreased viral clearance and increased mortality (Liu *et al.*, 2000), while similar treatment during the chronic phase of the disease halted the progression of demyelination and increased remyelination (Liu *et al.*, 2001b). Both of these outcomes were correlated with decreased numbers of infiltrating CD4+ and CD8+ T-cells. Thus, CXCL10 is important for the recruitment of T-cells during established demyelinating disease. These results suggest that different chemokines may be important for the recruitment of T-cells in the healthy CNS as opposed to the damaged CNS. Along these lines, antibodies against CXCL9, a related lymphocyte chemoattractant that also binds CXCR3, resulted in worse encephalomyelitis post–MHV infection, suggesting that this chemokine is involved in the recruitment of a protective T-cell response (Liu *et al.*, 2001a). These results also mirror what is observed clinically in many human viral encephalitides, where mononuclear cells are important for viral clearance but their persistence within the CNS also leads to immune-mediated neuronal injury and death.

While the mechanism of injury that occurs as a result of inflammatory infiltration into the CNS during viral infections is unclear, the extent of CNS dysfunction appears to correlate with increased expression of chemokines by activated microglia and astrocytes. As discussed above, various chemokine receptors are expressed by subpopulations of neurons in many brain regions. Thus, increased chemokine expression during neuroinflammation may directly impact neuronal function and survival. In HIV-1 encephalitis, various chemokines and chemokine receptors have been shown to play roles in each aspect of the pathologic process. During initial infection, HIV-1 infects T-cells and macrophages via the coreceptor CCR5 (Berger *et al.*, 1999). An accessory HIV-1 gene product, Nef, then induces the production of CCL3 and CCL4 by macrophages. By attracting and activating T-cells, this CCL3 and CCL4 production perpetuates the infection (Swingler *et al.*, 1999). HIV-1-infected macrophages traffic into the CNS as part of the normal turnover of perivascular monocytic cells, producing productive infection within parenchymal microglia and latent infection in astrocytes (Persidsky *et al.*, 2000; Kim *et al.*, 2003). The upregulation of a variety of mononuclear cell chemoattractants leads to the infiltration of leukocytes, which are believed to control the extent of infection and limit the numbers of virally infected microglia (Zink *et al.*, 1999). As immune function wanes, uncontrolled infection of microglia and astrocytes leads to increased chemokine expression, increasing the recruitment of HIV-infected monocytes/macrophages. Indeed, macrophage and microglial activation is the best neuropathological correlate of HIV dementia, and the appearance of the monocyte chemoattractant CCL2 in patient CSF has been shown to precede and predict encephalitis (Cinque *et al.*, 1998). In addition, CCL3 expression and CCR5 expression localize to microglial nodules, suggesting that this chemokine and receptor play a role in the local formation of these lesions (Bonwetsch *et al.*, 1999). Although HIV does not infect neurons, these cells begin to undergo apoptosis, leading to progressive neurological deficits and death. The main hypothesis has been that infected microglia are the source of neurotoxic substances that produce the observed extensive damage.

Initially, it was believed that the HIV envelope protein gp120, which is known to bind and signal through CXCR4, was the neurotoxic culprit. Indeed, gp120 proved to induce neuronal apoptosis when added to neuronal cultures, injected into the brain, or overexpressed in transgenic mice (Brenneman *et al.*, 1994). In addition, gp120 signaling through astrocyte CXCR4 was shown to increase glutamate release by this cell type (Bezzi *et al.*, 2001). Thus, it was postulated that gp120 was inducing excitotoxicity through this indirect mechanism. Given that glutamate receptor antagonists failed to prove efficacious in treating HIV dementia and that chemokine ligands for neuronal CXCR4 and CCR5 have been shown to induce apoptosis *in vitro* (Kaul *et al.*, 2001), direct mechanisms for the neurotoxic effect seemed likely. Interestingly, CXCL12, which should ideally compete with gp120 in binding to CXCR4, appears to have both neuroprotective and neurotoxic effects on its own (Kaul and Lipton, 1999; Meucci *et al.*, 1998). Recent studies have demonstrated that HIV-1-infected microglia secrete matrix metalloproteinase-2, which processes CXCL12 into a highly neurotoxic protein (Zhang *et al.*, 2003). Given that chemokines are expressed by both healthy and inflamed CNS, they could potentially be the substrate of a variety of peptide-processing enzymes both intracellularly and extracellularly.

Investigations of other viral encephalitides have observed patterns of chemokine upregulation consistent with the development of innate and adaptive immune responses. For example, upregulation of CNS neutrophil chemoattractants was observed in encephalitides caused by HSV-1, Japanese encephalitis virus (JEV), and Murray Valley encephalitis virus (MVEV) (Rosler *et al.*, 1998; Andrews *et al.*, 1999; Singh *et al.*, 2000). In JEV and MVEV encephalitides, levels of chemokines and subsequent neutrophil infiltration correlate with severity of disease. In HSV-1 encephalitits, CXCL10 expressed by virally infected microglia inhibited HSV-1 replication in neurons, demonstrating that these chemokines play a role in antiviral defense as well as in amplification of the immune response (Lokensgard *et al.*, 2001). CX3CL1, whose role in normal CNS function is unknown, is dramatically upregulated in a variety of infectious and autoimmune neuroinflammatory diseases (Kastenbauer *et al.*, 2003), suggesting that cytokine milieu plays a role in the expression of this chemokine.

V. CHEMOKINES IN CNS NEOPLASTIC DISEASES

The broad role that chemokines play in regulating development and function in the CNS appears to be recapitulated in their expression and activity in CNS tumors. CXCL12 (Barbero *et al.*, 2002; Rubin *et al.*, 2003), CXCL1

(Van Meir *et al.*, 1992; Tada *et al.*, 1993; Suliman *et al.*, 1999), and CCL2 (Desbaillets *et al.*, 1994; Takeshima *et al.*, 1994; Sato *et al.*, 1995), as well as CXCR3 (Goldberg *et al.*, 2001) and CXCR4 (Rempel *et al.*, 2000; Rubin *et al.*, 2003) are expressed in astrocytic, neuronal, or meningeal tumors. Tumor production of CXCL1 and CCL2 appears to be important for the recruitment of immune cells and may therefore promote antitumor immune responses (Tada *et al.*, 1993; Sato *et al.*, 1995; Oh *et al.*, 2001). It is interesting to note that CXCL12 may be an important regulator of CXCL1 and CCL1 expression in astroglioma cells.

CXCR3 expression in astrocytic tumors has not been correlated with stage or prognosis, and its potential role is inferred from the pattern of its expression and its activity in lymphocytes. CXCR3 is highly expressed on reactive astrocytes, and, by analogy to lymphocytes, it is proposed that it functions to mediate a chemotactic signal that recruits astrocytes to sites of inflammation. CXCR3 is expressed on a subset of gemistocytic astrocytes within high-grade glial neoplasms. It has been suggested that CXCR3 may mediate similar chemotactic responses and thereby contribute to the pattern of glioma growth (Goldberg *et al.*, 2001).

A significant role for CXCL12 and CXCR4 in brain tumor biology is more clearly established. Expression of CXCR4 in high-grade gliomas was first described by Rempel *et al.* (2000) and appeared to correlate with grade and be localized to areas of necrosis and angiogenesis. Its potential relevance to tumor biology was suggested by *in vitro* studies that demonstrated its ability to mediate proliferative signals (Bajetto *et al.*, 2001; Rubin *et al.*, 2003). Subsequent work has demonstrated that CXCR4 expression is more widespread in CNS tumors, including (in addition to high-grade gliomas) medulloblastoma, a tumor derived from cerebellar granule cells (Rubin *et al.*, 2003). Cerebellar granule cells are a neuronal population whose development is abnormal in the CXCR4- or CXCL12-deficient mice (Ma *et al.*, 1998; Zou *et al.*, 1998). CXCR4 is not known to be mutationally activated in these tumors, but the presence of CXCL12 in tumor-associated blood vessels suggests that a paracrine relationship for CXCR4 activation exists *in vivo*. *In vitro*, CXCL12 is a proliferative factor for both tumor types and also mediates chemotactic and survival signals. The importance of this to tumor biology was established when the systemic delivery of the CXCR4 antagonist AMD 3100 (Hendrix *et al.*, 2000) inhibited the growth of intracranial xenografts of medulloblastoma and glioblastoma multiforme (Rubin *et al.*, 2003).

The regulation of CXCR4 expression by factors known to be active in CNS tumors, such as Sonic hedgehog (Zhao *et al.*, 2002), vascular–endothelial growth factor (Lee *et al.*, 2002), the von Hippel–Lindau tumor suppressor pVHL (Staller *et al.*, 2003), NF-κB (Helbig *et al.*, 2003), and TNF-α

(Oh *et al.*, 2001), suggests that CXCR4 may be not only an important contributor to CNS tumor biology but one with broad potential as a therapeutic target.

VI. CONCLUDING REMARKS AND FUTURE DIRECTIONS

The chemokine system is involved in many important aspects of CNS functioning. In this review we have touched on some of the highlights from the recent literature that have revealed a role for chemokines in development, normal physiology, and pathological inflammatory and malignant processes of the CNS. As one might predict from other organ systems, chemokines orchestrate the trafficking of leukocytes into the CNS in homeostasis and in response to stress, injury, infection, and autoimmune diseases. However, chemokines also play a role in the positioning of neurons during CNS development and in the growth and movement of neuronal tumors. These studies suggest that chemokines represent novel therapeutic targets for inflammatory and malignant diseases of the CNS. While still in the early stages of study, there is also some indication that chemokines may signal adult neurons and may therefore play a role in neuronal signaling, which may lead to exciting and unanticipated findings that chemokines play a role in cognitive functions. Such findings might lead to novel therapies for degenerative mental illnesses.

References

Alt, C., Laschinger, M., and Engelhardt, B. (2002). Functional expression of the lymphoid chemokines CCL19 (ELC) and CCL 21 (SLC) at the blood-brain barrier suggests their involvement in G-protein-dependent lymphocyte recruitment into the central nervous system during experimental autoimmune encephalomyelitis. *Eur. J. Immunol.* **32,** 2133–2144.

Andrews, D. M., Matthews, V. B., Sammels, L. M., Carrello, A. C., and McMinn, P. C. (1999). The severity of murray valley encephalitis in mice is linked to neutrophil infiltration and inducible nitric oxide synthase activity in the central nervous system. *J. Virol.* **73,** 8781–8790.

Arakawa, Y., Bito, H., Furuyashiki, T., Tsuji, T., Takemoto-Kimura, S., Kimura, K., Nozaki, K., Hashimoto, N., and Narumiya, S. (2003). Control of axon elongation via an SDF-1alpha/Rho/mDia pathway in cultured cerebellar granule neurons. *J. Cell Biol.* **161,** 381–391.

Bagri, A., Gurney, T., He, X., Zou, Y. R., Littman, D. R., Tessier-Lavigne, M., and Pleasure, S. J. (2002). The chemokine SDF1 regulates migration of dentate granule cells. *Development* **129,** 4249–4260.

Bajetto, A., Bonavia, R., Barbero, S., Piccioli, P., Costa, A., Florio, T., and Schettini, G. (1999). Glial and neuronal cells express functional chemokine receptor CXCR4 and its natural ligand stromal cell-derived factor 1. *J. Neurochem.* **73,** 2348–2357.

Bajetto, A., Barbero, S., Bonavia, R., Piccioli, P., Pirani, P., Florio, T., and Schettini, G. (2001). Stromal cell-derived factor-1alpha induces astrocyte proliferation through the activation of extracellular signal-regulated kinases 1/2 pathway. *J. Neurochem.* **77,** 1226–1236.

Banisadr, G., Fontanges, P., Haour, F., Kitabgi, P., Rostene, W., and Melik Parsadaniantz, S. (2002a). Neuroanatomical distribution of CXCR4 in adult rat brain and its localization in cholinergic and dopaminergic neurons. *Eur. J. Neurosci.* **16,** 1661–1671.

Banisadr, G., Queraud-Lesaux, F., Boutterin, M. C., Pelaprat, D., Zalc, B., Rostene, W., Haour, F., and Parsadaniantz, S. M. (2002b). Distribution, cellular localization and functional role of CCR2 chemokine receptors in adult rat brain. *J. Neurochem.* **81,** 257–269.

Banisadr, G., Skrzydelski, D., Kitabgi, P., Rostene, W., and Parsadaniantz, S. M. (2003). Highly regionalized distribution of stromal cell-derived factor-1/CXCL12 in adult rat brain: constitutive expression in cholinergic, dopaminergic and vasopressinergic neurons. *Eur. J. Neurosci.* **18,** 1593–1606.

Barbero, S., Bajetto, A., Bonavia, R., Porcile, C., Piccioli, P., Pirani, P., Ravetti, J. L., Zona, G., Spaziante, R., Florio, T., and Schettini, G. (2002). Expression of the chemokine receptor CXCR4 and its ligand stromal cell-derived factor 1 in human brain tumors and their involvement in glial proliferation *in vitro*. *Ann. NY Acad. Sci.* **973,** 60–69.

Berger, E. A., Murphy, P. M., and Farber, J. M. (1999). Chemokine receptors as HIV-1 coreceptors: roles in viral entry, tropism, and disease. *Annu. Rev. Immunol.* **17,** 657–700.

Bernasconi, S., Cinque, P., Peri, G., Sozzani, S., Crociati, A., Torri, W., Vicenzi, E., Vago, L., Lazzarin, A., Poli, G., and Mantovani, A. (1996). Selective elevation of monocyte chemotactic protein-1 in the cerebrospinal fluid of AIDS patients with cytomegalovirus encephalitis. *J. Infect. Dis.* **174,** 1098–1101.

Bezzi, P., Domercq, M., Brambilla, L., Galli, R., Schols, D., De Clercq, E., Vescovi, A., Bagetta, G., Kollias, G., Meldolesi, J., and Volterra, A. (2001). CXCR4-activated astrocyte glutamate release via TNFalpha: Amplification by microglia triggers neurotoxicity. *Nat. Neurosci.* **4,** 702–710.

Bleul, C. C., Schultze, J. L., and Springer, T. A. (1998). B lymphocyte chemotaxis regulated in association with microanatomic localization, differentiation state, and B cell receptor engagement. *J. Exp. Med.* **187,** 753–762.

Bonavia, R., Bajetto, A., Barbero, S., Pirani, P., Florio, T., and Schettini, G. (2003). Chemokines and their receptors in the CNS: Expression of CXCL12/SDF-1 and CXCR4 and their role in astrocyte proliferation. *Toxicol. Lett.* **139,** 181–189.

Bonwetsch, R., Croul, S., Richardson, M. W., Lorenzana, C., Valle, L. D., Sverstiuk, A. E., Amini, S., Morgello, S., Khalili, K., and Rappaport, J. (1999). Role of HIV-1 Tat and CC chemokine MIP-1alpha in the pathogenesis of HIV associated central nervous system disorders. *J. Neurovirol.* **5,** 685–694.

Borghesani, P. R., Peyrin, J. M., Klein, R., Rubin, J., Carter, A. R., Schwartz, P. M., Luster, A., Corfas, G., and Segal, R. A. (2002). BDNF stimulates migration of cerebellar granule cells. *Development* **129,** 1435–1442.

Brenneman, D. E., McCune, S. K., Mervis, R. F., and Hill, J. M. (1994). gp120 as an etiologic agent for NeuroAIDS: neurotoxicity and model systems. *Adv. Neuroimmunol.* **4,** 157–165.

Carrasco, E., Blum, M., Weickert, C. S., and Casper, D. (2003). Epidermal growth factor receptor expression is related to post-mitotic events in cerebellar development: Regulation by thyroid hormone. *Brain Res. Dev. Brain Res.* **140,** 1–13.

Chalasani, S. H., Baribaud, F., Coughlan, C. M., Sunshine, M. J., Lee, V. M., Doms, R. W., Littman, D. R., and Raper, J. A. (2003a). The chemokine stromal cell-derived factor-1 promotes the survival of embryonic retinal ganglion cells. *J. Neurosci.* **23,** 4601–4612.

Chalasani, S. H., Sabelko, K. A., Sunshine, M. J., Littman, D. R., and Raper, J. A. (2003b). A chemokine, SDF-1, reduces the effectiveness of multiple axonal repellents and is required for normal axon pathfinding. *J. Neurosci.* **23,** 1360–1371.

Chen, S. C., Leach, M. W., Chen, Y., Cai, X. Y., Sullivan, L., Wiekowski, M., Dovey-Hartman, B. J., Zlotnik, A., and Lira, S. A. (2002). Central nervous system inflammation and neurological disease in transgenic mice expressing the CC chemokine CCL21 in oligodendrocytes. *J. Immunol.* **168,** 1009–1017.

Cinque, P., Vago, L., Mengozzi, M., Torri, V., Ceresa, D., Vicenzi, E., Transidico, P., Vagani, A., Sozzani, S., Mantovani, A., Lazzarin, A., and Poli, G. (1998). Elevated cerebrospinal fluid levels of monocyte chemotactic protein-1 correlate with HIV-1 encephalitis and local viral replication. *Aids* **12,** 1327–1332.

Combadiere, C., Gao, J., Tiffany, H. L., and Murphy, P. M. (1998). Gene cloning, RNA distribution, and functional expression of mCX3CR1, a mouse chemotactic receptor for the CX3C chemokine fractalkine. *Biochem. Biophys. Res. Commun.* **253,** 728–732.

Cook, D. N., Chen, S. C., Sullivan, L. M., Manfra, D. J., Wiekowski, M. T., Prosser, D. M., Vassileva, G., and Lira, S. A. (2001). Generation and analysis of mice lacking the chemokine fractalkine. *Mol. Cell. Biol.* **21,** 3159–3165.

Cotter, R., Williams, C., Ryan, L., Erichsen, D., Lopez, A., Peng, H., and Zheng, J. (2002). Fractalkine (CX3CL1) and brain inflammation: Implications for HIV-1-associated dementia. *J. Neurovirol.* **8,** 585–598.

Cowell, R. M., and Silverstein, F. S. (2003). Developmental changes in the expression of chemokine receptor CCR1 in the rat cerebellum. *J. Comp. Neurol.* **457,** 7–23.

Crazzolara, R., Kreczy, A., Mann, G., Heitger, A., Eibl, G., Fink, F. M., Mohle, R., and Meister, B. (2001). High expression of the chemokine receptor CXCR4 predicts extramedullary organ infiltration in childhood acute lymphoblastic leukaemia. *Br. J. Haematol.* **115,** 545–553.

Croitoru-Lamoury, J., Guillemin, G. J., Boussin, F. D., Mognetti, B., Gigout, L. I., Cheret, A., Vaslin, B., Le Grand, R., Brew, B. J., and Dormont, D. (2003). Expression of chemokines and their receptors in human and simian astrocytes: Evidence for a central role of TNF alpha and IFN gamma in CXCR4 and CCR5 modulation. *Glia* **41,** 354–370.

Danik, M., Puma, C., Quirion, R., and Williams, S. (2003). Widely expressed transcripts for chemokine receptor CXCR1 in identified glutamatergic, gamma-aminobutyric acidergic, and cholinergic neurons and astrocytes of the rat brain: A single-cell reverse transcription-multiplex polymerase chain reaction study. *J. Neurosci. Res.* **74,** 286–295.

Desbaillets, I., Tada, M., de Tribolet, N., Diserens, A. C., Hamou, M. F., and Van Meir, E. G. (1994). Human astrocytomas and glioblastomas express monocyte chemoattractant protein-1 (MCP-1) *in vivo* and *in vitro*. *Int. J. Cancer* **58,** 240–247.

DeVries, M. E., Ran, L., and Kelvin, D. J. (1999). On the edge: The physiological and pathophysiological role of chemokines during inflammatory and immunological responses. *Semin. Immunol.* **11,** 95–104.

Dubois-Dalcq, M., and Murray, K. (2000). Why are growth factors important in oligodendrocyte physiology? *Pathol. Biol. (Paris)* **48,** 80–86.

Dwinell, M. B., Ogawa, H., Barrett, K. E., and Kagnoff, M. F. (2003). Stromal cell-derived factor-1/CXCL12 regulates cyclic AMP production and ion transport in intestinal epithelial cells via CXCR4. *Am. J. Physiol. Gastrointest. Liver Physiol.* G844–G850.

Fife, B. T., Huffnagle, G. B., Kuziel, W. A., and Karpus, W. J. (2000). CC chemokine receptor 2 is critical for induction of experimental autoimmune encephalomyelitis. *J. Exp. Med.* **192,** 899–905.

Fife, B. T., Kennedy, K. J., Paniagua, M. C., Lukacs, N. W., Kunkel, S. L., Luster, A. D., and Karpus, W. J. (2001a). CXCL10 (IFN-gamma-inducible protein-10) control of encephalitogenic CD4+ T cell accumulation in the central nervous system during experimental autoimmune encephalomyelitis. *J. Immunol.* **166,** 7617–7624.

Fife, B. T., Paniagua, M. C., Lukacs, N. W., Kunkel, S. L., and Karpus, W. J. (2001b). Selective CC chemokine receptor expression by central nervous system-infiltrating encephalitogenic T cells during experimental autoimmune encephalomyelitis. *J. Neurosci. Res.* **66,** 705–714.

Franciotta, D., Martino, G., Zardini, E., Furlan, R., Bergamaschi, R., Andreoni, L., and Cosi, V. (2001). Serum and CSF levels of MCP-1 and IP-10 in multiple sclerosis patients with acute and stable disease and undergoing immunomodulatory therapies. *J. Neuroimmunol.* **115,** 192–198.

Gillard, S. E., Lu, M., Mastracci, R. M., and Miller, R. J. (2002). Expression of functional chemokine receptors by rat cerebellar neurons. *J. Neuroimmunol.* **124,** 16–28.

Giovannelli, A., Limatola, C., Ragozzino, D., Mileo, A. M., Ruggieri, A., Ciotti, M. T., Mercanti, D., Santoni, A., and Eusebi, F. (1998). CXC chemokines interleukin-8 (IL-8) and growth-related gene product alpha (GROalpha) modulate Purkinje neuron activity in mouse cerebellum. *J. Neuroimmunol.* **92,** 122–132.

Giunti, D., Borsellino, G., Benelli, R., Marchese, M., Capello, E., Valle, M. T., Pedemonte, E., Noonan, D., Albini, A., Bernardi, G., Mancardi, G. L., Battistini, L., and Uccelli, A. (2003). Phenotypic and functional analysis of T cells homing into the CSF of subjects with inflammatory diseases of the CNS. *J. Leukoc. Biol.* **73,** 584–590.

Glass, W. G., Chen, B. P., Liu, M. T., and Lane, T. E. (2002). Mouse hepatitis virus infection of the central nervous system: Chemokine-mediated regulation of host defense and disease. *Viral Immunol.* **15,** 261–272.

Gleichmann, M., Gillen, C., Czardybon, M., Bosse, F., Greiner-Petter, R., Auer, J., and Muller, H. W. (2000). Cloning and characterization of SDF-1gamma, a novel SDF-1 chemokine transcript with developmentally regulated expression in the nervous system. *Eur. J. Neurosci.* **12,** 1857–1866.

Goldberg, S. H., van der Meer, P., Hesselgesser, J., Jaffer, S., Kolson, D. L., Albright, A. V., Gonzalez-Scarano, F., and Lavi, E. (2001). CXCR3 expression in human central nervous system diseases. *Neuropathol. Appl. Neurobiol.* **27,** 127–138.

Gould, E., and Cameron, H. A. (1996). Regulation of neuronal birth, migration and death in the rat dentate gyrus. *Dev. Neurosci.* **18,** 22–35.

Handa, S. (1992). Concentrations of interleukin-1 beta, interleukin-6, interleukin-8 and TNF-alpha in cerebrospinal fluid from children with septic or aseptic meningitis. *Kurume. Med. J.* **39,** 257–265.

Harrison, J. K., Jiang, Y., Chen, S., Xia, Y., Maciejewski, D., McNamara, R. K., Streit, W. J., Salafranca, M. N., Adhikari, S., Thompson, D. A., Botti, P., Bacon, K. B., and Feng, L. (1998). Role for neuronally derived fractalkine in mediating interactions between neurons and CX3CR1-expressing microglia. *Proc. Natl. Acad. Sci. USA* **95,** 10896–10901.

Haskell, C. A., Hancock, W. W., Salant, D. J., Gao, W., Csizmadia, V., Peters, W., Faia, K., Fituri, O., Rottman, J. B., and Charo, I. F. (2001). Targeted deletion of CX(3)CR1 reveals a role for fractalkine in cardiac allograft rejection. *J. Clin. Invest.* **108,** 679–688.

Hatori, K., Nagai, A., Heisel, R., Ryu, J. K., and Kim, S. U. (2002). Fractalkine and fractalkine receptors in human neurons and glial cells. *J. Neurosci. Res.* **69,** 418–426.

Helbig, G., Christopherson, K. W., 2nd, Bhat-Nakshatri, P., Kumar, S., Kishimoto, H., Miller, K. D., Broxmeyer, H. E., and Nakshatri, H. (2003). NF-kappaB promotes breast cancer cell migration and metastasis by inducing the expression of the chemokine receptor CXCR4. *J. Biol. Chem.* **278,** 21631–21638.

Hendrix, C. W., Flexner, C., MacFarland, R. T., Giandomenico, C., Fuchs, E. J., Redpath, E., Bridger, G., and Henson, G. W. (2000). Pharmacokinetics and safety of AMD-3100, a novel antagonist of the CXCR-4 chemokine receptor, in human volunteers. *Antimicrob. Agents Chemother.* **44,** 1667–1673.

Horuk, R., Martin, A. W., Wang, Z., Schweitzer, L., Gerassimides, A., Guo, H., Lu, Z., Hesselgesser, J., Perez, H. D., Kim, J., Parker, J., Hadley, T. J., and Peiper, S. C. (1997). Expression of chemokine receptors by subsets of neurons in the central nervous system. *J. Immunol.* **158,** 2882–2890.

Huang, D. R., Han, Y., Rani, M. R., Glabinski, A., Trebst, C., Sorensen, T., Tani, M., Wang, J., Chien, P., O'Bryan, S., Bielecki, B., Zhou, Z. L., Majumder, S., and Ransohoff, R. M. (2000). Chemokines and chemokine receptors in inflammation of the nervous system: Manifold roles and exquisite regulation. *Immunol. Rev.* **177,** 52–67.

Huang, D. R., Wang, J., Kivisakk, P., Rollins, B. J., and Ransohoff, R. M. (2001). Absence of monocyte chemoattractant protein 1 in mice leads to decreased local macrophage recruitment and antigen-specific T helper cell type 1 immune response in experimental autoimmune encephalomyelitis. *J. Exp. Med.* **193,** 713–726.

Hulshof, S., van Haastert, E. S., Kuipers, H. F., van den Elsen, P. J., De Groot, C. J., van der Valk, P., Ravid, R., and Biber, K. (2003). CX3CL1 and CX3CR1 expression in human brain tissue: Noninflammatory control versus multiple sclerosis. *J. Neuropathol. Exp. Neurol.* **62,** 899–907.

Izikson, L., Klein, R. S., Charo, I. F., Weiner, H. L., and Luster, A. D. (2000). Resistance to experimental autoimmune encephalomyelitis in mice lacking the CC chemokine receptor (CCR)2. *J. Exp. Med.* **192,** 1075–1080.

Karpus, W. J., and Kennedy, K. J. (1997). MIP-1alpha and MCP-1 differentially regulate acute and relapsing autoimmune encephalomyelitis as well as Th1/Th2 lymphocyte differentiation. *J. Leukoc. Biol.* **62,** 681–687.

Kastenbauer, S., Koedel, U., Wick, M., Kieseier, B. C., Hartung, H. P., and Pfister, H. W. (2003). CSF and serum levels of soluble fractalkine (CX3CL1) in inflammatory diseases of the nervous system. *J. Neuroimmunol.* **137,** 210–217.

Kaul, M., and Lipton, S. A. (1999). Chemokines and activated macrophages in HIV gp120-induced neuronal apoptosis. *Proc. Natl. Acad. Sci. USA* **96,** 8212–8216.

Kaul, M., Garden, G. A., and Lipton, S. A. (2001). Pathways to neuronal injury and apoptosis in HIV-associated dementia. *Nature* **410,** 988–994.

Khan, M. Z., Brandimarti, R., Musser, B. J., Resue, D. M., Fatatis, A., and Meucci, O. (2003). The chemokine receptor CXCR4 regulates cell-cycle proteins in neurons. *J. Neurovirol.* **9,** 300–314.

Kim, C. H., and Broxmeyer, H. E. (1999). Chemokines: Signal lamps for trafficking of T and B cells for development and effector function. *J. Leukoc. Biol.* **65,** 6–15.

Kim, W. K., Corey, S., Alvarez, X., and Williams, K. (2003). Monocyte/macrophage traffic in HIV and SIV encephalitis. *J. Leukoc. Biol.* **74,** 650–656.

Klein, R. S., Williams, K. C., Alvarez-Hernandez, X., Westmoreland, S., Force, T., Lackner, A. A., and Luster, A. D. (1999). Chemokine receptor expression and signaling in macaque and human fetal neurons and astrocytes: Implications for the neuropathogenesis of AIDS. *J. Immunol.* **163,** 1636–1646.

Klein, R. S., Rubin, J. B., Gibson, H. D., DeHaan, E. N., Alvarez-Hernandez, X., Segal, R. A., and Luster, A. D. (2001). SDF-1 alpha induces chemotaxis and enhances Sonic hedgehog-induced proliferation of cerebellar granule cells. *Development* **128,** 1971–1981.

Klein, R. S., Izikson, L., Means, T., Gibson, H. D., Lin, E., Sobel, R. A., Weiner, H. L., and Luster, A. D. (2004). IFN-inducible protein 10/CXC chemokine ligand 10-independent induction of experimental autoimmune encephalomyelitis. *J. Immunol.* **172,** 550–559.

Lahrtz, F., Piali, L., Nadal, D., Pfister, H. W., Spanaus, K. S., Baggiolini, M., and Fontana, A. (1997). Chemotactic activity on mononuclear cells in the cerebrospinal fluid of patients with viral meningitis is mediated by interferon-gamma inducible protein-10 and monocyte chemotactic protein-1. *Eur. J. Immunol.* **27,** 2484–2489.

Lane, T. E., Liu, M. T., Chen, B. P., Asensio, V. C., Samawi, R. M., Paoletti, A. D., Campbell, I. L., Kunkel, S. L., Fox, H. S., and Buchmeier, M. J. (2000). A central role for CD4(+) T cells and RANTES in virus-induced central nervous system inflammation and demyelination. *J. Virol.* **74,** 1415–1424.

Lavi, E., Strizki, J. M., Ulrich, A. M., Zhang, W., Fu, L., Wang, Q., O'Connor, M., Hoxie, J. A., and Gonzalez-Scarano, F. (1997). CXCR-4 (Fusin), a co-receptor for the type 1 human immunodeficiency virus (HIV-1), is expressed in the human brain in a variety of cell types, including microglia and neurons. *Am. J. Pathol.* **151,** 1035–1042.

Lavi, E., Kolson, D. L., Ulrich, A. M., Fu, L., and Gonzalez-Scarano, F. (1998). Chemokine receptors in the human brain and their relationship to HIV infection. *J. Neurovirol.* **4,** 301–311.

Lee, T. H., Avraham, H., Lee, S. H., and Avraham, S. (2002). Vascular endothelial growth factor modulates neutrophil transendothelial migration via up-regulation of interleukin-8 in human brain microvascular endothelial cells. *J. Biol. Chem.* **277,** 10445–10451.

Limatola, C., Ciotti, M. T., Mercanti, D., Vacca, F., Ragozzino, D., Giovannelli, A., Santoni, A., Eusebi, F., and Miledi, R. (2000a). The chemokine growth-related gene product beta protects rat cerebellar granule cells from apoptotic cell death through alpha-amino-3-hydroxy-5-methyl-4-isoxazolepropionate receptors. *Proc. Natl. Acad. Sci. USA* **97,** 6197–6201.

Limatola, C., Giovannelli, A., Maggi, L., Ragozzino, D., Castellani, L., Ciotti, M. T., Vacca, F., Mercanti, D., Santoni, A., and Eusebi, F. (2000b). SDF-1alpha-mediated modulation of synaptic transmission in rat cerebellum. *Eur. J. Neurosci.* **12,** 2497–2504.

Limatola, C., Ciotti, M. T., Mercanti, D., Santoni, A., and Eusebi, F. (2002). Signaling pathways activated by chemokine receptor CXCR2 and AMPA-type glutamate receptors and involvement in granule cells survival. *J. Neuroimmunol.* **123,** 9–17.

Liu, M. T., Chen, B. P., Oertel, P., Buchmeier, M. J., Armstrong, D., Hamilton, T. A., and Lane, T. E. (2000). The T cell chemoattractant IFN-inducible protein 10 is essential in host defense against viral-induced neurologic disease. *J. Immunol.* **165,** 2327–2330.

Liu, M. T., Armstrong, D., Hamilton, T. A., and Lane, T. E. (2001a). Expression of Mig (monokine induced by interferon-gamma) is important in T lymphocyte recruitment and host defense following viral infection of the central nervous system. *J. Immunol.* **166,** 1790–1795.

Liu, M. T., Keirstead, H. S., and Lane, T. E. (2001b). Neutralization of the chemokine CXCL10 reduces inflammatory cell invasion and demyelination and improves neurological function in a viral model of multiple sclerosis. *J. Immunol.* **167,** 4091–4097.

Liu, Z., Geng, L., Li, R., He, X., Zheng, J. Q., and Xie, Z. (2003). Frequency modulation of synchronized Ca2+ spikes in cultured hippocampal networks through G-protein-coupled receptors. *J. Neurosci.* **23,** 4156–4163.

Lokensgard, J. R., Hu, S., Sheng, W., vanOijen, M., Cox, D., Cheeran, M. C., and Peterson, P. K. (2001). Robust expression of TNF-alpha, IL-1beta, RANTES, and IP-10 by human microglial cells during nonproductive infection with herpes simplex virus. *J. Neurovirol.* **7,** 208–219.

Lu, M., Grove, E. A., and Miller, R. J. (2002). Abnormal development of the hippocampal dentate gyrus in mice lacking the CXCR4 chemokine receptor. *Proc. Natl. Acad. Sci. USA* **99,** 7090–7095.

Lu, Q., Sun, E. E., Klein, R. S., and Flanagan, J. G. (2001). Ephrin-B reverse signaling is mediated by a novel PDZ-RGS protein and selectively inhibits G protein-coupled chemoattraction. *Cell* **105,** 69–79.

Ma, Q., Jones, D., Borghesani, P. R., Segal, R. A., Nagasawa, T., Kishimoto, T., Bronson, R. T., and Springer, T. A. (1998). Impaired B-lymphopoiesis, myelopoiesis, and derailed cerebellar neuron migration in CXCR4- and SDF-1-deficient mice. *Proc. Natl. Acad. Sci. USA* **95,** 9448–9453.

MacLean, A. G., Orandle, M. S., Alvarez, X., Williams, K. C., and Lackner, A. A. (2001). Rhesus macaque brain microvessel endothelial cells behave in a manner phenotypically distinct from umbilical vein endothelial cells. *J. Neuroimmunol.* **118,** 223–232.

McGrath, K. E., Koniski, A. D., Maltby, K. M., McGann, J. K., and Palis, J. (1999). Embryonic expression and function of the chemokine SDF-1 and its receptor, CXCR4. *Dev. Biol.* **213,** 442–456.

Meng, S. Z., Oka, A., and Takashima, S. (1999). Developmental expression of monocyte chemoattractant protein-1 in the human cerebellum and brainstem. *Brain Dev.* **21,** 30–35.

Meucci, O., Fatatis, A., Simen, A. A., Bushell, T. J., Gray, P. W., and Miller, R. J. (1998). Chemokines regulate hippocampal neuronal signaling and gp120 neurotoxicity. *Proc. Natl. Acad. Sci. USA* **95,** 14500–14505.

Meucci, O., Fatatis, A., Simen, A. A., and Miller, R. J. (2000). Expression of CX3CR1 chemokine receptors on neurons and their role in neuronal survival. *Proc. Natl. Acad. Sci. USA* **97,** 8075–8080.

Miller, D. H., Khan, O. A., Sheremata, W. A., Blumhardt, L. D., Rice, G. P., Libonati, M. A., Willmer-Hulme, A. J., Dalton, C. M., Miszkiel, K. A., and O'Connor, P. W. (2003). A controlled trial of natalizumab for relapsing multiple sclerosis. *N. Engl. J. Med.* **348,** 15–23.

Misu, T., Onodera, H., Fujihara, K., Matsushima, K., Yoshie, O., Okita, N., Takase, S., and Itoyama, Y. (2001). Chemokine receptor expression on T cells in blood and cerebrospinal fluid at relapse and remission of multiple sclerosis: Imbalance of Th1/Th2-associated chemokine signaling. *J. Neuroimmunol.* **114,** 207–212.

Miyagishi, R., Kikuchi, S., Fukazawa, T., and Tashiro, K. (1995). Macrophage inflammatory protein-1 alpha in the cerebrospinal fluid of patients with multiple sclerosis and other inflammatory neurological diseases. *J. Neurol. Sci.* **129,** 223–227.

Miyamoto, Y., and Kim, S. U. (1999). Cytokine-induced production of macrophage inflammatory protein-1alpha (MIP-1alpha) in cultured human astrocytes. *J. Neurosci. Res.* **55,** 245–251.

Mori, S., Nakano, H., Aritomi, K., Wang, C. R., Gunn, M. D., and Kakiuchi, T. (2001). Mice lacking expression of the chemokines CCL21-ser and CCL19 (plt mice) demonstrate delayed but enhanced T cell immune responses. *J. Exp. Med.* **193,** 207–218.

Nagasawa, T., Kikutani, H., and Kishimoto, T. (1994). Molecular cloning and structure of a pre-B-cell growth-stimulating factor. *Proc. Natl. Acad. Sci. USA* **91,** 2305–2309.

Nansen, A., Marker, O., Bartholdy, C., and Thomsen, A. R. (2000). CCR2+ and CCR5+ CD8+ T cells increase during viral infection and migrate to sites of infection. *Eur. J. Immunol.* **30,** 1797–1806.

Nicot, A., Lelievre, V., Tam, J., Waschek, J. A., and DiCicco-Bloom, E. (2002). Pituitary adenylate cyclase-activating polypeptide and sonic hedgehog interact to control cerebellar granule precursor cell proliferation. *J. Neurosci.* **22,** 9244–9254.

Nishiyori, A., Minami, M., Ohtani, Y., Takami, S., Yamamoto, J., Kawaguchi, N., Kume, T., Akaike, A., and Satoh, M. (1998). Localization of fractalkine and CX3CR1 mRNAs in rat brain: Does fractalkine play a role in signaling from neuron to microglia? *FEBS Lett.* **429,** 167–172.

Oh, J. W., Drabik, K., Kutsch, O., Choi, C., Tousson, A., and Benveniste, E. N. (2001). CXC chemokine receptor 4 expression and function in human astroglioma cells. *J. Immunol.* **166,** 2695–2704.

Oh, S. B., Endoh, T., Simen, A. A., Ren, D., and Miller, R. J. (2002). Regulation of calcium currents by chemokines and their receptors. *J. Neuroimmunol.* **123,** 66–75.

Pashenkov, M., Teleshova, N., Kouwenhoven, M., Smirnova, T., Jin, Y. P., Kostulas, V., Huang, Y. M., Pinegin, B., Boiko, A., and Link, H. (2002). Recruitment of dendritic cells to the cerebrospinal fluid in bacterial neuroinfections. *J. Neuroimmunol.* **122,** 106–116.

Pashenkov, M., Soderstrom, M., and Link, H. (2003). Secondary lymphoid organ chemokines are elevated in the cerebrospinal fluid during central nervous system inflammation. *J. Neuroimmunol.* **135,** 154–160.

Persidsky, Y., Zheng, J., Miller, D., and Gendelman, H. E. (2000). Mononuclear phagocytes mediate blood-brain barrier compromise and neuronal injury during HIV-1-associated dementia. *J. Leukoc. Biol.* **68,** 413–422.

Ragozzino, D., Giovannelli, A., Mileo, A. M., Limatola, C., Santoni, A., and Eusebi, F. (1998). Modulation of the neurotransmitter release in rat cerebellar neurons by GRO beta. *Neuroreport* **9,** 3601–3606.

Ragozzino, D., Renzi, M., Giovannelli, A., and Eusebi, F. (2002). Stimulation of chemokine CXC receptor 4 induces synaptic depression of evoked parallel fibers inputs onto Purkinje neurons in mouse cerebellum. *J. Neuroimmunol.* **127,** 30–36.

Rempel, S. A., Dudas, S., Ge, S., and Gutierrez, J. A. (2000). Identification and localization of the cytokine SDF1 and its receptor, CXC chemokine receptor 4, to regions of necrosis and angiogenesis in human glioblastoma. *Clin. Cancer Res.* **6,** 102–111.

Rezaie, P., Trillo-Pazos, G., Everall, I. P., and Male, D. K. (2002). Expression of beta-chemokines and chemokine receptors in human fetal astrocyte and microglial co-cultures: potential role of chemokines in the developing CNS. *Glia* **37,** 64–75.

Robinson, S., and Franic, L. A. (2001). Chemokine GRO1 and the spatial and temporal regulation of oligodendrocyte precursor proliferation. *Dev. Neurosci.* **23,** 338–345.

Rosler, A., Pohl, M., Braune, H. J., Oertel, W. H., Gemsa, D., and Sprenger, H. (1998). Time course of chemokines in the cerebrospinal fluid and serum during herpes simplex type 1 encephalitis. *J. Neurol. Sci.* **157,** 82–89.

Rossi, D. L., Hardiman, G., Copeland, N. G., Gilbert, D. J., Jenkins, N., Zlotnik, A., and Bazan, J. F. (1998). Cloning and characterization of a new type of mouse chemokine. *Genomics* **47,** 163–170.

Rottman, J. B., Ganley, K. P., Williams, K., Wu, L., Mackay, C. R., and Ringler, D. J. (1997). Cellular localization of the chemokine receptor CCR5. Correlation to cellular targets of HIV-1 infection. *Am. J. Pathol.* **151,** 1341–1351.

Rubin, J. B., Kung, A. L., Klein, R. S., Chan, J. A., Sun, Y., Schmidt, K., Kieran, M. W., Luster, A. D., and Segal, R. A. (2003). A small-molecule antagonist of CXCR4 inhibits intracranial growth of primary brain tumors. *Proc. Natl. Acad. Sci. USA* **100,** 13513–13518.

Sato, K., Kuratsu, J., Takeshima, H., Yoshimura, T., and Ushio, Y. (1995). Expression of monocyte chemoattractant protein-1 in meningioma. *J. Neurosurg.* **82,** 874–878.

Sievers, J., Pehlemann, F. W., Gude, S., and Berry, M. (1994). A time course study of the alterations in the development of the hamster cerebellar cortex after destruction of the overlying meningeal cells with 6 hydroxydopamine on the day of birth. *J. Neurocytol.* **23,** 117–134.

Singh, A., Kulshreshtha, R., and Mathur, A. (2000). Secretion of the chemokine interleukin-8 during Japanese encephalitis virus infection. *J. Med. Microbiol.* **49,** 607–612.

Sorensen, T. L., Tani, M., Jensen, J., Pierce, V., Lucchinetti, C., Folcik, V. A., Qin, S., Rottman, J., Sellebjerg, F., Strieter, R. M., Frederiksen, J. L., and Ransohoff, R. M. (1999). Expression of specific chemokines and chemokine receptors in the central nervous system of multiple sclerosis patients. *J. Clin. Invest.* **103,** 807–815.

Spanaus, K. S., Nadal, D., Pfister, H. W., Seebach, J., Widmer, U., Frei, K., Gloor, S., and Fontana, A. (1997). C-X-C and C-C chemokines are expressed in the cerebrospinal fluid in bacterial meningitis and mediate chemotactic activity on peripheral blood-derived polymorphonuclear and mononuclear cells *in vitro*. *J. Immunol.* **158,** 1956–1964.

Staller, P., Sulitkova, J., Lisztwan, J., Moch, H., Oakeley, E. J., and Krek, W. (2003). Chemokine receptor CXCR4 downregulated by von Hippel-Lindau tumour suppressor pVHL. *Nature* **425,** 307–311.

Stumm, R. K., Rummel, J., Junker, V., Culmsee, C., Pfeiffer, M., Krieglstein, J., Hollt, V., and Schulz, S. (2002). A dual role for the SDF-1/CXCR4 chemokine receptor system in adult brain: Isoform-selective regulation of SDF-1 expression modulates CXCR4-dependent neuronal plasticity and cerebral leukocyte recruitment after focal ischemia. *J. Neurosci.* **22,** 5865–5878.

Stumm, R. K., Zhou, C., Ara, T., Lazarini, F., Dubois-Dalcq, M., Nagasawa, T., Hollt, V., and Schulz, S. (2003). CXCR4 regulates interneuron migration in the developing neocortex. *J. Neurosci.* **23,** 5123–5130.

Suliman, M. E., Royds, J. A., Baxter, L., Timperley, W. R., Cullen, D. R., and Jones, T. H. (1999). IL-8 mRNA expression by *in situ* hybridisation in human pituitary adenomas. *Eur. J. Endocrinol.* **140,** 155–158.

Swingler, S., Mann, A., Jacque, J., Brichacek, B., Sasseville, V. G., Williams, K., Lackner, A. A., Janoff, E. N., Wang, R., Fisher, D., and Stevenson, M. (1999). HIV-1 Nef mediates lymphocyte chemotaxis and activation by infected macrophages. *Nat. Med.* **5,** 997–1003.

Tada, M., Suzuki, K., Yamakawa, Y., Sawamura, Y., Sakuma, S., Abe, H., van Meir, E., and de Tribolet, N. (1993). Human glioblastoma cells produce 77 amino acid interleukin-8 (IL-8(77)). *J. Neurooncol.* **16,** 25–34.

Takeshima, H., Kuratsu, J., Takeya, M., Yoshimura, T., and Ushio, Y. (1994). Expression and localization of messenger RNA and protein for monocyte chemoattractant protein-1 in human malignant glioma. *J. Neurosurg.* **80,** 1056–1062.

Tao, Y., Black, I. B., and DiCicco-Bloom, E. (1996). Neurogenesis in neonatal rat brain is regulated by peripheral injection of basic fibroblast growth factor (bFGF). *J. Comp. Neurol.* **376,** 653–663.

ten Donkelaar, H. J., Lammens, M., Wesseling, P., Thijssen, H. O., and Renier, W. O. (2003). Development and developmental disorders of the human cerebellum. *J. Neurol.* **250,** 1025–1036.

Tham, T. N., Lazarini, F., Franceschini, I. A., Lachapelle, F., Amara, A., and Dubois-Dalcq, M. (2001). Developmental pattern of expression of the alpha chemokine stromal cell-derived factor 1 in the rat central nervous system. *Eur. J. Neurosci.* **13,** 845–856.

Tong, N., Perry, S. W., Zhang, Q., James, H. J., Guo, H., Brooks, A., Bal, H., Kinnear, S. A., Fine, S., Epstein, L. G., Dairaghi, D., Schall, T. J., Gendelman, H. E., Dewhurst, S., Sharer, L. R., and Gelbard, H. A. (2000). Neuronal fractalkine expression in HIV-1 encephalitis: Roles for macrophage recruitment and neuroprotection in the central nervous system. *J. Immunol.* **164,** 1333–1339.

Tran, E. H., Kuziel, W. A., and Owens, T. (2000). Induction of experimental autoimmune encephalomyelitis in C57BL/6 mice deficient in either the chemokine macrophage inflammatory protein-1alpha or its CCR5 receptor. *Eur. J. Immunol.* **30,** 1410–1415.

Trysberg, E., Carlsten, H., and Tarkowski, A. (2000). Intrathecal cytokines in systemic lupus erythematosus with central nervous system involvement. *Lupus* **9,** 498–503.

Tsai, H. H., Frost, E., To, V., Robinson, S., Ffrench-Constant, C., Geertman, R., Ransohoff, R. M., and Miller, R. H. (2002). The chemokine receptor CXCR2 controls positioning of oligodendrocyte precursors in developing spinal cord by arresting their migration. *Cell* **110,** 373–383.

Vajkoczy, P., Laschinger, M., and Engelhardt, B. (2001). Alpha4-integrin-VCAM-1 binding mediates G protein-independent capture of encephalitogenic T cell blasts to CNS white matter microvessels. *J. Clin. Invest.* **108,** 557–565.

Van Der Meer, P., Goldberg, S. H., Fung, K. M., Sharer, L. R., Gonzalez-Scarano, F., and Lavi, E. (2001). Expression pattern of CXCR3, CXCR4, and CCR3 chemokine receptors in the developing human brain. *J. Neuropathol. Exp. Neurol.* **60,** 25–32.

Van Meir, E., Ceska, M., Effenberger, F., Walz, A., Grouzmann, E., Desbaillets, I., Frei, K., Fontana, A., and de Tribolet, N. (1992). Interleukin-8 is produced in neoplastic and infectious diseases of the human central nervous system. *Cancer Res.* **52,** 4297–4305.

Wechsler-Reya, R. J., and Scott, M. P. (1999). Control of neuronal precursor proliferation in the cerebellum by Sonic Hedgehog. *Neuron* **22,** 103–114.

Westmoreland, S. V., Rottman, J. B., Williams, K. C., Lackner, A. A., and Sasseville, V. G. (1998). Chemokine receptor expression on resident and inflammatory cells in the brain of macaques with simian immunodeficiency virus encephalitis. *Am. J. Pathol.* **152,** 659–665.

Westmoreland, S. V., Alvarez, X., de Bakker, C., Aye, P., Wilson, M. L., Williams, K. C., and Lackner, A. A. (2002). Developmental expression patterns of CCR5 and CXCR4 in the rhesus macaque brain. *J. Neuroimmunol.* **122,** 146–158.

Xia, M. Q., Bacskai, B. J., Knowles, R. B., Qin, S. X., and Hyman, B. T. (2000). Expression of the chemokine receptor CXCR3 on neurons and the elevated expression of its ligand IP-10 in reactive astrocytes: *In vitro* ERK1/2 activation and role in Alzheimer's disease. *J. Neuroimmunol.* **108,** 227–235.

Xia, M., Qin, S., McNamara, M., Mackay, C., and Hyman, B. T. (1997). Interleukin-8 receptor B immunoreactivity in brain and neuritic plaques of Alzheimer's disease. *Am. J. Pathol.* **150,** 1267–1274.

Xiang, Y., Li, Y., Zhang, Z., Cui, K., Wang, S., Yuan, X. B., Wu, C. P., Poo, M. M., and Duan, S. (2002). Nerve growth cone guidance mediated by G protein-coupled receptors. *Nat. Neurosci.* **5,** 843–848.

Ye, P., Xing, Y., Dai, Z., and D'Ercole, A. J. (1996). *In vivo* actions of insulin-like growth factor-I (IGF-I) on cerebellum development in transgenic mice: Evidence that IGF-I increases proliferation of granule cell progenitors. *Brain Res. Dev. Brain Res.* **95,** 44–54.

Zhang, K., McQuibban, G. A., Silva, C., Butler, G. S., Johnston, J. B., Holden, J., Clark-Lewis, I., Overall, C. M., and Power, C. (2003). HIV-induced metalloproteinase processing of the chemokine stromal cell derived factor-1 causes neurodegeneration. *Nat. Neurosci.* **6,** 1064–1071.

Zhao, Q., Kho, A., Kenney, A. M., Yuk Di, D. I., Kohane, I., and Rowitch, D. H. (2002). Identification of genes expressed with temporal-spatial restriction to developing cerebellar neuron precursors by a functional genomic approach. *Proc. Natl. Acad. Sci. USA* **99,** 5704–5709.

Zhu, Y., Yu, T., Zhang, X. C., Nagasawa, T., Wu, J. Y., and Rao, Y. (2002). Role of the chemokine SDF-1 as the meningeal attractant for embryonic cerebellar neurons. *Nat. Neurosci.* **5,** 719–720.

Zink, W. E., Zheng, J., Persidsky, Y., Poluektova, L., and Gendelman, H. E. (1999). The neuropathogenesis of HIV-1 infection. *FEMS Immunol. Med. Microbiol.* **26,** 233–241.

Zou, Y. R., Kottmann, A. H., Kuroda, M., Taniuchi, I., and Littman, D. R. (1998). Function of the chemokine receptor CXCR4 in haematopoiesis and in cerebellar development. *Nature* **393,** 595–599.

Zwijnenburg, P. J., de Bie, H. M., Roord, J. J., van der Poll, T., and van Furth, A. M. (2003). Chemotactic activity of CXCL5 in cerebrospinal fluid of children with bacterial meningitis. *J. Neuroimmunol.* **145,** 148–153.

CHAPTER 8

Chemokines and Chemokine Receptors in Pulmonary Disease

Molly S. Thomas, Allison L. Miller, and Nicholas W. Lukacs
Department of Pathology, University of Michigan Medical School, Ann Arbor, Michigan 48109

I. INTRODUCTION

Inflammatory mediators released at the site of an injury stimulate a cascade of events necessary for the repair and resolution of injury. A number of cells participate in the inflammatory response, including monocytes/macrophages, neutrophils, and lymphocytes. Structural cells such as fibroblasts and epithelial cells also play critical roles in directing the inflammatory response. Cellular injury results in the release of reactive oxygen species, arachadonic-acid metabolites, inflammatory cytokines, and chemokines that generate an inflammatory response. These paracrine mediators orchestrate complex interactions between the surrounding cell populations and direct the inflammatory responses necessary to restore homeostasis. The chemokine family members appear to play a particularly important role in directing the movement and localization of immune and inflammatory cells.

Current Topics in Membranes, Volume 55 1063-5823/05 $35.00

DOI: 10.1016/S1063-5823(04)55008-1

Chemokines are low-molecular-weight chemotactic cytokines that bind to seven-transmembrane G-protein-coupled chemokine receptors. Chemokines control the directional migration of specific inflammatory-cell populations across a concentration gradient and thereby direct these cellular populations to the site of inflammation (Rossi and Zlotnik, 2000). Chemokines are secreted from a variety of cells and can bind heparin-rich proteoglycans, a process that may serve as a solid phase for leukocyte chemotaxis. Chemokines can also be presented on the cell surface in coordination with the carbohydrate moieties glycosaminoglycans (GAGs) (Koopmann and Krangel, 1997). GAG-associated chemokine presentation may be especially important in endothelial cells, where the cell-surface expression of chemokines may function to recruit cells from the vasculature in one of the primary steps of leukocyte recruitment to the site of inflammation. Chemokines are classified into four groups based on the sequence of conserved cysteine amino-acid motifs, CC, CXC, C, and CX3C. CXC chemokines are further classified based on the presence of and ELR (glutamate–leucine–arginine) amino-acid motif. To date there are approximately 20 identified chemokine receptors and over 40 known chemokines (Table I). Chemokine receptors are classified based upon the classification of their chemokine ligand, CC chemokine receptors (CCR) or CXC chemokine receptors (CXCR). Chemokines are produced by a wide variety of cells, including leukocytes, fibroblasts, epithelial cells, endothelial cells, and smooth-muscle cells, each of which expresses distinct patterns of chemokine receptors (Rossi and Zlotnik, 2000). Chemokines perform a wide variety of functions in addition to their role in chemotaxis. Chemokines are also active in angiogenesis, antipathogen immunity, immunoregulation, inflammatory-cell activation, and the maintenance of overall homeostasis (Rossi and Zlotnik, 2000). ELR^+ CXC chemokines have been shown to be proangiogenic, whereas ELR^- CXC chemokines have been shown to be antiangiogenic (Keane *et al.*, 1998). Chemokine association with GAGs on the surface of structural cells can function to maintain organ-specific cellular populations, such as resident alveolar macrophages, Langerhans cells in the skin, and oligodendrocytes in the brain. In addition, chemokine receptors such as CXCR4 have also been implicated in the maintenance developmental processes evidenced by the embryonic lethality of $CXCR4^{-/-}$ mice (Gonzalo *et al.*, 2000).

Recent results have established that chemokines have a critical role in immunoregulation and inflammatory-cell activation using multiple mechanisms. Chemokine expression on the endothelial surface can stimulate the activation of cellular adhesion molecules on circulating leukocytes, and in so doing it can regulate leukocyte trafficking to the site of inflammation (Campbell *et al.*, 1996). Selective chemokine expression directs the preferential recruitment of specific leukocyte populations that express the

TABLE I

Chemokines and Their Effect on Allergic Airway Disease in Murine Models

Chemokine	Old nomenclature	Chemokine receptor	Effect on allergic response	Effector cell(s) altered
CCL22	MDC	CCR4	+	eosinophil
CCL17	Tarc	CCR4	+	lymphocyte
CCL5	RANTES	CCR1,3,5	+	eosinophil, lymphocyte
CCL11	Eotaxin	CCR3	+	eosinophil
CCL2	MCP-1	CCR2	+	monocyte
CCL7	MCP-3	CCR2	+	monocyte
CXCL10	IP-10	CXCR3	+/−	lymphocyte, eosinophil
CXCL9	MIG	CXCR3	−	monocyte
CCL3	MIP 1α	CCR1	+	monocyte, eosinophil
CCL20	MIP 3α	CCR6	+	dendritic cell
CCL6	C10	CCR1 (?)	+	eosinophil
CXCL12	SDF-1	CXCR4	+	monocyte, eosinophil
CXCL8	MIP 2 (IL-8)	CXCR2	+	monocyte, neutrophil

(+), aggravates the allergic response; (−), attenuates the allergic response.

corresponding chemokine receptors, resulting in a fine-tuned immune response tailored to the specific inflammatory stimulus. Chemokines and chemokine receptors also function in immunoregulation by altering the trafficking of antigen-presenting cells (APCs) to the lymph node and by influencing the activation of effector T- and B-lymphocytes. The chemokine receptors CCR7, CXCR3, CXCR4, and CXCR5 along with their chemokine ligands CCL21, CXCL9, and CXCL10 are of particular importance in regulating the trafficking of APCs to the lymph node for the generation of effector T-cells and effective B-cell responses (Janatpour *et al.*, 2001; Park *et al.*, 2002). Ligation of a chemokine receptor by its chemokine ligand stimulates the activation of signal transduction pathways, such as those involving PI3 kinase and various mitogen-activated protein (MAP) kinases, leading to transcriptional activation and cellular degranulation (Boehme *et al.*, 1999; Kampen *et al.*, 2000; Adachi *et al.*, 2001). Thus, the regulation of the immune response may be controlled by specific G-protein-mediated pathways that differ depending upon the ligand and receptor activated.

There are two general types of immune responses to foreign antigens: the cell-mediated immune response (type 1 or Th1) and the antibody-mediated immune response (type 2 or Th2). The cell-mediated response involves macrophages as well as NK cells, CD4 (Th1) lymphocytes, and $CD8^+$ (T-cytotoxic, Tc) lymphocytes and is primarily directed against viral and bacterial pathogens. The antibody-mediated immune response involves $CD4^+$ (Th2) and B-lymphocytes directed against chronic antigens such as parasite infections and other persistent antigens. These responses are characterized by functional activity as well as by the cytokine profile of the inflammatory response. Typically, cell-mediated responses produce higher levels of IFN-γ, whereas antibody-mediated responses produce higher levels of IL-4. Type-1 and type-2 responses are antagonistic to each other, as IL-4 and IL-12 direct the effector phenotype of naïve T-lymphocytes (Biedermann *et al.*, 2001; Cohn *et al.*, 2001) (Santana and Rosenstein, 2003). IL-12 stimulation induces IFN-γ expression and inhibits the expression of IL-4, thereby directing the lymphocyte towards a Th1 phenotype. Type-1 responses are characterized by increased levels of IFN-γ-inducible chemokines, CXCL9, CXCL10, and CCL3, and thereby recruit lymphocytes expressing the corresponding chemokine receptors CCR5 and CXCR3 (Bonecchi *et al.*, 1998). IL-4 stimulation induces increased development of T-cells expressing IL-4 while simultaneously inhibiting Th1 responses, thereby directing the lymphocyte towards a Th2 phenotype (Elser *et al.*, 2002; Santana and Rosenstein, 2003). Type-2 responses are characterized by increased levels of the chemokines CCL22, CCL17, and CCL11 and thereby recruit type-2 lymphocytes expressing the chemokine receptors CCR3, CCR4, and CCR8 (Bonecchi *et al.*, 1998). Th1-type lymphocytes have been associated with several pathological disease processes, including rheumatoid arthritis, systemic sclerosis, and inflammatory bowel disease; Th2-type lymphocytes have been associated with disease processes such as asthma and allergy (O'Byrne, 1996; Crimi *et al.*, 1997; Gu *et al.*, 1997; Qin *et al.*, 1998; Patel *et al.*, 2001; Yuan *et al.*, 2001). Thus, the chemokines produced during specific immune responses relate to the types of cells that migrate into the tissue.

II. CHEMOKINES AND PULMONARY DISEASE

A. Chemokines and Allergic Airway Disease

The incidence of allergic asthma has steadily increased over the last decade. According to the Centers of Disease Control, nearly 20.3 million Americans suffered from asthma in 2001; 12 million of them reported having asthma attacks. It is estimated that asthma costs the health-care system $13.8 billion annually. In addition, approximately 14 million school days are missed each year that are attributable to asthma-related complications.

Asthma-associated morbidity and mortality have also risen dramatically. From 1993 to 1998, over 5,000 people died annually from asthma. Several allergens have been identified to be clinically significant in allergic asthma: secondhand smoke, house dust-mite antigen, cockroach antigen, respiratory syncytial virus, mold, and fungal antigens. Nearly one in four inner-city children with asthma is also allergic to cockroaches. Allergic asthma is characterized by an increase in bronchial hyperreactivity, eosinophil accumulation, $CD4^+$ Th2 lymphocyte responses, goblet-cell hyperplasia, mucus production, and an increase in IgE antibody production (Corrigan and Kay, 1992). It has also been reported that asthmatic airways have extensive epithelial-cell damage and have undergone airway remodeling, evidenced by airway smooth-muscle-cell hyperplasia and collagen deposition (Djukanovic, 2000). The characteristics of allergic airway disease are mediated by complex interactions between multiple effector cells, including lymphocytes, eosinophils, epithelial cells, and fibroblasts that produce the inflammatory mediators that participate in the development of disease. Resident cells in and around the airways, such as alveolar macrophages and mast cells, respond rapidly to inhaled allergens by releasing a plethora of chemokines and other inflammatory mediators that further stimulate chemokine release from the surrounding airway epithelial cells. These chemokines then recruit lymphocytes, neutrophils, and eosinophils into the airway, where they become activated and perform effector functions responsible for the characteristics of allergic airway disease.

Due to the inherent difficulties associated with studying human asthma patients, animal models of allergic disease have offered much of what is known about the mechanism of allergic airway disease. Multiple models of antigen-induced allergic disease have been developed to investigate the pathological progression of asthma, including ovalbumin (OVA), cockroach antigen (CRA), aspergillus, and a number of other clinically relevant allergens (Lukacs *et al.*, 1996; Wiley *et al.*, 2001; Thomas *et al.*, 2002). These models have been shown to induce disease in mice that closely parallels several aspects of human asthma. For example, allergen-sensitized and allergen-challenged mice exhibit intense pulmonary lymphocytic and eosinophilic influx into the lung, which is also characteristic of human asthma (Fig. 1). In addition, these models exhibit increased airway hyperreactivity and overproduction of mucus, and some researchers have established chronic models of allergen responses that have significant airway remodeling.

Although multiple cell populations are involved in allergic airway disease, $CD4^+$ T-cells have been demonstrated to have an integral role in directing the allergic inflammatory response. It has been shown that transfer of allergen-specific $CD4^+$ Th2 cells into naïve mice is sufficient to generate an allergic response following allergen challenge (Li *et al.*, 1998). Similarly, depletion of $CD4^+$ T-cells prevents the development of allergic disease

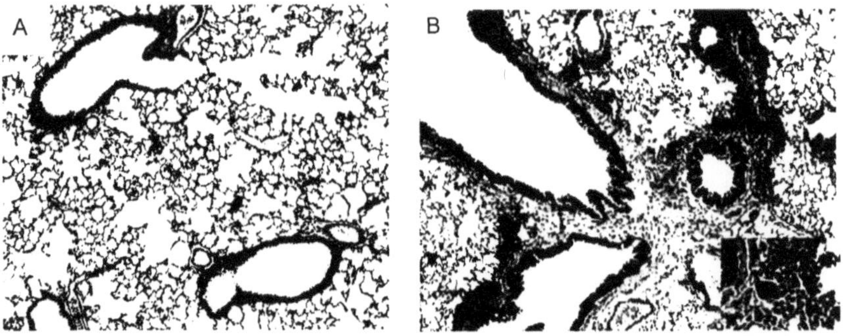

FIGURE 1 CRA induced allergic airway inflammation. Representative histologic sections of lungs from unchallenged mice (A) versus that from mice challenged with cockroach allergen (B) demonstrate that cockroach Allergen is able to induce allergic airway disease exhibiting characteristic intense lymphocytic and eosinophilic inflammation. Arrow depicts area of inset in panel B.

(Crimi *et al.*, 1997; Haddad el *et al.*, 2002). Th2 $CD4^+$ lymphocytes produce a number of cytokines that mediate the production of chemokines and other inflammatory mediators that regulate allergic airway disease. IL-4 is an apical Th2-associated cytokine that promotes the development of type-2 responses by inhibiting the production of IFN-γ (Elser *et al.*, 2002). IL-4 and tumor necrosis factor (TNF)-α are capable of inducing the production of chemokines that recruit additional Th2-type lymphocytes into the lung as well as activating those cells upon arrival to perpetuate the allergic inflammatory response (Chen *et al.*, 2000). IL-13 is another important Th2-associated cytokine that has been implicated in the development of airway hyperreactivity and eosinophilia, production of mucus, and (in the isotype) a switch to IgE-antibody production (McKenzie *et al.*, 1998; Ford *et al.*, 2001; Foster *et al.*, 2001). IL-4 and IL-13 together can induce a number of allergy-associated chemokines; CCL2, CCL5, CCL11, and CCL22 can all appear in both immune and structural cells. Of these chemokines, CCL2 and CCL5 appeared to be especially pertinent in allergen-induced disease. CCL2 recruits mononuclear cells into the lung and airway via the chemokine receptor CCR2 (Chen *et al.*, 2000), whereas CCL5 functions to recruit mononuclear cells and eosinophils into the lung and airway via the chemokine receptors CCR1, CCR3, and CCR5. Using models of allergic airway inflammation, neutralization of CCL2 dramatically reduces total leukocyte accumulation in the lung and attenuates airway hyperreactivity following antigen challenge (Lukacs *et al.*, 1997), while CCL5 contributes to allergic airway disease through its ability to recruit eosinophils into the lung and to stimulate eosinophil respiratory burst and degranulation (Beck *et al.*, 1997; Chen *et al.*, 2000; Linden, 2001). A number of other lymphocyte-derived chemokines have also demonstrated a role in allergic airway disease. CCL3 recruits lymphocytes and eosinophils into allergen-challenged lungs via CCR1 and CCR5 and has been demonstrated to induce eosinophil degranulation (Lukacs *et al.*, 1997; Gonzalo *et al.*, 1998; Oliveira *et al.*, 2002). The CCR4 ligands CCL17 and CCL22 also participate in the progression of allergic inflammation through the recruitment of CCR4-expressing Th2-type leukocytes. It has been reported that CCL17 levels in the serum of asthmatic patients correlate with acute asthmatic exacerbations (Sekiya *et al.*, 2002; Leung *et al.*, 2003), while CCL22 has been demonstrated to stimulate eosinophil degranulation (Pinho *et al.*, 2003). Thus, the coordinated production of Th2-type cytokines followed by the induction of particular chemokines appears to contribute to the progression of the responses and severity of the disease.

Through their production of IFN-γ, Th1-type lymphocytes play an important regulatory role in allergic airway disease and provide a critical inhibitory signal, resulting in the resolution of allergic inflammation. It has previously been demonstrated that introduction of IFN-γ into an allergic lung is able to weaken the allergic inflammatory response (Ford *et al.*, 2001). IFN-γ induces the expression of the type-1-associated chemokines CXCL9 and CXCL10 in monocytes, neutrophils, and bronchial epithelial cells (Farber, 1997). Lymph-node-associated CXCL10 has been shown to be important for the generation of an effector Th1-type response and the negative regulation of allergic inflammation (Wiley *et al.*, 2001; Dufour *et al.*, 2002; Thomas *et al.*, 2002). However, CXCL10 levels are increased in the airways of asthmatic patients (Medoff *et al.*, 2002), and recent studies using animal models of allergic disease suggest that CXCL10 may also play a role in the exacerbation of allergic airway inflammation, possibly by increasing eosinophil accumulation in the lung following allergen challenge (Medoff *et al.*, 2002; Thomas *et al.*, 2002). These latter studies suggest that local expression of CXCL10 may drive the recruitment of eosinophils and other CXCR3-expressing cells into the lung, leading to an overall exacerbation of the allergic response. Thus, IFNγ-inducible chemokines may play a protective role in allergic disease by mediating lymph-node localization of Th1-type lymphocytes. However, local effects of these chemokines allow for the exacerbation of disease, suggesting that the site of chemokine production determines function.

Most resting effector lymphocytes express specific chemokine receptor patterns that facilitate homing to the lymph node and/or inflamed tissues, especially CXCR4 and CCR7. Th2-type effector lymphocytes express higher levels of the chemokine receptors CCR3, CCR4, and CCR8, while Th1-type effector lymphocytes express higher levels of the chemokine receptors CCR2, CCR5, and CXCR3 (Annunziato *et al.*, 1999; Chung *et al.*, 2003). Several studies have used chemokine receptor knockouts to determine the overall importance of a number of chemokine receptors in the development of

TABLE II

Phenotype of Allergic Response in Chemokine Receptor Knockout Mice

Chemokine receptor	Phenotype of receptor knockout in allergic response
CCR1	Decreased airway remodeling
CCR2	Increased AHR, eosinophil accumulation, IL-13 expression
CCR3	Failure to develop AHR, decreased eosinophil accumulation
CCR4	No effect on allergic response
CCR5	Decreased AHR and airway remodeling
CCR6	Decreased AHR, eosinophil recruitment, IgE levels
CCR8	Decreased eosinophil accumulation, Th2 cytokine production
CXCR2	Decreased AHR, Th2 associated cytokines, eotaxin

AHR, airway hyperreactivity.

allergic airway disease (Table II). While the use of these animals may not be optimal, it allows for the elucidation of potential mechanisms of disease pathogenesis. CCR3 knockout mice fail to develop airway hyperreactivity and eosinophilia following antigen challenge, demonstrating the importance of CCR3 in eosinophil migration and disease progression (Ma *et al.*, 2002). CCR6-deficient mice exhibit reduced airway hyperreactivity and eosinophil accumulation following allergen challenge (Lukacs *et al.*, 2001) through a mechanism involving both T-lymphocytes and pulmonary dendritic cells (Lukacs, 2005, unpublished data). Studies from CCR8-deficient mice argue this chemokine receptor–ligand pair may also play a role in allergen-induced airway eosinophilia (Chensue *et al.*, 2001). Using models of chronic allergic fungal disease, it has been demonstrated that $CCR1^{-/-}$ mice exhibit reduced airway remodeling, while $CCR5^{-/-}$ mice exhibit reduced airway hyperreactivity, goblet-cell hyperplasia, and peribronchial fibrosis following chronic allergen challenge (Blease *et al.*, 2000; Schuh *et al.*, 2002a). CXCR2-deficient mice were also shown to exhibit a reduced allergic response in the lung following chronic allergen challenge (Schuh *et al.*, 2002b). These studies suggest that CCR1-, CCR5-, and CXCR2-expressing cells, such as eosinophils, neutrophils, and T-lymphocytes, may significantly contribute to airway remodeling observed in chronic allergen-induced disease. Results from these studies using chemokine receptor knockout mice demonstrate the importance of specific chemokine receptors in different phases of the development and progression of allergic airway disease and also highlight the role of the specific chemokine receptor ligands in the pathogenesis of allergic disease by illustrating how the absence of chemokine receptor–stimulated cellular activation contributes to the allergic response.

During chronic asthmatic responses, eosinophils appear to act as important effector cells that correlate to the severity of the disease (Boyce, 1997). Eosinophils can express multiple chemokine receptors, including CCR1, CCR3, CCR5, CCR6, and CCR8 (Hochstetter *et al.*, 2000; Jinquan *et al.*, 2000; Nagase *et al.*, 2001). This profile of chemokine receptor expression facilitates the recruitment and activation of eosinophils into inflammatory environments producing the Th2-associated chemokines CCL3, CCL5, CCL11, and CCL6 (Bandeira-Melo *et al.*, 2001). Eosinophils can also express and secrete several chemokines, such as CCL5, CCL6, and CCL11, as well as Th2 cytokines (IL-4, IL-5, IL-13); they therefore perpetuate allergic inflammatory responses (Simon *et al.*, 1995; Chihara *et al.*, 1997; Hertz *et al.*, 2001; Woerly *et al.*, 2002). In addition, several eosinophil degranulation products have been directly linked to allergic airway hyperreactivity. Leukotrienes have been shown to stimulate smooth-muscle-cell contraction, leading to bronchial hyperreactivity and mucus hypersecretion (Bandeira-Melo *et al.*, 2001, 2002). Eosinophil peroxidase and eosinophil cationic protein can facilitate protein nitration and can damage epithelial cells, leading to the induction of further inflammatory mediators (Duguet *et al.*, 2001). Furthermore, eosinophil major basic protein can stimulate CXCL8 production in neutrophils and has also been shown to prolong the stability of CXCL8 mRNA, which may contribute to increased neutrophil recruitment (Page *et al.*, 1999; Linden, 2001). Thus, the localization and activation of eosinophils has been and should continue to be considered a disease-associated event that has a detrimental role in asthma.

Although neutrophils are important effector cells in the innate immune response, they have also been linked to the early inflammatory response in allergic airway disease. Neutrophils have been associated with acute asthmatic exacerbations in humans and have been correlated with fatal asthma (Gibson *et al.*, 2001). Neutrophil numbers and IL-8 level were significantly elevated in patients suffering from acute severe asthma that required ventilation and were significantly higher in the sputum of patients with severe asthma than in patients with moderate asthma or control subjects (Jatakanon *et al.*, 1999; Ordonez *et al.*, 2000). Neutrophils contain granular proteins such as neutrophil peroxidase and myeloperoxidase that, when released, can trigger the release of inflammatory mediators from nearby monocytes and epithelial cells. In addition, neutrophil elastase secretion can result in the degradation of basement membrane components, allowing for the increased influx of inflammatory cells. Release of oxygen-free radials, as well as elastase by neutrophils, can increase mucus secretion from epithelial cells through the transactivation of EGF (Linden, 2001). Neutrophils express the chemokine receptors CCR1, CCR3, CXCR1, and CXCR2 and are recruited into allergic inflammatory sites via local production of CCL3, CCL5, CCL7, and

CCL11. Interestingly, CCR1 expression has been shown to correlate with elastase$^+$ neutrophil influx in allergen-induced late-phase cutaneous reactions (Ying *et al.*, 2001). Neutrophil-derived TNF-α expression can stimulate the production of a number of chemokines associated with allergic airway disease, including CCL3, CXCL8 (MIP 2), CXCL9, and CXCL10, in the surrounding inflammatory and structural cells (Gasperini *et al.*, 1999; Linden, 2001). Neutrophil-derived chemokines, such as CCL3 and CXCL10, may be important in other allergic events, such as eosinophil and mast-cell degranulation (Oliveira *et al.*, 2002).

Mast-cell activation during allergic disease is initially dependent upon ligation and cross-linking of IgE and participates in the inflammatory responses during asthma (Rossi and Olivieri, 1997). Mast cells often localize to airway smooth muscle in the lung and are poised to rapidly respond to allergen stimulation. Through their expression of the chemokine receptors CCR1, CCR3, CCR5, CXCR2, CXCR4, and CX3CR1, mast cells are recruited to inflammatory lesions expressing the corresponding chemokine ligands CCL3, CCL11, CCL5, CXCL12, CXCL8, and CX3CL1 (Brightling *et al.*, 2003). Mast cells produce a number of chemokines, including CCL11, CCL8, CCL7 and CXCL8, that function in the recruitment and activation of additional effector cells that are involved in the perpetuation of the allergic response, such as eosinophils and neutrophils (Brightling *et al.*, 2003). CCL11 is a potent eosinophil chemoattractant and can induce eosinophil degranulation (Stellato *et al.*, 1997). In addition, CCL3 and CCL7 released from mast cells can stimulate the influx of monocytes into the airway. Furthermore, a number of chemokines, such as CCL3 and CCL5, can directly induce mast-cell degranulation. Activated mast cells also secrete mediators that can directly contribute to several characteristics of allergic airway disease. Upon degranulation, mast cells release several mediators that perpetuate the inflammatory process, such as TNF-α and IL-4, as well as leukotrienes that can directly induce smooth-muscle-cell contraction and airway hyperreactivity (Lukacs *et al.*, 1998; Brightling *et al.*, 2003). In addition, mast cells secrete other Th2-associated cytokines, IL-9 and IL-13, which have been shown to induce mucus production in epithelial cells (Romagnani, 2002). Furthermore, several mast-cell products can also directly influence airway remodeling, including transforming growth factor (TGF)-β, histamine, and tryptase, that can act as smooth-muscle-cell mitogens and induce smooth-muscle hypertrophy (Brightling *et al.*, 2003). Additionally, TGF-β has been shown to induce fibroblast proliferation and collagen deposition, which may contribute to asthma-associated airway remodeling. Thus, chemokine production by mast cells and chemokine-induced activation of mast cells both likely participate in several aspects of allergic asthma.

Alveolar macrophages are the most abundant cell population in the airway and respond rapidly to inhaled allergens through the release of mediators that also drive the allergic inflammatory response. Alveolar macrophages produce a number of chemokines, including CCL3, CCL5, CCL22, CXCL9, and CXCL10. In addition, local alveolar-macrophage populations express the chemokine receptors CCR2, CCR5, and CXCR4, which may regulate their activation (Rodenburg *et al.*, 1998; Janatpour *et al.*, 2001; Katschke *et al.*, 2001; Mariani *et al.*, 2002; Park *et al.*, 2002). One macrophage chemokine receptor that appears particularly important in allergic disease is CCR2. CCR2 knockout mice display enhanced airway hyperreactivity, eosinophil accumulation, and IL-13 expression following OVA challenge, suggesting that CCR2 and CCR2-expressing cells play an important regulatory role in the development of allergic asthma (Kim *et al.*, 2001). It has been demonstrated that monocytes isolated from asthmatics are more highly activated than monocytes isolated from control subjects (Hallsworth *et al.*, 1994). IL-1 and TNF-α produced by activated macrophages can induce a number of chemokines, such as CCL22, CCL17, and CCL11, that regulate the migration and activation of infiltrating inflammatory cells, which all contributes to the pathogenesis of allergic airway disease (Chen *et al.*, 2000).

In addition to circulating inflammatory cells, structural cells such as epithelial cells and fibroblasts also serve important effector functions in the regulation of allergic airway inflammation. Epithelial cells express the chemokine receptors CCR2, CCR3, and CXCR4 (Murdoch *et al.*, 1999; Stellato *et al.*, 2001; Eddleston *et al.*, 2002) and release multiple chemokines, including CCL2, CCL3, CCL5, CCL7, CCL11, CCL13, CCL17, and CCL22 (Lundien *et al.*, 2002; Michalec *et al.*, 2002). Several of these epithelial cell–derived chemokines contribute to increased severity of allergic responses by influencing lymphocyte and eosinophil recruitment into the airway and their subsequent activation. Panina-Bordignon *et al.* (2001) have demonstrated that the expression of the CCR4 ligands CCL22 and CCL17 is upregulated on epithelial cells following allergen challenge. Epithelial cells produce a number of chemokines in response to Th2 cytokines. For example, IL-13 can induce CXCL8 expression in airway epithelium, leading to increased neutrophil recruitment into the airway (Shim *et al.*, 2001). Epithelial cells also express the IFN-γ-inducible chemokines CXCL9 and CXCL10 (Sauty *et al.*, 1999). Epithelial-cell expression of CCR3 can activate airway epithelial cells, but it may also serve as a regulatory mechanism by acting as a sink for excess CCL11 as well as other CCR3 ligands and may function to downregulate the allergic inflammatory response. The ability of epithelial cells to produce both proinflammatory and anti-inflammatory mediators suggests that these cells play a modulatory role in the development of allergic airway inflammation.

In addition to epithelial-cell damage, it has become apparent that two significant consequences of chronic asthmatic responses are airway remodeling and peribronchial fibrosis. Fibroblast proliferation and activation therefore also play a role in the pathogenesis of allergic airway disease. Fibroblasts can express and secrete a number of chemokines associated with the allergic response, such as CCL2, CCL3, CCL5, CCL6, and CCL11 (Hogaboam *et al.*, 1999; Li *et al.*, 1999). These chemokines have demonstrated roles in the recruitment and activation of multiple leukocyte populations that participate in the inflammation and amplify the allergic responses (White *et al.*, 1997; Gonzalo *et al.*, 1998; Hogaboam *et al.*, 1999).

B. Chemokines and Chronic Pulmonary Disease

1. Chronic Obstructive Pulmonary Disease

Chronic obstructive pulmonary disease (COPD) is the sixth-highest cause of death worldwide and accounts for 13% of hospital admissions annually (Shapiro, 1999). COPD can be divided into three distinct pulmonary disorders: chronic bronchitis, emphysema, and small airway disease–bronchiolitis. COPD is characterized by a robust influx of neutrophils, monocytes, and $CD8^+$ T-lymphocytes that mediate goblet-cell hyperplasia, mucus hypersecretion, and the destruction of alveolar septae (De Boer, 2002). The most common acute exacerbations of COPD result from viral or bacterial infection (Aaron *et al.*, 2001). Researchers have encountered a considerable amount of difficulty in developing animal models that sufficiently mimic human disease. Therefore, our understanding of the mechanisms related to chemokine and chemokine-receptor involvement in the pathogenesis of COPD are limited. Much of the current knowledge of the disease mechanisms involved in COPD has been gleaned through the examination of clinical specimens. Interestingly, $CXCR3^+CD8^+$ IFN-γ-producing lymphocytes are associated with severe COPD responses (Panina-Bordignon and D'Ambrosio, 2003). Exacerbations of COPD have also been associated with increased neutrophil recruitment and increased sputum levels of TNF-α, CXCL8, and CCL2 (Aaron *et al.*, 2001; Traves *et al.*, 2002). In COPD patients, CCL2 mRNA was localized primarily to inflammatory cells, suggesting that the infiltrating cells act to perpetuate the responses (De Boer *et al.*, 2000). Increased production of the inflammatory cytokines IL-1 and TNF-α induces the expression of the chemokines CXCL8 and CXCL5 (ENA 78), leading to further neutrophil recruitment and activation (Panina-Bordignon and D'Ambrosio, 2003). The correlation of CCR2 and CXCR2 ligand expression (CCL2 and CXCL8, respectively) with disease severity suggests that CCR2- and CXCR2-expressing inflammatory cells play prominent roles in the pathogenesis of COPD. Neutrophils and monocytes can secrete a variety of mediators, such as elastases and matrix metalloproteinases (MMPs), in response to CCL2 and CXCL8 that degrade various components of the alveolar wall, leading to increased inflammation and enhanced disease. It is the release of specific MMPs that may lead to the most severe disease progression in COPD. In particular, by use of a murine model of COPD, it has been demonstrated that MMP-12-deficient mice fail to develop disease, suggesting that leukocyte MMP production may significantly contribute to the pathogenesis of COPD (Aaron *et al.*, 2001). A number of chemokines may induce the release of MMPs from inflammatory cells and therefore contribute to the progression of disease. Defining this and other mechanisms of COPD progression may allow targeting of a particular chemokine pathway for therapeutic intervention.

2. Chemokines and Pulmonary Fibrosis

Chemokines and chemokine receptors have also been identified that contribute to the pathogenesis of various etiologies of pulmonary fibrotic diseases, or idiopathic pulmonary fibrosis (IPF). IPF is characterized by the excessive accumulation of matrix proteins, which leads to the derangement of the alveolar wall, loss of airway elasticity, and overall decreased pulmonary function. Although it is still controversial, there appear to be three phases of pulmonary fibrosis. The first is an initial triggering event, such as particulate matter or a viral infection, that results in epithelial-cell injury, leading to the second phase, the inflammatory response (Razzaque and Taguchi, 2003). The inflammatory phase of fibrosis is characterized by an increased monocyte, neutrophil, T-lymphocyte, and eosinophil influx into the lung, along with increased levels of the chemokines CXCL8, CCL2, and CCL5 (Petrek *et al.*, 1997; Boomars *et al.*, 1998; Razzaque and Taguchi, 2003). Using *in situ* hybridization, CCL5 mRNA expression in airway cells isolated from IPF patients was localized to alveolar macrophages and eosinophils, suggesting that these cells are involved in the persistent inflammation associated with progressive disease (Petrek *et al.*, 1997). Neovascularization is necessary for fibroblast proliferation and deposition of the extracellular matrix (Keane *et al.*, 1999a). CXCL8 is an ELR^+ CXC chemokine that possesses potent angiogenic activity, and its levels in BAL fluid samples correlate to disease progression (Keane *et al.*, 1999b). CXCL10, on the other hand, is an ELR^- CXC chemokine and possesses angiostatic activity. Interestingly, IPF patients have higher levels of CXCL8 and lower levels of CXCL10 in their BAL fluid than control patients. It has been suggested that CXCL8 and CXCL10 regulate angiogenic activity in IPF and that the balance of these two chemokines determines disease progression versus disease resolution (Keane *et al.*, 1999b). Thus, targeting these angiogenic/angiostatic pathways may allow researchers to control severe disease.

A murine model of pulmonary fibrosis has been developed using Bleomycin as the initial triggering event to elicit epithelial-cell damage and instigate a fibrotic response (Smith *et al.*, 1994). Although not a perfect model, it has given us much of our insight into the pathological mechanism of idiopathic pulmonary fibrosis. Bleomycin-induced pulmonary fibrosis exhibits an increase in lymphocyte accumulation, fibroblast proliferation, and collagen deposition characteristic of the human disease. It has been suggested that $CD4^+$ Th2-type lymphocytes participate in the pathogenesis of IPF while Th1-type lymphocytes mediate resolution of the fibrotic response. Lungs of mice challenged with bleomycin have decreased levels of INF-γ induced chemokine CXCL10 and hydroxyproline content in the lung, indicative of an increased fibrotic response. It has also been demonstrated that exogenous CXCL10 administered to bleomycin-challenged mice can weaken pulmonary fibrosis, presumably via its negative regulation of angiogenesis (Keane *et al.*, 1999a, 2005). Several other chemokines, including CCL2 and CCL5, have been implicated in the pathogenesis of IPF. CCL5 stimulates the CCR3- and CCR5-mediated recruitment of eosinophils into the lung, whereas CCL2 can facilitate the CCR2-mediated accumulation of monocytes in the lung. In addition, CCL2 can directly induce collagen production in fibroblasts, whereas several studies have demonstrated that CCR2-deficient mice have decreased fibrotic responses (Gharaee-Kermani *et al.*, 1996). Thus, CCL2 and CCR2 interactions may be crucial for the progression of fibrotic disease. Corresponding to observations in human IPF samples, neutralization of CXCL8 (MIP-2) resulted in (1) a decreased fibrotic response independent of fibroblast proliferation or neutrophil recruitment and (2) activation that was dependent on its ability to negatively regulate angiogenesis (Keane *et al.*, 1999b). Other chemokines, such as CCL3, also appear to be important for localization of leukocytes to the lung during the fibrotic response (Smith *et al.*, 1994, 1996, 1998). In conjunction with these mediators, levels of IL-13 and CCL6 are also increased in the lungs of bleomycin-challenged mice (Hogaboam *et al.*, 1999). Neutralization of either IL-13 or CCL6 during bleomycin-induced injury decreases monocyte accumulation and reduces the overall fibrotic response in the lung (Belperio *et al.*, 2002). Although a number of chemokines are associated with disease severity, it is not clear whether targeting chemokines or chemokine receptors will have any effect on established fibrotic disease.

C. Chemokines and Pulmonary Viral Infection

Respiratory viral infections, including adenovirus, influenza, metapneumovirus, parainfluenza, rhinovirus, and respiratory syncytial virus (RSV), are the primary cause of asthma exacerbations in both children and adults

(Wilson, 2003). Up to 80% of asthma exacerbations in school-aged children and 50% of asthma exacerbations in adults are associated with viral infections (Yamaya and Sasaki, 2003). Furthermore, there is some evidence that suggests that in the case of RSV, a severe infection during infancy may predispose an individual to develop more severe asthma later in life (Pullan and Hey, 1982; Stein *et al.*, 1999; Sigurs *et al.*, 2000; Oddy *et al.*, 2002). Despite the heterogeneity in their taxonomy and genomes, these viruses share many similarities, including the cells they infect, the chemokine response they induce, and the disease phenotype they elicit (Table III). Because chemokines and their receptors are intermediates between viral infection and disease, it is interesting to consider their role in viral induction or exacerbation of asthma. Respiratory viral infections induce similar profiles of chemokine and chemokine-receptor production. Consequently, it may be possible to target these molecules during disease to alleviate symptoms of primary viral infection as well as to prevent asthma complications.

Some postulate that viral infections during childhood may help to protect against asthma by skewing the lung milieu to a Th1-type environment. This idea stems from the "hygiene hypothesis," which contends that childhood infections help to prevent against asthma by creating a Th1-type environment in the lung that antagonizes the development of the Th2-type environment characteristic of asthma. However, not all viral infections may experience a homogeneous Th1-type immune response. For example, RSV infection has been shown in both rodents and humans to induce IL-4 and IL-13 secretion (Alwan *et al.*, 1992; Anderson *et al.*, 1994; Bright *et al.*, 1995; Bendelja *et al.*, 2000; Tekkanat *et al.*, 2001; de Waal *et al.*, 2003). In the case where there is viral induction of Th2 cytokines, it is not clear whether the production of these cytokines can lead to an asthma phenotype or if a host response to a viral infection that includes IL-4 and IL-13 is reflective of a genetic predisposition to viral infections and/or asthma.

1. Viruses Implicated in the Induction and/or Exacerbation of Asthma

a. Adenovirus. Adenovirus infections account for a significant portion of viral infections in children during their first year of life. RSV and adenovirus induce a similar disease with two distinct long-term consequences. RSV infection is thought to be associated with chronic lung changes and the later development of asthma, while adenovirus is often associated with chronic lung disease and has been associated with acute episodes of near-fatal asthma (Simila *et al.*, 1981; Spigelblatt and Rosenfeld, 1983). Latent infection with adenovirus has been shown to result in steroid-resistant asthma (Macek *et al.*, 1994), and more recently, this steroid resistance has been associated with the persistence of the virally encoded E1A protein

TABLE III

Chemokines and Pulmonary Viral Infection

	Genome	Target cells	Chemokines	Disease	Effects on asthma
Adenovirus					
Family: *Adenoviridae*	Nonenveloped dsDNA	Airway epithelium, lymphocytes	CXCL8	URTI, pneumonia, wheezing	1. Exacerbates asthma 2. Contributes to steroid-resistant asthma
Genus: *Mastadenovirus*					
Influenza					
Family: *Orthomyxoviridae*	Enveloped, negative-sense ssRNA fragments	Airway epithelium, monocytes/ macrophages	CCL2, CCL3, CCL4, CCL5, CCL7, CCL11 CXCL8, CXCL10	Bronchiolitis, pneumonia	1. Exacerbates asthma
Genus: *Influenzavirus A,B*					
Metapneumovirus					
Family: *Paramyxoviridae*	Enveloped, negative-sense ssRNA	Unknown	CXCL8	URTI, bronchiolitis, pneumonitis, wheezing	1. Exacerbates asthma 2. May contribute to the severity of RSV-induced disease

Genus: *Metapneumovirus*					
Parainfluenza					
Family: *Paramyxoviridae*	Enveloped, negative-sense ssRNA	Airway epithelium	CXCL8	Bronchiolitis, laryngo-tracheitis (croup), pneumonia, wheezing (<2 years of age)	1. Exacerbates asthma
Genus: *Paramyxovirus*					
Respiratory Syncytial Virus					
Family: *Paramyxoviridae*	Enveloped, negative-sense ssRNA	Airway epithelium, lymphocytes, monocytes/macrophages	CCL2, CCL3, CCL5, CXCL8	Bronchiolitis, pneumonia, wheezing (<2 years of age)	1. May predispose to or exacerbate asthma
Genus: *Pneumovirus*					
Rhinovirus					
Family: *Picornaviridae*	Nonenveloped, positive-sense ssRNA	Airway epithelium, basophils, mast cells, macrophages	CCL3, CCL5, CCL11, CCL24, CXCL5, CXCL8	URTI, bronchiolitis, wheezing (>2 years of age)	1. Exacerbates asthma
Genus: *Rhinovirus*					

URTI, upper respiratory tract infection.

(Yamada *et al.*, 2002). Interestingly, Marin *et al.* (2000) showed that adenovirus DNA was present in the nasal passages of 78.4% of asthmatic children, but in only 5% of the healthy controls; this suggested a link between adenovirus infection and asthma.

Adenovirus primarily infects airway epithelial cells. Infection alone is sufficient to trigger CXCL8 production, because infection with either replication-deficient or replication-sufficient virus is able to induce CXCL8 (Rothenberg *et al.*, 1995; Leland Booth and Metcalf, 1999). However, it appears that different strains of adenovirus display different capacities to upregulate CXCL8 (Leland Booth and Metcalf, 1999). Further investigation is necessary to determine whether CXCL8 production correlates with adenovirus-mediated disease severity or adenovirus-mediated asthma exacerbations. As CXCL8 has been demonstrated to exaggerate the allergic inflammatory response, increased CXCL8 production following viral infection may contribute to an inflammatory environment in the lung, leading to exacerbated allergic responses.

b. Influenza. Influenza outbreaks occur annually during the winter months. A recent study implicated influenza virus infections in 9% of wheezing episodes in children with asthma (Isaacs and Joshi, 2002). Additionally, influenza has been associated with acute exacerbations of acute asthma and COPD (Tan *et al.*, 2003). Influenza virus targets airway epithelial cells; however, it also can infect monocytes and macrophages. Infection of airway epithelial cells and macrophages results in a cell-specific profile of chemokine induction. Airway epithelial cells upregulate CCL2, CCL5, CCL11, and CXCL8 in response to influenza (Julkunen *et al.*, 2000; Kawaguchi *et al.*, 2000). In contrast, influenza infection of monocytes and macrophages results in the upregulation of CCL2, CCL3, CCL5, CCL7, and CXCL10 and the suppression of CXCL1 and CXCL8 (Sprenger *et al.*, 1996; Julkunen *et al.*, 2000). Examination of nasopharyngeal secretions from patients with Influenza A infection revealed elevated levels of CCL2, CCL3, CCL4, and CXCL8 (Julkunen *et al.*, 2000), while CXCL8 levels have been shown to correlate well with viral shedding and disease symptoms in experimentally infected adults (Skoner *et al.*, 1999; Gern *et al.*, 2002). However, CXCL8 levels are not elevated in asthmatic patients with influenza relative to nonasthmatics with influenza (Oh *et al.*, 2002), suggesting that influenza-mediated asthma exacerbations may not be due to a viral amplification of CXCL8 secretion. CCL5 and CCL11 are important eosinophil chemoattractants, and viral induction of these chemokines may aid in enhanced eosinophil recruitment into the lung during subsequent allergic responses.

c. Metapneumovirus. In 2001, van den Hoogen *et al.* reported that they had isolated a new human respiratory viral pathogen that induced symptoms similar to those of RSV. This virus, given the name human metapneumovirus (hMPV), was initially isolated from children in the Netherlands. Studies done in the Netherlands found that almost all children had been infected with this virus by the age of 5 and that this pathogen was ubiquitous and had been circulating for over 45 years (van den Hoogen *et al.*, 2001). hMPV infection can cause upper-respiratory-tract infection, bronchiolitis, and pneumonia. It has been associated with an increase in CXCL8 levels and low levels of CCL5 in nasal secretions (Jartti *et al.*, 2002). A report from Finland documented that 8% of 132 children who were admitted to the hospital for wheezing were positive for hMPV (Jartti *et al.*, 2002), suggesting that hMPV can induce asthma-like symptoms. hMPV did not appear to exacerbate asthma (Rawlinson *et al.*, 2003), but in one study of 10 infants who presented with severe RSV disease in the absence of any predisposing factors, 90% were positive for hMPV (Greensill *et al.*, 2003). Thus, hMPV may be a factor that dictates the severity of RSV-induced disease, and as such it may be implicated in the precipitation of wheezing and childhood asthma.

d. Parainfluenza. Approximately 9% of asthma exacerbations in children are a result of parainfluenza infection (Isaacs and Joshi, 2002). Parainfluenza is similar to RSV in the disease phenotype it elicits (Counihan *et al.*, 2001). Both viral infections induce a cell-mediated immune response to viral antigens and induce virus-specific IgE production. Although infants in their first year of life are the patients most susceptible to infection with these viruses, parainfluenza affects infants from 6 to 12 months old whereas RSV appears to be more problematic in infants between 1 and 3 months of age (Welliver, 2003). At present, parainfluenza infection has only been reported to induce CXCL8 production in humans; however, this chemokine was not found to correlate with disease severity (Gern *et al.*, 2002).

e. Respiratory Syncytial Virus. In developed countries, bronchiolitis is the most common cause of infant hospitalization (Shay *et al.*, 1999). Although many respiratory viruses have the potential to induce bronchiolitis, RSV is responsible for approximately 70% of these hospitalizations (Hall, 1998). RSV infection is characterized by epithelial-cell necrosis, lymphocytic inflammation, submucosal edema, and obstruction of the bronchioles by mucus, cellular debris, and fibrin (Lugo and Nahata, 1993). Infants between 1 and 3 months of age appear to be the patients most susceptible to RSV infection (Welliver, 2003), due at least partly to their small, easily occludable airways.

A great deal of research has been done on RSV to illuminate the disease caused by primary infection as well as to gain insight into how RSV may impact asthma. As is the case for many respiratory viral infections, RSV has been shown to exacerbate asthma. One study showed that 5% of asthma exacerbations in children are caused by RSV infection (Isaacs and Joshi, 2002). Some also argue that RSV is important in the pathogenesis of asthma. Stein *et al.* (1999) reported that a severe RSV infection during infancy was responsible for significant increases in wheezing during childhood that declined with age and were absent by age 13. A study by Sigurs *et al.* (2000) showed that RSV bronchiolitis was the strongest independent risk factor for asthma and atopic sensitization in children up to age 7. Similarly, another report that analyzed data from six studies reported that RSV infection during the first year of life was associated with recurrent wheezing and asthma in children up to 5 years postinfection, but that this correlation was not significant after 5 years (Kneyber *et al.*, 2000). Together these studies indicate that the RSV does have an association with childhood asthma. However, it is still not clear whether RSV infection can predispose an infant to childhood asthma or whether the development of a severe RSV infection during infancy reflects a genetic predisposition to childhood asthma.

RSV preferentially infects and replicates in airway epithelial cells; however, RSV can also infect alveolar macrophages. Airway epithelial cells have been shown to upregulate the chemokines CCL2, CCL5, CXCL8, CXCL10, and KC in response to RSV infection (Saito *et al.*, 1997; Olszewska-Pazdrak *et al.*, 1998; Harrison *et al.*, 1999; Zhang *et al.*, 2001; Miller *et al.*, 2003). Others have shown that RSV induces CCL5 and CXCL8 secretion in macrophages (Thomas *et al.*, 2000; Miller *et al.*, 2003). Many of the same chemokines observed to be upregulated *in vitro* have also been documented *in vivo* in murine models of RSV infection, as well as in human nasal secretions (Bonville *et al.*, 1999; Sheeran *et al.*, 1999; Domachowske *et al.*, 2000; Noah and Becker, 2000; Power *et al.*, 2001; Noah *et al.*, 2002; Miller *et al.*, 2003). Of these, CCL5 and CXCL8 have been linked to disease severity in humans (Hornsleth *et al.*, 2001; Noah *et al.*, 2002; Smyth *et al.*, 2002). CCL5, CXCL8, and CXCL10 have been linked to severity of allergic airway disease as well. Further investigation into chemokine levels during RSV-induced exacerbations of human asthma will be necessary to confirm these chemokines' role in the various aspects of an asthmatic response.

f. Rhinovirus. Rhinovirus infections account for over 60% of virally induced asthma exacerbations (Isaacs and Joshi, 2002). It has been reported that 32.4% of asthmatic children (versus 0% of nonasthmatic children) had rhinovirus genetic material in their nasal passages (Marin *et al.*, 2000). Experimental infection with rhinovirus in humans increases airway hyperresponsiveness (Cheung *et al.*, 1995). Furthermore, this increase in airway hyperresponsiveness was more severe in patients with allergic rhinitis than in control subjects (Gern *et al.*, 1997). This may be due to the fact that eosinophil numbers remain elevated in rhinovirus-infected asthmatics during the convalescence of the viral infection, which is the time when eosinophil numbers return to baseline in nonasthmatics (Fraenkel *et al.*, 1995). Both eosinophils and eosinophil cationic protein (ECP) were positively correlated to rhinovirus-mediated airway hyperresponsiveness in asthmatic individuals (Grunberg *et al.*, 1997). It has been demonstrated that experimental rhinovirus infection of asthmatic individuals resulted in increases in ECP, IL-6, and CXCL8 (Grunberg *et al.*, 1997). Rhinoviruses have been shown to upregulate, in addition to CXCL8, CCL3, CCL5, CCL11, CCL24, and CXCL5 (Subauste *et al.*, 1995; Schroth *et al.*, 1999; Papadopoulos *et al.*, 2001; Donninger *et al.*, 2003). Of these, CCL11, CCL24, and CXCL8 seem likely candidates in mediating rhinovirus-induced disease. CCL11 and CCL24 are potent eosinophil chemoattractants, and as such they may be responsible for the recruitment and retention of eosinophils to the lung, which cause airway hyperreactivity, whereas CXCL8 levels have been found to correspond to rhinovirus-induced disease severity (Gern *et al.*, 2002).

2. Animal Models of Allergic Disease Exacerbated by Respiratory Viruses

Data from humans have provided scientists, both at the desk and in the clinic, tremendous insight into the relationship between respiratory viral infections and asthma. However, interpretation of these data is often complicated by factors that cannot be properly controlled for in a clinical setting. In an attempt to circumvent these obstacles, animal models have been developed that will help researchers examine the impact of viral infection on asthma under defined conditions. The cytokine response following a viral infection is critical because of both the direct effects of the cytokines and the downstream effects that the cytokines may have on polarizing the immune response and directing chemokine production. Recent studies have demonstrated that influenza or adenoviral infection prevented Th2 cellular recruitment into the lungs following OVA challenge (Stampfli *et al.*, 1998; Wohlleben *et al.*, 2003). In contrast, RSV infection followed by an allergen challenge appears to exacerbate the asthmatic response. Specifically, RSV infection prior to OVA sensitization and challenge was found to result in an increase in inflammatory cell recruitment, including recruitment of neutrophils and eosinophils, to the lungs during the asthmatic response (Schwarze *et al.*, 1997). In addition, we have shown that RSV infections can exacerbate cockroach-allergen-induced airway hyperreactivity and inflammation via an IL-13-dependent mechanism (Lukacs *et al.*, 2001). These latter responses may reflect similar responses to those observed in the clinic in children with severe RSV infections.

To address the mechanism of virally induced exacerbations of asthma, Stephens *et al.* infected mice with a parainfluenza virus and then adoptively transferred allergen-specific Th2 cells. The viral infection, even in the absence of allergen challenge, caused allergen-specific Th2 cellular recruitment to the lungs. Upon further analysis, this same group found that upon OVA challenge of mice that had received OVA-specific Th1 cells and HEL-specific Th2 cells, both Th1 and Th2 cell populations were recruited to the lungs, and, furthermore, this recruitment was associated with an increase in CCL3 and CCL5 (Stephens *et al.*, 2002). Thus, viral infections, or the inflammatory response mounted against viral infections, may induce chemokine production that results in the recruitment of both virus-specific and allergen-specific T-cells to the lungs. The resulting leukocytic inflammation may play a role in the exacerbation of asthma. As is shown in Table III, CXCL8 and CCL5 are two chemokines that are upregulated by almost all clinically important respiratory viruses that have been shown to predispose patients to, or exacerbate, asthma. In the case of CXCL8, levels are upregulated in all of the viruses mentioned above; however, there is a correlation with disease severity only in the case of RSV and rhinovirus. It is interesting to consider that the differences that exist among the diseases elicited by these viruses may be a result of differences in CXCR2, the receptor for CXCL8. However, at present, there is a paucity of data regarding the chemokine receptor profiles during a primary viral infection or during viral exacerbations of asthma.

Our lab has begun to investigate the role of chemokines and chemokine receptors in primary RSV infection, as well as in a combined RSV and allergen model. So far, we have demonstrated that CXCR2, the CXCL8 receptor, is upregulated in a murine model of RSV infection. Furthermore, we have demonstrated that CXCR2 plays a role in RSV-induced airway hyperresponsiveness and mucus overproduction (Miller *et al.*, 2003). In the case of CCL5, we have shown that the chemokine can modulate RSV-induced airway hyperresponsiveness (Tekkanat *et al.*, 2002). A recent publication has gone one step further and has examined the role of RSV-induced CCL5 production in allergic asthma. The data from this study showed that neutralization of CCL5 during RSV infection prevented cockroach-allergen-induced eosinophilia, chemokine production, leukotriene release, and inflammatory-cell recruitment to the bronchoalveolar space but had little effect on the pathophysiological response (John *et al.*, 2003). Thus, data collected in rodents and humans suggest that there may be common chemokines and/or receptors, for example, CCL5 or CXCL8, that could be targeted to cause the weakening of a number of virally induced asthmatic responses. However, further research is necessary to correlate findings in animal models with clinical data and conclusively prove a role for these molecules in the exacerbation of asthma.

III. SUMMARY

Many of the current therapies available to treat pulmonary diseases are centered on the use of pharmacological agents designed to inhibit the overall initiation of an inflammatory response. Inhaled glucocorticoids and β_2 antagonists have long been used to treat allergic airway disease, and it is not clear whether these treatments have long-term benefits. The use of glucocorticoids in the treatment of COPD and pulmonary fibrosis has met with limited success. Recently, several leukotriene synthesis inhibitors have also been developed and used with some success. Additionally, it has been reported that the use of cyclooxygenase (COX-1 or COX-2) inhibitors increased airway hyperreactivity and allergic inflammation in the lung and that these inhibitors specifically increased CCR1, CCR2, CCR3, CCR4, and CCR5 expression in allergen-challenged lungs (Peebles *et al.*, 2002). They therefore may eventually worsen disease outcome. The relative paucity of successful therapies in pulmonary disease mandates continued research into new therapeutic targets in this area.

References

Aaron, S. D., Angel, J. B., Lunau, M., Wright, K., Fex, C., Le Saux, N., and Dales, R. E. (2001). Granulocyte inflammatory markers and airway infection during acute exacerbation of chronic obstructive pulmonary disease. *Am. J. Respir. Crit. Care Med.* **163,** 349.

Adachi, T., Vita, R., Sannohe, S., Stafford, S., Alam, R., Kayaba, H., and Chihara, J. (2001). The functional role of rho and rho-associated coiled-coil forming protein kinase in eotaxin signaling of eosinophils. *J. Immunol.* **167,** 4609.

Alwan, W. H., Record, F. M., and Openshaw, P. J. (1992). CD4+ T cells clear virus but augment disease in mice infected with respiratory syncytial virus. Comparison with the effects of CD8+ T cells. *Clin. Exp. Immunol.* **88,** 527.

Anderson, L. J., Tsou, C., Potter, C., Keyserling, H. L., Smith, T. F., Ananaba, G., and Bangham, C. R. (1994). Cytokine response to respiratory syncytial virus stimulation of human peripheral blood mononuclear cells. *J. Infect. Dis.* **170,** 1201.

Annunziato, F., Cosmi, L., Galli, G., Beltrame, C., Romagnani, P., Manetti, R., Romagnani, S., and Maggi, E. (1999). Assessment of chemokine receptor expression by human Th1 and Th2 cells *in vitro* and *in vivo*. *J. Leukoc. Biol.* **65,** 691.

Bandeira-Melo, C., Bozza, P. T., and Weller, P. F. (2002). The cellular biology of eosinophil eicosanoid formation and function. *J. Allergy Clin. Immunol.* **109,** 393.

Bandeira-Melo, C., Phoofolo, M., and Weller, P. F. (2001). Extranuclear lipid bodies, elicited by CCR3-mediated signaling pathways, are the sites of chemokine-enhanced leukotriene C4 production in eosinophils and basophils. *J. Biol. Chem.* **276,** 22779.

Beck, L. A., Dalke, S., Leiferman, K. M., Bickel, C. A., Hamilton, R., Rosen, H., Bochner, B. S., and Schleimer, R. P. (1997). Cutaneous injection of RANTES causes eosinophil recruitment: Comparison of nonallergic and allergic human subjects. *J. Immunol.* **159,** 2962.

Belperio, J. A., Dy, M., Burdick, M. D., Xue, Y. Y., Li, K., Elias, J. A., and Keane, M. P. (2002). Interaction of IL-13 and C10 in the pathogenesis of bleomycin-induced pulmonary fibrosis. *Am. J. Respir. Cell. Mol. Biol.* **27,** 419.

Bendelja, K., Gagro, A., Bace, A., Lokar-Kolbas, R., Krsulovic-Hresic, V., Drazenovic, V., Mlinaric-Galinovic, G., and Rabatic, S. (2000). Predominant type-2 response in infants with respiratory syncytial virus (RSV) infection demonstrated by cytokine flow cytometry. *Clin. Exp. Immunol.* **121,** 332.

Biedermann, T., Mailhammer, R., Mai, A., Sander, C., Ogilvie, A., Brombacher, F., Maier, K., Levine, A. D., and Rocken, M. (2001). Reversal of established delayed type hypersensitivity reactions following therapy with IL-4 or antigen-specific Th2 cells. *Eur. J. Immunol.* **31,** 1582.

Blease, K., Mehrad, B., Standiford, T. J., Lukacs, N. W., Kunkel, S. L., Chensue, S. W., Lu, B., Gerard, C. J., and Hogaboam, C. M. (2000). Airway remodeling is absent in CCR1−/− mice during chronic fungal allergic airway disease. *J. Immunol.* **165,** 1564.

Boehme, S. A., Sullivan, S. K., Crowe, P. D., Santos, M., Conlon, P. J., Sriramarao, P., and Bacon, K. B. (1999). Activation of mitogen-activated protein kinase regulates eotaxin-induced eosinophil migration. *J. Immunol.* **163,** 1611.

Bonecchi, R., Bianchi, G., Bordignon, P. P., D'Ambrosio, D., Lang, R., Borsatti, A., Sozzani, S., Allavena, P., Gray, P. A., Mantovani, A., and Sinigaglia, F. (1998). Differential expression of chemokine receptors and chemotactic responsiveness of type 1 T helper cells (Th1s) and Th2s. *J. Exp. Med.* **187,** 129.

Bonville, C. A., Rosenberg, H. F., and Domachowske, J. B. (1999). Macrophage inflammatory protein-1alpha and RANTES are present in nasal secretions during ongoing upper respiratory tract infection. *Pediatr. Allergy Immunol.* **10,** 39.

Boomars, K. A., Schweizer, R. C., Zanen, P., van den Bosch, J. M., Lammers, J. W., and Koenderman, L. (1998). Eosinophil chemotactic activity in bronchoalveolar lavage from idiopathic pulmonary fibrosis is dependent on cytokine priming of eosinophils. *Eur. Respir. J.* **11,** 1009.

Boyce, J. A. (1997). The pathobiology of eosinophilic inflammation. *Allergy Asthma Proc.* **18,** 293.

Bright, H., Turnbull, T., Toms, G. L., and Scott, R. (1995). Comparison of the T helper cell response induced by respiratory syncytial virus and its fusion protein in BALB/c mice. *Vaccine* **13,** 915.

Brightling, C. E., Bradding, P., Pavord, I. D., and Wardlaw, A. J. (2003). New insights into the role of the mast cell in asthma. *Clin. Exp. Allergy* **33,** 550.

Campbell, J. J., Qin, S., Bacon, K. B., Mackay, C. R., and Butcher, E. C. (1996). Biology of chemokine and classical chemoattractant receptors: Differential requirements for adhesion-triggering versus chemotactic responses in lymphoid cells. *J. Cell Biol.* **134,** 255.

Chen, Y., Song, A., and Krensky, A. M. (2000). Transcriptional Regulation of Chemokines. *In* "Chemokines in Allergic Disease" (M. E. Rothenberg, ed.), p. 117. Marcel Dekker, Inc, New York.

Chensue, S. W., Lukacs, N. W., Yang, T. Y., Shang, X., Frait, K. A., Kunkel, S. L., Kung, T., Wiekowski, M. T., Hedrick, J. A., Cook, D. N., Zingoni, A., Narula, S. K., Zlotnik, A., Barrat, F. J., O'Garra, A., Napolitano, M., and Lira, S. A. (2001). Aberrant *in vivo* T helper type 2 cell response and impaired eosinophil recruitment in CC chemokine receptor 8 knockout mice. *J. Exp. Med.* **193,** 573.

Cheung, D., Dick, E. C., Timmers, M. C., de Klerk, E. P., Spaan, W. J., and Sterk, P. J. (1995). Rhinovirus inhalation causes long-lasting excessive airway narrowing in response to methacholine in asthmatic subjects *in vivo*. *Am. J. Respir. Crit. Care Med.* **152,** 1490.

Chihara, J., Oyamada, H., Yamada, H., Tsuda, A., Kamada, Y., and Urayama, O. (1997). Expression of mRNA for RANTES in human eosinophils. *Int. Arch. Allergy Immunol* **114** (Suppl. 1), 33.

Chung, C. D., Kuo, F., Kumer, J., Motani, A. S., Lawrence, C. E., Henderson, W. R., Jr., and Venkataraman, C. (2003). CCR8 is not essential for the development of inflammation in a mouse model of allergic airway disease. *J. Immunol.* **170,** 581.

Cohn, L., Herrick, C., Niu, N., Homer, R., and Bottomly, K. (2001). IL-4 promotes airway eosinophilia by suppressing IFN-gamma production: Defining a novel role for IFN-gamma in the regulation of allergic airway inflammation. *J. Immunol.* **166,** 2760.

Corrigan, C. J., and Kay, A. B. (1992). T cells and eosinophils in the pathogenesis of asthma. *Immunol. Today* **13,** 501.

Counihan, M. E., Shay, D. K., Holman, R. C., Lowther, S. A., and Anderson, L. J. (2001). Human parainfluenza virus-associated hospitalizations among children less than five years of age in the United States. *Pediatr. Infect. Dis. J.* **20,** 646.

Crimi, E., Gaffi, D., Frittoli, E., Borgonovo, B., and Burastero, S. E. (1997). Depletion of circulating allergen-specific TH2 T lymphocytes after allergen exposure in asthma. *J. Allergy Clin. Immunol.* **99,** 788.

De Boer, W. I. (2002). Cytokines and therapy in COPD: A promising combination? *Chest* **121,** 209S.

De Boer, W. I., Sont, J. K., van Schadewijk, A., Stolk, J., van Krieken, J. H., and Hiemstra, P. S. (2000). Monocyte chemoattractant protein-1, interleukin-8, and chronic airways inflammation in COPD. *J. Pathol.* **190,** 619.

de Waal, L., Koopman, L. P., van Benten, I. J., Brandenburg, A. H., Mulder, P. G., de Swart, R. L., Fokkens, W. J., Neijens, H. J., and Osterhaus, A. D. (2003). Moderate local and systemic respiratory syncytial virus-specific T-cell responses upon mild or subclinical RSV infection. *J. Med. Virol.* **70,** 309.

Djukanovic, R. (2000). Asthma: A disease of inflammation and repair. *J. Allergy Clin. Immunol.* **105,** S522.

Domachowske, J. B., Bonville, C. A., Gao, J. L., Murphy, P. M., Easton, A. J., and Rosenberg, H. F. (2000). MIP-1alpha is produced but it does not control pulmonary inflammation in response to respiratory syncytial virus infection in mice. *Cell Immunol.* **206,** 1.

Donninger, H., Glashoff, R., Haitchi, H. M., Syce, J. A., Ghildyal, R., van Rensburg, E., and Bardin, P. G. (2003). Rhinovirus induction of the CXC chemokine epithelial-neutrophil activating peptide-78 in bronchial epithelium. *J. Infect. Dis.* **187,** 1809.

Dufour, J. H., Dziejman, M., Liu, M. T., Leung, J. H., Lane, T. E., and Luster, A. D. (2002). IFN-gamma-inducible protein 10 (IP-10; CXCL10)-deficient mice reveal a role for IP-10 in effector T cell generation and trafficking. *J. Immunol.* **168,** 3195.

Duguet, A., Iijima, H., Eum, S. Y., Hamid, Q., and Eidelman, D. H. (2001). Eosinophil peroxidase mediates protein nitration in allergic airway inflammation in mice. *Am. J. Respir. Crit. Care Med.* **164,** 1119.

Eddleston, J., Christiansen, S. C., and Zuraw, B. L. (2002). Functional expression of the C-X-C chemokine receptor CXCR4 by human bronchial epithelial cells: Regulation by proinflammatory mediators. *J. Immunol.* **169,** 6445.

Elser, B., Lohoff, M., Kock, S., Giaisi, M., Kirchhoff, S., Krammer, P. H., and Li-Weber, M. (2002). IFN-gamma represses IL-4 expression via IRF-1 and IRF-2. *Immunity* **17,** 703.

Farber, J. M. (1997). Mig and IP-10: CXC chemokines that target lymphocytes. *J. Leukoc. Biol.* **61,** 246.

Ford, J. G., Rennick, D., Donaldson, D. D., Venkayya, R., McArthur, C., Hansell, E., Kurup, V. P., Warnock, M., and Grunig, G. (2001). Il-13 and IFN-gamma: Interactions in lung inflammation. *J. Immunol.* **167,** 1769.

Foster, P. S., Mould, A. W., Yang, M., Mackenzie, J., Mattes, J., Hogan, S. P., Mahalingam, S., McKenzie, A. N., Rothenberg, M. E., Young, I. G., Matthaei, K. I., and Webb, D. C.

(2001). Elemental signals regulating eosinophil accumulation in the lung. *Immunol. Rev.* **179,** 173.

Fraenkel, D. J., Bardin, P. G., Sanderson, G., Lampe, F., Johnston, S. L., and Holgate, S. T. (1995). Lower airways inflammation during rhinovirus colds in normal and in asthmatic subjects. *Am. J. Respir. Crit. Care Med.* **151,** 879.

Gasperini, S., Marchi, M., Calzetti, F., Laudanna, C., Vicentini, L., Olsen, H., Murphy, M., Liao, F., Farber, J., and Cassatella, M. A. (1999). Gene expression and production of the monokine induced by IFN-gamma (MIG), IFN-inducible T cell alpha chemoattractant (I-TAC), and IFN-gamma-inducible protein-10 (IP-10) chemokines by human neutrophils. *J. Immunol.* **162,** 4928.

Gern, J. E., Calhoun, W., Swenson, C., Shen, G., and Busse, W. W. (1997). Rhinovirus infection preferentially increases lower airway responsiveness in allergic subjects. *Am. J. Respir. Crit. Care Med.* **155,** 1872.

Gern, J. E., Martin, M. S., Anklam, K. A., Shen, K., Roberg, K. A., Carlson-Dakes, K. T., Adler, K., Gilbertson-White, S., Hamilton, R., Shult, P. A., Kirk, C. J., Da Silva, S. A., Sund, S. A., Kosorok, M. R., and Lemanske, R. F., Jr. (2002). Relationships among specific viral pathogens, virus-induced interleukin-8, and respiratory symptoms in infancy. *Pediatr. Allergy Immunol.* **13,** 386.

Gharaee-Kermani, M., Denholm, E. M., and Phan, S. H. (1996). Costimulation of fibroblast collagen and transforming growth factor beta1 gene expression by monocyte chemoattractant protein-1 via specific receptors. *J. Biol. Chem.* **271,** 17779.

Gibson, P. G., Simpson, J. L., and Saltos, N. (2001). Heterogeneity of airway inflammation in persistent asthma: Evidence of neutrophilic inflammation and increased sputum interleukin-8. *Chest* **119,** 1329.

Gonzalo, J. A., Lloyd, C. M., Peled, A., Delaney, T., Coyle, A. J., and Gutierrez-Ramos, J. C. (2000). Critical involvement of the chemotactic axis CXCR4/stromal cell-derived factor-1 alpha in the inflammatory component of allergic airway disease. *J. Immunol.* **165,** 499.

Gonzalo, J. A., Lloyd, C. M., Wen, D., Albar, J. P., Wells, T. N., Proudfoot, A., Martinez, A. C., Dorf, M., Bjerke, T., Coyle, A. J., and Gutierrez-Ramos, J. C. (1998). The coordinated action of CC chemokines in the lung orchestrates allergic inflammation and airway hyperresponsiveness. *J. Exp. Med.* **188,** 157.

Greensill, J., McNamara, P. S., Dove, W., Flanagan, B., Smyth, R. L., and Hart, C. A. (2003). Human metapneumovirus in severe respiratory syncytial virus bronchiolitis. *Emerg. Infect. Dis.* **9,** 372.

Grunberg, K., Smits, H. H., Timmers, M. C., de Klerk, E. P., Dolhain, R. J., Dick, E. C., Hiemstra, P. S., and Sterk, P. J. (1997). Experimental rhinovirus 16 infection. Effects on cell differentials and soluble markers in sputum in asthmatic subjects. *Am. J. Respir. Crit. Care Med.* **156,** 609.

Gu, L., Rutledge, B., Fiorillo, J., Ernst, C., Grewal, I., Flavell, R., Gladue, R., and Rollins, B. (1997). *In vivo* properties of monocyte chemoattractant protein-1. *J. Leukoc. Biol.* **62,** 577.

Haddad el, B., Underwood, S. L., Dabrowski, D., Birrell, M. A., McCluskie, K., Battram, C. H., Pecoraro, M., Foster, M. L., and Belvisi, M. G. (2002). Critical role for T cells in sephadex-induced airway inflammation: Pharmacological and immunological characterization and molecular biomarker identification. *J. Immunol.* **168,** 3004.

Hall, C. B. (1998). "Respiratory Syncytial Virus." Lippincott Williams & Wilkins, Philadelphia.

Hallsworth, M. P., Soh, C. P., Lane, S. J., Arm, J. P., and Lee, T. H. (1994). Selective enhancement of GM-CSF, TNF-alpha, IL-1 beta and IL-8 production by monocytes and macrophages of asthmatic subjects. *Eur. Respir. J.* **7,** 1096.

Harrison, A. M., Bonville, C. A., Rosenberg, H. F., and Domachowske, J. B. (1999). Respiratory syncytical virus-induced chemokine expression in the lower airways: Eosinophil recruitment and degranulation. *Am. J. Respir. Crit. Care Med.* **159,** 1918.

Hertz, M., Mahalingam, S., Dalum, I., Klysner, S., Mattes, J., Neisig, A., Mouritsen, S., Foster, P. S., and Gautam, A. (2001). Active vaccination against IL-5 bypasses immunological tolerance and ameliorates experimental asthma. *J. Immunol.* **167,** 3792.

Hochstetter, R., Dobos, G., Kimmig, D., Dulkys, Y., Kapp, A., and Elsner, J. (2000). The CC chemokine receptor 3 CCR3 is functionally expressed on eosinophils but not on neutrophils. *Eur. J. Immunol.* **30,** 2759.

Hogaboam, C. M., Gallinat, C. S., Taub, D. D., Strieter, R. M., Kunkel, S. L., and Lukacs, N. W. (1999). Immunomodulatory role of C10 chemokine in a murine model of allergic bronchopulmonary aspergillosis. *J. Immunol.* **162,** 6071.

Hornsleth, A., Loland, L., and Larsen, L. B. (2001). Cytokines and chemokines in respiratory secretion and severity of disease in infants with respiratory syncytial virus (RSV) infection. *J. Clin. Virol.* **21,** 163.

Isaacs, D., and Joshi, P. (2002). Respiratory infections and asthma. *Med. J. Aust.* **177**(Suppl.), S50.

Janatpour, M. J., Hudak, S., Sathe, M., Sedgwick, J. D., and McEvoy, L. M. (2001). Tumor necrosis factor-dependent segmental control of MIG expression by high endothelial venules in inflamed lymph nodes regulates monocyte recruitment. *J. Exp. Med.* **194,** 1375.

Jartti, T., van den Hoogen, B., Garofalo, R. P., Osterhaus, A. D., and Ruuskanen, O. (2002). Metapneumovirus and acute wheezing in children. *Lancet* **360,** 1393.

Jatakanon, A., Uasuf, C., Maziak, W., Lim, S., Chung, K. F., and Barnes, P. J. (1999). Neutrophilic inflammation in severe persistent asthma. *Am. J. Respir. Crit. Care Med.* **160,** 1532.

Jinquan, T., Jing, C., Jacobi, H. H., Reimert, C. M., Millner, A., Quan, S., Hansen, J. B., Dissing, S., Malling, H. J., Skov, P. S., and Poulsen, L. K. (2000). CXCR3 expression and activation of eosinophils: Role of IFN-gamma-inducible protein-10 and monokine induced by IFN-gamma. *J. Immunol.* **165,** 1548.

John, A. E., Berlin, A. A., and Lukacs, N. W. (2003). Respiratory syncytial virus-induced CCL5/RANTES contributes to exacerbation of allergic airway inflammation. *Eur. J. Immunol.* **33,** 1677.

Julkunen, I., Melen, K., Nyqvist, M., Pirhonen, J., Sareneva, T., and Matikainen, S. (2000). Inflammatory responses in influenza A virus infection. *Vaccine* **19**(Suppl. 1), S32.

Kampen, G. T., Stafford, S., Adachi, T., Jinquan, T., Quan, S., Grant, J. A., Skov, P. S., Poulsen, L. K., and Alam, R. (2000). Eotaxin induces degranulation and chemotaxis of eosinophils through the activation of ERK2 and p38 mitogen-activated protein kinases. *Blood* **95,** 1911.

Katschke, K. J., Jr., Rottman, J. B., Ruth, J. H., Qin, S., Wu, L., LaRosa, G., Ponath, P., Park, C. C., Pope, R. M., and Koch, A. E. (2001). Differential expression of chemokine receptors on peripheral blood, synovial fluid, and synovial tissue monocytes/macrophages in rheumatoid arthritis. *Arthritis Rheum* **44,** 1022.

Kawaguchi, M., Kokubu, F., Kuga, H., Tomita, T., Matsukura, S., Kadokura, M., and Adachi, M. (2000). Expression of eotaxin by normal airway epithelial cells after influenza virus A infection. *Int. Arch. Allergy Immunol.* **122**(Suppl. 1), 44.

Keane, M. P., Arenberg, D. A., Lynch, I.,P., J., Whyte, R. I., Iannettoni, M. D., Burdick, M. D., Wilke, C. A., Morris, S. B., Glass, M. C., DiGiovine, B., Kunkel, S. L., and Strieter, R. M. (2005). The CXC chemokines, IL-8 and IP-10, regulate angiogenic activity in idiopathic pulmonary fibrosis. *J. Immunol.* In press.

Keane, M. P., Arenberg, D. A., Moore, B. B., Addison, C. L., and Strieter, R. M. (1998). CXC chemokines and angiogenesis/angiostasis. *Proc. Assoc. Am. Physicians* **110,** 288.

Keane, M. P., Belperio, J. A., Arenberg, D. A., Burdick, M. D., Xu, Z. J., Xue, Y. Y., and Strieter, R. M. (1999a). IFN-gamma-inducible protein-10 attenuates bleomycin-induced pulmonary fibrosis via inhibition of angiogenesis. *J. Immunol.* **163,** 5686.

Keane, M. P., Belperio, J. A., Moore, T. A., Moore, B. B., Arenberg, D. A., Smith, R. E., Burdick, M. D., Kunkel, S. L., and Strieter, R. M. (1999b). Neutralization of the CXC chemokine, macrophage inflammatory protein-2, attenuates bleomycin-induced pulmonary fibrosis. *J. Immunol.* **162,** 5511.

Kim, Y., Sung, S., Kuziel, W. A., Feldman, S., Fu, S. M., and Rose, C. E., Jr. (2001). Enhanced airway Th2 response after allergen challenge in mice deficient in CC chemokine receptor-2 (CCR2). *J. Immunol.* **166,** 5183.

Kneyber, M. C. J., Steyerberg, E. W., de Groot, R., and Moll, H. A. (2000). Long-term effects of respiratory syncytial virus (RSV) bronchiolitis in infants and young children: A quantitative review. *Acta. Paediatr.* **89,** 654.

Koopmann, W., and Krangel, M. S. (1997). Identification of a glycosaminoglycan-binding site in chemokine macrophage inflammatory protein-1 alpha. *J. Biol. Chem.* **272,** 10103.

Leland Booth, J., and Metcalf, J. P. (1999). Type-specific induction of interleukin-8 by adenovirus. *Am. J. Respir. Cell Mol. Biol.* **21,** 521.

Leung, T. F., Wong, C. K., Lam, C. W., Li, A. M., Ip, W. K., Wong, G. W., and Fok, T. F. (2003). Plasma TARC concentration may be a useful marker for asthmatic exacerbation in children. *Eur. Respir. J.* **21,** 616.

Li, L., Xia, Y., Nguyen, A., Feng, L., and Lo, D. (1998). Th2-induced eotaxin expression and eosinophilia coexist with Th1 responses at the effector stage of lung inflammation. *J. Immunol.* **161,** 3128.

Li, L., Xia, Y., Nguyen, A., Lai, Y. H., Feng, L., Mosmann, T. R., and Lo, D. (1999). Effects of Th2 cytokines on chemokine expression in the lung: IL-13 potently induces eotaxin expression by airway epithelial cells. *J. Immunol.* **162,** 2477.

Linden, A. (2001). Role of interleukin-17 and the neutrophil in asthma. *Int. Arch. Allergy Immunol.* **126,** 179.

Lugo, R. A., and Nahata, M. C. (1993). Pathogenesis and treatment of bronchiolitis. *Clin. Pharm.* **12,** 95.

Lukacs, N. W., Hogaboam, C. M., Kunkel, S. L., Chensue, S. W., Burdick, M. D., Evanoff, H. L., and Strieter, R. M. (1998). Mast cells produce ENA-78, which can function as a potent neutrophil chemoattractant during allergic airway inflammation. *J. Leukoc. Biol.* **63,** 746.

Lukacs, N. W., Prosser, D. M., Wiekowski, M., Lira, S. A., and Cook, D. N. (2001). Requirement for the chemokine receptor CCR6 in allergic pulmonary inflammation. *J. Exp. Med.* **194,** 551.

Lukacs, N. W., Standiford, T. J., Chensue, S. W., Kunkel, R. G., Strieter, R. M., and Kunkel, S. L. (1996). C-C chemokine-induced eosinophil chemotaxis during allergic airway inflammation. *J. Leukoc. Biol.* **60,** 573.

Lukacs, N. W., Strieter, R. M., Warmington, K., Lincoln, P., Chensue, S. W., and Kunkel, S. L. (1997). Differential recruitment of leukocyte populations and alteration of airway hyperreactivity by C-C family chemokines in allergic airway inflammation. *J. Immunol.* **158,** 4398.

Lukacs, N. W., Tekkanat, K. K., Berlin, A., Hogaboam, C. M., Miller, A., Evanoff, H., Lincoln, P., and Maassab, H. (2001). Respiratory syncytial virus predisposes mice to augmented allergic airway responses via IL-13-mediated mechanisms. *J. Immunol.* **167,** 1060.

Lundien, M. C., Mohammed, K. A., Nasreen, N., Tepper, R. S., Hardwick, J. A., Sanders, K. L., Van Horn, R. D., and Antony, V. B. (2002). Induction of MCP-1 expression in airway epithelial cells: Role of CCR2 receptor in airway epithelial injury. *J. Clin. Immunol.* **22,** 144.

Ma, W., Bryce, P. J., Humbles, A. A., Laouini, D., Yalcindag, A., Alenius, H., Friend, D. S., Oettgen, H. C., Gerard, C., and Geha, R. S. (2002). CCR3 is essential for skin eosinophilia and airway hyperresponsiveness in a murine model of allergic skin inflammation. *J. Clin. Invest.* **109,** 621.

Macek, V., Sorli, J., Kopriva, S., and Marin, J. (1994). Persistent adenoviral infection and chronic airway obstruction in children. *Am. J. Respir. Crit. Care Med.* **150,** 7.

Mariani, E., Pulsatelli, L., Neri, S., Dolzani, P., Meneghetti, A., Silvestri, T., Ravaglia, G., Forti, P., Cattini, L., and Facchini, A. (2002). RANTES and MIP-1alpha production by T lymphocytes, monocytes and NK cells from nonagenarian subjects. *Exp. Gerontol.* **37,** 219.

Marin, J., Jeler-Kacar, D., Levstek, V., and Macek, V. (2000). Persistence of viruses in upper respiratory tract of children with asthma. *J. Infect.* **41,** 69.

McKenzie, G. J., Emson, C. L., Bell, S. E., Anderson, S., Fallon, P., Zurawski, G., Murray, R., Grencis, R., and McKenzie, A. N. (1998). Impaired development of Th2 cells in IL-13-deficient mice. *Immunity* **9,** 423.

Medoff, B. D., Sauty, A., Tager, A. M., Maclean, J. A., Smith, R. N., Mathew, A., Dufour, J. H., and Luster, A. D. (2002). IFN-gamma-inducible protein 10 (CXCL10) contributes to airway hyperreactivity and airway inflammation in a mouse model of asthma. *J. Immunol.* **168,** 5278.

Michalec, L., Choudhury, B. K., Postlethwait, E., Wild, J. S., Alam, R., Lett-Brown, M., and Sur, S. (2002). CCL7 and CXCL10 orchestrate oxidative stress-induced neutrophilic lung inflammation. *J. Immunol.* **168,** 846.

Miller, A. L., Bowlin, T. L., and Lukacs, N. W. (2003). RSV-Induced Chemokine Production: Linking Viral Infection and Replication to Chemokine Production *in vitro* and *in vivo*. *J. Infect. Dis.* **189,** 1419–1430.

Miller, A. L., Strieter, R. M., Gruber, A. D., Ho, S. B., and Lukacs, N. W. (2003). CXCR2 regulates respiratory syncytial virus-induced airway hyperreactivity and mucus overproduction. *J. Immunol.* **170,** 3348.

Murdoch, C., Monk, P. N., and Finn, A. (1999). Functional expression of chemokine receptor CXCR4 on human epithelial cells. *Immunology* **98,** 36.

Nagase, H., Kudo, K., Izumi, S., Ohta, K., Kobayashi, N., Yamaguchi, M., Matsushima, K., Morita, Y., Yamamoto, K., and Hirai, K. (2001). Chemokine receptor expression profile of eosinophils at inflamed tissue sites: Decreased CCR3 and increased CXCR4 expression by lung eosinophils. *J. Allergy Clin. Immunol.* **108,** 563.

Noah, T. L., Ivins, S. S., Murphy, P., Kazachkova, I., Moats-Staats, B., and Henderson, F. W. (2002). Chemokines and inflammation in the nasal passages of infants with respiratory syncytial virus bronchiolitis. *Clin. Immunol.* **104,** 86.

Noah, T. L., and Becker, S. (2000). Chemokines in Nasal Secretions of Normal Adults Experimentally Infected with Respiratory Syncytial Virus. *Clin. Immunol.* **97,** 43.

O'Byrne, P. M. (1996). Airway inflammation and asthma. *Aliment. Pharmacol. Ther.* **10** (Suppl. 2), 18.

Oddy, W. H., de Klerk, N. H., Sly, P. D., and Holt, P. G. (2002). The effects of respiratory infections, atopy, and breastfeeding on childhood asthma. *Eur. Respir. J.* **19,** 899.

Oh, J. W., Lee, H. B., Park, I. K., and Kang, J. O. (2002). Interleukin-6, interleukin-8, interleukin-11, and interferon-gamma levels in nasopharyngeal aspirates from wheezing children with respiratory syncytial virus or influenza A virus infection. *Pediatr. Allergy Immunol.* **13,** 350.

Oliveira, S. H., Lira, S., Martinez, A. C., Wiekowski, M., Sullivan, L., and Lukacs, N. W. (2002). Increased responsiveness of murine eosinophils to MIP-1beta (CCL4) and TCA-3 (CCL1) is mediated by their specific receptors, CCR5 and CCR8. *J. Leukoc. Biol.* **71,** 1019.

Olszewska-Pazdrak, B., Casola, A., Saito, T., Alam, R., Crowe, S. E., Mei, F., Ogra, P. L., and Garofalo, R. P. (1998). Cell-specific expression of RANTES, MCP-1, and MIP-1alpha by lower airway epithelial cells and eosinophils infected with respiratory syncytial virus. *J. Virol.* **72,** 4756.

Ordonez, C. L., Shaughnessy, T. E., Matthay, M. A., and Fahy, J. V. (2000). Increased neutrophil numbers and IL-8 levels in airway secretions in acute severe asthma: Clinical and biologic significance. *Am. J. Respir. Crit. Care Med.* **161,** 1185.

Page, S. M., Gleich, G. J., Roebuck, K. A., and Thomas, L. L. (1999). Stimulation of neutrophil interleukin-8 production by eosinophil granule major basic protein. *Am. J. Respir. Cell Mol. Biol.* **21,** 230.

Panina-Bordignon, P., Papi, A., Mariani, M., Di Lucia, P., Casoni, G., Bellettato, C., Buonsanti, C., Miotto, D., Mapp, C., Villa, A., Arrigoni, G., Fabbri, L. M., and Sinigaglia, F. (2001). The C-C chemokine receptors CCR4 and CCR8 identify airway T cells of allergen-challenged atopic asthmatics. *J. Clin. Invest.* **107,** 1357.

Panina-Bordignon, P., and D'Ambrosio, D. (2003). Chemokines and their receptors in asthma and chronic obstructive pulmonary disease. *Curr. Opin. Pulm. Med.* **9,** 104.

Papadopoulos, N. G., Papi, A., Meyer, J., Stanciu, L. A., Salvi, S., Holgate, S. T., and Johnston, S. L. (2001). Rhinovirus infection up-regulates eotaxin and eotaxin-2 expression in bronchial epithelial cells. *Clin. Exp. Allergy* **31,** 1060.

Park, M. K., Amichay, D., Love, P., Wick, E., Liao, F., Grinberg, A., Rabin, R. L., Zhang, H. H., Gebeyehu, S., Wright, T. M., Iwasaki, A., Weng, Y., DeMartino, J. A., Elkins, K. L., and Farber, J. M. (2002). The CXC chemokine murine monokine induced by IFN-gamma (CXC chemokine ligand 9) is made by APCs, targets lymphocytes including activated B cells, and supports antibody responses to a bacterial pathogen *in vivo*. *J. Immunol.* **169,** 1433.

Patel, D. D., Zachariah, J. P., and Whichard, L. P. (2001). CXCR3 and CCR5 ligands in rheumatoid arthritis synovium. *Clin. Immunol.* **98,** 39.

Peebles, R. S., Jr., Hashimoto, K., Morrow, J. D., Dworski, R., Collins, R. D., Hashimoto, Y., Christman, J. W., Kang, K. H., Jarzecka, K., Furlong, J., Mitchell, D. B., Talati, M., Graham, B. S., and Sheller, J. R. (2002). Selective cyclooxygenase-1 and -2 inhibitors each increase allergic inflammation and airway hyperresponsiveness in mice. *Am. J. Respir. Crit. Care Med.* **165,** 1154.

Petrek, M., Pantelidis, P., Southcott, A. M., Lympany, P., Safranek, P., Black, C. M., Kolek, V., Weigl, E., and du Bois, R. M. (1997). The source and role of RANTES in interstitial lung disease. *Eur. Respir. J.* **10,** 1207.

Pinho, V., Oliveira, S. H., Souza, D. G., Vasconcelos, D., Alessandri, A. L., Lukacs, N. W., and Teixeira, M. M. (2003). The role of CCL22 (MDC) for the recruitment of eosinophils during allergic pleurisy in mice. *J. Leukoc. Biol.* **73,** 356.

Power, U. F., Huss, T., Michaud, V., Plotnicky-Gilquin, H., Bonnefoy, J. Y., and Nguyen, T. N. (2001). Differential histopathology and chemokine gene expression in lung tissues following respiratory syncytial virus (RSV) challenge of formalin-inactivated RSV- or BBG2Na-immunized mice. *J. Virol.* **75,** 12421.

Pullan, C. R., and Hey, E. N. (1982). Wheezing, asthma, and pulmonary dysfunction 10 years after infection with respiratory syncytial virus in infancy. *Br. Med. J. (Clin Res Ed)* **284,** 1665.

Qin, S., Rottman, J. B., Myers, P., Kassam, N., Weinblatt, M., Loetscher, M., Koch, A. E., Moser, B., and Mackay, C. R. (1998). The chemokine receptors CXCR3 and CCR5 mark subsets of T cells associated with certain inflammatory reactions. *J. Clin. Invest.* **101,** 746.

Rawlinson, W. D., Waliuzzaman, Z., Carter, I. W., Belessis, Y. C., Gilbert, K. M., and Morton, J. R. (2003). Asthma exacerbations in children associated with rhinovirus but not human metapneumovirus infection. *J. Infect. Dis.* **187,** 1314.

Razzaque, M. S., and Taguchi, T. (2003). Pulmonary fibrosis: Cellular and molecular events. *Pathol. Int.* **53,** 133.

Rodenburg, R. J., Brinkhuis, R. F., Peek, R., Westphal, J. R., Van Den Hoogen, F. H., van Venrooij, W. J., and van de Putte, L. B. (1998). Expression of macrophage-derived chemokine (MDC) mRNA in macrophages is enhanced by interleukin-1beta, tumor necrosis factor alpha, and lipopolysaccharide. *J. Leukoc. Biol.* **63,** 606.

Romagnani, S. (2002). Cytokines and chemoattractants in allergic inflammation. *Mol. Immunol.* **38,** 881.

Rossi, D., and Zlotnik, A. (2000). The biology of chemokines and their receptors. *Annu. Rev. Immunol.* **18,** 217.

Rossi, G. L., and Olivieri, D. (1997). Does the mast cell still have a key role in asthma? *Chest* **112,** 523.

Rothenberg, M. E., Luster, A. D., Lilly, C. M., Drazen, J. M., and Leder, P. (1995). Constitutive and allergen-induced expression of eotaxin mRNA in the guinea pig lung. *J. Exp. Med.* **181,** 1211.

Saito, T., Deskin, R. W., Casola, A., Haeberle, H., Olszewska, B., Ernst, P. B., Alam, R., Ogra, P. L., and Garofalo, R. (1997). Respiratory syncytial virus induces selective production of the chemokine RANTES by upper airway epithelial cells. *J. Infect. Dis.* **175,** 497.

Santana, M. A., and Rosenstein, Y. (2003). What it takes to become an effector T cell: The process, the cells involved, and the mechanisms. *J. Cell. Physiol.* **195,** 392.

Sauty, A., Dziejman, M., Taha, R. A., Iarossi, A. S., Neote, K., Garcia-Zepeda, E. A., Hamid, Q., and Luster, A. D. (1999). The T cell-specific CXC chemokines IP-10, Mig, and I-TAC are expressed by activated human bronchial epithelial cells. *J. Immunol.* **162,** 3549.

Schroth, M. K., Grimm, E., Frindt, P., Galagan, D. M., Konno, S. I., Love, R., and Gern, J. E. (1999). Rhinovirus replication causes RANTES production in primary bronchial epithelial cells. *Am. J. Respir. Cell Mol. Biol.* **20,** 1220.

Schuh, J. M., Blease, K., and Hogaboam, C. M. (2002a). The role of CC chemokine receptor 5 (CCR5) and RANTES/CCL5 during chronic fungal asthma in mice. *FASEB J.* **16,** 228.

Schuh, J. M., Blease, K., and Hogaboam, C. M. (2002b). CXCR2 is necessary for the development and persistence of chronic fungal asthma in mice. *J. Immunol.* **168,** 1447.

Schwarze, J., Hamelmann, E., Bradley, K. L., Takeda, K., and Gelfand, E. W. (1997). Respiratory syncytial virus infection results in airway hyperresponsiveness and enhanced airway sensitization to allergen. *J. Clin. Invest.* **100,** 226.

Sekiya, T., Yamada, H., Yamaguchi, M., Yamamoto, K., Ishii, A., Yoshie, O., Sano, Y., Morita, A., Matsushima, K., and Hirai, K. (2002). Increased levels of a TH2-type CC chemokine thymus and activation-regulated chemokine (TARC) in serum and induced sputum of asthmatics. *Allergy* **57,** 173.

Shapiro, S. D. (1999). The macrophage in chronic obstructive pulmonary disease. *Am. J. Respir. Crit. Care Med.* **160,** S29.

Shay, D. K., Holman, R. C., Newman, R. D., Liu, L. L., Stout, J. W., and Anderson, L. J. (1999). Bronchiolitis-associated hospitalizations among US children, 1980–1996. *Jama* **282,** 1440.

Sheeran, P., Jafri, H., Carubelli, C., Saavedra, J., Johnson, C., Krisher, K., Sanchez, P. J., and Ramilo, O. (1999). Elevated cytokine concentrations in the nasopharyngeal and tracheal secretions of children with respiratory syncytial virus disease. *Pediatr. Infect. Dis. J.* **18,** 115.

Shim, J. J., Dabbagh, K., Ueki, I. F., Dao-Pick, T., Burgel, P. R., Takeyama, K., Tam, D. C., and Nadel, J. A. (2001). IL-13 induces mucin production by stimulating epidermal growth factor receptors and by activating neutrophils. *Am. J. Physiol. Lung Cell Mol. Physiol.* **280,** L134.

Sigurs, N., Bjarnason, R., Sigurbergsson, F., and Kjellman, B. (2000). Respiratory syncytial virus bronchiolitis in infancy is an important risk factor for asthma and allergy at age 7. *Am. J. Respir. Crit. Care Med.* **161,** 1501.

Simila, S., Linna, O., Lanning, P., Heikkinen, E., and Ala-Houhala, M. (1981). Chronic lung damage caused by adenovirus type 7: A ten-year follow-up study. *Chest* **80,** 127.

Simon, H. U., Yousefi, S., Weber, M., Simon, D., Holzer, C., Hartung, K., and Blaser, K. (1995). Human peripheral blood eosinophils express and release interleukin-8. *Int. Arch. Allergy Immunol.* **107,** 124.

Skoner, D. P., Gentile, D. A., Patel, A., and Doyle, W. J. (1999). Evidence for cytokine mediation of disease expression in adults experimentally infected with influenza A virus. *J. Infect. Dis.* **180,** 10.

Smith, R. E., Strieter, R. M., Phan, S. H., Lukacs, N. W., Huffnagle, G. B., Wilke, C. A., Burdick, M. D., Lincoln, P., Evanoff, H., and Kunkel, S. L. (1994). Production and function of murine macrophage inflammatory protein-1 alpha in bleomycin-induced lung injury. *J. Immunol.* **153,** 4704.

Smith, R. E., Strieter, R. M., Phan, S. H., Lukacs, N., and Kunkel, S. L. (1998). TNF and IL-6 mediate MIP-1alpha expression in bleomycin-induced lung injury. *J. Leukoc. Biol.* **64,** 528.

Smith, R. E., Strieter, R. M., Phan, S. H., and Kunkel, S. L. (1996). C-C chemokines: Novel mediators of the profibrotic inflammatory response to bleomycin challenge. *Am. J. Respir. Cell Mol. Biol.* **15,** 693.

Smyth, R. L., Mobbs, K. J., O'Hea, U., Ashby, D., and Hart, C. A. (2002). Respiratory syncytial virus bronchiolitis: Disease severity, interleukin-8, and virus genotype. *Pediatr. Pulmonol.* **33,** 339.

Spigelblatt, L., and Rosenfeld, R. (1983). Hyperlucent lung: Long-term complication of adenovirus type 7 pneumonia. *Can. Med. Assoc. J.* **128,** 47.

Sprenger, H., Meyer, R. G., Kaufmann, A., Bussfeld, D., Rischkowsky, E., and Gemsa, D. (1996). Selective induction of monocyte and not neutrophil-attracting chemokines after influenza A virus infection. *J. Exp. Med.* **184,** 1191.

Stampfli, M. R., Ritz, S. A., Neigh, G. S., Sime, P. J., Lei, X. F., Xing, Z., Croitoru, K., and Jordana, M. (1998). Adenoviral infection inhibits allergic airways inflammation in mice. *Clin. Exp. Allergy* **28,** 1581.

Stein, R. T., Sherrill, D., Morgan, W. J., Holberg, C. J., Halonen, M., Taussig, L. M., Wright, A. L., and Martinez, F. D. (1999). Respiratory syncytial virus in early life and risk of wheeze and allergy by age 13 years. *Lancet* **354,** 541.

Stein, R. T., Sherrill, D., Morgan, W. J., Holberg, C. J., Halonen, M., Taussig, L. M., Wright, A. L., and Martinez, F. D. (1999). Respiratory syncytial virus in early life and risk of wheeze and allergy by age 13 years. *Lancet* **354,** 541.

Stellato, C., Brummet, M. E., Plitt, J. R., Shahabuddin, S., Baroody, F. M., Liu, M. C., Ponath, P. D., and Beck, L. A. (2001). Expression of the C-C chemokine receptor CCR3 in human airway epithelial cells. *J. Immunol.* **166,** 1457.

Stellato, C., Collins, P., Ponath, P. D., Soler, D., Newman, W., La Rosa, G., Li, H., White, J., Schwiebert, L. M., Bickel, C., Liu, M., Bochner, B. S., Williams, T., and Schleimer, R. P. (1997). Production of the novel C-C chemokine MCP-4 by airway cells and comparison of its biological activity to other C-C chemokines. *J. Clin. Invest.* **99,** 926.

Stephens, R., Randolph, D. A., Huang, G., Holtzman, M. J., and Chaplin, D. D. (2002). Antigen-nonspecific recruitment of Th2 cells to the lung as a mechanism for viral infection-induced allergic asthma. *J. Immunol.* **169,** 5458.

Subauste, M. C., Jacoby, D. B., Richards, S. M., and Proud, D. (1995). Infection of a human respiratory epithelial cell line with rhinovirus. Induction of cytokine release and modulation of susceptibility to infection by cytokine exposure. *J. Clin. Invest.* **96,** 549.

Tan, W. C., Xiang, X., Qiu, D., Ng, T. P., Lam, S. F., and Hegele, R. G. (2003). Epidemiology of respiratory viruses in patients hospitalized with near-fatal asthma, acute exacerbations of asthma, or chronic obstructive pulmonary disease. *Am. J. Med.* **115,** 272.

Tekkanat, K. K., Maassab, H. F., Cho, D. S., Lai, J. J., John, A., Berlin, A., Kaplan, M. H., and Lukacs, N. W. (2001). IL-13-induced airway hyperreactivity during respiratory syncytial virus infection is STAT6 dependent. *J. Immunol.* **166,** 3542.

Tekkanat, K. K., Maassab, H., Miller, A., Berlin, A. A., Kunkel, S. L., and Lukacs, N. W. (2002). RANTES (CCL5) production during primary respiratory syncytial virus infection exacerbates airway disease. *Eur. J. Immunol.* **32,** 3276.

Thomas, L. H., Wickremasinghe, M. I., Sharland, M., and Friedland, J. S. (2000). Synergistic upregulation of interleukin-8 secretion from pulmonary epithelial cells by direct and monocyte-dependent effects of respiratory syncytial virus infection. *J. Virol.* **74,** 8425.

Thomas, M. S., Kunkel, S. L., and Lukacs, N. W. (2002). Differential role of IFN-gamma-inducible protein 10 kDa in a cockroach antigen-induced model of allergic airway hyperreactivity: Systemic versus local effects. *J. Immunol.* **169,** 7045.

Traves, S. L., Culpitt, S. V., Russell, R. E., Barnes, P. J., and Donnelly, L. E. (2002). Increased levels of the chemokines GROalpha and MCP-1 in sputum samples from patients with COPD. *Thorax* **57,** 590.

van den Hoogen, B. G., de Jong, J. C., Groen, J., Kuiken, T., de Groot, R., Fouchier, R. A., and Osterhaus, A. D. (2001). A newly discovered human pneumovirus isolated from young children with respiratory tract disease. *Nat. Med.* **7,** 719.

Welliver, R. C. (2003). Respiratory syncytial virus and other respiratory viruses. *Pediatr. Infect. Dis. J.* **22,** S6.

White, J. R., Imburgia, C., Dul, E., Appelbaum, E., O'Donnell, K., O'Shannessy, D. J., Brawner, M., Fornwald, J., Adamou, J., Elshourbagy, N. A., Kaiser, K., Foley, J. J., Schmidt, D. B., Johanson, K., Macphee, C., Moores, K., McNulty, D., Scott, G. F., Schleimer, R. P., and Sarau, H. M. (1997). Cloning and functional characterization of a novel human CC chemokine that binds to the CCR3 receptor and activates human eosinophils. *J. Leukoc. Biol.* **62,** 667.

Wiley, R., Palmer, K., Gajewska, B., Stampfli, M., Alvarez, D., Coyle, A., Gutierrez-Ramos, J., and Jordana, M. (2001). Expression of the Th1 chemokine IFN-gamma-inducible protein 10 in the airway alters mucosal allergic sensitization in mice. *J. Immunol.* **166,** 2750.

Wilson, N. M. (2003). Virus infections, wheeze and asthma. *Paediatr. Respir. Rev.* **4,** 184.

Woerly, G., Lacy, P., Younes, A. B., Roger, N., Loiseau, S., Moqbel, R., and Capron, M. (2002). Human eosinophils express and release IL-13 following CD28-dependent activation. *J. Leukoc. Biol.* **72,** 769.

Wohlleben, G., Muller, J., Tatsch, U., Hambrecht, C., Herz, U., Renz, H., Schmitt, E., Moll, H., and Erb, K. J. (2003). Influenza A virus infection inhibits the efficient recruitment of Th2 cells into the airways and the development of airway eosinophilia. *J. Immunol.* **170,** 4601.

Yamada, K., Elliott, W. M., Brattsand, R., Valeur, A., Hogg, J. C., and Hayashi, S. (2002). Molecular mechanisms of decreased steroid responsiveness induced by latent adenoviral infection in allergic lung inflammation. *J. Allergy Clin. Immunol.* **109,** 35.

Yamaya, M., and Sasaki, H. (2003). Rhinovirus and asthma. *Viral Immunol.* **16,** 99.

Ying, S., Meng, Q., Barata, L. T., and Kay, A. B. (2001). Macrophage inflammatory protein-1alpha and C-C chemokine receptor-1 in allergen-induced skin late-phase reactions: Relationship to macrophages, neutrophils, basophils, eosinophils and T lymphocytes. *Clin. Exp. Allergy* **31,** 1724.

Yuan, Y. H., ten Hove, T., The, F. O., Slors, J. F., van Deventer, S. J., and te Velde, A. A. (2001). Chemokine receptor CXCR3 expression in inflammatory bowel disease. *Inflamm. Bowel Dis.* **7,** 281.

Zhang, Y., Luxon, B. A., Casola, A., Garofalo, R. P., Jamaluddin, M., and Brasier, A. R. (2001). Expression of respiratory syncytial virus-induced chemokine gene networks in lower airway epithelial cells revealed by cDNA microarrays. *J. Virol.* **75,** 9044.

CHAPTER 9

Chemokines, Chemokine Receptors and Atherosclerosis

Alison E. John,* Keith M. Channon,† and David R. Greaves*
*Sir William Dunn School of Pathology, University of Oxford, Oxford OX1 3RE, United Kingdom
†Department of Cardiovascular Medicine, University of Oxford, Oxford OX1 3RE, United Kingdom

I. INTRODUCTION

Atherosclerosis is the leading cause of morbidity and mortality in Westernized societies and is the underlying basis of myocardial infarction, stroke, and peripheral arterial disease (Libby, 2002). Atherosclerosis is a

Current Topics in Membranes, Volume 55
1063-5823/05 $35.00
DOI: 10.1016/S1063-5823(04)55009-3

multifactorial disease associated with a number of genetic and environmental risk factors. However, it has become increasingly evident that inflammation plays a prominent role in both the initiation and the progression of atherosclerotic lesion development as well as in the complications related to plaque destabilization and rupture (Ross, 1999). Early atherosclerotic lesions consist primarily of monocyte-derived macrophages and T-cells (Stary *et al.*, 1994), but these cells persist even in advanced lesions. Given the complexity of the inflammatory processes involved in disease pathogenesis, enormous amounts of research have been directed towards identifying the key mediators of this inflammation so that new targets for therapeutic intervention can be discovered. This brings us to chemokines. Chemokines play essential roles in inflammation and disease (D'Ambrosio *et al.*, 2003; John and Lukacs, 2003; Proudfoot *et al.*, 2000; Sheikine and Hansson, 2004). This family of inflammatory mediators has diverse roles in both immune and inflammatory responses, including regulation of cellular recruitment, activation, and differentiation (Gerard and Rollins, 2001). Given the importance of chemokines in the regulation of inflammatory processes, numerous strategies have been devised to modulate chemokine-receptor interactions in an attempt to identify novel therapeutic candidates for intervention in inflammatory disease.

II. INFLAMMATION AND ATHEROGENESIS

Atherosclerotic lesions are typically found at arterial bifurcations, branch points, or curvatures, as these regions are subject to disturbed or turbulent blood flow patterns and altered shear stresses at the vascular wall (Gimbrone *et al.*, 2000). The initiating trigger for lesion development within these regions is still not clear, but it is most likely due to endothelial-cell activation by a combination of altered shear stresses at the vessel wall (Gimbrone *et al.*, 2000; Kinlay *et al.*, 2001) and accumulation of low-density lipoprotein (LDL) cholesterol in the subendothelial region (Frostegard *et al.*, 1991). Once inside the vessel wall, the LDL undergoes oxidation or glycation and the resulting modified lipoproteins induce further activation of the endothelial cells lining the vasculature in these regions. As a consequence of endothelial-cell activation, adhesion molecules such as ICAM-1, VCAM-1, E-selectin, and P-selectin and chemoattractants such as CCL2 are upregulated (Hansson, 2001; Libby, 2002; Osterud and Bjorklid, 2003). Leukocytes are then able to attach, roll, transmigrate through the endothelium, and subsequently accumulate in subintimal regions of the blood vessel. Initial leukocyte recruitment is believed to be a protective inflammatory response to remove excess lipid from the blood-vessel wall. In early atherosclerosis, the inflammatory lesions' content is more than 80% monocytes; the remaining cells are primarily T-cells (Jonasson *et al.*, 1986; Moreno *et al.*, 1994; Weber *et al.*, 2004).

Transmigration of monocytes into the extravascular compartment triggers maturation and differentiation of these cells into macrophages that express scavenger receptors on their cell surfaces (Gough *et al.*, 1999; Li and Glass, 2002; Libby, 2002; Osterud and Bjorklid, 2003). Modified LDL particles trapped in the intima are internalized into macrophages via these scavenger receptors, inducing the formation of foam cells, and the presence of multiple foam cells within the vessel wall leads to the formation of a fatty streak, a characteristic feature of the earliest atherosclerotic lesions (Faggiotto *et al.*, 1984; Napoli *et al.*, 1997). Macrophages are the key cell type within these early lesions, as they are able to secrete numerous growth factors, cytokines, and chemokines. These mediators contribute to lesion progression not only by promoting additional leukocyte recruitment but by activating resident endothelial and smooth muscle cells. This leads to further development of the lesions to form a more complex next stage of lesion formation, intermediate or fibro-fatty lesions (Ross, 1999).

The role of activated T-cells in atherosclerosis is still unclear. The presence of this cell type in early lesions and at the site of lesion growth and rupture indicates that they may play an important role in lesion development (Hansson, 2001). Studies in chow-fed mice deficient in apolipoprotein E (ApoE) and RAG-1, a gene essential for normal T- and B-lymphocyte development, exhibit lesions that are 40% smaller than those seen in immunocompetent mice deficient in ApoE alone (Dansky *et al.*, 1997). These data suggest that lesion development is at least partly dependent on the presence of functional T- and B-lymphocytes. However, in mice fed a high-fat diet, lesion size and the progression to advanced lesions with fibrous caps and necrotic cores were identical in ApoE/RAG-1$^{-/-}$and immunocompetent ApoE$^{-/-}$mice (Dansky *et al.*, 1997; Daugherty *et al.*, 1997). Additional experiments with ApoE/scid mice that lack T-cells and B-cells demonstrated a 73% reduction in fatty streak lesion development (Zhou *et al.*, 2000). Transfer of CD4+ve T-cells from ApoE$^{-/-}$ mice into these immunocompromised mice promoted a 164% increase in lesion size, following infiltration of the transferred T-cell into the developing lesions. These studies in ApoE/scid$^{-/-}$ mice also involved animals that were fed a normal chow diet. The contrast between the effects of different diets may indicate that T-cells play a key role in early fatty streak formation but are not essential for the development of more complex lesions.

CD4-positive Th1 lymphocytes are considered to be the predominant T-cell within human and murine lesions, although both CD8-positive and memory T-cells have also been identified (Jonasson *et al.*, 1986; Stemme *et al.*, 1992; Hansson, 2001). The Th1 cytokine, IFN-γ, is a key mediator of cellular immune responses, and it has been demonstrated to have a direct role in lesion development with ApoE/IFN-γ receptor knockout mice exhibiting lesions that are approximately 60% smaller in size than control ApoE$^{-/-}$ mice (Gupta *et al.*, 1997). IFN-γ is an important regulator of chemokine production, and its importance in atherosclerosis may be at least in part due to its effects on this group of inflammatory mediators.

Mast cells have also been found in significant numbers within atherosclerotic plaques, particularly at sites of plaque rupture (Kovanen *et al.*, 1995). The ability of these cells to secrete matrix-degrading enzymes and their detection in unstable plaques indicates that they may facilitate plaque rupture by damaging the fibrous cap (Kaartinen *et al.*, 1994).

Other immune cells that have been implicated in atherogenesis are natural killer (NK) T-cells (NKTs) and dendritic cells (DCs). NKTs have a key role in innate immunity and have been detected in human and murine atherosclerotic lesions. LDL receptor (LDLR)–deficient mice depleted of functional NKTs exhibit a reduction in lesion formation, indicating that recruitment of these cells into the vessel wall contributes to the development of atherosclerosis (Whitman *et al.*, 2004). DCs have a crucial role in the initiation of immune responses and are key antigen-presenting cells. Although their role in atherosclerosis is still unclear, they have been identified in the intima of human lesions and at arterial branch points in animal models, where they localize with T-cells and macrophages (Lord and Bobryshev, 2002).

For many years, lesion progression was believed to be a gradual process, but clinical studies have determined that lesions actually develop discontinuously (Davies, 1996; Yokoya *et al.*, 1999). In advanced atherosclerotic lesions, a fibrous cap is formed that separates the thrombogenic lipid-rich core of the atheroma from the circulation (Ross, 1999). Damage or destabilization of this cap exposes the lipid core and induces a thrombotic event, which is countered by the triggering of a healing response within the vessel. During the healing response, the thrombus is reabsorbed and inflammatory mediators induce SMC accumulation and proliferation as well as promoting collagen deposition, all in an attempt to stabilize the plaque. Repeated destabilization and healing of atherosclerotic lesions and the associated inflammation lead to an increase in lesion size (Ross, 1999).

III. ANIMAL MODELS OF ATHEROSCLEROSIS

Unambiguous demonstration of a nonredundant role for individual inflammatory mediators in atherosclerosis requires an *in vivo* model of the disease that closely mimics the human condition. The generation of small-animal models of atherosclerosis has revolutionized the study of the disease, and the availability of transgenic mice has proven to be a particularly useful tool in investigating the function of individual genes. Although there is no perfect model of the human disease, studies utilizing mouse strains deficient in either ApoE (Piedrahita *et al.*, 1992; Plump *et al.*, 1992) or LDL receptor (LDLR) (Ishibashi *et al.*, 1993) have enabled scientists to gain insights into the role of individual molecules in the initiation and progression of the disease (Knowles and Maeda, 2000; Smith and Breslow, 1997). ApoE is the major protein ligand on the surface of murine LDL, and it is recognized by the LDLR. In mice lacking a functional ApoE gene, plasma LDL-cholesterol levels are dramatically elevated, leading to hypercholesterolemia. Similar effects are seen in mice deficient in LDLR, and these mice mimic the situation seen in homozygous familial hypercholesterolemia in humans (Ishibashi *et al.*, 1993).

Both of these mouse strains develop atherosclerotic lesions that share many of the features of early human lesions, including the presence of large numbers of macrophage-derived foam cells (Smith and Breslow, 1997; Knowles and Maeda, 2000). The plaques develop in a time-dependent manner, starting at the proximal aorta and spreading towards the distal aorta. In advanced murine lesions, there is also evidence of T-cell recruitment as well as of smooth muscle cell migration and proliferation (Knowles and Maeda, 2000). The absence of spontaneous plaque rupture, and the thrombotic occlusion of the artery that sometimes accompanics it, is considered to be the major difference between the pathologies seen in human and murine atherosclerotic disease (Cullen *et al.*, 2003). However, as the process of plaque rupture is inherently difficult to study in humans, attempts have been made to generate animal models in which plaque destabilization occurs either spontaneously or through surgical or experimental intervention. Evidence of plaque rupture is not normally seen within the timeframe of experiments using murine models to examine atherosclerotic lesion development.

IV. CHEMOKINES AND ATHEROSCLEROSIS

As with other diseases characterized by a major inflammatory component, numerous chemokine–chemokine receptor interactions have been hypothesized to play a key role in the pathogenesis of atherosclerosis (Reape and Groot, 1999; Burke-Gaffney *et al.*, 2002; Sheikine and Hansson, 2004). A wide range of chemokines, including those from the CC, CXC, and CX3C families, have been detected in atherosclerotic lesions (Table I). Many of these chemokines have been shown to regulate leukocyte trafficking *in vitro*

TABLE I

Cellular Localization of Chemokine Expression in Atherosclerosis

Chemokine	Cellular source	Receptor	Target cells
CCL1	EC (Haque *et al.*, 2000)	CCR8	M
CCL2	M (Nelken *et al.*, 1991), SMC (Cushing *et al.*, 1990; Nelken *et al.*, 1991), EC (Cushing *et al.*, 1990)	CCR2	M, T, EC
CCL3	T (Wilcox *et al.*, 1994)	CCR1, CCR5	M, T
CCL4	T (Wilcox *et al.*, 1994)	CCR5	M, aT, SMC
CCL5	M, T, EC (Wilcox *et al.*, 1994; Pattison *et al.*, 1996)	CCR1, CCR3, CCR5	M, SMC, aT, DC
CCL11	SMC (Haley *et al.*, 2000)	CCR3	T, B, DC
CCL13	M, EC (Berkhout *et al.*, 1997)	CCR2, CCR3	M, aT
CCL17	M (Greaves *et al.*, 2001)	CCR4	T, DC
CCL18	M (Reape *et al.*, 1999)	Unknown	T
CCL19	M, SMC (Reape *et al.*, 1999)	CCR7	aT
CCL22	M (Greaves *et al.*, 2001)	CCR4, CCR8	aT
CXCL8	M (Wang *et al.*, 1996a)	CXCR1, CXCR2	M, EC, SMC
CXCL9	M, EC (Mach *et al.*, 1999)	CXCR3	aT
CXCL10	EC, M, SMC (Mach *et al.*, 1999)	CXCR3	aT
CXCL11	M, EC (Mach *et al.*, 1999)	CXCR3	aT
CXCL12	M, SMC, EC (Abi-Younes *et al.*, 2000)	CXCR4	T, progenitor
CXCL16	M (Minami *et al.*, 2001; Wuttge *et al.*, 2004)	CXCR6	T
CX_3CL1	SMC, M (Greaves *et al.*, 2001; Wong *et al.*, 2002; Lucas *et al.*, 2003)	CX3CR1	M, T

M, monocyte/macrophage; T, T-cell; aT, activated T-cell; EC, endothelial cell; SMC, smooth muscle cell; DC, dendritic cell; B, B-cell.

as well as regulating inflammatory processes in disease models. Although it seems likely that many chemokines may play a role in atherosclerosis, there have been very few of these chemokines whose function in lesion initiation, progression, or plaque stabilization has been confirmed *in vivo*. This chapter will present the currently available evidence that suggests a key role for chemokines in atherosclerosis. It will focus on the chemokines for which the most extensive data exists but will also consider the potential role of some new chemokine/receptor interactions that have only recently been implicated in the disease.

A. CCL2 and CCR2

CCL2/MCP-1 is a key regulator of monocyte recruitment via interactions with its receptor CCR2. Initial studies identified smooth muscle cells and the luminal endothelium as two sources of CCL2 in early atherosclerotic lesions (Takeya *et al.*, 1993). Although it is not possible to conclude that these cells are the initial source of CCL2 in the developing lesion, it is possible that CCL2 release from these cells triggers the first wave of monocyte recruitment into the vessel wall, a prerequisite for lesion formation (Reape and Groot, 1999). CCL2 also directly induces scavenger-receptor expression and promotes monocytic differentiation in foam cells (Tabata *et al.*, 2003).

In more established lesions, CCL2 is detected in macrophage-rich regions of atherosclerotic plaques (Nelken *et al.*, 1991; Yla-Herttuala *et al.*, 1991; Takeya *et al.*, 1993), and macrophage-derived CCL2 is therefore believed to reinforce the recruitment of additional monocytes into developing lesions (Reape and Groot, 1999). The CCL2 receptor, CCR2, is expressed on circulating monocytes, and the level of expression is increased in patients with hypercholesterolemia (Han *et al.*, 1998) and hypertension (Ishibashi *et al.*, 2004). This finding indicates that monocyte recruitment via CCL2–CCR2 interactions may play a more important role in patients with elevated lipid levels and hypertension and further demonstrates the potential importance of this chemokine in lesion development. Monocytes treated with oxLDL *in vitro* exhibit lower CCR2 levels than cells treated with unmodified forms of LDL (Han *et al.*, 2000). CCR2 is important for the recruitment of monocytes to developing lesions, and, once it is inside the lesion, exposure to modified LDL leads to loss of cell-surface CCR2 expression, which may aid retention of the monocyte within the vessel wall.

The CCL2–CCR2 interaction was one of the first to be investigated for a potential role in the pathogenesis of atherosclerosis due to the availability of viable knockout mice. Both $LDLR^{-/-}$ and $ApoE^{-/-}$ mouse models have been used to investigate the roles of CCL2 and CCR2 in atherosclerosis. When $CCR2^{-/-}$ mice were crossed with $ApoE^{-/-}$ mice, the lesions in the double-knockout mice were reduced in size and contained significantly lower numbers of macrophages than the $ApoE^{-/-}$ mice with wild-type CCR2 expression (Boring *et al.*, 1998; Dawson *et al.*, 1999). Circulating lipid levels were similar in the $CCR2^{+/+}$ and $CCR2^{-/-}$ mouse strains. The study by Boring *et al.* (1998) was the first to demonstrate that CCR2 was an important regulator of monocyte/macrophage recruitment in atherosclerosis and that the functional consequence of reduced macrophage accumulation was a decrease in lesion size. Similar findings were obtained in $LDLR^{-/-}$ mice crossed with mice deficient in CCL2 (Gu *et al.*, 1998).

The importance of monocyte/macrophage–derived CCL2 in atherosclerosis was also examined in irradiated ApoE$^{-/-}$ mice receiving bone marrow from mice overexpressing the chemokine (Aiello *et al.*, 1999). Following transfer, macrophages expressing the CCL2 transgene under the control of the mouse mammary tumor virus long-terminal repeat (MMTV-LTR) promoter were detected within atherosclerotic lesions, and the area covered by the lesions in these mice was increased in size by 50% compared to those found in ApoE$^{-/-}$ mice that received transfer of nontransgenic bone marrow. Lesional macrophage numbers and lipid levels were also increased in mice overexpressing CCL2.

In addition to its role in lesion formation, CCL2's importance in the progression of preexisting atherosclerotic lesions has also been examined using gene therapy to introduce a mutant form of CCL2 into hypercholesteremic mice (Inoue *et al.*, 2002). This mutated CCL2 inhibits normal CCL2 and has previously been shown to reduce lesion formation (Ni *et al.*, 2001). The mutant CCL2 slowed lesion progression and altered the composition of the remaining lesion to a more stable phenotype characterized by decreased leukocyte accumulation, lower lipid levels, increased smooth muscle cell presence, and increased collagen content. Taken together, these two studies demonstrate not only that CCL2 is a key mediator of lesion progression but that it appears to be involved in lesion destabilization.

All of the studies investigating the role of CCL2–CCR2 interactions in atherosclerosis have demonstrated that this chemokine ligand–receptor pair is important in the process of atherosclerotic lesion development. However, it is important to note that in the absence of either of these genes or in situations where normal chemokine–receptor interactions are disrupted (e.g., in introduction of mutant CCL2 [Inoue *et al.*, 2002; Ni *et al.*, 2001]), lesions containing macrophages are still detected in murine models of atherosclerosis. This is confirmation that, although CCL2 plays a key role in disease pathogenesis, other chemokines are also likely to contribute to this process.

B. CXCL8, CXCR1, and CXCR2

CXCL8/IL-8, and its murine homologue CXCL1/GRO-α, were initially characterized as neutrophil chemoattractants and as such have been shown to play an important role in neutrophil-mediated acute inflammation (Harada *et al.*, 1994). However, it is now known to have much more wide-ranging effects on both non-leukocyte and leukocyte subtypes of cells (Harada *et al.*, 1994). The CXCL8/CXCL1 receptor, CXCR2, is also expressed on the surface of monocytes (Moser *et al.*, 1993; Chuntharapai *et al.*,

1994), and the finding that CXCR2 ligands can promote monocyte adhesion (Schwartz *et al.*, 1994; Gerszten *et al.*, 1999) led to the hypothesis that this chemokine may play a role in atherosclerosis. CXCR2 expression patterns were examined in both early and advanced atherosclerotic lesions, and the receptor appears to be expressed predominantly on lesional macrophages (Boisvert *et al.*, 1998). The role of CXCL8/CXCL1–CXCR2 interaction in lesion development was examined in irradiated $ApoE^{-/-}$mice that were reconstituted with leukocytes deficient in CXCR2 (Boisvert *et al.*, 1998). In these mice, the absence of leukocyte-associated CXCR2 led to a significant reduction in lesion size that was accompanied by a decrease in macrophage accumulation within the intima. This study suggested that CXCR2 and its ligands play a role in trafficking of monocytes to the intima and that the subsequent reduction in macrophage numbers attenuated lesion development.

C. CX_3CL1 and CX_3CR1

CX_3CL1/fractalkine is the only member of the CX_3C subfamily, the most recent class of chemokines to be identified. The unusual structure of CX_3CL1, an N-terminal chemoattractant domain that is attached to the cell membrane via a mucin-like stalk (Bazan *et al.*, 1997), allows it to function as both a soluble chemoattractant following cleavage of the stalk (Tsou *et al.*, 2001) and an adhesion molecule that rapidly mediates tethering and arrest of monocytes and T-cells under flow conditions (Imai *et al.*, 1997; Fong *et al.*, 1998; Haskell *et al.*, 1999). The effects of CX_3CL1 are mediated via interactions with its receptor, CX_3CR1. Soluble CX_3CL1 mediates chemotaxis of monocytes, T-cells, and NK cells (Bazan *et al.*, 1997) while membrane-bound CX_3CL1 mediates adhesion of monocytes, endothelial cells, and smooth muscle cells (Imai *et al.*, 1997; Harrison *et al.*, 1999). CX_3CL1 is not expressed in normal arteries but has been detected in human atherosclerotic lesions (Greaves *et al.*, 2001; Wong *et al.*, 2002; Lucas *et al.*, 2003). Expression seems to be associated with mononuclear leukocytes, including macrophages, foam cells, and smooth muscle cells (Greaves *et al.*, 2001; Wong *et al.*, 2002). The receptor, CX_3CR1, is present in both normal arteries and atherosclerotic lesions, and the smooth muscle cells are the primary site of expression in diseased arteries (Wong *et al.*, 2002; Lucas *et al.*, 2003). *In vitro*, CX_3CL1 has been shown to induce migration of smooth muscle cells (Lucas *et al.*, 2003).

Studies of genetic association indicated that polymorphisms of the CX_3CR1 gene may be associated with significantly lower risk of developing coronary artery disease and endothelial-cell dysfunction (McDermott *et al.*, 2001; Moatti *et al.*, 2001). These polymorphisms have also been shown to have functional consequences. Reduction of binding sites for CX_3CL1 on the cell surface reduces the ability of cells to adhere and respond to chemotactic signals.

Two recent studies further confirmed a potential role for CX_3CL1–CX_3CR1 interactions in formation of atherosclerotic lesions (Combadiere *et al.*, 2003; Lesnik *et al.*, 2003). These groups independently generated $ApoE^{-/-}$ mouse strains that were deficient in CX_3CR1 in order to examine the role of this receptor in lesion development. The studies differed in the choice of diet; Lesnik *et al.* (2003) used a high-fat Western diet, and Combadiere *et al.* (2003) chose to use normal chow. Lesnik *et al.* reported a significant reduction in mean lesion area and lesion thickness as well as a 40% reduction in lesional monocytes/macrophages in $CX_3CR1^{-/-}$/$ApoE^{-/-}$. CX_3CL1 was detected only in lesional areas of the aorta and was detected on smooth muscle cells located directly beneath macrophages. Although the macrophages were only weakly positive, the detection of CX_3CL1 in regions of macrophage infiltration was in agreement with previous findings in human atherosclerotic lesions (Greaves *et al.*, 2001). CX_3CL1 expression was also detected on endothelial cells in areas of lesion development.

Combadiere *et al.* (2003) addressed the role of CX_3CR1 in a less severe murine model of hypercholesterolemia. The study determined that the ligand CX_3CL1 was expressed by leukocytes within early fatty streak lesions. In more advanced lesions, some smooth muscle and endothelial cells were positive for CX_3CL1, although acellular regions surrounding macrophages within the lipid core also showed reactivity for the chemokine. The double ApoE/CX_3CR1 knockout mice exhibited a significant reduction in lesion size over $CX_3CR1^{+/+}$ mice, and this reduction was accompanied by a 50% reduction in the number of monocyte/macrophages found within the lesion. This study also reported that the plaques in the $CX_3CR1^{-/-}$ mice were not only significantly smaller but also appeared to be stabilized by substantial accumulation of smooth muscle cells and collagen. The changes in lesion size and composition in atherosclerosis were not just seen in homozygous null CX_3CR1 mice—this phenotype was present even in mice that were heterozygous for the CX_3CR1 gene. It appears that only one mutated allele is sufficient to alter disease pathogenesis, further indicating the potential importance of the CX_3CL1–CX_3CR1 interactions in lesion development.

D. CCL5 and CCR5

CCL5/RANTES is secreted by a wide range of cell types, many of which play a key role in atherosclerotic lesion development, including endothelial cells, smooth muscle cells, activated T-cells, macrophages (Pattison *et al.*,

1996; Hayes *et al.*, 1998), and platelets (von Hundelshausen *et al.*, 2001; Huo *et al.*, 2003). CCL5 expression is undetectable in normal arteries but has been detected in macrophages and T-cells in human atherosclerotic plaques (Wilcox *et al.*, 1994; Pattison *et al.*, 1996). In addition, CCL5 has been detected on the luminal surface of atherosclerotic endothelium and on neointimal lesions found in $ApoE^{-/-}$ mice following arterial injury (von Hundelshausen *et al.*, 2001).

Elevated CCL5 levels have been detected in patients suffering from cardiovascular disease (Parissis *et al.*, 2002) and have also been implicated in monocyte arrest on atherosclerotic endothelium (von Hundelshausen *et al.*, 2001). CCL5 mediates a range of cellular responses via interactions with CCR1, CCR3, and CCR5. Two of the CCL5 receptors, CCR5 (Schecter *et al.*, 2000) and CCR3 (Haley *et al.*, 2000), have been detected in atherosclerotic lesions, but there is currently little information regarding expression patterns of CCR1 in lesions. CCR5 is the CCL5 receptor that has been most extensively studied with regard to a role in atherosclerosis. CCR5 is expressed on isolated macrophages, T-cells, endothelial cells, and aortic smooth muscle cells (Rottman *et al.*, 1997; Berger *et al.*, 1999; Schecter *et al.*, 2000; Mack *et al.*, 2001). A recent genetic screening study indicated that a common 32bp-deletion mutation in CCR5 (CCR5Δ32) is associated with a decreased risk of severe coronary artery disease (Szalai *et al.*, 2001). This deletion abolishes expression of the receptor from the cell surface, and in the homozygous state there is no functional CCR5 expressed in cells.

A recent study utilizing an $ApoE^{-/-}/CCR5^{-/-}$-deficient mouse attempted to determine the contribution of CCR5 to atherosclerotic lesion formation and progression (Kuziel *et al.*, 2003). These knockout mice, fed on normal chow, did not show any significant differences in lesion area or lesion composition up to 18 weeks of age. In the absence of CCR5, lesions developed normally and leukocyte infiltration into the plaques was unaffected. This finding was disappointing, as $CCR5^{-/-}$ mice have previously been shown to exhibit defects in macrophage migration and function (Zhou *et al.*, 1998; Huffnagle *et al.*, 1999). The absence of an effect from CCR5 does not completely rule out a nonredundant role for this receptor in lesion progression, but it does appear to indicate that the receptor plays no role in formation of early lesions. Additional studies in which mice are fed a high-fat diet to induce formation of more complex lesions will be needed to conclusively rule out the possibility that a role is played by CCR5 in atherosclerosis.

The potential of a role by CCL5 in lesion development was recently suggested by Veillard *et al.* (2004) in a study in which the authors tested the efficacy of the chemokine receptor antagonist met-RANTES in a murine model of atherosclerosis. When recombinant forms of CCL5 are synthesized in prokaryotic cells, an initiating methionine residue is retained. This met-RANTES form of the chemokine has been shown to function as a selective antagonist for the murine chemokine receptors CCR1 and CCR5, blocking calcium mobilization and chemotaxis at nanomolar potency (Proudfoot *et al.*, 1996). Previous studies have demonstrated that, in several animal models, met-RANTES can significantly reduce inflammatory responses such as rheumatoid arthritis (Plater-Zyberk *et al.*, 1997) and airway inflammation (Gonzalo *et al.*, 1998). Veillard *et al.* (2004) demonstrated that antagonism of the CCL5 receptors, CCR1 and CCR5, using met-RANTES in fat-fed LDLR$^{-/-}$ mice reduced atherosclerotic lesion formation (Veillard *et al.*, 2004). The reduction in lesion size was associated with a decrease in foam cell numbers, and changes in lesion conformation with a reduced number of macrophage foam cells and T-cells in the plaque. Treatment with met-RANTES also led to an increase in the number of smooth muscle cells that contributed to a thicker fibrous cap and a seemingly more stable lesion. The collagen content of lesions was also increased following met-RANTES treatment, whilst MMP-9 levels were reduced. Both of these factors would also contribute to a more stable plaque phenotype.

Although met-RANTES is known to inhibit RANTES binding, the antagonist is also considered to be a functional antagonist of CCL3/MIP-1α-CCR1 interactions (Bonville *et al.*, 2004). The effects of met-RANTES in the atherosclerosis model can therefore not be assumed to reflect only CCL5's role in the disease, as it is possible that CCL3 may also be mediating some effects. It is possible that antagonism of both CCR1 and CCR5 to produce a significant effect with met-RANTES.

E. CXCL16/CXCR6

CXCL16 is one of the most recently identified chemokines (Matloubian *et al.*, 2000). Like CX_3CL1, it exists in both transmembrane and soluble forms and is expressed by macrophages (Minami *et al.*, 2001; Wuttge *et al.*, 2004), T-cells (Shashkin *et al.*, 2003), smooth muscle cells (Hofnagel *et al.*, 2002), and DCs (Matloubian *et al.*, 2000). CXCL16 has been identified as a T-cell chemoattractant and mediates its effects via interactions with CXCR6. It was also independently identified as the scavenger receptor SR-PSOX (Shimaoka *et al.*, 2000). Scavenger receptors play a crucial role in the pathogenesis of atherosclerosis due to their role in the uptake of oxLDL by macrophages and the subsequent foam-cell development. SR-PSOX was identified as a scavenger receptor for phosphatidylserine and oxLDL. It is expressed on macrophages and mediates internalization of oxLDL (Shimaoka *et al.*, 2000). CXCL16's dual role as both a chemokine and a scavenger receptor has led to great interest in the potential role of this chemokine in atherosclerosis.

CXCL16 and its receptor CXCR6 have both been detected in human and murine atherosclerotic lesions, particularly in regions rich in macrophages and T-cells (Wuttge *et al.*, 2004). CXCL16 expression has been shown to be upregulated by inflammatory cytokines, including IFN-γ, and as a result of increased SR-PSOX expression there is likely to be increased uptake of oxLDL into these cells (Abel *et al.*, 2004; Wuttge *et al.*, 2004). The chemokine has also been shown to mediate smooth muscle cell proliferation and promote cell–cell adhesion (Chandrasekar *et al.*, 2004). This suggests that CXCL16 may play an important role in foam-cell formation and T-cell recruitment into the developing atherosclerotic plaque, but we must await experiments with knockout mice before we can speculate on the relative contribution of this ligand–receptor pairing in atherogenesis.

F. CXCR3 Ligands

The CXCR3 receptor binds a group of IFN-γ-inducible chemokine ligands, CXCL9/MIG, CXCL10/IP-10, and CXCL11/ITAC. These ligands are differentially expressed in human atherosclerotic lesions: CXCL10 is expressed by endothelial cells, smooth muscle cells, and macrophages while CXCL9 and CXCL11 are expressed in endothelial cells and macrophages (Mach *et al.*, 1999). All of the T-cells in the lesions expressed CXCR3, suggesting that this ligand–receptor interaction may be extremely important in regulating T-cell recruitment into the lesions. Although many of the effects of the CXCR3 ligands appear to be directed towards T-cells, CXCL10 is also known to be a monocyte chemoattractant (Taub *et al.*, 1993). In addition, CXCL10 can also induce vascular smooth muscle cell migration and proliferation (Wang *et al.*, 1996b).

V. THERAPEUTIC STRATEGIES FOR MODULATING CHEMOKINE–RECEPTOR INTERACTIONS

Antagonism of chemokines and their receptors represents an attractive and potentially effective means of modulating inflammatory responses, and it may prove to be a useful therapeutic strategy for atherosclerosis and many other inflammatory diseases. Successful therapeutic intervention is based on selective disease modification with minimal impact on the normal physiological processes so as to limit the side effects of treatment. Due to the complexity of the chemokine–chemokine receptor system, the inherent redundancy that has evolved within it, and the widespread involvement of chemokines in many physiological processes, the task of generating therapeutically useful chemokine or chemokine-receptor inhibitors is likely to be difficult. However, there is already clinical evidence that shows that successful targeting of the proinflammatory cytokine TNF-α via soluble receptors and antibodies has proven to be a successful treatment for both rheumatoid arthritis and Crohn's disease (Feldmann *et al.*, 2001; Foxwell *et al.*, 2003; Korzenik, 2004). Given the success of targeting a specific cytokine for therapy in a chronic inflammatory pathology, it is hoped that the same may also be possible for chemoattractant cytokines.

In an ideal situation, inhibitors of chemokine–chemokine receptor interactions would be designed to act on a single chemokine receptor that is expressed in only a single cell type during disease development. Given the complex nature of inflammatory diseases and the evidence suggesting that multiple chemokine–receptor interactions are involved in the inflammatory process, a more powerful approach might be to target multiple chemokine receptors at once in order to produce an effective drug treatment.

A. Chemokine Receptor Antagonists

There are two main categories of chemokine receptor antagonists: those that specifically target one or a small number of receptors (e.g., small-molecule antagonists or modified chemokines) and those that are broad-spectrum chemokine inhibitors, which target a range of chemokine receptors (e.g., virally encoded chemokine antagonist proteins). Arrays of neutralizing anti-chemokine receptor antibodies have also been generated and tested in animal models of disease, including those targeted against CXCR2 (Ness *et al.*, 2003), CXCR3 (Xie *et al.*, 2003), and CCR2 (Horvath *et al.*, 2002). To date, the only humanized monoclonal antibody to be produced is one targeted to CXCL8 (Yang *et al.*, 1999; Huang *et al.*, 2002).

B. Specific Low-Molecular-Weight Inhibitors

A number of potent antagonists to chemokine–chemokine receptor interactions have been designed in recent years, and the majority of them are either peptide or nonpeptide small-molecular-weight inhibitors of chemokine receptors. Receptor antagonists have been reported for CCR1 (Ng *et al.*, 1999; Liang *et al.*, 2000; Gladue *et al.*, 2003), CCR2 (Mirzadegan *et al.*, 2000), CCR5 (Baba *et al.*, 1999; Strizki *et al.*, 2001), CXCR2 (White *et al.*, 1998; Widdowson *et al.*, 2004), and CXCR4 (De Clercq, 2000).

Although there is currently no information available regarding the ability of these antagonists to modulate lesion progression in atherosclerosis, they have been shown to be effective in some other models of inflammatory disease, which suggests that they do have therapeutic potential. CXCR2 antagonists have proven effective in reducing accumulation of leukocytes and secretion of proinflammatory mediators in synovial fluid in models of acute or chronic arthiritis (Podolin *et al.*, 2002) and in reducing inflammatory-cell recruitment and liver damage in a model of ischemia-reperfusion injury (Bertini *et al.*, 2004). An MCP-1 small-peptide antagonist has also been shown to reduce inflammation and lesion severity in a murine model of arthritis (Gong *et al.*, 1997). CCR1 antagonists have received considerable interest recently because of a double-blind, placebo-controlled clinical study that suggested that treatment with the antagonist for 14 days resulted in an improvement in clinical symptoms for patients with rheumatoid arthritis. The improvement in symptoms was associated with decreased mononuclear leukocyte accumulation and reduced swelling and pain in joints (Haringman *et al.*, 2003).

The best studied example of a modified chemokine that functions as a potent chemokine receptor antagonist is met-RANTES. Met-RANTES is a functional inhibitor of both CCR1- and CCR5-mediated signaling and has been widely tested in a number of animal models of disease (Proudfoot *et al.*, 2000). This antagonist has been shown to significantly reduce inflammatory responses in several animal disease models, including those for rheumatoid arthritis (Plater-Zyberk *et al.*, 1997) and airway inflammation (Gonzalo *et al.*, 1998). As discussed earlier in this chapter, met-RANTES has also proven to be effective in reducing atherosclerotic lesion formation in a murine model of disease (Veillard *et al.*, 2004).

C. Broad-Spectrum Blockade of Chemokine-Chemokine Receptor Interactions

Although the development of specific antagonists of chemokine receptors may be desirable in terms of allowing for specific targeting of one type of chemokine–receptor interaction, the efficacy of this approach to chemokine-receptor blockade may be limited by the redundancy in chemokine signaling pathways. A more rational and potentially more successful therapeutic strategy may be broad-spectrum blockade of the function of a whole group of chemokine receptors at one time. Viruses have already evolved a mechanism for evading the chemokine arm of the immune system by synthesizing proteins that bind and inactivate multiple chemokines simultaneously (Alcami, 2003; Alcami and Koszinowski, 2000). The vaccinia virus (strain

Lister) expresses a 35-Kda protein termed 35K, vCKBP (Alcami *et al.*, 1998), or VCCI (Carfi *et al.*, 1999) that binds with high affinity to almost all CC chemokines but does not interact with other classes of chemokines (Smith *et al.*, 1997; Burns *et al.*, 2002). The 35K protein has previously been demonstrated to potently inhibit CC chemokine–induced cell migration and signaling (Bursill *et al.*, 2003).

Bursill *et al.* (2004) used adenoviral-mediated gene transfer to express a soluble broad spectrum CC-chemokine inhibitor, 35K, and they examined its efficacy in mediating atherosclerotic lesion formation. In mice expressing the soluble 35K inhibitor, atherosclerotic lesion size was reduced by 55%, and this was accompanied by an 85% decrease in macrophage recruitment. This report confirmed the involvement of CC chemokines in the initiation and progression of lesion development and demonstrated that broad-spectrum CC-chemokine inhibition is a rational therapeutic target.

Another virally encoded chemokine binding protein synthesized by the murine γ-herpesvirus (MHV-68) is a 44-Kda protein termed M3. The M3 protein has been shown to bind to members of all four classes of chemokines, with particularly high affinities to CCL2, CCL5, CXCL1, CXCL10, and CX_3CL1 (Parry *et al.*, 2000; van Berkel *et al.*, 2000). The efficacy of the M3 protein in modulating atherosclerotic-lesion development has not been tested, although a transgenic mouse that conditionally expresses the M3 protein exhibits reduced intimal hyperplasia following vascular injury (Pyo *et al.*, 2004). Additional viral proteins with chemokine-binding properties have been described, including a new group of α-herpesvirus glycoprotein-G isoforms that interact with a number of CC, CXC, and CX_3C chemokines (Alcami, 2003). Due to the fact that 35K (CC chemokines only) and M3 (CC chemokines plus CX_3CL1) viral proteins appear to bind to a more restricted group of chemokines, they appear to be well suited for use therapeutically in the treatment of inflammatory disease.

D. Statins

The statins, which are 3-hydrozy-3-methyl-glutaryl-coenzyme A (HMG-CoA)-reductase inhibitors, are a group of widely used cholesterol-lowering drugs that have proved very effective in the prevention and treatment of atherosclerosis (Bhatnagar, 1998). In recent years, clinical trials have concluded that although the primary effect of statin administration is a reduction in circulating cholesterol levels, some of the benefits of statin therapy may be ascribed to effects other than those on LDL levels, including the modulation of mediators involved in inflammation (Diomede *et al.*, 2001; Schonbeck and Libby, 2004). *In vitro* experiments examining the effect of statins on cells that contribute to atherosclerotic lesion development demonstrate that members of this class of drugs inhibit adhesion-molecule expression on endothelial cells (Weber *et al.*, 1997; Rasmussen *et al.*, 2001; Weitz-Schmidt *et al.*, 2001).

A number of the statins (atorvastatin, lovastatin, fluvastatin, pravastatin, and simvastatin) have also been shown to regulate expression of chemokines *in vitro*. In isolated monocytes (Romano and Slee, 2001; Rezaie-Majd *et al.*, 2002), endothelial cells (Pasceri *et al.*, 2001), and smooth muscle cells (Bustos *et al.*, 1998; Ito *et al.*, 2002), expression of CCL2, CCL5, or CXCL8 is reduced by treatment with statins. HMG-CoA reductase inhibitors have been shown to inhibit lesion development in animal models of atherosclerosis. The decrease in lesion size is associated with a reduction in chemokine production within the lesions (Bustos *et al.*, 1998; Martinez-Gonzalez *et al.*, 2001; Mitani *et al.*, 2003) and in levels of macrophage accumulation (Bustos *et al.*, 1998), as well as with increased stability for the lesion, suggested by elevated smooth muscle cell and collagen content (Mitani *et al.*, 2003).

E. Peroxisome Proliferator-Activated Receptor γ Agonists

Peroxisome proliferator-activated receptor γ (PPARγ) is a member of the nuclear receptor superfamily and functions as a ligand-dependent transcription factor. PPARγ is abundantly expressed in adipose tissue and plays a key role in lipid metabolism. It is also highly expressed on a number of cell types involved in atherogenesis, including monocyte/macrophages, endothelial cells, and vascular smooth muscle cells (Ricote *et al.*, 1998; Law *et al.*, 2000; Sueyoshi *et al.*, 2001). PPARγ agonists have been implicated in the regulation of inflammatory and proliferative responses, and treatment of hypercholesteremic mice with PPARγ ligand reduces atherosclerotic-lesion development (Chen *et al.*, 2001; Collins *et al.*, 2001). Although the exact anti-inflammatory mechanism of these ligands is not completely clear, some of the effects appear to involve regulation of chemokine signaling. In particular, PPARγ ligands have been shown to substantially reduce both CCR2 protein expression and mRNA expression in circulating monocytes (Han *et al.*, 2000; Ishibashi *et al.*, 2002), resulting in blockade of CCL2-mediated chemotaxis (Han and Quehenberger, 2000). In endothelial cells, PPARγ agonists inhibited expression of CXCL9, CXCL10, and CXCL11, although CCL2 expression was unaffected (Han and Quehenberger, 2000). The anti-inflammatory effects of PPARγ ligands are likely to result from a combination of effects on cytokine signaling, lipid metabolism, and adhesion-molecule expression, but the modulation of chemokine-mediated inflammatory responses is also likely to be important.

VI. UNANSWERED QUESTIONS AND FUTURE DIRECTIONS

Although this review has highlighted a number of chemokines from the CC, CXC, and CX_3C families believed to contribute (to a greater or lesser extent) to the pathogenesis of atherosclerosis, there are potentially a number of chemokines yet to be identified that also play roles in the disease. Much of the evidence for the functional role of chemokine–chemokine receptor interactions in the disease has been determined in animal models of atherosclerosis, in which mice overexpressing or deficient in individual chemokines or chemokine receptors have been studied. In future years, it is likely that new strains of chemokine or chemokine-receptor knockouts and transgenic animals will be generated and that additional chemokines will be implicated in the inflammatory process that characterizes the disease. Of particular interest may be the use of conditional "knock-in" or knockout mice, which would allow genes encoding chemokines or their receptors to be switched on or off at different stages of lesion development and allow the role of chemokines in early lesion formation to be distinguished from their role in later lesion progression. If manipulation of chemokine–chemokine receptor interactions is ever to become a useful therapeutic strategy for reducing atherosclerotic lesion formation, the kinetics of ligand and receptor expression during the disease process will need to be more clearly defined for the most useful targets to be determined. The use of animal models will continue to be a major investigative tool in this effort, in spite of some of the differences highlighted in this review between human disease and animal models of atherosclerosis.

Much of the research on chemokines in atherosclerosis is concentrated on the role of these inflammatory mediators as leukocyte chemoattractants and their importance in regulating accumulation of monocytes and T-lymphocytes to the vessel wall. Less is known about the effect of chemokines on the other cell types present in the vessel wall, including smooth muscle cells and endothelial cells. Although for many years smooth muscle cells were considered to function only as structural cells, they are now recognized to both secrete and respond to a number of cell–cell signaling proteins, including chemokines released from recruited leukocytes or other cells within the vascular wall. Vascular smooth muscle cells have been shown to express a number of chemokine receptors on their cell surfaces, both *in vitro in vivo* (Hayes *et al.*, 1998; Schecter *et al.*, 2003). However, information regarding the mechanisms by which chemokine-receptor expression is regulated by cytokines or other inflammatory mediators is limited, as is evidence of the functional effects of chemokine–chemokine receptor interactions on smooth muscle cells. Recent evidence has demonstrated that CX_3CR1 (Lucas *et al.*, 2003) and CCR8 (Haque *et al.*, 2004) are both able to mediate vascular smooth muscle cell migration in response to their respective chemokine ligands, CX_3CL1 and CCL1, while these cells also possess a functional CCR5 receptor (Schecter *et al.*, 2000). The promigratory effect of chemokines on smooth muscle cells suggests that this class of inflammatory mediators may have an important role in vascular remodeling, although their relative contribution to this process in comparison to those of mediators such as PDGF (Bernstein *et al.*, 1982) or VEGF (Grosskreutz *et al.*, 1999) is still to be elucidated.

The potential to use circulating chemokine levels or genetic screening for chemokines or chemokine receptors as predictors of coronary heart disease has gained attention in recent years. Elevated levels of the CCR2 ligand, CCL2, have been suggested to be predictors of coronary artery disease (Blaha *et al.*, 2004) in addition to being described as a potentially useful marker to determine the efficacy of LDL elimination (Okopien *et al.*, 2002). It remains to be determined whether the circulating levels of a single chemokine, or, indeed, a panel of chemokines, will prove to be a useful tool for either predicting the likelihood of lesion development or marking the progression of lesions. Genetic association studies have also identified mutations in the chemokine receptor genes that appear to have atheroprotective effects. Polymorphisms in CCR2, CCR5, and CX_3CR1 have all been linked to a reduced risk of severe coronary disease. The relative importance of each of these genes in lesion development and progression may prove difficult to define in a highly complex and chronic disease like atherosclerosis, for which a number of genetic and environmental risk factors have already been identified. It is unlikely that polymorphisms in chemokines or chemokine receptors alone will ever prove to be useful diagnostic tools for identifying patients at risk of coronary artery disease. However, combining information about chemokine-receptor polymorphisms with assessments of other risk factors may aid in the identification of the patients most at risk of severe atherosclerosis.

A major goal in the field of chemokine research is the identification of antagonists of chemokine–chemokine receptor interactions that would selectively reduce inflammation. A number of selective receptor antagonists (White *et al.*, 1998; Ng *et al.*, 1999; De Clercq *et al.*, 2000; Liang *et al.*, 2000; Mirzadegan *et al.*, 2000; Gladue *et al.*, 2003; Widdowson *et al.*, 2004) are currently available, along with some modified chemokines, although only met-RANTES (Proudfoot *et al.*, 2000) has so far been shown to reduce atherosclerotic lesion development in a murine model of the disease (Veillard *et al.*, 2004). In addition to identifying new types of inhibitors and assessing the efficacy of the available chemokine-receptor antagonists in the treatment of atherosclerosis, future studies will also need to address the potential difficulties posed by systemically targeting chemokine receptors that may have important roles in processes other than inflammation (such as immune function). It will be important to establish whether chemokine-receptor antagonists are effective in the treatment of chronic inflammatory diseases such as atherosclerosis in humans and to determine the time course of any such treatment. Although chemokine-receptor antagonists may be effective in reducing lesion development in animal models over the course of a few months, administration of these drugs to humans will most likely be over a much longer time period. Chemokines and chemokine receptors are expressed in a wide range of normal tissues, and many have homeostatic roles in the functioning of the immune system or of other organ systems in addition to the role they play in the regulation of inflammatory responses. Long-term blockade of normal chemokine–chemokine receptor interactions may interfere with these normal processes. As a result, chemokine antagonists may be a feasible treatment option only in the short term, where, for example, local or systemic delivery at the time of angioplasty may serve to limit the inflammatory response, which significantly contributes to restenosis.

The global interest in the role of inflammation in the pathogenesis of atherosclerosis has led to a considerable research effort to understand the role of inflammatory mediators in the disease. Although it seems clear that chemokines play a fundamental role in both the formation of early fatty streak lesions and the subsequent development into more complex lesions, a considerable amount of research is still required before the manipulation of chemokine–chemokine receptor interactions can be considered a viable treatment option in coronary artery disease.

Acknowledgments

Research in our laboratory is funded by the British Heart Foundation.

References

Abel, S., Hundhausen, C., Mentlein, R., Schulte, A., Berkhout, T. A., Broadway, N., Hartmann, D., Sedlacek, R., Dietrich, S., Muetze, B., Schuster, B., Kallen, K. J., Saftig, P., Rose-John, S., and Ludwig, A. (2004). The transmembrane CXC-chemokine ligand 16 is induced by IFN-gamma and TNF-alpha and shed by the activity of the disintegrin-like metalloproteinase ADAM10. *J. Immunol.* **172,** 6362–6372.

Abi-Younes, S., Sauty, A., Mach, F., Sukhova, G. K., Libby, P., and Luster, A. D. (2000). The stromal cell-derived factor-1 chemokine is a potent platelet agonist highly expressed in atherosclerotic plaques. *Circ. Res.* **86,** 131–138.

Aiello, R. J., Bourassa, P. A., Lindsey, S., Weng, W., Natoli, E., Rollins, B. J., and Milos, P. M. (1999). Monocyte chemoattractant protein-1 accelerates atherosclerosis in apolipoprotein E-deficient mice. *Arterioscler Thromb. Vasc. Biol.* **19,** 1518–1525.

Alcami, A. (2003). Viral mimicry of cytokines, chemokines and their receptors. *Nat. Rev. Immunol.* **3,** 36–50.

Alcami, A., and Koszinowski, U. H. (2000). Viral mechanisms of immune evasion. *Immunol. Today* **21,** 447–455.

Alcami, A., Symons, J. A., Collins, P. D., Williams, T. J., and Smith, G. L. (1998). Blockade of chemokine activity by a soluble chemokine binding protein from vaccinia virus. *J. Immunol.* **160,** 624–633.

Baba, M., Nishimura, O., Kanzaki, N., Okamoto, M., Sawada, H., Iizawa, Y., Shiraishi, M., Aramaki, Y., Okonogi, K., Ogawa, Y., Meguro, K., and Fujino, M. (1999). A small-molecule, nonpeptide CCR5 antagonist with highly potent and selective anti-HIV-1 activity. *Proc. Natl. Acad. Sci. USA* **96,** 5698–5703.

Bazan, J. F., Bacon, K. B., Hardiman, G., Wang, W., Soo, K., Rossi, D., Greaves, D. R., Zlotnik, A., and Schall, T. J. (1997). A new class of membrane-bound chemokine with a CX3C motif. *Nature* **385,** 640–644.

Berger, O., Gan, X., Gujuluva, C., Burns, A. R., Sulur, G., Stins, M., Way, D., Witte, M., Weinand, M., Said, J., Kim, K. S., Taub, D., Graves, M. C., and Fiala, M. (1999). CXC and CC chemokine receptors on coronary and brain endothelia. *Mol. Med.* **5,** 795–805.

Berkhout, T. A., Sarau, H. M., Moores, K., White, J. R., Elshourbagy, N., Appelbaum, E., Reape, R. J., Brawner, M., Makwana, J., Foley, J. J., Schmidt, D. B., Imburgia, C., McNulty, D., Matthews, J., O'Donnell, K., O'Shannessy, D., Scott, M., Groot, P. H., and Macphee, C. (1997). Cloning, *in vitro* expression, and functional characterization of a novel human CC chemokine of the monocyte chemotactic protein (MCP) family (MCP-4) that binds and signals through the CC chemokine receptor 2B. *J. Biol. Chem.* **272,** 16404–16413.

Bernstein, L. R., Antoniades, H., and Zetter, B. R. (1982). Migration of cultured vascular cells in response to plasma and platelet-derived factors. *J. Cell Sci.* **56,** 71–82.

Bertini, R., Allegretti, M., Bizzarri, C., Moriconi, A., Locati, M., Zampella, G., Cervellera, M. N., Di Cioccio, V., Cesta, M. C., Galliera, E., Martinez, F. O., Di Bitondo, R., Troiani, G., Sabbatini, V., D'Anniballe, G., Anacardio, R., Cutrin, J. C., Cavalieri, B., Mainiero, F., Strippoli, R., Villa, P., Di Girolamo, M., Martin, F., Gentile, M., Santoni, A., Corda, D., Poli, G., Mantovani, A., Ghezzi, P., and Colotta, F. (2004). Noncompetitive allosteric inhibitors of the inflammatory chemokine receptors CXCR1 and CXCR2: Prevention of reperfusion injury. *Proc. Natl. Acad. Sci. USA* **101,** 11791–11796.

Bhatnagar, D. (1998). Lipid-lowering drugs in the management of hyperlipidaemia. *Pharmacol. Ther.* **79,** 205–230.

Blaha, M., Krejsek, J., Blaha, V., Andrys, J., Vokurkova, D., Maly, J., Blazek, M., and Skorepova, M. (2004). Selectins and monocyte chemotactic peptide as the markers of atherosclerosis activity. *Physiol. Res.* **53,** 273–278.

Boisvert, W. A., Santiago, R., Curtiss, L. K., and Terkeltaub, R. A. (1998). A leukocyte homologue of the IL-8 receptor CXCR-2 mediates the accumulation of macrophages in atherosclerotic lesions of LDL receptor-deficient mice. *J. Clin. Invest.* **101,** 353–363.

Bonville, C. A., Lau, V. K., DeLeon, J. M., Gao, J. L., Easton, A. J., Rosenberg, H. F., and Domachowske, J. B. (2004). Functional antagonism of chemokine receptor CCR1 reduces mortality in acute pneumovirus infection *in vivo*. *J. Virol.* **78,** 7984–7989.

Boring, L., Gosling, J., Cleary, M., and Charo, I. F. (1998). Decreased lesion formation in CCR2-/- mice reveals a role for chemokines in the initiation of atherosclerosis. *Nature* **394,** 894–897.

Burke-Gaffney, A., Brooks, A. V., and Bogle, R. G. (2002). Regulation of chemokine expression in atherosclerosis. *Vascul. Pharmacol.* **38,** 283–292.

Burns, J. M., Dairaghi, D. J., Deitz, M., Tsang, M., and Schall, T. J. (2002). Comprehensive mapping of poxvirus vCCI chemokine-binding protein. Expanded range of ligand interactions and unusual dissociation kinetics. *J. Biol. Chem.* **277,** 2785–2789.

Bursill, C. A., Cai, S., Channon, K. M., and Greaves, D. R. (2003). Adenoviral-mediated delivery of a viral chemokine binding protein blocks CC-chemokine activity *in vitro* and *in vivo*. *Immunobiology* **207,** 187–196.

Bursill, C., Choudhury, R., Ali, Z., and Greaves, D. R. (2004). Broad-spectrum CC-chemokine blockade by gene transfer inhibits macrophage recruitment and athersclerotic plaque formation in apolipoprotein E-knockout mice. *Circulation* **110,** 2460–2466.

Bustos, C., Hernandez-Presa, M. A., Ortego, M., Tunon, J., Ortega, L., Perez, F., Diaz, C., Hernandez, G., and Egido, J. (1998). HMG-CoA reductase inhibition by atorvastatin reduces neointimal inflammation in a rabbit model of atherosclerosis. *J. Am. Coll. Cardiol.* **32,** 2057–2064.

Carfi, A., Smith, C. A., Smolak, P. J., McGrew, J., and Wiley, D. C. (1999). Structure of a soluble secreted chemokine inhibitor vCCI (p35) from cowpox virus. *Proc. Natl. Acad. Sci. USA* **96,** 12379–12383.

Chandrasekar, B., Bysani, S., and Mummidi, S. (2004). CXCL16 signals via Gi, phosphatidylinositol 3-kinase, Akt, I kappa B kinase, and nuclear factor-kappa B and induces cell-cell adhesion and aortic smooth muscle cell proliferation. *J. Biol. Chem.* **279,** 3188–3196.

Chen, Z., Ishibashi, S., Perrey, S., Osuga, Ji., Gotoda, T., Kitamine, T., Tamura, Y., Okazaki, H., Yahagi, N., Iizuka, Y., Shionoiri, F., Ohashi, K., Harada, K., Shimano, H., Nagai, R., and Yamada, N. (2001). Troglitazone inhibits atherosclerosis in apolipoprotein E-knockout mice: Pleiotropic effects on CD36 expression and HDL. *Arterioscler Thromb. Vasc. Biol.* **21,** 372–377.

Chuntharapai, A., Lee, J., Hebert, C. A., and Kim, K. J. (1994). Monoclonal antibodies detect different distribution patterns of IL-8 receptor A and IL-8 receptor B on human peripheral blood leukocytes. *J. Immunol.* **153,** 5682–5688.

Collins, A. R., Meehan, W. P., Kintscher, U., Jackson, S., Wakino, S., Noh, G., Palinski, W., Hsueh, W. A., and Law, R. E. (2001). Troglitazone inhibits formation of early atherosclerotic lesions in diabetic and nondiabetic low density lipoprotein receptor-deficient mice. *Arterioscler Thromb. Vasc. Biol.* **21,** 365–371.

Combadiere, C., Potteaux, S., Gao, J. L., Esposito, B., Casanova, S., Lee, E. J., Debre, P., Tedgui, A., Murphy, P. M., and Mallat, Z. (2003). Decreased atherosclerotic lesion formation in CX3CR1/apolipoprotein E double knockout mice. *Circulation* **107,** 1009–1016.

Cullen, P., Baetta, R., Bellosta, S., Bernini, F., Chinetti, G., Cignarella, A., von Eckardstein, A., Exley, A., Goddard, M., Hofker, M., Hurt-Camejo, E., Kanters, E., Kovanen, P., Lorkowski, S., McPheat, W., Pentikainen, M., Rauterberg, J., Ritchie, A., Staels, B., Weitkamp, B., and de Winther, M. (2003). MAFAPS Consortium. Rupture of the atherosclerotic plaque: does a good animal model exist? *Arterioscler Thromb. Vasc. Biol.* **23,** 535–542.

Cushing, S. D., Berliner, J. A., Valente, A. J., Territo, M. C., Navab, M., Parhami, F., Gerrity, R., Schwartz, C. J., and Fogelman, A. M. (1990). Minimally modified low density lipoprotein induces monocyte chemotactic protein 1 in human endothelial cells and smooth muscle cells. *Proc. Natl. Acad. Sci. USA* **87,** 5134–5138.

D'Ambrosio, D., Panina-Bordignon, P., and Sinigaglia, F. (2003). Chemokine receptors in inflammation: An overview. *J. Immunol. Methods* **273,** 3–13.

Dansky, H. M., Charlton, S. A., Harper, M. M., and Smith, J. D. (1997). T and B lymphocytes play a minor role in atherosclerotic plaque formation in the apolipoprotein E-deficient mouse. *Proc. Natl. Acad. Sci. USA* **94,** 4642–4646.

Daugherty, A., Pure, E., Delfel-Butteiger, D., Chen, S., Leferovich, J., Roselaar, S. E., and Rader, D. J. (1997). The effects of total lymphocyte deficiency on the extent of atherosclerosis in apolipoprotein E-/- mice. *J. Clin. Invest.* **100,** 1575–1580.

Davies, M. J. (1996). Stability and instability: two faces of coronary atherosclerosis. The Paul Dudley White Lecture 1995. *Circulation* **94,** 2013–2020.

Dawson, T. C., Kuziel, W. A., Osahar, T. A., and Maeda, N. (1999). Absence of CC chemokine receptor-2 reduces atherosclerosis in apolipoprotein E-deficient mice. *Atherosclerosis* **143,** 205–211.

De Clercq, E. (2000). Inhibition of HIV infection by bicyclams, highly potent and specific CXCR4 antagonists. *Mol. Pharmacol.* **57,** 833–839.

Diomede, L., Albani, D., Sottocorno, M., Donati, M. B., Bianchi, M., Fruscella, P., and Salmona, M. (2001). *In vivo* anti-inflammatory effect of statins is mediated by nonsterol mevalonate products. *Arterioscler Thromb. Vasc. Biol.* **21,** 1327–1332.

Faggiotto, A., Ross, R., and Harker, L. (1984). Studies of hypercholesterolemia in the non-human primate. I. Changes that lead to fatty streak formation. *Arteriosclerosis* **4,** 323–340.

Feldmann, M., Maini, R. N., Bondeson, J., Taylor, P., Foxwell, B. M., and Brennan, F. M. (2001). Cytokine blockade in rheumatoid arthritis. *Adv. Exp. Med. Biol.* **490,** 119–127.

Fong, A. M., Robinson, L. A., Steeber, D. A., Tedder, T. F., Yoshie, O., Imai, T., and Patel, D. D. (1998). Fractalkine and CX3CR1 mediate a novel mechanism of leukocyte capture, firm adhesion, and activation under physiologic flow. *J. Exp. Med.* **188,** 1413–1419.

Foxwell, B., Andreakos, E., Brennan, F., Feldmann, M., Smith, C., and Conron, M. (2003). Prospects for the development of small molecular weight compounds to replace anti-tumour necrosis factor biological agents. *Ann. Rheum. Dis.* **62**(Suppl 2.), ii90–93.

Frostegard, J., Haegerstrand, A., Gidlund, M., and Nilsson, J. (1991). Biologically modified LDL increases the adhesive properties of endothelial cells. *Atherosclerosis* **90,** 119–126.

Gerard, C., and Rollins, B. J. (2001). Chemokines and disease. *Nat. Immunol.* **2,** 108–115.

Gerszten, R. E., Garcia-Zepeda, E. A., Lim, Y. C., Yoshida, M., Ding, H. A., Gimbrone, M. A., Jr., Luster, A. D., Luscinskas, F. W., and Rosenzweig, A. (1999). MCP-1 and IL-8 trigger firm adhesion of monocytes to vascular endothelium under flow conditions. *Nature* **398,** 718–723.

Gimbrone, M. A., Jr., Topper, J. N., Nagel, T., Anderson, K. R., and Garcia-Cardena, G. (2000). Endothelial dysfunction, hemodynamic forces, and atherogenesis. *Ann. New York Acad. Sci.* **902,** 230–240.

Gladue, R. P., Tylaska, L. A., Brissette, W. H., Lira, P. D., Kath, J. C., Poss, C. S., Brown, M. F., Paradis, T. J., Conklyn, M. J., Ogborne, K. T., McGlynn, M. A., Lillie, B. M., DiRico, A. P., Mairs, E. N., McElroy, E. B., Martin, W. H., Stock, I. A., Shepard, R. M., Showell, H. J., and Neote, K. (2003). CP-481,715, a potent and selective CCR1 antagonist with potential therapeutic implications for inflammatory diseases. *J. Biol. Chem.* **278,** 40473–40480.

Gong, J. H., Ratkay, L. G., Waterfield, J. D., and Clark-Lewis, I. (1997). An antagonist of monocyte chemoattractant protein 1 (MCP-1) inhibits arthritis in the MRL-lpr mouse model. *J. Exp. Med.* **186,** 131–137.

Gonzalo, J. A., Lloyd, C. M., Wen, D., Albar, J. P., Wells, T. N., Proudfoot, A., Martinez-A, C., Dorf, M., Bjerke, T., Coyle, A. J., and Gutierrez-Ramos, J. C. (1998). The coordinated action of CC chemokines in the lung orchestrates allergic inflammation and airway hyperresponsiveness. *J. Exp. Med.* **188,** 157–167.

Gough, P. J., Greaves, D. R., Suzuki, H., Hakkinen, T., Hiltunen, M. O., Turunen, M., Herttuala, S. Y., Kodama, T., and Gordon, S. (1999). Analysis of macrophage scavenger receptor (SR-A) expression in human aortic atherosclerotic lesions. *Arterioscler Thromb. Vasc. Biol.* **19,** 461–471.

Greaves, D. R., Hakkinen, T., Lucas, A. D., Liddiard, K., Jones, E., Quinn, C. M., Senaratne, J., Green, F. R., Tyson, K., Boyle, J., Shanahan, C., Weissberg, P. L., Gordon, S., and Yla-Hertualla, S. (2001). Linked chromosome 16q13 chemokines, macrophage-derived chemokine, fractalkine, and thymus- and activation-regulated chemokine, are expressed in human atherosclerotic lesions. *Arterioscler Thromb. Vasc. Biol.* **21,** 923–929.

Grosskreutz, C. L., Anand-Apte, B., Duplaa, C., Quinn, T. P., Terman, B. I., Zetter, B., and D'Amore, P. A. (1999). Vascular endothelial growth factor-induced migration of vascular smooth muscle cells *in vitro*. *Microvasc. Res.* **58,** 128–136.

Gu, L., Okada, Y., Clinton, S. K., Gerard, C., Sukhova, G. K., Libby, P., and Rollins, B. J. (1998). Absence of monocyte chemoattractant protein-1 reduces atherosclerosis in low density lipoprotein receptor-deficient mice. *Mol. Cell* **2,** 275–281.

Gupta, S., Pablo, A. M., Jiang, X., Wang, N., Tall, A. R., and Schindler, C. (1997). IFN-gamma potentiates atherosclerosis in ApoE knock-out mice. *J. Clin. Invest.* **99,** 2752–2761.

Haley, K. J., Lilly, C. M., Yang, J. H., Feng, Y., Kennedy, S. P., Turi, T. G., Thompson, J. F., Sukhova, G. H., Libby, P., and Lee, R. T. (2000). Overexpression of eotaxin and the CCR3 receptor in human atherosclerosis: Using genomic technology to identify a potential novel pathway of vascular inflammation. *Circulation* **102,** 2185–2189.

Han, K. H., and Quehenberger, O. (2000). Ligands for peroxisome proliferator-activated receptor inhibit monocyte CCR2 expression stimulated by plasma lipoproteins. *Trends. Cardiovasc. Med.* **10,** 209–216.

Han, K. H., Chang, M. K., Boullier, A., Green, S. R., Li, A., Glass, C. K., and Quehenberger, O. (2000). Oxidized LDL reduces monocyte CCR2 expression through pathways involving peroxisome proliferator-activated receptor gamma. *J. Clin. Invest.* **106,** 793–802.

Han, K. H., Tangirala, R. K., Green, S. R., and Quehenberger, O. (1998). Chemokine receptor CCR2 expression and monocyte chemoattractant protein-1-mediated chemotaxis in human monocytes. A regulatory role for plasma LDL. *Arterioscler Thromb. Vasc. Biol.* **18,** 1983–1991.

Hansson, G. K. (2001). Immune mechanisms in atherosclerosis. *Arterioscler. Thromb. Vasc. Biol.* **21,** 1876–1890.

Haque, N. S., Fallon, J. T., Pan, J. J., Taubman, M. B., and Harpel, P. C. (2004). Chemokine receptor-8 (CCR8) mediates human vascular smooth muscle cell chemotaxis and metalloproteinase-2 secretion. *Blood* **103,** 1296–1304.

Haque, N. S., Zhang, X., French, D. L., Li, J., Poon, M., Fallon, J. T., Gabel, B. R., Taubman, M. B., Koschinsky, M., and Harpel, P. C. (2000). CC chemokine I-309 is the principal monocyte chemoattractant induced by apolipoprotein(a) in human vascular endothelial cells. *Circulation* **102,** 786–792.

Harada, A., Sekido, N., Akahoshi, T., Wada, T., Mukaida, N., and Matsushima, K. (1994). Essential involvement of interleukin-8 (IL-8) in acute inflammation. *J. Leukoc. Biol.* **56,** 559–564.

Haringman, J. J., Kraan, M. C., Smeets, T. J., Zwinderman, K. H., and Tak, P. P. (2003). Chemokine blockade and chronic inflammatory disease: Proof of concept in patients with rheumatoid arthritis. *Ann. Rheum. Dis.* **62,** 715–721.

Harrison, J. K., Jiang, Y., Wees, E. A., Salafranca, M. N., Liang, H. X., Feng, L., and Belardinelli, L. (1999). Inflammatory agents regulate *in vivo* expression of fractalkine in endothelial cells of the rat heart. *J. Leukoc. Biol.* **66,** 937–944.

Haskell, C. A., Cleary, M. D., and Charo, I. F. (1999). Molecular uncoupling of fractalkine-mediated cell adhesion and signal transduction. Rapid flow arrest of CX3CR1-expressing cells is independent of G-protein activation. *J. Biol. Chem.* **274,** 10053–10058.

Hayes, I. M., Jordan, N. J., Towers, S., Smith, G., Paterson, J. R., Earnshaw, J. J., Roach, A. G., Westwick, J., and Williams, R. J. (1998). Human vascular smooth muscle cells express receptors for CC chemokines. *Arterioscler Thromb. Vasc. Biol.* **18,** 397–403.

Hofnagel, O., Luechtenborg, B., Plenz, G., and Robenek, H. (2002). Expression of the novel scavenger receptor SR-PSOX in cultured aortic smooth muscle cells and umbilical endothelial cells. *Arterioscler Thromb. Vasc. Biol.* **22,** 710–711.

Horvath, C., Welt, F. G., Nedelman, M., Rao, P., and Rogers, C. (2002). Targeting CCR2 or CD18 inhibits experimental in-stent restenosis in primates: Inhibitory potential depends on type of injury and leukocytes targeted. *Circ. Res.* **90,** 488–494.

Huang, S., Mills, L., Mian, B., Tellez, C., McCarty, M., Yang, X. D., Gudas, J. M., and Bar-Eli, M. (2002). Fully humanized neutralizing antibodies to interleukin-8 (ABX-IL8) inhibit angiogenesis, tumor growth, and metastasis of human melanoma. *Am. J. Pathol.* **161,** 125–134.

Huffnagle, G. B., McNeil, L. K., McDonald, R. A., Murphy, J. W., Toews, G. B., Maeda, N., and Kuziel, W. A. (1999). Cutting edge: Role of C-C chemokine receptor 5 in organ-specific and innate immunity to Cryptococcus neoformans. *J. Immunol.* **163,** 4642–4646.

Huo, Y., Schober, A., Forlow, S. B., Smith, D. F., Hyman, M. C., Jung, S., Littman, D. R., Weber, C., and Ley, K. (2003). Circulating activated platelets exacerbate atherosclerosis in mice deficient in apolipoprotein E. *Nat. Med.* **9,** 61–67.

Imai, T., Hieshima, K., Haskell, C., Baba, M., Nagira, M., Nishimura, M., Kakizaki, M., Takagi, S., Nomiyama, H., Schall, T. J., and Yoshie, O. (1997). Identification and molecular characterization of fractalkine receptor CX3CR1, which mediates both leukocyte migration and adhesion. *Cell* **91,** 521–530.

Inoue, S., Egashira, K., Ni, W., Kitamoto, S., Usui, M., Otani, K., Ishibashi, M., Hiasa, K., Nishida, K., and Takeshita, A. (2002). Anti-monocyte chemoattractant protein-1 gene therapy limits progression and destabilization of established atherosclerosis in apolipoprotein E-knockout mice. *Circulation* **106,** 2700–2706.

Ishibashi, M., Egashira, K., Hiasa, K., Inoue, S., Ni, W., Zhao, Q., Usui, M., Kitamoto, S., Ichiki, T., and Takeshita, A. (2002). Antiinflammatory and antiarteriosclerotic effects of pioglitazone. *Hypertension* **40,** 687–693.

Ishibashi, M., Hiasa, K., Zhao, Q., Inoue, S., Ohtani, K., Kitamoto, S., Tsuchihashi, M., Sugaya, T., Charo, I. F., Kura, S., Tsuzuki, T., Ishibashi, T., Takeshita, A., and Egashira, K. (2004). Critical role of monocyte chemoattractant protein-1 receptor CCR2 on monocytes in hypertension-induced vascular inflammation and remodeling. *Circ. Res.* **94,** 1203–1210.

Ishibashi, S., Brown, M. S., Goldstein, J. L., Gerard, R. D., Hammer, R. E., and Herz, J. (1993). Hypercholesterolemia in low density lipoprotein receptor knockout mice and its reversal by adenovirus-mediated gene delivery. *J. Clin. Invest.* **92,** 883–893.

Ito, T., Ikeda, U., Yamamoto, K., and Shimada, K. (2002). Regulation of interleukin-8 expression by HMG-CoA reductase inhibitors in human vascular smooth muscle cells. *Atherosclerosis* **165,** 51–55.

John, A. E., and Lukacs, N. W. (2003). Chemokines and asthma. *Sarcoidosis Vasc. Diffuse. Lung. Dis.* **20,** 180–189.

Jonasson, L., Holm, J., Skalli, O., Bondjers, G., and Hansson, G. K. (1986). Regional accumulations of T cells, macrophages, and smooth muscle cells in the human atherosclerotic plaque. *Arteriosclerosis* **6,** 131–138.

Kaartinen, M., Penttila, A., and Kovanen, P. T. (1994). Accumulation of activated mast cells in the shoulder region of human coronary atheroma, the predilection site of atheromatous rupture. *Circulation* **90,** 1669–1678.

Kinlay, S., Libby, P., and Ganz, P. (2001). Endothelial function and coronary artery disease. *Curr. Opin. Lipidol.* **12,** 383–389.

Knowles, J. W., and Maeda, N. (2000). Genetic modifiers of atherosclerosis in mice. *Arterioscler Thromb. Vasc. Biol.* **20,** 2336–2345.

Korzenik, J. R. (2004). Crohn's disease: Future anti-tumor necrosis factor therapies beyond infliximab. *Gastroenterol Clin. North Am* **33**(ix), 285–301.

Kovanen, P. T., Kaartinen, M., and Paavonen, T. (1995). Infiltrates of activated mast cells at the site of coronary atheromatous erosion or rupture in myocardial infarction. *Circulation* **92,** 1084–1088.

Kuziel, W. A., Dawson, T. C., Quinones, M., Garavito, E., Chenaux, G., Ahuja, S. S., Reddick, R. L., and Maeda, N. (2003). CCR5 deficiency is not protective in the early stages of atherogenesis in apoE knockout mice. *Atherosclerosis* **167,** 25–32.

Law, R. E., Goetze, S., Xi, X. P., Jackson, S., Kawano, Y., Demer, L., Fishbein, M. C., Meehan, W. P., and Hsueh, W. A. (2000). Expression and function of PPARgamma in rat and human vascular smooth muscle cells. *Circulation* **101,** 1311–1318.

Lesnik, P., Haskell, C. A., and Charo, I. F. (2003). Decreased atherosclerosis in CX3CR1-/- mice reveals a role for fractalkine in atherogenesis. *J. Clin. Invest.* **111,** 333–340.

Li, A. C., and Glass, C. K. (2002). The macrophage foam cell as a target for therapeutic intervention. *Nat. Med.* **8,** 1235–1242.

Liang, M., Mallari, C., Rosser, M., Ng, H. P., May, K., Monahan, S., Bauman, J. G., Islam, I., Ghannam, A., Buckman, B., Shaw, K., Wei, G. P., Xu, W., Zhao, Z., Ho, E., Shen, J., Oanh, H., Subramanyam, B., Vergona, R., Taub, D., Dunning, L., Harvey, S., Snider, R. M., Hesselgesser, J., Morrissey, M. M., and Perez, H. D. (2000). Identification and characterization of a potent, selective, and orally active antagonist of the CC chemokine receptor-1. *J. Biol. Chem.* **275,** 19000–19008.

Libby, P. (2002). Inflammation in atherosclerosis. *Nature* **420,** 868–874.

Lord, R. S., and Bobryshev, Y. V. (2002). Hallmarks of atherosclerotic lesion development with special reference to immune inflammatory mechanisms. *Cardiovasc. Surg.* **10,** 405–414.

Lucas, A. D., Bursill, C., Guzik, T. J., Sadowski, J., Channon, K. M., and Greaves, D. R. (2003). Smooth muscle cells in human atherosclerotic plaques express the fractalkine receptor CX3CR1 and undergo chemotaxis to the CX3C chemokine fractalkine (CX3CL1). *Circulation* **108,** 2498–2504.

Mach, F., Sauty, A., Iarossi, A. S., Sukhova, G. K., Neote, K., Libby, P., and Luster, A. D. (1999). Differential expression of three T lymphocyte-activating CXC chemokines by human atheroma-associated cells. *J. Clin. Invest.* **104,** 1041–1050.

Mack, M., Cihak, J., Simonis, C., Luckow, B., Proudfoot, A. E., Plachy, J., Bruhl, H., Frink, M., Anders, H. J., Vielhauer, V., Pfirstinger, J., Stangassinger, M., and Schlondorff, D. (2001). Expression and characterization of the chemokine receptors CCR2 and CCR5 in mice. *J. Immunol.* **166,** 4697–4704.

Martinez-Gonzalez, J., Alfon, J., Berrozpe, M., and Badimon, L. (2001). HMG-CoA reductase inhibitors reduce vascular monocyte chemotactic protein-1 expression in early lesions from hypercholesterolemic swine independently of their effect on plasma cholesterol levels. *Atherosclerosis* **159,** 27–33.

Matloubian, M., David, A., Engel, S., Ryan, J. E., and Cyster, J. G. (2000). A transmembrane CXC chemokine is a ligand for HIV-coreceptor Bonzo. *Nat. Immunol.* **1,** 298–304.

McDermott, D. H., Halcox, J. P., Schenke, W. H., Waclawiw, M. A., Merrell, M. N., Epstein, N., Quyyumi, A. A., and Murphy, P. M. (2001). Association between polymorphism in the chemokine receptor CX3CR1 and coronary vascular endothelial dysfunction and atherosclerosis. *Circ. Res.* **89,** 401–407.

Minami, M., Kume, N., Shimaoka, T., Kataoka, H., Hayashida, K., Akiyama, Y., Nagata, I., Ando, K., Nobuyoshi, M., Hanyuu, M., Komeda, M., Yonehara, S., and Kita, T. (2001). Expression of SR-PSOX, a novel cell-surface scavenger receptor for phosphatidylserine and oxidized LDL in human atherosclerotic lesions. *Arterioscler Thromb. Vasc. Biol.* **21,** 1796–1800.

Mirzadegan, T., Diehl, F., Ebi, B., Bhakta, S., Polsky, I., McCarley, D., Mulkins, M., Weatherhead, G. S., Lapierre, J. M., Dankwardt, J., Morgans, D., Jr, Wilhelm, R., and Jarnagin, K. (2000). Identification of the binding site for a novel class of CCR2b chemokine receptor antagonists: Binding to a common chemokine receptor motif within the helical bundle. *J. Biol. Chem.* **275,** 25562–25571.

Mitani, H., Egashira, K., and Kimura, M. (2003). HMG-CoA reductase inhibitor, fluvastatin, has cholesterol-lowering independent "direct" effects on atherosclerotic vessels in high cholesterol diet-fed rabbits. *Pharmacol. Res.* **48,** 417–427.

Moatti, D., Faure, S., Fumeron, F., Amara, Mel-W., Seknadji, P., McDermott, D. H., Debre, P., Aumont, M. C., Murphy, P. M., de Prost, D., and Combadiere, C. (2001).

Polymorphism in the fractalkine receptor CX3CR1 as a genetic risk factor for coronary artery disease. *Blood* **97,** 1925–1928.

Moreno, P. R., Falk, E., Palacios, I. F., Newell, J. B., Fuster, V., and Fallon, J. T. (1994). Macrophage infiltration in acute coronary syndromes. Implications for plaque rupture. *Circulation* **90,** 775–778.

Moser, B., Barella, L., Mattei, S., Schumacher, C., Boulay, F., Colombo, M. P., and Baggiolini, M. (1993). Expression of transcripts for two interleukin-8 receptors in human phagocytes, lymphocytes and melanoma cells. *Biochem. J.* **294 (Pt 1),** 285–292.

Napoli, C., D'Armiento, F. P., Mancini, F. P., Postiglione, A., Witztum, J. L., Palumbo, G., and Palinski, W. (1997). Fatty streak formation occurs in human fetal aortas and is greatly enhanced by maternal hypercholesterolemia. Intimal accumulation of low density lipoprotein and its oxidation precede monocyte recruitment into early atherosclerotic lesions. *J. Clin. Invest.* **100,** 2680–2690.

Nelken, N. A., Coughlin, S. R., Gordon, D., and Wilcox, J. N. (1991). Monocyte chemoattractant protein-1 in human atheromatous plaques. *J. Clin. Invest.* **88,** 1121–1127.

Ness, T. L., Hogaboam, C. M., Strieter, R. M., and Kunkel, S. L. (2003). Immunomodulatory role of CXCR2 during experimental septic peritonitis. *J. Immunol.* **171,** 3775–3784.

Ng, H. P., May, K., Bauman, J. G., Ghannam, A., Islam, I., Liang, M., Horuk, R., Hesselgesser, J., Snider, R. M., Perez, H. D., and Morrissey, M. M. (1999). Discovery of novel non-peptide CCR1 receptor antagonists. *J. Med. Chem.* **42,** 4680–4694.

Ni, W., Egashira, K., Kitamoto, S., Kataoka, C., Koyanagi, M., Inoue, S., Imaizumi, K., Akiyama, C., Nishida, K. I., and Takeshita, A. (2001). New anti-monocyte chemoattractant protein-1 gene therapy attenuates atherosclerosis in apolipoprotein E-knockout mice. *Circulation* **103,** 2096–2101.

Okopien, B., Haberka, M., Cwalina, L., Kowalski, J., Belowski, D., Madej, A., Zielinski, M., Krysiak, R., Labuzek, K., Kalina, Z., and Herman, Z. S. (2002). Plasma cytokines as predictors of coronary heart disease. *Res. Commun. Mol. Pathol. Pharmacol.* **112,** 5–15.

Osterud, B., and Bjorklid, E. (2003). Role of monocytes in atherogenesis. *Physiol. Rev.* **83,** 1069–1112.

Parissis, J. T., Adamopoulos, S., Venetsanou, K. F., Mentzikof, D. G., Karas, S. M., and Kremastinos, D. T. (2002). Serum profiles of C-C chemokines in acute myocardial infarction: Possible implication in postinfarction left ventricular remodeling. *J. Interferon. Cytokine. Res.* **22,** 223–229.

Parry, C. M., Simas, J. P., Smith, V. P., Stewart, C. A., Minson, A. C., Efstathiou, S., and Alcami, A. (2000). A broad spectrum secreted chemokine binding protein encoded by a herpesvirus. *J. Exp. Med.* **191,** 573–578.

Pasceri, V., Cheng, J. S., Willerson, J. T., Yeh, E. T., and Chang, J. (2001). Modulation of C-reactive protein-mediated monocyte chemoattractant protein-1 induction in human endothelial cells by anti-atherosclerosis drugs. *Circulation* **103,** 2531–2534.

Pattison, J. M., Nelson, P. J., Huie, P., Sibley, R. K., and Krensky, A. M. (1996). RANTES chemokine expression in transplant-associated accelerated atherosclerosis. *J. Heart Lung Transplant* **15,** 1194–1199.

Piedrahita, J. A., Zhang, S. H., Hagaman, J. R., Oliver, P. M., and Maeda, N. (1992). Generation of mice carrying a mutant apolipoprotein E gene inactivated by gene targeting in embryonic stem cells. *Proc. Natl. Acad. Sci. USA* **89,** 4471–4475.

Plater-Zyberk, C., Hoogewerf, A. J., Proudfoot, A. E., Power, C. A., and Wells, T. N. (1997). Effect of a CC chemokine receptor antagonist on collagen induced arthritis in DBA/1 mice. *Immunol. Lett.* **57,** 117–120.

Plump, A. S., Smith, J. D., Hayek, T., Aalto-Setala, K., Walsh, A., Verstuyft, J. G., Rubin, E. M., and Breslow, J. L. (1992). Severe hypercholesterolemia and atherosclerosis in apolipoprotein E-deficient mice created by homologous recombination in ES cells. *Cell* **71,** 343–353.

Podolin, P. L., Bolognese, B. J., Foley, J. J., Schmidt, D. B., Buckley, P. T., Widdowson, K. L., Jin, Q., White, J. R., Lee, J. M., Goodman, R. B., Hagen, T. R., Kajikawa, O., Marshall, L. A., Hay, D. W., and Sarau, H. M. (2002). A potent and selective nonpeptide antagonist of CXCR2 inhibits acute and chronic models of arthritis in the rabbit. *J. Immunol.* **169,** 6435–6444.

Proudfoot, A. E., Power, C. A., Hoogewerf, A. J., Montjovent, M. O., Borlat, F., Offord, R. E., and Wells, T. N. (1996). Extension of recombinant human RANTES by the retention of the initiating methionine produces a potent antagonist. *J. Biol. Chem.* **271,** 2599–2603.

Proudfoot, A. E., Power, C. A., and Wells, T. N. (2000). The strategy of blocking the chemokine system to combat disease. *Immunol. Rev.* **177,** 246–256.

Pyo, R., Jensen, K. K., Wiekowski, M. T., Manfra, D., Alcami, A., Taubman, M. B., and Lira, S. A. (2004). Inhibition of intimal hyperplasia in transgenic mice conditionally expressing the chemokine-binding protein M3. *Am. J. Pathol.* **164,** 2289–2297.

Rasmussen, L. M., Hansen, P. R., Nabipour, M. T., Olesen, P., Kristiansen, M. T., and Ledet, T. (2001). Diverse effects of inhibition of 3-hydroxy-3-methylglutaryl-CoA reductase on the expression of VCAM-1 and E-selectin in endothelial cells. *Biochem. J.* **360,** 363–370.

Reape, T. J., and Groot, P. H. (1999). Chemokines and atherosclerosis. *Atherosclerosis* **147,** 213–225.

Reape, T. J., Rayner, K., Manning, C. D., Gee, A. N., Barnette, M. S., Burnand, K. G., and Groot, P. H. (1999). Expression and cellular localization of the CC chemokines PARC and ELC in human atherosclerotic plaques. *Am. J. Pathol.* **154,** 365–374.

Rezaie-Majd, A., Maca, T., Bucek, R. A., Valent, P., Muller, M. R., Husslein, P., Kashanipour, A., Minar, E., and Baghestanian, M. (2002). Simvastatin reduces expression of cytokines interleukin-6, interleukin-8, and monocyte chemoattractant protein-1 in circulating monocytes from hypercholesterolemic patients. *Arterioscler Thromb. Vasc. Biol.* **22,** 1194–1199.

Ricote, M., Huang, J., Fajas, L., Li, A., Welch, J., Najib, J., Witztum, J. L., Auwerx, J., Palinski, W., and Glass, C. K. (1998). Expression of the peroxisome proliferator-activated receptor gamma (PPARgamma) in human atherosclerosis and regulation in macrophages by colony stimulating factors and oxidized low density lipoprotein. *Proc. Natl. Acad. Sci. USA* **95,** 7614–7619.

Romano, S. J., and Slee, D. H. (2001). Targeting selectins for the treatment of respiratory diseases. *Curr. Opin. Investig. Drugs* **2,** 907–913.

Ross, R. (1999). Atherosclerosis–an inflammatory disease. *N. Engl. J. Med.* **340,** 115–126.

Rottman, J. B., Ganley, K. P., Williams, K., Wu, L., Mackay, C. R., and Ringler, D. J. (1997). Cellular localization of the chemokine receptor CCR5. Correlation to cellular targets of HIV-1 infection. *Am. J. Pathol.* **151,** 1341–1351.

Schecter, A. D., Berman, A. B., and Taubman, M. B. (2003). Chemokine receptors in vascular smooth muscle. *Microcirculation* **10,** 265–272.

Schecter, A. D., Calderon, T. M., Berman, A. B., McManus, C. M., Fallon, J. T., Rossikhina, M., Zhao, W., Christ, G., Berman, J. W., and Taubman, M. B. (2000). Human vascular smooth muscle cells possess functional CCR5. *J. Biol. Chem.* **275,** 5466–5471.

Schonbeck, U., and Libby, P. (2004). Inflammation, immunity, and HMG-CoA reductase inhibitors: Statins as antiinflammatory agents? *Circulation* **109,** II18–26.

Schwartz, D., Andalibi, A., Chaverri-Almada, L., Berliner, J. A., Kirchgessner, T., Fang, Z. T., Tekamp-Olson, P., Lusis, A. J., Gallegos, C., and Fogelman, A. M. (1994). Role of the GRO family of chemokines in monocyte adhesion to MM-LDL-stimulated endothelium. *J. Clin. Invest.* **94,** 1968–1973.

Shashkin, P., Simpson, D., Mishin, V., Chesnutt, B., and Ley, K. (2003). Expression of CXCL16 in human T cells. *Arterioscler Thromb. Vasc. Biol.* **23,** 148–149.

Sheikine, Y., and Hansson, G. K. (2004). Chemokines and atherosclerosis. *Ann. Med.* **36,** 98–118.

Shimaoka, T., Kume, N., Minami, M., Hayashida, K., Kataoka, H., Kita, T., and Yonehara, S. (2000). Molecular cloning of a novel scavenger receptor for oxidized low density lipoprotein, SR-PSOX, on macrophages. *J. Biol. Chem.* **275,** 40663–40666.

Smith, C. A., Smith, T. D., Smolak, P. J., Friend, D., Hagen, H., Gerhart, M., Park, L., Pickup, D. J., Torrance, D., Mohler, K., Schooley, K., and Goodwin, R. G. (1997). Poxvirus genomes encode a secreted, soluble protein that preferentially inhibits beta chemokine activity yet lacks sequence homology to known chemokine receptors. *Virology* **236,** 316–327.

Smith, J. D., and Breslow, J. L. (1997). The emergence of mouse models of atherosclerosis and their relevance to clinical research. *J. Intern. Med.* **242,** 99–109.

Stary, H. C., Chandler, A. B., Glagov, S., Guyton, J. R., Insull, W., Jr., Rosenfeld, M. E., Schaffer, S. A., Schwartz, C. J., Wagner, W. D., and Wissler, R. W. (1994). A definition of initial, fatty streak, and intermediate lesions of atherosclerosis. A report from the Committee on Vascular Lesions of the Council on Arteriosclerosis, American Heart Association. *Circulation* **89,** 2462–2478.

Stemme, S., Holm, J., and Hansson, G. K. (1992). T lymphocytes in human atherosclerotic plaques are memory cells expressing CD45RO and the integrin VLA-1. *Arterioscler Thromb.* **12,** 206–211.

Strizki, J. M., Xu, S., Wagner, N. E., Wojcik, L., Liu, J., Hou, Y., Endres, M., Palani, A., Shapiro, S., Clader, J. W., Greenlee, W. J., Tagat, J. R., McCombie, S., Cox, K., Fawzi, A. B., Chou, C. C., Pugliese-Sivo, C., Davies, L., Moreno, M. E., Ho, D. D., Trkola, A., Stoddart, C. A., Moore, J. P., Reyes, G. R., and Baroudy, B. M. (2001). SCH-C (SCH 351125), an orally bioavailable, small molecule antagonist of the chemokine receptor CCR5, is a potent inhibitor of HIV-1 infection *in vitro* and *in vivo*. *Proc. Natl. Acad. Sci. USA* **98,** 12718–12723.

Sueyoshi, S., Yamada, T., Niihasi, M., Kusumi, Y., Oinuma, T., Esumi, M., Tsuru, K., Imai, S., Nemoto, N., Sakura, I., and Mitsumata, M. (2001). Expression of peroxisome proliferator-activated receptor subtypes in human atherosclerosis. *Ann. New York Acad. Sci.* **947,** 429–432.

Szalai, C., Duba, J., Prohaszka, Z., Kalina, A., Szabo, T., Nagy, B., Horvath, L., and Csaszar, A. (2001). Involvement of polymorphisms in the chemokine system in the susceptibility for coronary artery disease (CAD). Coincidence of elevated Lp(a) and MCP-1 -2518 G/G genotype in CAD patients. *Atherosclerosis* **158,** 233–239.

Tabata, T., Mine, S., Kawahara, C., Okada, Y., and Tanaka, Y. (2003). Monocyte chemoattractant protein-1 induces scavenger receptor expression and monocyte differentiation into foam cells. *Biochem. Biophys. Res. Commun.* **305,** 380–385.

Takeya, M., Yoshimura, T., Leonard, E. J., and Takahashi, K. (1993). Detection of monocyte chemoattractant protein-1 in human atherosclerotic lesions by an anti-monocyte chemoattractant protein-1 monoclonal antibody. *Hum. Pathol.* **24,** 534–539.

Taub, D. D., Lloyd, A. R., Conlon, K., Wang, J. M., Ortaldo, J. R., Harada, A., Matsushima, K., Kelvin, D. J., and Oppenheim, J. J. (1993). Recombinant human interferon-inducible protein 10 is a chemoattractant for human monocytes and T lymphocytes and promotes T cell adhesion to endothelial cells. *J. Exp. Med.* **177,** 1809–1814.

Tsou, C. L., Haskell, C. A., and Charo, I. F. (2001). Tumor necrosis factor-alpha-converting enzyme mediates the inducible cleavage of fractalkine. *J. Biol. Chem.* **276,** 44622–44626.

van Berkel, V., Barrett, J., Tiffany, H. L., Fremont, D. H., Murphy, P. M., McFadden, G., Speck, S. H., and Virgin, H. W., IV. (2000). Identification of a gammaherpesvirus selective chemokine binding protein that inhibits chemokine action. *J. Virol.* **74,** 6741–6747.

Veillard, N. R., Kwak, B., Pelli, G., Mulhaupt, F., James, R. W., Proudfoot, A. E., and Mach, F. (2004). Antagonism of RANTES receptors reduces atherosclerotic plaque formation in mice. *Circ. Res.* **94,** 253–261.

von Hundelshausen, P., Weber, K. S., Huo, Y., Proudfoot, A. E., Nelson, P. J., Ley, K., and Weber, C. (2001). RANTES deposition by platelets triggers monocyte arrest on inflamed and atherosclerotic endothelium. *Circulation* **103,** 1772–1777.

Wang, N., Tabas, I., Winchester, R., Ravalli, S., Rabbani, L. E., and Tall, A. (1996a). Interleukin 8 is induced by cholesterol loading of macrophages and expressed by macrophage foam cells in human atheroma. *J. Biol. Chem.* **271,** 8837–8842.

Wang, X., Yue, T. L., Ohlstein, E. H., Sung, C. P., and Feuerstein, G. Z. (1996b). Interferon-inducible protein-10 involves vascular smooth muscle cell migration, proliferation, and inflammatory response. *J. Biol. Chem.* **271,** 24286–24293.

Weber, C., Erl, W., Weber, K. S., and Weber, P. C. (1997). HMG-CoA reductase inhibitors decrease CD11b expression and CD11b-dependent adhesion of monocytes to endothelium and reduce increased adhesiveness of monocytes isolated from patients with hypercholesterolemia. *J. Am. Coll. Cardiol.* **30,** 1212–1217.

Weber, C., Schober, A., and Zernecke, A. (2004). Chemokines. Key Regulators of Mononuclear Cell Recruitment in Atherosclerotic Vascular Disease. *Arterioscler Thromb. Vasc. Biol.* **24,** 1997–2008.

Weitz-Schmidt, G., Welzenbach, K., Brinkmann, V., Kamata, T., Kallen, J., Bruns, C., Cottens, S., Takada, Y., and Hommel, U. (2001). Statins selectively inhibit leukocyte function antigen-1 by binding to a novel regulatory integrin site. *Nat. Med.* **7,** 687–692.

White, J. R., Lee, J. M., Young, P. R., Hertzberg, R. P., Jurewicz, A. J., Chaikin, M. A., Widdowson, K., Foley, J. J., Martin, L. D., Griswold, D. E., and Sarau, H. M. (1998). Identification of a potent, selective non-peptide CXCR2 antagonist that inhibits interleukin-8-induced neutrophil migration. *J. Biol. Chem.* **273,** 10095–10098.

Whitman, S. C., Rateri, D. L., Szilvassy, S. J., Yokoyama, W., and Daugherty, A. (2004). Depletion of natural killer cell function decreases atherosclerosis in low-density lipoprotein receptor null mice. *Arterioscler Thromb. Vasc. Biol.* **24,** 1049–1054.

Widdowson, K. L., Elliott, J. D., Veber, D. F., Nie, H., Rutledge, M. C., McCleland, B. W., Xiang, J. N., Jurewicz, A. J., Hertzberg, R. P., Foley, J. J., Griswold, D. E., Martin, L., Lee, J. M., White, J. R., and Sarau, H. M. (2004). Evaluation of potent and selective small-molecule antagonists for the CXCR2 chemokine receptor. *J. Med. Chem.* **47,** 1319–1321.

Wilcox, J. N., Nelken, N. A., Coughlin, S. R., Gordon, D., and Schall, T. J. (1994). Local expression of inflammatory cytokines in human atherosclerotic plaques. *J. Atheroscler Thromb.* **1**(Suppl. 1), S10–S13.

Wong, B. W., Wong, D., and McManus, B. M. (2002). Characterization of fractalkine (CX3CL1) and CX3CR1 in human coronary arteries with native atherosclerosis, diabetes mellitus, and transplant vascular disease. *Cardiovasc. Pathol.* **11,** 332–338.

Wuttge, D. M., Zhou, X., Sheikine, Y., Wagsater, D., Stemme, V., Hedin, U., Stemme, S., Hansson, G. K., and Sirsjo, A. (2004). CXCL16/SR-PSOX is an interferon-gamma-regulated chemokine and scavenger receptor expressed in atherosclerotic lesions. *Arterioscler Thromb. Vasc. Biol.* **24,** 750–755.

Xie, J. H., Nomura, N., Lu, M., Chen, S. L., Koch, G. E., Weng, Y., Rosa, R., Di Salvo, J., Mudgett, J., Peterson, L. B., Wicker, L. S., and DeMartino, J. A. (2003). Antibody-mediated blockade of the CXCR3 chemokine receptor results in diminished recruitment of T helper 1 cells into sites of inflammation. *J. Leukoc. Biol.* **73,** 771–780.

Yang, X. D., Corvalan, J. R., Wang, P., Roy, C. M., and Davis, C. G. (1999). Fully human anti-interleukin-8 monoclonal antibodies: Potential therapeutics for the treatment of inflammatory disease states. *J. Leukoc. Biol.* **66,** 401–410.

Yla-Herttuala, S., Lipton, B. A., Rosenfeld, M. E., Sarkioja, T., Yoshimura, T., Leonard, E. J., Witztum, J. L., and Steinberg, D. (1991). Expression of monocyte chemoattractant protein-1 in macrophage-rich areas of human and rabbit atherosclerotic lesions. *Proc. Natl. Acad. Sci. USA* **88,** 5252–5256.

Yokoya, K., Takatsu, H., Suzuki, T., Hosokawa, H., Ojio, S., Matsubara, T., Tanaka, T., Watanabe, S., Morita, N., Nishigaki, K., Takemura, G., Noda, T., Minatoguchi, S., and Fujiwara, H. (1999). Process of progression of coronary artery lesions from mild or moderate stenosis to moderate or severe stenosis: A study based on four serial coronary arteriograms per year. *Circulation* **100,** 903–909.

Zhou, X., Nicoletti, A., Elhage, R., and Hansson, G. K. (2000). Transfer of CD4(+) T cells aggravates atherosclerosis in immunodeficient apolipoprotein E knockout mice. *Circulation* **102,** 2919–2922.

Zhou, Y., Kurihara, T., Ryseck, R. P., Yang, Y., Ryan, C., Loy, J., Warr, G., and Bravo, R. (1998). Impaired macrophage function and enhanced T cell-dependent immune response in mice lacking CCR5, the mouse homologue of the major HIV-1 coreceptor. *J. Immunol.* **160,** 4018–4025.

CHAPTER 10

CXC Chemokines in Cancer

Robert M. Strieter,* Marie D. Burdick,† Ammar Sakkour,† Nilo O. Arnaiz,† John A. Belperio,† and Michael P. Keane†

*Departments of †Medicine, Pathology, and Pediatrics, Division of Pulmonary and Critical Care Medicine, School of Medicine, University of California, Los Angeles, California 90024

Current Topics in Membranes, Volume 55

1063-5823/05 $35.00
DOI: 10.1016/S1063-5823(04)55010-X

I. INTRODUCTION

CXC chemokines can enhance innate and adaptive immunity, regulate angiogenesis, and mediate tumor-cell metastases. The functions of CXC chemokines have a direct impact on both the biology of cancer and the host's response to the tumor. Tumor growth, invasion, and metastasis to distant organs are dependent on a highly orchestrated series of events that includes numerous steps: preneoplastic-to-neoplastic cellular transformation; establishment of a proangiogenic environment; local tumor-cell growth; separation from adjacent cells and/or extracellular matrix (ECM) followed by local invasion through ECM/vascular basement membrane and entry into the circulation; tumor-cell embolization and extravasation; and metastatic tumor growth in distant organs based on Paget's theory of "seed and soil" regarding the homing of tumor cells (the "seed") to a specific organ (the "soil") (Paget, 1889). In addition, tumor cells within a local microenvironment are capable of eliciting the infiltration of leukocytes, which may play a role in either mediating tumoricidal activity or enhancing tumorigenicity. The list of events that destine a tumor cell to invade and metastasize to distant organs in a specific manner is analogous to the process of leukocyte maturation, subsequent entry into the circulation, and eventual homing to specific tissue sites. Over the last 20 years, it has been recognized that a specific group of cytokines, the chemokines, have had an increasingly important role in mediating the homing of populations of leukocytes under conditions of both homeostasis and inflammatory/immunological responses. However, numerous studies over the last decade have demonstrated that specific expressions of chemokines and their receptors in the context of cancer are essential events that appear to be important in either promoting tumor growth and metastasis or inhibiting tumor growth. These studies highlight the expanding role that chemokines play in promoting autocrine, paracrine, and hormonal influence for successful tumor growth, invasion, and metastasis to distant organs. In this chapter we will review the role that CXC chemokines play in mediating various events that are important and relevant to cancer.

II. EVIDENCE THAT CXC CHEMOKINES ARE INVOLVED IN CELLULAR TRANSFORMATION RELEVANT TO PRENEOPLASTIC-TO-NEOPLASTIC TRANSFORMATION

The pathogenesis of cancer is a multistep process that involves sequential morphological and molecular changes that precede invasive lung cancer; these events are not too dissimilar from other neoplasms. For example, the human Kaposi's sarcoma herpesvirus (KSHV), which mediates the pathogenesis of Kaposi's sarcoma (KS), encodes a seven-transmembrane G-protein-coupled receptor (GPCR) (known in the case of KSHV as "KSHV-GPCR") that is homologous to the CXC chemokine receptor, CXCR2 (Gershengorn *et al.*, 1998; Burger *et al.*, 1999). Seven-transmembrane GPCR activation leads to dissociation of the heterotrimeric protein complex ($G_{\alpha\beta\gamma}$) to α and $\beta\gamma$ subunits that mediate downstream regulation of several intracellular signaling pathways (e.g., cAMP/protein kinase A [PKA], protein kinase C [PKC], phospholipase C [PLC], phosphoinositide 3-kinase [PI3K]/AKT/mTOR, Ras/Raf/MEK/JNK/p38/ERK1/ERK2) and activate NF-kB pathways (Pawson and Scott, 1997; Sugden and Clerk, 1997; Shyamala and Khoja, 1998; Couty and Gershengorn, 2004). Some of these signaling pathways are mechanisms identical to signal transduction by protein tyrosine kinase receptors (PTKRs), which are important for cellular proliferation, migration, and regulation of apoptosis (Pawson and Scott, 1997; Sugden and Clerk, 1997; Shyamala and Khoja, 1998). These findings support the notion that seven-transmembrane GPCRs, like CXCR2, may be involved in preneoplastic-to-neoplastic transformation and the development of cancer.

The KSHV-GPCR has been found to constitutively signal-couple, and signal-coupling of this receptor can be further augmented with CXC-chemokine ligand binding (e.g., binding of IL-8/CXCL8 and/or GRO-α/CXCL1) (Arvanitakis *et al.*, 1997; Bais *et al.*, 1998; Geras-Raaka *et al.*, 1998a,b). To ascertain the relevance of KSHV-GPCR in promoting the pathogenesis of KS, researchers have studied transgenic mice that overexpress KSHV-GPCR, which have been found to constitutively produce tumors similar to KS (Yang *et al.*, 2000; Guo *et al.*, 2003). The mice develop angioproliferative lesions in multiple organs that morphologically resemble KS lesions (Yang *et al.*, 2000; Guo *et al.*, 2003). These findings suggest that the expression of only one viral chemokine receptor–like gene can lead to the histopathological recapitulation of KS, along with cellular transformation and the development of a lesion that resembles an angiosarcoma. This supports the notion that a CXCR2-like receptor facilitates preneoplastic-to-neoplastic cellular transformation. The KSHV-GPCR has been found to signal-couple a number of signaling pathways relevant to cellular transformation.

In further support of this contention is the finding that a point mutation of CXCR2, but not of CXCR1, results in constitutive signaling of the receptor and cellular transformation of transfected cells in a similar manner to KSHV-GPCR (Burger *et al.*, 1999). Furthermore, the persistent activation of CXCR2 by specific CXC-chemokine ligands can lead to a cellular transformation similar to that seen either with the point mutation of CXCR2 or with KSHV-GPCR (Burger *et al.*, 1999). CXCR2 has been further implicated in the cellular transformation of melanocytes into melanoma

(Luan *et al.*, 1997; Owen *et al.*, 1997). Therefore, the expression of CXCR2 on certain cells in the presence of persistent autocrine and paracrine stimulation with specific CXC chemokine ligands has important implications in promoting cellular transformation that may be relevant to the preneoplastic-to-neoplastic transformation of lung cancer.

III. EVIDENCE THAT CHEMOKINES ARE INVOLVED IN THE REGULATION OF ANGIOGENESIS IN TUMORS

Preneoplastic-to-neoplastic transformation, tumor growth, invasion, and metastases are dependent on the establishment of a proangiogenic environment. Net angiogenesis in the local microenvironment is determined by an imbalance in the ratio of expression of angiogenic, as compared to angiostatic, factors. Various factors have been described that promote angiogenesis (Zetter, 1998; Belperio *et al.*, 2000; Folkman, 2002). In parallel, there is a growing list of angiogenesis inhibitors that have been identified (Belperio *et al.*, 2000; Folkman, 2003). The CXC chemokine family represents a unique group of cytokines that have the ability to function as either promoters or inhibitors of angiogenesis.

CXC chemokines are heparin-binding proteins that display disparate roles in the regulation of angiogenesis (Belperio *et al.*, 2000; Strieter *et al.*, 2004a, b). CXC chemokines are characteristically heparin-binding proteins. On a structural level, they have four highly conserved cysteine amino-acid residues, and the first two cysteines are separated by one nonconserved amino-acid residue, hence the name CXC (Belperio *et al.*, 2000; Romagnani *et al.*, 2004; Strieter *et al.*, 2004a,b). Although the CXC motif distinguishes this family from other chemokine families, a second structural domain within this family dictates their angiogenic potential. The NH_2-terminus of the majority of the CXC chemokines contains three amino-acid residues (Glu–Leu–Arg: the " 'ELR" motif), which precede the first cysteine amino-acid residue of the primary structure of these cytokines (Belperio *et al.*, 2000; Strieter *et al.*, 2004a,b). The family members that contain the ELR motif (such members are designated "ELR+") are potent promoters of angiogenesis in physiological concentrations of 1 to 100 nM (Strieter *et al.*, 1995). In contrast, members of the family that lack the ELR motif (designated "non–ELR+"), which are, in general, interferon-inducible, are potent inhibitors of angiogenesis in physiological concentrations of 500 pM to 100 nM (Strieter *et al.*, 1995, 2004a,b; Belperio *et al.*, 2000; Romagnani *et al.*, 2004).

A. ELR+ CXC Chemokines Promote Angiogenesis

The angiogenic members of the CXC chemokine family include CXCL1, CXCL2, CXCL3, CXCL5, CXCL6, CXCL7, and CXCL8 (Table I) (Strieter *et al.*, 1995, 2004a,b; Belperio *et al.*, 2000; Romagnani *et al.*, 2004). While angiogenic factors can behave in a redundant manner, recent studies have shown that serial interactions may ultimately contribute to perpetuation of the angiogenic phenotype of the endothelium. For example, vascular endothelial cell growth factor (VEGF) activation of endothelial cells leads to upregulation of the antiapoptotic molecule, Bcl-2, which in turn is associated with the expression of endothelial-cell CXCL8, which functions in an autocrine and paracrine manner to maintain the angiogenic phenotype of the endothelial cell (Nor *et al.*, 2001). This autocrine and paracrine function of VEGF, promoting the expression of CXCL8, has been further substantiated in a recent finding of neutrophils in the promotion of angiogenesis by a paracrine feed-forward mechanism involving endothelial cell–derived CXCL8 (Schruefer *et al.*, 2004). In this study, activation of neutrophils with N-formyl-Met-Leu-Phe (fMLP) resulted in the generation of proangiogenic activity that was related to VEGF and CXCL8 (Schruefer *et al.*, 2004). Moreover, VEGF induced the expression of CXCL8 that was associated with formation of CXCL8-dependent capillary-like structures from endothelial cells.

TABLE I

CXC Chemokine Family Members That Regulate Angiogenesis

Angiogenic ELR+ CXC chemokines	Angiostatic non-ELR+ CXC chemokines
-IL-8 (CXCL8)	-PF-4 (CXCL4 and CXCL4L1)
-ENA-78 (CXCL5)	-CXCL14
-GRO-α (CXCL1)	-Interferon-Inducible non-ELR+ CXC Chemokines
-GRO-β (CXCL2)	—IP-10 (CXCL10)
-GRO-γ (CXCL3)	—MIG (CXCL9)
-GCP-2 (CXCL6)	—ITAC (CXCL11)
-PBP	
—CTAP-III	
—β-TG	
—NAP-2 (CXCL7)	

Note: CXCR2 is the putative receptor for chemokines biological activity; CXCR3 is the putative receptor for chemokines biological activity, with the exception of CXCL14.

The ability of CXCL8 to promote enhanced endothelial-cell survival and proliferation has been substantiated by other studies (Li *et al.*, 2003). Moreover, this pathway *in vitro* can be reproduced *in vivo* using human tumor cells, which normally do not form tumors, mixed with endothelial cells overexpressing Bcl-2, which can lead to CXCL8-dependent tumorigenesis (Nor *et al.*, 2001). In addition, other serial pathways that promote CXC chemokine–mediated angiogenesis are operative that include receptor activation of seven-transmembrane GPCR and PTKR that contribute to the expression of angiogenic CXC chemokines via NF-kB activation in cancer cells and enhanced tumor-associated angiogenesis (Dong *et al.*, 2001; Hirata *et al.*, 2002; Richmond, 2002; Levine *et al.*, 2003). These results demonstrate the existence of novel paracrine and autocrine signal pathways that lead to enhanced tumor microvasculature and tumor growth related to angiogenic ELR+ CXC chemokines.

B. CXCR2 is the Receptor for Angiogenic ELR+ CXC Chemokine-Mediated Angiogenesis

The fact that all ELR+ CXC chemokines mediate angiogenesis highlights the importance of identifying a common receptor that mediates their biological function in the promotion of angiogenesis. While the candidate CXC chemokine receptors are CXCR1 and/or CXCR2, only CXCL8 and CXCL6 specifically bind to CXCR1, whereas all ELR+ CXC chemokines bind to CXCR2 (Addison *et al.*, 2000b). The ability of all ELR+ CXC chemokine ligands to bind to CXCR2 supports the notion that this receptor mediates the angiogenic activity of ELR+ CXC chemokines.

While CXCR1 and CXCR2 are detected in endothelial cells (Murdoch *et al.*, 1999; Addison *et al.*, 2000b; Salcedo *et al.*, 2000), the expression of CXCR2, not CXCR1, has been found to be the primary functional chemokine receptor in mediating endothelial-cell chemotaxis (Murdoch *et al.*, 1999; Addison *et al.*, 2000b). Heidemann *et al.* (2003) have further confirmed the importance of CXCR2 in mediating the effects of angiogenesis in human microvascular endothelial cells. They found that endothelial cells respond to CXCL8 with rapid stress-fiber assembly, chemotaxis, enhanced proliferation, and phosphorylation of extracellular signal-regulated protein kinase 1/2 (ERK 1/2) related to activation of CXCR2 (Heidemann *et al.*, 2003). Blocking the function of CXCR2 either with specific neutralizing antibodies or by inhibiting downstream signaling using specific inhibitors of ERK1/2 and PI3K impaired CXCL8-induced stress fiber assembly, chemotaxis, and endothelial tube formation in endothelial cells (Heidemann *et al.*, 2003).

What is the fate of CXCR2 after activation by ELR+ CXC chemokines? CXCR2 activation leads to receptor internalization, recycling of the receptor back to the cell membrane, or targeting of CXCR2 for degradation. ELR+ CXC chemokine activation of CXCR2 under conditions in which the receptor is transiently exposed or stimulated with less than saturable concentrations results in movement of CXCR2 into clathrin-coated pits, which leads to movement into the early endosome, the sorting endosome, and on to the recycling endosome, which leads to trafficking back to the plasma membrane compartment and reexpression on the cell surface (Richmond *et al.*, 2004). However, if CXCR2 is exposed to prolonged saturating concentrations of ELR+ CXC chemokines, a significant proportion of CXCR2 will move into the late endosome and on to the lysosome for degradation (Richmond *et al.*, 2004). Interestingly, CXCR2 internalization is necessary for generating a chemotactic response. Mutation of the CXCR2, which impairs receptor internalization by altering the binding of adaptor proteins AP-2 or beta arrestin to the receptor, results in a marked reduction in the chemotactic response (Richmond *et al.*, 2004).

The importance of CXCR2 in mediating ELR+ CXC chemokine–induced angiogenesis has been shown *in vivo* using the cornea micropocket assay of angiogenesis in CXCR2+/+ and −/− animals. ELR+ CXC chemokine–mediated angiogenesis was inhibited in the corneas of CXCR2−/− mice and in the presence of neutralizing antibodies to CXCR2 in the rat corneal micropocket assay (Addison *et al.*, 2000b). These studies have been further substantiated using CXCR2−/− mice in a wound-repair model system (Devalaraja *et al.*, 2000). Devalaraja *et al.* (2000) have examined the significance of CXC chemokines in wound healing. In this study, full excisional wounds were created on CXCR2+/+, heterozygous +/−, or CXCR2−/− mice. Significant delays in wound-healing parameters were found in CXCR2−/− mice, including decreased neovascularization (Devalaraja *et al.*, 2000). These studies have now been extended to a lung-cancer syngeneic tumor model system in CXCR2−/−, as compared to CXCR2+/+, mice. Lung cancer in CXCR2−/− mice demonstrated reduced growth, increased tumor-associated necrosis, and inhibited tumor-associated angiogenesis and metastatic potential (Keane *et al.*, 2004). These *in vitro* and *in vivo* studies establish that CXCR2 is an important receptor that mediates ELR+ CXC chemokine–dependent angiogenic activity.

C. In General, Non-ELR+ CXC Chemokines are Inhibitors of Angiogenesis

The angiostatic members of the CXC chemokine family include CXCL4, CXCL4L1, CXCL9, CXCL10, CXCL11, and CXCL14 (Strieter *et al.*, 1995, 2002; Belperio *et al.*, 2000; Struyf *et al.*, 2004; Shellenberger *et al.*, in press) (Table I). Platelet factor-4 (PF-4)/CXCL4 was the first chemokine described to inhibit neovascularization (Maione *et al.*, 1990). However, the product of the nonallelic variant gene of CXCL4, PF-4var1/PF-4alt (designated CXCL4L1) was recently isolated from thrombin-stimulated human platelets and purified to homogeneity (Struyf *et al.*, 2004). Although secreted CXCL4 and CXCL4L1 differ in only three amino acids, CXCL4L1 is more potent for inhibiting angiogenesis in response to angiogenic factors in both *in vitro* and *in vivo* models of angiogenesis (Struyf *et al.*, 2004). Thus, activated platelets release CXCL4L1, a potent regulator of endothelial cell biology, which affects angiogenesis.

CXCL9, CXCL10, and CXCL11 are induced by both type-I and type-II interferons (type-I interferons are IFN-α and -β; the type-II interferon is IFN-γ) (Rollins, 1997; Balkwill, 1998; Luster, 1998; Strieter *et al.*, 2004a,b). The IFN-inducible non-ELR+ members of the CXC chemokine family inhibit angiogenesis in response to ELR+ CXC chemokines, bFGF, and VEGF (Strieter *et al.*, 2002). These findings suggest that all IFN-inducible ELR− CXC chemokines are potent inhibitors of angiogenesis. Moreover, this interrelationship of interferons and IFN-inducible CXC chemokines and their biological functions are directly relevant to the function of other cytokines, such as Th1/type-1 cytokines that stimulate the expression of interferons. Therefore, cytokines such as IL-23, IL-21, IL-18, IL-15, IL-12, and IL-2 and chemokines such as CCL19 and CCL21 will, via the induction of interferons? have profound effects on the production of CXCL9, CXCL10, and CXCL11. The subsequent expression of IFN-inducible CXC chemokines represents the final common pathway and may, in part, explain the mechanisms for the attenuation of angiogenesis related to interferons. Furthermore, this cytokine cascade interconnects with Th1/type-1 cytokine-mediated immunity toward tumor-associated antigens and creates the concept of "immunoangiostasis" (see the next section).

Breast- and kidney-expressed chemokine (BRAK)/CXCL14 is another non-ELR+ CXC chemokine and has been recently identified to inhibit angiogenesis *in vitro* and *in vivo* (Shellenberger *et al.*, 2005). CXCL14 was first identified by differential display of normal oral epithelial cells and head and neck squamous-cell carcinoma (SCCA) (Frederick *et al.*, 2000). CXCL14 was found to be downregulated in tumor specimens more than normal adjacent tissue was (Frederick *et al.*, 2000). The biological significance of the absence of CXCL14 in these tumors was not elucidated until Shellenberger *et al.* (2005) discovered that CXCL14 inhibited microvascular endothelial-cell chemotaxis *in vitro* in response to CXCL8, bFGF, and VEGF and inhibited neovascularization *in vivo* in response to the same angiogenic agonists. These findings support the notion that the loss of CXCL14 constitutive expression is associated with the transformation of normal epithelial cells to cancer cells and the promotion of a proangiogenic microenvironment.

D. Alternative Mechanisms to Receptor Interaction Related to the Angiostatic Activity of CXCL4 That May be Applicable to Other Angiostatic Non-ELR+ CXC Chemokines

Eloquent studies have delineated potential alternative mechanisms to receptor interaction for CXCL4 and its ability to inhibit angiogenesis that may be relevant to other non-ELR+ CXC chemokines. The ability of CXCL4 to bind to glycosaminoglycans (GAG: heparin and heparan sulfate) with high affinity appears to be important in several of its biological functions. CXCL4 inhibits endothelial-cell migration, proliferation, and *in vivo* angiogenesis in response to bFGF or VEGF (Maione *et al.*, 1990; Gupta and Singh, 1994). Moreover, FITC-labeled CXCL4, injected systemically, selectively binds to the endothelium only in areas of active angiogenesis (Hansell *et al.*, 1995; Borgstrom *et al.*, 1998). This suggests that the microvasculature is the major target for the biological effects of CXCL4 during angiogenesis. CXCL4 has been shown to inhibit bFGF and $VEGF_{165}$ binding to their respective receptors (Sato *et al.*, 1990; Gengrinovitch *et al.*, 1995; Perollet *et al.*, 1998). One mechanism for this effect is related to the generation of CXCL4-bFGF or CXCL4-$VEGF_{165}$ heterodimeric complexes, which impairs bFGF or $VEGF_{165}$ binding to their respective receptors (Gengrinovitch *et al.*, 1995; Perollet *et al.*, 1998; Jouan *et al.*, 1999). Basic FGF must undergo dimerization in the presence of endogenous heparin in order to bind to its receptor (Perollet *et al.*, 1998; Jouan *et al.*, 1999). CXCL4 complexes to bFGF and prevents bFGF dimerization followed by impaired receptor binding and internalization (Perollet *et al.*, 1998). $VEGF_{165}$ possesses a heparin-binding ability similar to that of bFGF. CXCL4 impairs $VEGF_{165}$ binding to its receptors on the endothelium via a mechanism similar to what has been reported for its ability to inhibit bFGF (Gengrinovitch *et al.*, 1995). Although the ability of CXCL4 to form heterodimers with bFGF and $VEGF_{165}$ is one potential mechanism by which to inhibit bFGF and $VEGF_{165}$ biological activity, it appears that CXCL4 may inhibit angiogenesis through additional mechanisms.

While CXCL4 inhibits specific $VEGF_{165}$ binding, it does not inhibit $VEGF_{121}$ binding to VEGF receptors on endothelial cells. In contrast to $VEGF_{165}$, $VEGF_{121}$ is not a heparin-binding protein (Houck *et al.*, 1991, 1992; Gengrinovitch *et al.*, 1995). CXCL4 neither forms heterodimers with $VEGF_{121}$ nor competitively interferes with $VEGF_{121}$ binding to its receptor. However, CXCL4 directly inhibits $VEGF_{121}$-induced endothelial-cell proliferation (Gengrinovitch *et al.*, 1995). These findings suggest that other mechanisms must be operative for CXCL4 inhibition of mitogen stimulation of endothelial cells, perhaps mediated through its own independent biological signal. Studies have suggested that CXCL4 inhibits the endothelial-cell cycle by preventing cell entry into S phase (Gupta and Singh, 1994). In a model system of endothelial-cell stimulation independent of interaction with cell-surface GAGs, CXCL4 inhibits epidermal growth factor (EGF)-stimulated endothelial-cell proliferation by causing a decrease in cyclin E-cyclin-dependent kinase 2 (cdk2) activity that results in a decrease in the rate of retinoblastoma protein (pRb) phosphorylation (Gentilini *et al.*, 1999). The mechanism is related to CXCL4-dependent sustained increase in the levels and binding of the cyclin-dependent kinase inhibitor (CKI), $p21^{Cip1/WAF1}$, to the cyclin E-cdk2 complex. This inhibits cell-cycle progression by preventing the downregulation of $p21^{Cip1/WAF1}$, leading to inhibition of both cyclin E-cdk2 activity and phosphorylation of pRb (Gentilini *et al.*, 1999). These mechanisms suggest that CXCL4 can inhibit a variety of endothelial-cell mitogens at multiple levels and may be relevant to the other angiostatic non-ELR+ CXC chemokines, including specific receptor binding.

E. CXCR3 Appears to be the Major Receptor for CXC Chemokines That Inhibit Angiogenesis

The major receptor that has been identified for angiostatic CXC chemokines is CXCR3, and it exists in alternative splice forms (i.e., CXCR3A, CXCR3B, and CXCR3-alt) that are involved in mediating recruitment of Th1 cells and act as the receptors for inhibition of angiogenesis (Rollins, 1997; Balkwill, 1998; Loetscher *et al.*, 1998; Luster, 1998; Ehlert *et al.*, 2004). CXCR3A is the major chemokine receptor found on Th1 effector T-cells, cytotoxic CD8 T-cells, activated B-cells, and NK cells (Loetscher *et al.*, 1996, 1998; Luster, 1998; Qin *et al.*, 1998; Rabin *et al.*, 1999; Moser and Loetscher, 2001; Beider *et al.*, 2003). IL-2 is a major agonist for the expression of CXCR3A on these cells (Loetscher *et al.*, 1996, 1998; Beider *et al.*, 2003). CXCR3 was originally identified on murine endothelial cells (Soto *et al.*, 1998); subsequent studies demonstrated that CXCR3 ligands could block both migration and proliferation of human microvascular endothelial cells in response to a variety of angiogenic factors (Salcedo *et al.*, 2000; Romagnani *et al.*, 2001). Further clarification of the role of CXCR3 in mediating angiostatic activity has come from the discovery that CXCR3 exists in two alternative splice forms (Lasagni *et al.*, 2003). These variants have been termed CXCR3A and CXCR3B (Lasagni *et al.*, 2003). CXCR3A mediates the CXCR3 ligand–dependent chemotactic activity of mononuclear cells (Lasagni *et al.*, 2003). CXCR3B mediates the angiostatic activity of CXCL4, CXCL9, CXCL10, and CXCL11 on human microvascular endothelial cells (Lasagni *et al.*, 2003). Moreover, specific antibodies to CXCR3B immunolocalize to endothelial cells within neoplastic tissues (Lasagni *et al.*, 2003). This supports the notion that if CXCR3 ligands can be spatially expressed within the tumor, then CXCR3B activation can inhibit tumor-associated angiogenesis (Lasagni *et al.*, 2003). To add to the complexity of CXCR3 biology, a variant of human CXCR3 has been recently identified that is generated by posttranscriptional exon skipping and is referred to as CXCR3-alt (Ehlert *et al.*, 2004). This receptor is expressed and responds to CXCL11 more than CXCL9 and CXCL10 (Ehlert *et al.*, 2004). These findings support the notion that augmenting CXCR3/CXCR3 ligand biology will be a therapeutic strategy to enhance angiostasis within the tumor.

While these studies have supported CXCR3 as the putative receptor for CXCL4, CXCL9, CXCL10, and CXCL11, it has remained unclear *in vivo* whether these CXCR3 ligands use CXCR3 on endothelium to mediate their angiostatic effect. Yang and Richmond (2004) have recently demonstrated that CXCL10 mediates its angiostatic activity *in vivo* by binding to CXCR3 and not by binding to glycosaminoglycans. To clarify this issue, they created expression constructs for mutants of CXCL10 that exhibit partial or total loss of binding to CXCR3 or loss of binding to glycosaminoglycans. They transfected a human melanoma cell line with these expression vectors, and stable clones were selected and inoculated into immunoincompetent mice (Yang and Richmond, 2004). Tumor cells expressing wild-type CXCL10 showed remarkably less tumor growth than the control vector–transfected tumor cells. Surprisingly, mutation of CXCL10 resulting in partial loss of receptor binding (IP-10C), or loss of glycosaminoglycans binding (IP-10H), did not significantly alter the ability to inhibit tumor growth. The reduction in tumor growth was associated with a reduction in tumor-associated angiogenesis, leading to the observed increase in both apoptosis and necrosis of tumor cells (Yang and Richmond, 2004). In contrast, expression of the CXCL10 mutant that fails to bind to CXCR3 failed to inhibit tumor growth (Yang and Richmond, 2004). This study has been confirmed in another *in vivo* angiogenesis-dependent model. Burdick and associates have found that CXCL11 in a CXCR3-dependent manner inhibits angiogenesis in a murine model of pulmonary fibrosis (Burdick *et al.*, 2004). These data suggest that CXCR3 receptor binding, but not glycosaminoglycan binding, is essential for the tumor-angiostatic activity of CXCR3 ligands (Fig. 1).

IV. ELR+ CXC CHEMOKINES PROMOTE ANGIOGENESIS ASSOCIATED WITH TUMORIGENESIS

The ELR+ CXC chemokines are important mediators of tumorigenesis related to their angiogenic properties. Studies using melanoma tumor models support that CXCLs 1, 2, and 3 play a significant role in mediating

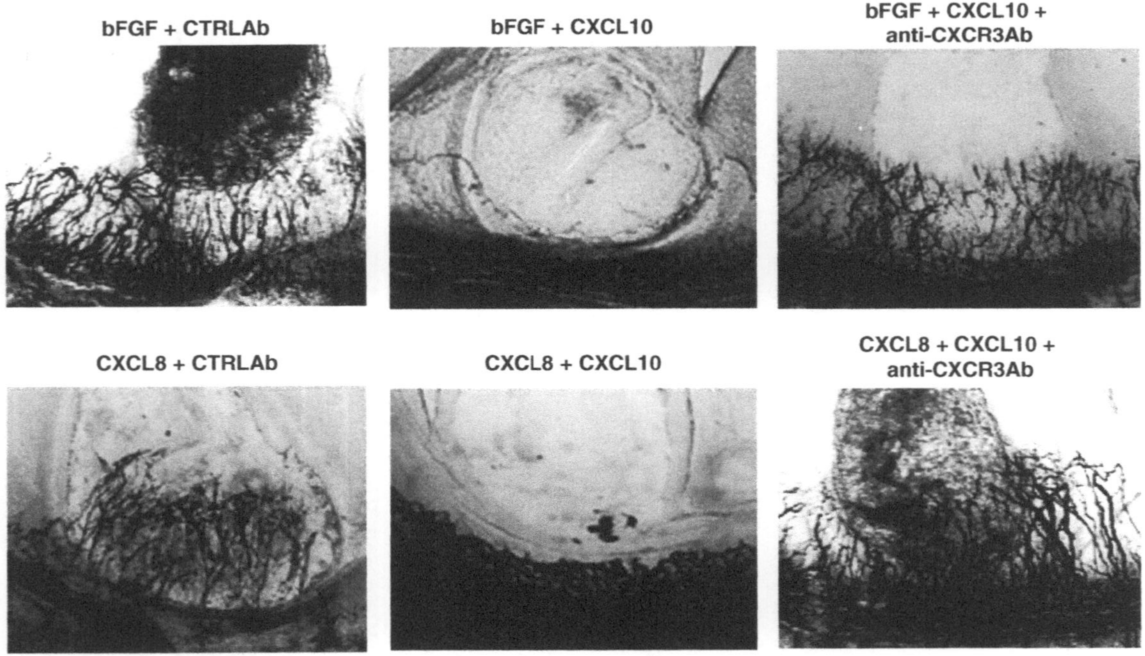

FIGURE 1 CXCR3 is the putative mediator of the angiostatic activity of CXCR3 ligands in the rat corneal micropocket assay of angiogenesis. CXCL8 and bFGF are shown to promote angiogenesis. The CXCR3 ligand, CXCL10, has been shown to inhibit both bFGF and CXCL8 mediated angiogenesis. Inhibition of CXCR3 results in reconstitution of bFGF and CXCL8-mediated angiogenesis in the presence of CXCL10.

tumorigenesis related to both their mitogenic and their angiogenic activities (Luan *et al.*, 1997). For example, CXCL1, CXCL2, and CXCL3 have all been found to be highly expressed in human melanoma (Luan *et al.*, 1997). To determine the biological significance of the presence of these ELR+ CXC chemokines in melanoma, human CXCL1, CXCL2, and CXCL3 genes were transfected into immortalized murine melanocytes that otherwise do not form tumors by themselves (Luan *et al.*, 1997; Owen *et al.*, 1997). The persistent expression of CXCLs 1, 2, or 3 in these cells transformed their phenotype to one with anchorage-independent growth *in vitro* and the ability to form tumors *in vivo* in immunoincompetent mice (Luan *et al.*, 1997; Owen *et al.*, 1997). The tumors were highly vascular and similar to the vascularity of B16 melanoma controls (Luan *et al.*, 1997; Owen *et al.*, 1997). When tumors were depleted of CXCLs 1, 2, or 3 there was a marked reduction of tumor-derived angiogenesis directly related to inhibition of tumor growth (Luan *et al.*, 1997; Owen *et al.*, 1997). These findings support the notion that the ELR+ CXC chemokines have the ability to act both as autocrine growth factors for melanoma and as potent paracrine mediators of angiogenesis to promote tumorigenesis and metastases.

The progression and growth of ovarian carcinoma is also dependent on successful angiogenesis, and CXCL8 has been determined to play a significant role in mediating human ovarian carcinoma–derived angiogenesis and tumorigenesis (Yoneda *et al.*, 1998). The expression of CXCL8, bFGF, and VEGF was examined in five different human ovarian carcinoma cell lines (Yoneda *et al.*, 1998). All cell lines *in vitro* expressed similar levels of bFGF; however, these cells expressed either high or low levels of CXCL8 or VEGF. When implanted into the peritoneum of immunoincompetent mice, the high-expressing CXCL8 tumors were found in all animals dying at less than 51 days (Yoneda *et al.*, 1998). The expression of CXCL8 was directly correlated with neovascularization and inversely correlated with survival, whereas VEGF expression was only correlated with production of ascites. No correlation was found for bFGF with either tumor neovascularization or tumor survival (Yoneda *et al.*, 1998). This study has been substantiated in patients with ovarian cancer, where ascites fluid demonstrates angiogenic activity directly correlated to CXCL8 (Gawrychowski *et al.*, 1998). These findings support the notion that antigenic ELR+ CXC chemokines play a greater role than bFGF and VEGF in mediating angiogenesis associated with ovarian cancer.

CXCL8 is markedly elevated and contributes to the overall angiogenic activity of nonsmall cell lung cancer (NSCLC) (Smith *et al.*, 1994). Extending these studies to an *in vivo* model system of human tumorigenesis (i.e., human NSCLC/SCID mouse chimera) (Arenberg *et al.*, 1995), tumor-derived CXCL8 was found to be directly correlated with tumorigenesis (Arenberg *et al.*, 1995). Tumor-bearing animals depleted of IL-8/CXCL8 demonstrated a >40% reduction in tumor growth and a reduction in spontaneous metastases (Arenberg *et al.*, 1995). The weakening of tumor growth and metastases was directly correlated to reduced angiogenesis (Arenberg *et al.*, 1995). These findings have been further corroborated with the use of several human NSCLC cell lines grown in immunoincompetent mice. NSCLC cell lines that constitutively express CXCL8 display greater tumorigenicity that is directly correlated to angiogenesis (Yatsunami *et al.*, 1997).

While CXCL8 was the first angiogenic ELR+ CXC chemokine to be discovered in NSCLC, CXCL5 has now been determined to have a higher degree of correlation with NSCLC-derived angiogenesis (Arenberg *et al.*, 1998). Surgical specimens of NSCLC tumors demonstrate a direct correlation of CXCL5 with tumor angiogenesis. These studies were extended to a SCID mouse model of human NSCLC tumorigenesis. CXCL5 expression was directly correlated with tumor growth (Arenberg *et al.*, 1998). Moreover, when NSCLC tumor–bearing animals were depleted of CXCL5, both tumor growth and spontaneous metastases were markedly slowed (Arenberg *et al.*, 1998). The reduction of angiogenesis was also accompanied by an increase in tumor-cell apoptosis, consistent with the previous observation that inhibition of tumor-derived angiogenesis is associated with increased tumor-cell apoptosis (Arenberg *et al.*, 1998). Similarly, *in vivo* and *in vitro* proliferation of NSCLC cells was unaffected by the presence of CXCL5. While a significant correlation exists between CXCL5 and tumor-derived angiogenesis, tumor growth, and metastases, CXCL5 depletion does not completely inhibit tumor growth (Arenberg *et al.*, 1998). This displays that the angiogenic activity of NSCLC tumors is related to many overlapping and potentially redundant factors acting in a parallel or serial manner. Furthermore, when all ELR+ CXC chemokines are evaluated in human NSCLC, it appears that they correlate with patient mortality (Chen *et al.*, 2003; White *et al.*, 2003).

Prostate-cancer tumorigenesis and metastases are dependent on angiogenesis (Fregene *et al.*, 1993; Bostwick and Iczkowski, 1998). Serum levels of CXCL8 have been found to be markedly elevated in patients with prostate cancer. These levels are highly correlated with the stage of the disease and have been determined to be a variable that is independent of the ratio of free to total (free:total) prostate-specific antigen (PSA) (Fregene *et al.*, 1993). In fact, the combined use of f:t PSA and IL-8/CXCL8 levels was more effective in distinguishing prostate cancer from benign prostatic hypertrophy (Fregene *et al.*, 1993). This suggests that ELR+ CXC chemokines may be playing an important role in mediating prostate cancer–derived angiogenesis in support of tumorigenesis and metastases. This observation in patients has been substantiated in human/SCID mice chimeras of human prostate-cancer tumorigenesis (Moore *et al.*, 1999). Three human prostate-cancer cell lines were examined for constitutive production of angiogenic ELR+ CXC chemokines (Moore *et al.*, 1999). Tumorigenesis of the human prostate-cancer cell line, PC-3, was shown to be attributable, in part, to the production of the angiogenic CXC chemokine, CXCL8. Depletion of endogenous CXCL8 inhibited PC-3 tumor growth in SCID mice, which was entirely attributable to inhibition of PC-3 tumor-derived angiogenesis (Moore *et al.*, 1999). In contrast, the human prostate-cancer cell line, Du145, was found to utilize a different angiogenic CXC chemokine, CXCL1, to mediate tumor-derived angiogenesis (Moore *et al.*, 1999). Depletion of endogenous CXCL1, but not of CXCL8, reduced tumor growth that was directly related to attenuated angiogenic activity (Moore *et al.*, 1999). Thus, prostate-cancer cell lines can utilize distinct CXC chemokines to mediate their tumorigenic potential. Other studies have confirmed this observation in prostate-cancer models (Kim *et al.*, 2001), and similar findings have been shown in gastric carcinoma, breast, and head and neck cancer (Singh *et al.*, 1994; Cohen *et al.*, 1995; Richards *et al.*, 1997; Kitadai *et al.*, 1998; Miller *et al.*, 1998; Chen *et al.*, 1999).

Glioblastoma is a devastating tumor of the central nervous system, and its lethality approaches 80% in the first year postdiagnosis. The hallmark of these tumors is the marked presence of angiogenesis (Garkavtsev *et al.*, 2004), which suggests that it is a biomarker necessary for malignant progression of this tumor. However, the precise molecular mechanisms underlying the regulation of glioblastoma growth and angiogenesis remain to be elucidated. Garkavtsev *et al.* have recently identified a candidate tumor suppressor gene, ING4, that is involved in regulating glioblastoma tumor growth and angiogenesis (Garkavtsev *et al.*, 2004). In this study, the expression of ING4 was found to be significantly reduced in glioblastomas, as compared with normal human brain tissue, and the extent of reduction correlated with the progression from a lower to a higher grade of tumor (Garkavtsev *et al.*, 2004). Human glioblastomas that exhibited decreased expression of ING4 when engrafted into immunoincompetent mice grew markedly faster and displayed greater angiogenesis than control tumors (Garkavtsev *et al.*, 2004). The mechanism for increased tumorigenicity in glioblastomas that expressed lower levels of ING4 was related to ING4's physical ability to bind the p65 (RelA) subunit of NF-kB, impair its nuclear translocation, and subsequently inhibit transactivation of NF-kB-dependent genes (Garkavtsev *et al.*, 2004). In fact, the mechanism for the angiogenic activity of glioblastomas that expressed low levels of ING4 was CXCL8 dependent, as inhibition of CXCL8 *in vivo* markedly reduced their tumor growth and tumor-associated angiogenesis (Garkavtsev *et al.*, 2004). These results indicate that ING4 has an important role in glioblastoma pathogenesis related to ELR+ CXC chemokine expression in response to NF-kB transactivation. Furthermore, these findings link a tumor-suppressor gene to function and control of the expression of angiogenic ELR+ CXC chemokines in human tumors and provide the unique opportunity to consider targeting ELR+ CXC chemokine-mediated angiogenesis.

V. NON-ELR+ CXC CHEMOKINES WEAKEN ANGIOGENESIS ASSOCIATED WITH TUMORIGENESIS

Non-ELR+ CXC chemokines have been shown to inhibit angiogenesis in several model systems; for example, Burkitt's lymphoma cell lines form tumors in immunoincompetent mice (Gurtsevitch *et al.*, 1988). Angiogenesis is essential for tumorigenesis of these lymphomas, analogous to carcinomas. The expression of CXCL10 and CXCL9 was found to be higher in tumors that demonstrated spontaneous regression and was directly related to impaired angiogenesis (Sgadari *et al.*, 1996b). To determine whether this effect was attributable to CXCL10 or CXCL9, more virulent Burkitt's lymphoma cell lines were grown in immunoincompetent mice and subjected to intratumor inoculation with either CXCL10 or CXCL9. Both conditions resulted in marked reduction in tumor-associated angiogenesis (Sgadari *et al.*, 1996a,b, 1997; Teruya-Feldstein *et al.*, 1997). Although these CXCR3 ligands have been shown to bind to CXCR3 on mononuclear cells (Loetscher *et al.*, 1996, 1998; Luster, 1998; Qin *et al.*, 1998; Rabin *et al.*, 1999; Moser and Loetscher, 2001; Beider *et al.*, 2003), the ability of these non-ELR+ CXC chemokines to inhibit angiogenesis and induce lymphoma regression in immunoincompetent mice supports the notion that these chemokines can mediate their effects in a T-cell-independent manner.

To examine the role of CXCL10 in the regulation of angiogenesis in carcinoma, the level of CXCL10 from human surgical NSCLC tumor specimens was examined and found to be significantly higher in the tumor specimens than in normal adjacent lung tissue (Arenberg *et al.*, 1996b). The increase in CXCL10 from human NSCLC tissue was entirely attributable to the levels of CXCL10, which were higher in SCCA than in adenocarcinoma (Arenberg *et al.*, 1996b). Moreover, depletion of CXCL10 from SCCA surgical specimens resulted in augmented angiogenic activity (Arenberg *et al.*, 1996b). The marked difference in the levels and bioactivity of CXCL10 in SCCA and adenocarcinoma is clinically and pathophysiologically relevant and represents a possible mechanism for the biologic differences of these two cell-types of NSCLC. Patient survival is lower, metastatic potential is higher, and evidence of angiogenesis is greater for adenocarcinoma than for SCCA of the lung (Carney, 1988; Minna, 1991; Yuan *et al.*, 1995). These studies were extended to a SCID mouse system to examine the effect of CXCL10 on human NSCLC cell line tumor growth in a T- and B-cell independent manner. SCID mice were inoculated with either adenocarcinoma or SCCA cell lines (Arenberg *et al.*, 1996b). The production of CXCL10 from adenocarcinoma and SCCA tumors was inversely correlated with tumor growth (Arenberg *et al.*, 1996b). However, CXCL10 levels were significantly higher in the SCCA than in adenocarcinoma tumors (Arenberg *et al.*, 1996b). The appearance of spontaneous lung metastases in SCID mice bearing adenocarcinoma tumors occurred after CXCL10 levels from either the primary tumor or the plasma had reached a nadir (Arenberg *et al.*, 1996b). In subsequent experiments, SCID mice bearing SCCA tumors were depleted of CXCL10, whereas animals bearing adenocarcinoma tumors were treated with intratumor CXCL10 (Arenberg *et al.*, 1996b). Depletion of CXCL10 in SCCA tumors resulted in an increase in their size (Arenberg *et al.*, 1996b). In contrast, reconstitution of intratumor CXCL10 in adenocarcinoma tumors reduced both their size and their metastatic potential, which were unrelated to infiltrating neutrophils or mononuclear cells (e.g., macrophages or NK cells) and directly attributable to a reduction in tumor-associated angiogenesis (Arenberg *et al.*, 1996b). Similar strategies have been found for CXCL10 in melanoma by way of a gene-therapeutic strategy (Feldman *et al.*, 2002).

Similarly to CXCL10, CXCL9 plays a significant role in regulating angiogenesis of NSCLC. CXCL9 levels in human specimens of NSCLC were not significantly different from those found in normal lung tissue (Addison *et al.*, 2000a). However, these results suggested that the increased expression of ELR+ CXC chemokines and other angiogenic factors found in these tumors were not counter-regulated by a concomitant increase in the expression of the angiostatic CXC chemokine, CXCL9. Thus, this imbalance could promote a microenvironment that promotes angiogenesis. To alter this imbalance, studies were performed to overexpress CXCL9 by three different strategies, including gene transfer (Addison *et al.*, 2000a). These experiments resulted in the inhibition of NSCLC tumor growth and metastasis via a decrease in tumor-associated angiogenesis (Addison *et al.*, 2000a). These findings support the importance of the IFN-inducible non-ELR+ CXC chemokines in inhibiting NSCLC tumor growth by attenuation of tumor-derived angiogenesis. In addition, this study demonstrates the potential efficacy of gene therapy as an alternative means to deliver and overexpress a potent angiostatic CXC chemokine.

VI. THE CONCEPT OF 'IMMUNOANGIOSTASIS': THE ROLE OF CXCR3/ CXCR3 LIGAND BIOLOGICAL AXIS IN MEDIATING TH1 CELL-MEDIATED IMMUNITY AND ANGIOSTASIS

While CXCR3/CXCR3 ligands inhibit angiogenesis, CXCR3 ligands represent the major chemoattractants for the recruitment of Th1 cells expressing CXCR3 during cell-mediated immunity (Loetscher *et al.*, 1996; Luster, 1998; Qin *et al.*, 1998; Rabin *et al.*, 1999; Moser and Loetscher, 2001). Thus, CXCR3 ligands play a critical role in orchestrating Th1 cytokine-induced cell-mediated immunity via the recruitment of mononuclear cells expressing CXCR3 (Loetscher *et al.*, 1996; Luster, 1998; Qin *et al.*, 1998; Rabin *et al.*, 1999; Moser and Loetscher, 2001). We have linked CXCR3/CXCR3 ligands to Th1 cytokine-induced cell-mediated immunity and inhibition of angiogenesis, which lead to suppression of tumor growth in preclinical models of NSCLC (Sharma *et al.*, 2000, 2003). Intratumoral injection of a recombinant CC chemokine, CCL21, in tumors induced complete tumor eradication in several of the treated mice (Sharma *et al.*, 2000, 2003). While CCL21-mediated antitumor responses were lymphocyte-dependent, its biological effect was entirely dependent on the spatial generation of intratumor IFN-γ and CXCR3 ligands (Sharma *et al.*, 2000, 2003). In immunocompetent mice, intratumoral CCL21 injection led to significant increases in CD4, CD8, NK, and dendritic cells infiltrating both the tumor and the draining lymph nodes, and a reduction in angiogenesis (Sharma *et al.*, 2000, 2003). CCL21-treated tumor-bearing mice demonstrated enhanced cytolytic capacity, suggesting the generation of a systemic immune response to tumor-associated antigens (TAAs). The mononuclear cell infiltration into the tumor was associated with enhanced production of Th1 cytokines and CXCR3 ligands (Sharma *et al.*, 2000, 2003). To assess the importance of Th1 cytokines and CXCR3 ligands in mediating the effects of CCL21, depletion studies of IFN-γ or CXCR3 ligands in the presence of CCL21 treatment demonstrated that CXCL9 $\geq$ CXCL10 $\geq$IFN-γ weakened the antitumor effects of CCL21 (Sharma *et al.*, 2003). These findings support the notion that CCL21-mediated antitumor response was dependent on CXCR3 ligands. These findings are similar to the previously reported study of IL-12-mediated regression of renal-cell carcinoma in a murine model, where the antitumor effect of IL-12 was lost when CXCR3 ligands were depleted (Tannenbaum *et al.*, 1998). These findings provide evidence that CXCR3/CXCR3 ligand biology is critical for the development of antitumor immunity and inhibition of angiogenesis relevant to a variety of tumors.

On the basis of the ability of CXCR3/CXCR3 ligands to promote Th1-dependent immunity and at the same time inhibit angiogenesis, we have coined the term 'immunoangiostasis' for their biological role in promoting tumor regression (Fig. 2). Although the concept of immunoangiostasis (i.e., promoting Th1 immunity via Th1 mononuclear-cell recruitment and at the same time inducing angiostasis) may seem a paradox, the precedent for this concept exists: the host response to *Mycobacterium tuberculosis*. *M. tuberculosis* is an aerobic bacillus that requires the full development of host-mediated Th1-induced cell-mediated immunity to contain the microorganism. The response is characterized by granulomas with a rim of mononuclear cells, epithelioid cells, giant cells, fibroblasts, and endothelial cells surrounding a central area of caseating necrosis, which is devoid of vasculature—the concept angiostasis. The mononuclear-cell response in the rim of the granuloma, together with subsequent processing of *M. tuberculosis* antigen, leads to adaptive immunity—the concept of immunity. The microenvironment within avascular caseating necrosis is angiostatic and hypoxic in nature, which induces dormancy of *M. tuberculosis*. Further organizing fibrosis of the granuloma contains the microbe in a dormant state. Therefore, the full development of immunoangiostasis provides an optimal host response to promote eradication and/or dormancy of this aerobic microbe—our analogy in promoting tumor-related immunoangiostasis.

If CXCR3/CXCR3 ligands promote Th1 cytokine-induced cell-mediated immune response and angiostasis, then why is immunotherapy effective in only a minority of patients with certain cancers? Is it because not all cancer cells are immunogenic, or is due to patient differences in development of tumor-related immunity, ineffective temporal/spatial immunotherapy, or both in order to fully develop immunoangiostasis in cancer? We hypothesize that the reason that immunotherapy is effective in only a minority of patients with cancer is related to the systemic nature of the administration of the therapy, not necessarily the therapy itself. For example, if systemic IL-2 is given alone or with another systemically administered immune-enhancing cytokine, the effect will be systemic activation and induction of increased circulating levels of CXCR3 ligands, along with variable expression of CXCR3 on activated T-cells and NK cells in circulation. This scenario leads to the presence of activated cells in circulation that have *no* chemotactic gradient to specifically instruct them where to traffic and extravasate (e.g., intratumor). The result of systemic immunotherapy alone leads to a hit-or-miss scenario, and only a minority of patients respond with appropriate tumor infiltration of activated mononuclear cells, durable TAA immunity, and induction of an angiostatic microenvironment in this case—the full development of immunoangiostasis.

If immunoangiostasis is critical for the success of immunotherapy in cancer, then how do we temporally and spatially optimize this biology?

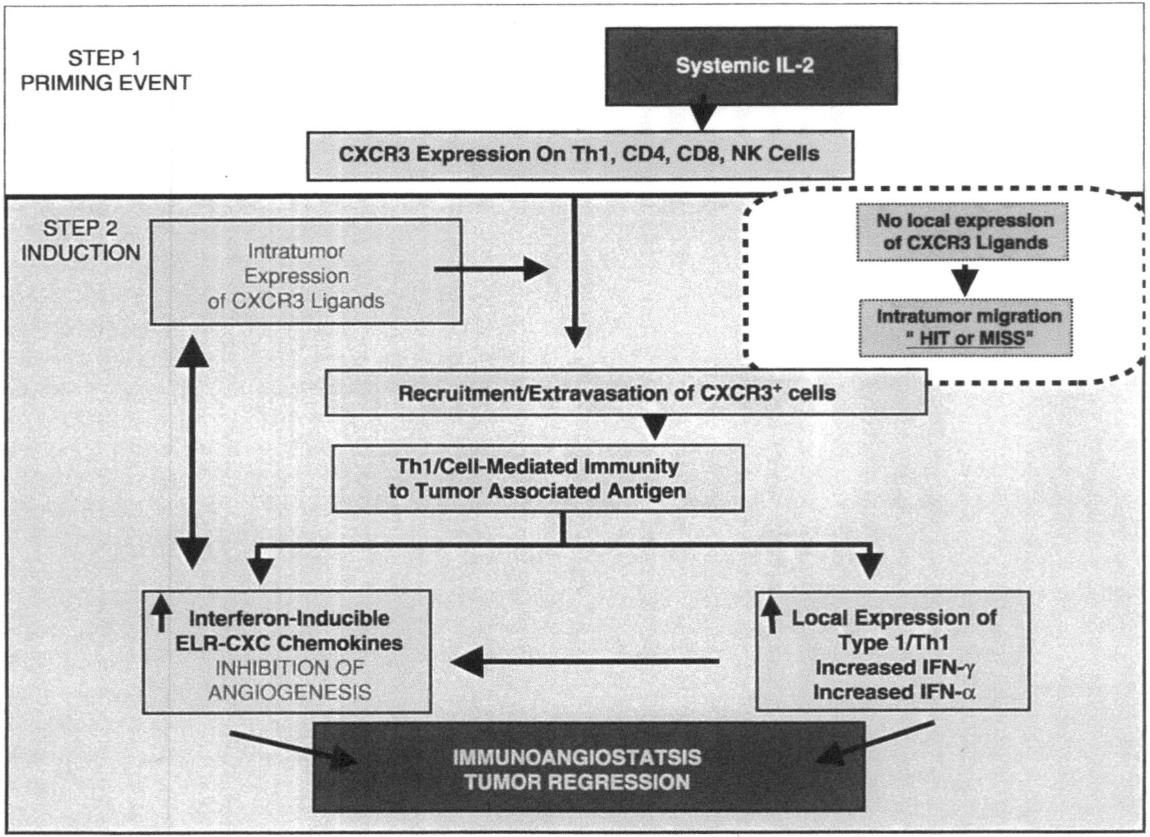

FIGURE 2 (*For legend, see opposite page.*)

We hypothesize that this can be achieved in two steps that may be temporally and spatially separated. The first step is dependent on systemic 'priming' of circulating CD4, CD8, and NK cells to optimally express CXCR3 (i.e., IL-2). Once this has been temporally achieved, step two would be to spatially (i.e., locally or intratumorally) induce the expression of CXCR3 ligands to take advantage of a maximal chemotactic gradient within the tumor and to fully optimize immunoangiostasis. The mechanism for inducing the expression of intratumor CXCR3 ligands can be direct (i.e., expression of CXCR3 ligands) or indirect (i.e., expression of IFN-α or -γ). The paradigm for effective immunoangiostasis would be to enhance selective and specific extravasation of Th1 cells into the tumor; enhance Th1-mediated immunity *in situ* (i.e., increase immunity to TAAs); increase the expression of local IFNs; further augment the local expression of CXCR3 ligands within the tumor; amplify further Th1-mediated immunity; and promote CXCR3 ligand–mediated angiostasis (Fig. 2).

VII. EVIDENCE THAT CHEMOKINES ARE INVOLVED IN TUMOR-CELL INVASION

Tumor-cell invasion through ECM and entry into the circulation is dependent on cellular adeherence, motility, ECM, and basement membrane degradation. Invasive tumor cells have the ability to secrete a variety of enzymes, including metalloproteinases (MMPs), serine proteinases, and cysteine proteinases. The generation of these proteinases allows for the migration of tumor cells through the ECM and penetration through the basement membrane and entry into the circulation. Chemokine activation of tumor cells is important in this process. The expression of CXCL8 by human melanoma cells upregulates MMP-2 activity and increases tumor growth and metastasis (Luca *et al.*, 1997). Luca and associates demonstrated that the expression of CXCL8 by human melanoma cell lines directly correlates with their metastatic potential. Although the angiogenic properties of

FIGURE 2 The two steps of priming and induction are necessary for the full induction of the concept of immunoangiostasis. Systemic IL-2 promotes the expression of CXCR3 on mononuclear cells; however, without the establishment of a chemotactic gradient from the circulation to within the tumor, mononuclear cells have no chemotactic gradient to extravasate into the tumor, the concept of "hit or miss." Induction of the local expression of CXCR3 ligands within the tumor leads to the establishment of a chemotactic gradient for mononuclear cells expressing CXCR3 and concomitantly promotes angiostasis. The combined steps of systemic priming for mononuclear cell expression of CXCR3 and local induction of CXCR3 ligand expression fully establish immunoangiostasis.

CXCL8 may be one explanation for this effect, they used nonmetastatic melanoma cells with negligible levels of IL-8/CXCL8 and transfected them to overexpress CXCL8 (Luca *et al.*, 1997). The overexpression of CXCL8 enhanced the tumorigenicity and metastatic potential of the melanoma cells *in vivo* (Luca *et al.*, 1997). In conjunction with this change in biological behavior, the CXCL8-transfected cells displayed upregulation in MMP-2 (Luca *et al.*, 1997). This expression was accompanied by heightened collagenase activity and increased invasiveness *in vitro* (Luca *et al.*, 1997). Moreover, they found that the effect of CXCL8 activation was at the level of induction of the promoter of the MMP-2 gene, suggesting that CXCL8 was involved in MMP-2 gene transcription (Luca *et al.*, 1997). These findings support the hypothesis that CXCL8 can active tumor cell–derived type IV collagenase activity that can lead to enhanced tumor-cell invasion into the host stroma and increase metastatic potential.

These findings have been further substantiated in prostate cancer, where CXCL8 expression regulates tumorigenicity and metastases in androgen-independent prostate cancer (Inoue *et al.*, 2000). Inoue and colleagues have found that prostate-cancer cell lines that overexpress CXCL8 are associated with a highly metastatic phenotype (Inoue *et al.*, 2000). They further evaluated this biology *in vivo* by implanting these human tumor cells in athymic mice (Inoue *et al.*, 2000). They transfected the overexpressing CXCL8 cell lines with full-sequence antisense CXCL8 cDNA, transfected the underexpressing CXCL8 cell lines to overexpress CXCL8, and engrafted these cells in immunoincompetent mice (Inoue *et al.*, 2000). The overexpression of CXCL8 *in vitro* resulted in the upregulation of MMP-9 in these cells (Inoue *et al.*, 2000). The effect was at the levels of mRNA, protein, and biological function; heightened collagenase activity resulted in increased invasiveness of the prostate cancer cell lines *in vitro* (Inoue *et al.*, 2000). Orthotopic implantation of the human prostate-cancer cells overexpressing CXCL8 that normally, in and of themselves, are not tumorigenic or metastatic in behavior, caused them to become highly tumorigenic and metastatic with associated increased angiogenesis, whereas the cells transfected with antisense CXCL8 were inhibited in their growth and metastatic potential (Inoue *et al.*, 2000). These findings have been further corroborated by Kim and associates (Kim *et al.*, 2001), who found that the expression of CXCL8 correlated with angiogenesis, tumorigenicity, and metastasis of human prostate cancer cells implanted orthotopically in immunoincompetent mice. These findings suggest that angiogenic ELR+ CXC chemokines like CXCL8 play a multifunctional role in aiding tumor-cell invasion by augmenting their local angiogenic environment and upregulating the expression of MMPs to aid tumor-cell invasion and entry into the circulation.

VIII. EVIDENCE THAT CHEMOKINES REGULATE THE PATTERN OF ORGAN-SPECIFIC METASTASIS OF CANCER

As described in the opening sections, in 1889, Paget described the theory of tumor cell metastasis based on the concept of a "seed" (the tumor cell) and the "soil" (the specific organ) for the nonrandom visceral metastases of breast cancer. While this theory has been debated, experimental data have demonstrated that sites of metastasis are determined not only by the characteristics of neoplastic cells but by the microenvironment of the specific organ (Hart and Fidler, 1980). However, the specific mechanisms that actually promote organ-specific metastasis have not been fully elucidated. Breast cancer, as characterized by the original observation of Paget (1889), has a distinct metastatic pattern preferentially involving the regional lymph nodes, bone marrow, lung, and liver. This distinctive metastatic pattern is associated with a number of cancers. Müller, Homey, and colleagues (Muller *et al.*, 2001) have provided new insight into potential mechanisms related to organ-specific metastasis of breast-cancer cells. They found that among all known chemokine receptors, CXCR4 and CCR7 are highly expressed in human breast-cancer cells, malignant breast tumors, and metastases. Their ligands for these receptors (CXCL12 for CXCR4 and CCL21 for CCR7) exhibited peak levels of expression in organs that are preferential destinations of breast-cancer metastasis (Muller *et al.*, 2001). In breast-cancer cells, signaling through CXCR4 or CCR7 mediated actin polymerization and pseudopodia formation and, subsequently, induced chemotactic and invasive responses at the local level (Muller *et al.*, 2001). Moreover, neutralization of CXCL12/CXCR4 interactions significantly inhibited metastasis of breast-cancer cells to regional lymph nodes and lung (Muller *et al.*, 2001). While this study suggested that the CXCL12/CXCR4 biological axis was important in mediating organ-specific metastases of breast cancer, this concept has remained somewhat controversial, as the findings did not take into consideration other potential biological effects of CXCL12/CXCR4 in the tumor.

For example, CXCL12 is a non-ELR+ CXC chemokine, but via CXCR4 it has been implicated in the promotion of angiogenesis (Salcedo *et al.*, 1999; Kijowski *et al.*, 2001; Bachelder *et al.*, 2002; Salcedo and Oppenheim, 2003). This has led to the speculation that the predominant function of this ligand/receptor pair in tumorigenesis is due to its angiogenic effect, not necessarily to its potential for mediating organ-specific metastases. However, in order for the biological axis of CXCL12/CXCR4 to mediate tumor-associated angiogenesis, both the ligand and the receptor should be temporally and spatially present within the tumor. Schrader and colleagues (2002) demonstrated in both renal-cell carcinoma cell lines and actual patient specimens that CXCR4 is expressed predominantly by the tumor cells and that its ligand CXCL12 is essentially absent within the tumor. These findings have been further substantiated in human breast cancer and NSCLC tumor specimens, in which CXCR4 was found expressed on the tumor cells, and does not mediate tumor-associated angiogenesis *in vivo* (Muller *et al.*, 2001; Phillips *et al.*, 2003). The studies demonstrated that when animals bearing breast or NSCLC tumors were treated with either neutralizing anti-CXCL12 or anti-CXCR4 antibodies, there was no change in the size of the primary tumor, nor was there any evidence of a decline in primary tumor-associated angiogenesis; A. Zlotnik, personal communication); however, there was a marked attenuation of tumor metastases in an organ-specific manner (Muller *et al.*, 2001; Phillips *et al.*, 2003). These studies support the notion that CXCL12/CXCR4 biology mediates metastases of the tumor cells in an angiogenesis-independent manner.

An explanation for the disparity between the tumor studies of CXCL12/CXCR4-mediated angiogenesis *in vivo* and those *in vitro* is that tumor cells expressing CXCR4 are themselves able to "out-compete" endothelial cells for CXCL12 if it is present (Fig. 3). In support of this contention, classical angiogenic factors are elevated in human tumors, whereas CXCL12 is not (Smith *et al.*, 1994; Arenberg., 1996a,b; Belperio *et al.*, 2000; Schrader *et al.*, 2002; Phillips *et al.*, 2003). Moreover, the depletion of classical angiogenic factors *in vivo* results in a net reduction of angiogenesis and a consequent reduction in primary tumor size and metastatic potential (Smith *et al.*, 1994; Arenberg *et al.*, 1996a,b; Belperio *et al.*, 2000). These findings suggest a disparity between the function for CXCL12 and that for classical angiogenic factors; angiogenic factors promote metastasis through their mediation of angiogenesis, whereas CXCL12 promotes metastasis in an angiogenesis-independent manner via CXCR4-dependent tumor-cell migration. This concept supports the notion that expression of CXCR4 on tumor cells may represent a critical biomarker for their propensity to metastasize. Therefore, further understanding the molecular mechanisms that are involved in the regulation of CXCR4 expression on tumor cells could lead to targets to modify expression of this receptor, which impacts metastases.

Recently, hypoxia and more specifically hypoxia-inducible factor-1α (HIF-1α) has been found to be a critical transcription factor for gene expression of CXCR4 on a variety of cells, including tumor cells (Schioppa *et al.*, 2003; Staller *et al.*, 2003). Moreover, the von Hippel–Lindau (VHL) tumor-suppressor gene can negatively regulate the expression of CXCR4, owing to its capacity to target HIF-1α for degradation under normoxic conditions (Schioppa *et al.*, 2003; Staller *et al.*, 2003). This process may be

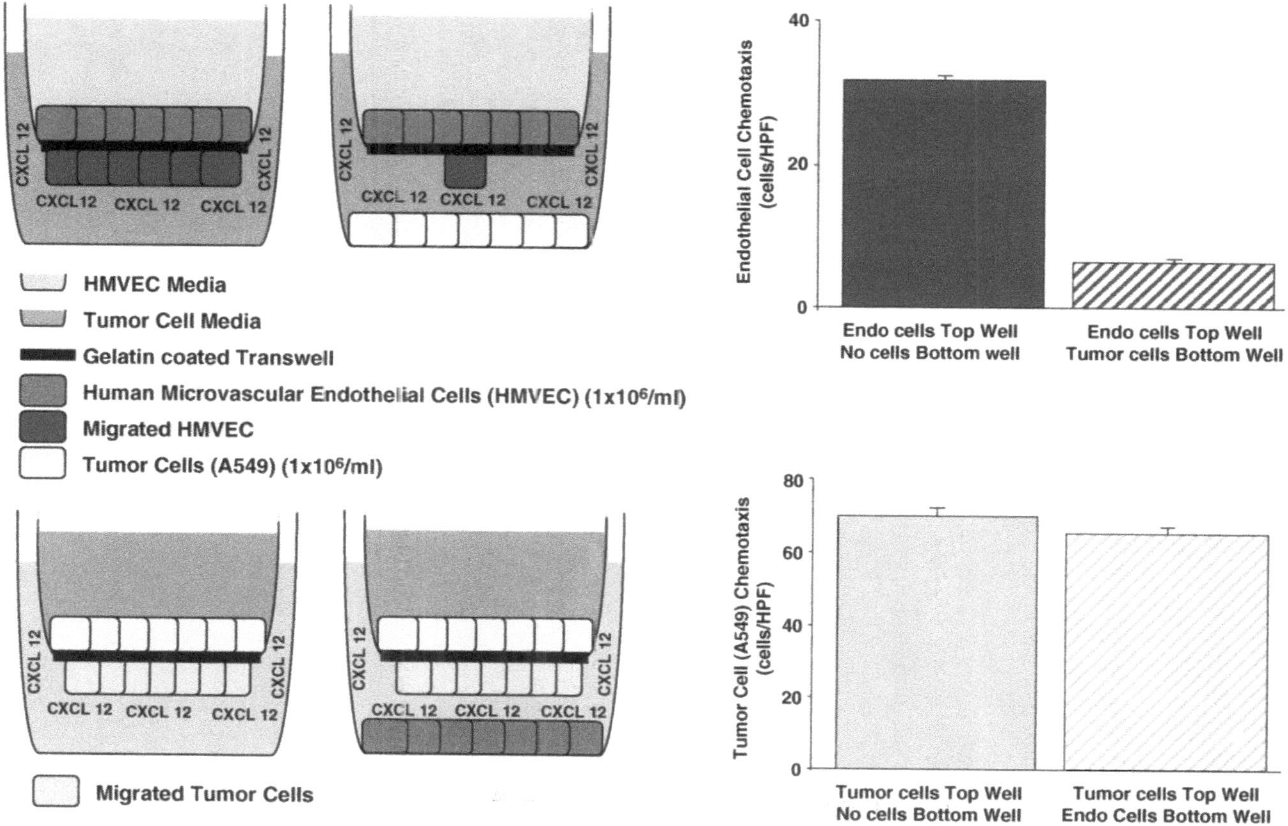

FIGURE 3 Transwell chemotaxis assay analogous to the tumor microenvironment. Chemotaxis of tumor and endothelial cells expressing CXCR4 in response to CXCL12. The presence of the tumor outcompetes endothelial cells for CXCL12-mediated cellular migration. (See Color Insert.)

TABLE II

CXC Chemokines Are Pleiotropic Cytokines in Cancer Biology

CXC chemokines are responsible for the following:
• Preneoplastic-to-neoplastic transformation.
• Regulation of angiogenesis.
• Recruitment of tumor-associated leukocytes and induction of immunoangiostasis.
• Promotion of tumor-cell invasion.
• Regulation of the pattern of organ-specific metastasis.

suppressed under hypoxic conditions in cells allowing HIF-1α-dependent induction of CXCR4 expression (Schioppa *et al.*, 2003; Staller *et al.*, 2003). In contrast, under normoxic conditions, receptor-activated PI3K/PTEN/AKT/mTor and ERK1/2/MAPkinase pathways can augment the expression of HIF-1α in a posttranscriptional manner (Zhong *et al.*, 2000; Semenza, 2002, 2003). However, the activation of these pathways has not been linked to the expression of CXCR4 on tumor cells. Further understanding of the molecular link between receptor- and hypoxia-induced HIF-1α and CXCR4 expression may spur the discovery of novel means to intervene and help reduce metastases of a variety of cancers.

IX. CONCLUSION

Although CXC chemokine biology was originally thought to be restricted to recruitment of subpopulations of leukocytes, it has become increasingly clear that these cytokines can display pleiotropic biology-mediating effects that go beyond their originally described function (Table II). There is no better human disease by which to study this diversity of function than tumor biology. Chemokines have an autocrine, a paracrine, and a hormonal role at every level related to primary tumor growth, invasion, and organ-specific metastases. The understanding of this expanded role of CXC chemokines in tumor biology will open new doors for novel therapeutic intervention

Acknowledgments

This work was supported, in part, by NIH grants HL66027, CA87879, P50 CA90388, and P50 HL67665 (RMS), HL04493 (JAB), and HL03906 and P50 HL67665 (MPK).

References

Addison, C. L., Arenberg, D. A., Morris, S. B., Xue, Y. Y., Burdick, M. D., Mulligan, M. S., Iannettoni, M. D., and Strieter, R. M. (2000a). The CXC chemokine, monokine induced by interferon-gamma, inhibits non-small cell lung carcinoma tumor growth and metastasis. *Hum. Gene Ther.* **11**, 247–261.

Addison, C. L., Daniel, T. O., Burdick, M. D., Liu, H., Ehlert, J. E., Xue, Y. Y., Buechi, L., Walz, A., Richmond, A., and Strieter, R. M. (2000b). The CXC chemokine receptor 2, CXCR2, is the putative receptor for ELR(+) CXC chemokine-induced angiogenic activity. *J. Immunol.* **165,** 5269–5277.

Arenberg, D. A., Kunkel, S. L., Burdick, M. D., Polverini, P. J., and Strieter, R. M. (1995). Treatment with anti-IL-8 inhibits non-small cell lung cancer tumor growth (Meeting abstract). *J. Investig. Med.* **43,** 479.

A. Arenberg, D. A., Kunkel, S. L., Polverini, P. J., Glass, M., Burdick, M. D., and Strieter, R. M. (1996a). Inhibition of interleukin-8 reduces tumorigenesis of human non-small cell lung cancer in SCID mice. *J. Clin. Invest.* **97,** 2792–2802.

Arenberg, D. A., Kunkel, S. L., Polverini, P. J., Morris, S. B., Burdick, M. D., Glass, M. C., Taub, D. T., Iannettoni, M. D., Whyte, R. I., and Strieter, R. M. (1996b). Interferon-gamma-inducible protein 10 (IP-10) is an angiostatic factor that inhibits human non-small cell lung cancer (NSCLC) tumorigenesis and spontaneous metastases. *J. Exp. Med.* **184,** 981–992.

Arenberg, D. A., Keane, M. P., DiGiovine, B., Kunkel, S. L., Morris, S. B., Xue, Y. Y., Burdick, M. D., Glass, M. C., Iannettoni, M. D., and Strieter, R. M. (1998). Epithelial-neutrophil activating peptide (ENA-78) is an important angiogenic factor in non-small cell lung cancer. *J. Clin. Invest.* **102,** 465–472.

Arvanitakis, L., Geras-Raaka, E., Varma, A., Gershengorn, M. C., and Cesarman, E. (1997). Human herpesvirus KSHV encodes a constitutively active G-protein-coupled receptor linked to cell proliferation. *Nature* **385,** 347–350.

Bachelder, R. E., Wendt, M. A., and Mercurio, A. M. (2002). Vascular endothelial growth factor promotes breast carcinoma invasion in an autocrine manner by regulating the chemokine receptor CXCR4. *Cancer Res.* **62,** 7203–7206.

Bais, C., Santomasso, B., Coso, O., Arvanitakis, L., Raaka, E. G., Gutkind, J. S., Asch, A. S., Cesarman, E., Gershengorn, M. C., Mesri, E. A., and Gerhengorn, M. C. (1998). G-protein-coupled receptor of Kaposi's sarcoma-associated herpesvirus is a viral oncogene and angiogenesis activator [published erratum appears in *Nature* (1998) **392** (6672), 210]. *Nature* **391,** 86–89.

Balkwill, F. (1998). The molecular and cellular biology of the chemokines. *J. Viral. Hepat.* **5,** 1–14.

Beider, K., Nagler, A., Wald, O., Franitza, S., Dagan-Berger, M., Wald, H., Giladi, H., Brocke, S., Hanna, J., Mandelboim, O., Darash-Yahana, M., Galun, E., and Peled, A. (2003). Involvement of CXCR4 and IL-2 in the homing and retention of human NK and NK T cells to the bone marrow and spleen of NOD/SCID mice. *Blood* **102,** 1951–1958.

Belperio, J. A., Keane, M. P., Arenberg, D. A., Addison, C. L., Ehlert, J. E., Burdick, M. D., and Strieter, R. M. (2000). CXC chemokines in angiogenesis. *J. Leukoc. Biol.* **68,** 1–8.

Borgstrom, P., Discipio, R., and Maione, T. E. (1998). Recombinant platelet factor 4, an angiogenic marker for human breast carcinoma. *Anticancer Res.* **18,** 4035–4041.

Bostwick, D. G., and Iczkowski, K. A. (1998). Microvessel density in prostate cancer: Prognostic and therapeutic utility. *Semin. Urol. Oncol.* **16,** 118–123.

Burdick, M. D., Murray, L. A., Keane, M. P., Xue, Y. Y., Zisman, D. A., Belperio, J. A., and Strieter, R. M. (2005). CXCL11 attenuates bleomycin-induced pulmonary fibrosis via inhibition of vascular remodeling. *Am. J. Respir. Crit. Care. Med.*

Burger, M., Burger, J. A., Hoch, R. C., Oades, Z., Takamori, H., and Schraufstatter, I. U. (1999). Point mutation causing constitutive signaling of CXCR2 leads to transforming activity similar to Kaposi's sarcoma herpesvirus-G protein-coupled receptor. *J. Immunol.* **163,** 2017–2022.

Carney, D. N. (1988). Cancers of the lungs. *In* "Pulmonary Diseases and Disorders" (A. P. Fishman, ed.), pp. 1885–2068. McGraw-Hill, New York.

Chen, J. J., Yao, P. L., Yuan, A., Hong, T. M., Shun, C. T., Kuo, M. L., Lee, Y. C., and Yang, P. C. (2003). Up-regulation of tumor interleukin-8 expression by infiltrating macrophages: Its correlation with tumor angiogenesis and patient survival in non-small cell lung cancer. *Clin. Cancer Res.* **9,** 729–737.

Chen, Z., Malhotra, P. S., Thomas, G. R., Ondrey, F. G., Duffey, D. C., Smith, C. W., Enamorado, I., Yeh, N. T., Kroog, G. S., Rudy, S., McCullagh, L., Mousa, S., Quezado, M., Herscher, L. L., and Van Waes, C. (1999). Expression of proinflammatory and proangiogenic cytokines in patients with head and neck cancer. *Clin. Cancer Res.* **5,** 1369–1379.

Cohen, R. F., Contrino, J., Spiro, J. D., Mann, E. A., Chen, L. L., and Kreutzer, D. L. (1995). Interleukin-8 expression by head and neck squamous cell carcinoma. *Arch. Otolaryngol. Head Neck Surg.* **121,** 202–209.

Couty, J. P., and Gershengorn, M. C. (2004). Insights into the viral G protein-coupled receptor encoded by human herpesvirus type 8 (HHV-8). *Biol. Cell* **96,** 349–354.

Devalaraja, R. M., Nanney, L. B., Qian, Q., Du, J., Yu, Y., Devalaraja, M. N., and Richmond, A. (2000). Delayed wound healing in CXCR2 knockout mice. *J. Invest. Dermatol.* **115,** 234–244.

Dong, G., Chen, Z., Li, Z. Y., Yeh, N. T., Bancroft, C. C., and Van Waes, C. (2001). Hepatocyte growth factor/scatter factor-induced activation of MEK and PI3K signal pathways contributes to expression of proangiogenic cytokines interleukin-8 and vascular endothelial growth factor in head and neck squamous cell carcinoma. *Cancer Res.* **61,** 5911–5918.

Ehlert, J. E., Addison, C. A., Burdick, M. D., Kunkel, S. L., and Strieter, R. M. (2004). Identification and partial characterization of a variant of human CXCR3 generated by posttranscriptional exon skipping. *J. Immunol.* **173,** 6234–6240.

Feldman, A. L., Friedl, J., Lans, T. E., Libutti, S. K., Lorang, D., Miller, M. S., Turner, E. M., Hewitt, S. M., and Alexander, H. R. (2002). Retroviral gene transfer of interferon-inducible protein 10 inhibits growth of human melanoma xenografts. *Int. J. Cancer* **99,** 149–153.

Folkman, J. (2002). Role of angiogenesis in tumor growth and metastasis. *Semin. Oncol.* **29,** 15–18.

Folkman, J. (2003). Angiogenesis inhibitors: A new class of drugs. *Cancer Biol. Ther.* **2,** S127–S133.

Frederick, M. J., Henderson, Y., Xu, X., Deavers, M. T., Sahin, A. A., Wu, H., Lewis, D. E., El-Naggar, A. K., and Clayman, G. L. (2000). *In vivo* expression of the novel CXC chemokine BRAK in normal and cancerous human tissue. *Am. J. Pathol.* **156,** 1937–1950.

Fregene, T. A., Khanuja, P. S., Noto, A. C., Gehani, S. K., Van Egmont, E. M., Luz, D. A., and Pienta, K. J. (1993). Tumor-associated angiogenesis in prostate cancer. *Anticancer Res.* **13,** 2377–2381.

Garkavtsev, I., Kozin, S. V., Chernova, O., Xu, L., Winkler, F., Brown, E., Barnett, G. H., and Jain, R. K. (2004). The candidate tumour suppressor protein ING4 regulates brain tumour growth and angiogenesis. *Nature* **428,** 328–332.

Gawrychowski, K., Skopinska-Rozewska, E., Barcz, E., Sommer, E., Szaniawska, B., Roszkowska-Purska, K., Janik, P., and Zielinski, J. (1998). Angiogenic activity and interleukin-8 content of human ovarian cancer ascites. *Eur. J. Gynaecol. Oncol.* **19,** 262–264.

Gengrinovitch, S., Greenberg, S. M., Cohen, T., Gitay-Goren, H., Rockwell, P., Maione, T. E., Levi, B. Z., and Neufeld, G. (1995). Platelet factor-4 inhibits the mitogenic activity of VEGF121 and VEGF165 using several concurrent mechanisms. *J. Biol. Chem.* **270,** 15059–15065.

Gentilini, G., Kirschbaum, N. E., Augustine, J. A., Aster, R. H., and Visentin, G. P. (1999). Inhibition of human umbilical vein endothelial cell proliferation by the CXC chemokine, platelet factor 4 (PF4), is associated with impaired downregulation of p21(Cip1/WAF1). *Blood* **93,** 25–33.

Geras-Raaka, E., Arvanitakis, L., Bais, C., Cesarman, E., Mesri, E. A., and Gershengorn, M. C. (1998a). Inhibition of constitutive signaling of Kaposi's sarcoma-associated herpesvirus G protein-coupled receptor by protein kinases in mammalian cells in culture. *J. Exp. Med.* **187,** 801–806.

Geras-Raaka, E., Varma, A., Ho, H., Clark-Lewis, I., and Gershengorn, M. C. (1998b). Human interferon-gamma-inducible protein 10 (IP-10) inhibits constitutive signaling of Kaposi's sarcoma-associated herpesvirus G protein-coupled receptor. *J. Exp. Med.* **188,** 405–408.

Gershengorn, M. C., Geras-Raaka, E., Varma, A., and Clark-Lewis, I. (1998). Chemokines activate Kaposi's sarcoma-associated herpesvirus G protein-coupled receptor in mammalian cells in culture [see comments]. *J. Clin. Invest.* **102,** 1469–1472.

Guo, H. G., Sadowska, M., Reid, W., Tschachler, E., Hayward, G., and Reitz, M. (2003). Kaposi's sarcoma-like tumors in a human herpesvirus 8 ORF74 transgenic mouse. *J. Virol.* **77,** 2631–2639.

Gupta, S. K., and Singh, J. P. (1994). Inhibition of endothelial cell proliferation by platelet factor-4 involves a unique action on S phase progression. *J. Cell. Biol.* **127,** 1121–1127.

Gurtsevitch, V. E., O'Conor, G. T., and Lenoir, G. M. (1988). Burkitt's lymphoma cell lines reveal different degrees of tumorigenicity in nude mice. *Int. J. Cancer* **41,** 87–95.

Hansell, P., Maione, T. E., and Borgstrom, P. (1995). Selective binding of platelet factor 4 to regions of active angiogenesis *in vivo*. *Am. J. Physiol.* **269,** H829–H836.

Hart, I. R., and Fidler, I. J. (1980). Role of organ selectivity in the determination of metastatic patterns of B16 melanoma. *Cancer Res.* **40,** 2281–2287.

Heidemann, J., Ogawa, H., Dwinell, M. B., Rafiee, P., Maaser, C., Gockel, H. R., Otterson, M. F., Ota, D. M., Lugering, N., Domschke, W., and Binion, D. G. (2003). Angiogenic effects of interleukin-8 (CXCL8) in human intestinal microvascular endothelial cells are mediated by CXCR2. *J. Biol. Chem.* **278,** 8508–8515.

Hirata, A., Ogawa, S., Kometani, T., Kuwano, T., Naito, S., Kuwano, M., and Ono, M. (2002). ZD1839 (Iressa) induces antiangiogenic effects through inhibition of epidermal growth factor receptor tyrosine kinase. *Cancer Res.* **62,** 2554–2560.

Houck, K. A., Ferrara, N., Winer, J., Cachianes, G., Li, B., and Leung, D. W. (1991). The vascular endothelial growth factor family: Identification of a fourth molecular species and characterization of alternative splicing of RNA. *Mol. Endocrinol.* **5,** 1806–1814.

Houck, K. A., Leung, D. W., Rowland, A. M., Winer, J., and Ferrara, N. (1992). Dual regulation of vascular endothelial growth factor bioavailability by genetic and proteolytic mechanisms. *J. Biol. Chem.* **267,** 26031–26037.

Inoue, K., Slaton, J. W., Eve, B. Y., Kim, S. J., Perrotte, P., Balbay, M. D., Yano, S., Bar-Eli, M., Radinsky, R., Pettaway, C. A., and Dinney, C. P. (2000). Interleukin-8 expression regulates tumorigenicity and metastases in androgen-independent prostate cancer. *Clin. Cancer Res.* **6,** 2104–2119.

Jouan, V., Canron, X., Alemany, M., Caen, J. P., Quentin, G., Plouet, J., and Bikfalvi, A. (1999). Inhibition of *in vitro* angiogenesis by platelet factor-4-derived peptides and mechanism of action. *Blood* **94,** 984–993.

Keane, M. P., Belperio, J. A., Xue, Y. Y., Burdick, M. D., and Strieter, R. M. (2004). Depletion of CXCR2 inhibits tumor growth and angiogenesis in a murine model of lung cancer. *J. Immunol.* **172,** 2853–2860.

Kijowski, J., Baj-Krzyworzeka, M., Majka, M., Reca, R., Marquez, L. A., Christofidou-Solomidou, M., Janowska-Wieczorek, A., and Ratajczak, M. Z. (2001). The SDF-1-CXCR4 axis stimulates VEGF secretion and activates integrins but does not affect proliferation and survival in lymphohematopoietic cells. *Stem Cells* **19,** 453–466.

Kim, S. J., Uehara, H., Karashima, T., McCarty, M., Shih, N., and Fidler, I. J. (2001). Expression of interleukin-8 correlates with angiogenesis, tumorigenicity, and metastasis of human prostate cancer cells implanted orthotopically in nude mice. *Neoplasia* **3,** 33–42.

Kitadai, Y., Haruma, K., Sumii, K., Yamamoto, S., Ue, T., Yokozaki, H , Yasui, W., Ohmoto, Y., Kajiyama, G., Fidler, I. J., and Tahara, E. (1998). Expression of interleukin-8 correlates with vascularity in human gastric carcinomas. *Am. J. Pathol.* **152,** 93–100.

Lasagni, L., Francalanci, M., Annunziato, F., Lazzeri, E., Giannini, S., Cosmi, L., Sagrinati, C., Mazzinghi, B., Orlando, C., Maggi, E., Marra, F., Romagnani, S., Serio, M., and Romagnani, P. (2003). An alternatively spliced variant of CXCR3 mediates the inhibition of endothelial cell growth induced by IP-10, Mig, and I-TAC, and acts as functional receptor for platelet factor 4. *J. Exp. Med.* **197,** 1537–1549.

Levine, L., Lucci, J. A., 3rd., Pazdrak, B., Cheng, J. Z., Guo, Y. S., Townsend, C. M., Jr., and Hellmich, M. R. (2003). Bombesin stimulates nuclear factor kappa B activation and expression of proangiogenic factors in prostate cancer cells. *Cancer Res.* **63,** 3495–3502.

Li, A., Dubey, S., Varney, M. L., Dave, B. J., and Singh, R. K. (2003). IL-8 directly enhanced endothelial cell survival, proliferation, and matrix metalloproteinases production and regulated angiogenesis. *J. Immunol.* **170,** 3369–3376.

Loetscher, M., Gerber, B., Loetscher, P., Jones, S. A., Piali, L., Clark-Lewis, I., Baggiolini, M., and Moser, B. (1996). Chemokine receptor specific for IP10 and mig: Structure, function, and expression in activated T-lymphocytes. *J. Exp. Med.* **184,** 963–969.

Loetscher, M., Loetscher, P., Brass, N., Meese, E., and Moser, B. (1998). Lymphocyte-specific chemokine receptor CXCR3: Regulation, chemokine binding and gene localization. *Eur. J. Immunol.* **28,** 3696–3705.

Luan, J., Shattuck-Brandt, R., Haghnegahdar, H., Owen, J. D., Strieter, R., Burdick, M., Nirodi, C., Beauchamp, D., Johnson, K. N., and Richmond, A. (1997). Mechanism and biological significance of constitutive expression of MGSA/GRO chemokines in malignant melanoma tumor progression. *J. Leukoc. Biol.* **62,** 588–597.

Luca, M., Huang, S., Gershenwald, J. E., Singh, R. K., Reich, R., and Bar-Eli, M. (1997). Expression of interleukin-8 by human melanoma cells up-regulates MMP-2 activity and increases tumor growth and metastasis. *Am. J. Pathol.* **151,** 1105–1113.

Luster, A. D. (1998). Chemokines–chemotactic cytokines that mediate inflammation. *N. Engl. J. Med.* **338,** 436–445.

Maione, T. E., Gray, G. S., Petro, J., Hunt, A. J., Donner, A. L., Bauer, S. I., Carson, H. F., and Sharpe, R. J. (1990). Inhibition of angiogenesis by recombinant human platelet factor-4 and related peptides. *Science* **247,** 77–79.

Miller, L. J., Kurtzman, S. H., Wang, Y., Anderson, K. H., Lindquist, R. R., and Kreutzer, D. L. (1998). Expression of interleukin-8 receptors on tumor cells and vascular endothelial cells in human breast cancer tissue. *Anticancer Res.* **18,** 77–81.

Minna, J. D. (1991). Neoplasms if the lung. *In* "Principles of Internal Medicine" (K. J. Isselbacher, ed.), pp. 1102–1110. McGraw-Hill, New York.

Moore, B. B., Arenberg, D. A., Stoy, K., Morgan, T., Addison, C. L., Morris, S. B., Glass, M., Wilke, C., Xue, Y. Y., Sitterding, S., Kunkel, S. L., Burdick, M. D., and Strieter, R. M. (1999). Distinct CXC chemokines mediate tumorigenicity of prostate cancer cells. *Am. J. Pathol.* **154,** 1503–1512.

Moser, B., and Loetscher, P. (2001). Lymphocyte traffic control by chemokines. *Nat. Immunol.* **2,** 123–128.

Muller, A., Homey, B., Soto, H., Ge, N., Catron, D., Buchanan, M. E., McClanahan, T., Murphy, E., Yuan, W., Wagner, S. N., Barrera, J. L., Mohar, A., Verastegui, E., and

Zlotnik, A. (2001). Involvement of chemokine receptors in breast cancer metastasis. *Nature* **410,** 50–56.

Murdoch, C., Monk, P. N., and Finn, A. (1999). Cxc chemokine receptor expression on human endothelial cells. *Cytokine* **11,** 704–712.

Nor, J. E., Christensen, J., Liu, J., Peters, M., Mooney, D. J., Strieter, R. M., and Polverini, P. J. (2001). Up-regulation of Bcl-2 in microvascular endothelial cells enhances intratumoral angiogenesis and accelerates tumor growth. *Cancer Res.* **61,** 2183–2188.

Owen, J. D., Strieter, R., Burdick, M., Haghnegahdar, H., Nanney, L., Shattuck-Brandt, R., and Richmond, A. (1997). Enhanced tumor-forming capacity for immortalized melanocytes expressing melanoma growth stimulatory activity/growth-regulated cytokine beta and gamma proteins. *Int. J. Cancer* **73,** 94–103.

Paget, S. L. (1889). The distribution of secondary growths in cancer of the breast. *Lancet* **1,** 571.

Pawson, T., and Scott, J. D. (1997). Signaling through scaffold, anchoring, and adaptor proteins. *Science* **278,** 2075–2080.

Perollet, C., Han, Z. C., Savona, C., Caen, J. P., and Bikfalvi, A. (1998). Platelet factor 4 modulates fibroblast growth factor 2 (FGF-2) activity and inhibits FGF-2 dimerization. *Blood* **91,** 3289–3299.

Phillips, R. J., Burdick, M. D., Lutz, M., Belperio, J. A., Keane, M. P., and Strieter, R. M. (2003). The stromal derived factor-1/CXCL12-CXC chemokine receptor 4 biological axis in non-small cell lung cancer metastases. *Am. J. Respir. Crit. Care. Med.* **167,** 1676–1686.

Qin, S., Rottman, J. B., Myers, P., Kassam, N., Weinblatt, M., Loetscher, M., Koch, A. E., Moser, B., and Mackay, C. R. (1998). The chemokine receptors CXCR3 and CCR5 mark subsets of T cells associated with certain inflammatory reactions. *J. Clin. Invest.* **101,** 746 754.

Rabin, R. L., Park, M. K., Liao, F., Swofford, R., Stephany, D., and Farber, J. M. (1999). Chemokine receptor responses on T cells are achieved through regulation of both receptor expression and signaling. *J. Immunol.* **162,** 3840–3850.

Richards, B. L., Eisma, R. J., Spiro, J. D., Lindquist, R. L., and Kreutzer, D. L. (1997). Coexpression of interleukin-8 receptors in head and neck squamous cell carcinoma. *Am. J. Surg.* **174,** 507–512.

Richmond, A. (2002). Nf-kappa B, chemokine gene transcription and tumour growth. *Nat. Rev. Immunol.* **2,** 664–674.

Richmond, A., Fan, G. H., Dhawan, P., and Yang, J. (2004). How do chemokine/chemokine receptor activations affect tumorigenesis? *Novartis Found Symp.* **256,** 74–89.

Rollins, B. J. (1997). Chemokines. *Blood* **90,** 909–928.

Romagnani, P., Annunziato, F., Lasagni, L., Lazzeri, E., Beltrame, C., Francalanci, M., Uguccioni, M., Galli, G., Cosmi, L., Maurenzig, L., Baggiolini, M., Maggi, E., Romagnani, S., and Serio, M. (2001). Cell cycle–dependent expression of CXC chemokine receptor 3 by endothelial cells mediates angiostatic activity. *J. Clin. Invest.* **107,** 53–63.

Romagnani, P., Lasagni, L., Annunziato, F., Serio, M., and Romagnani, S. (2004). CXC chemokines: The regulatory link between inflammation and angiogenesis. *Trends Immunol.* **25,** 201–209.

Salcedo, R., and Oppenheim, J. J. (2003). Role of chemokines in angiogenesis: CXCL12/SDF-1 and CXCR4 interaction, a key regulator of endothelial cell responses. *Microcirculation* **10,** 359–370.

Salcedo, R., Resau, J. H., Halverson, D., Hudson, E. A., Dambach, M., Powell, D., Wasserman, K., and Oppenheim, J. J. (2000). Differential expression and responsiveness of chemokine receptors (CXCR1-3) by human microvascular endothelial cells and umbilical vein endothelial cells. *FASEB J.* **14,** 2055–2064.

Salcedo, R., Wasserman, K., Young, H. A., Grimm, M. C., Howard, O. M., Anver, M. R., Kleinman, H. K., Murphy, W. J., and Oppenheim, J. J. (1999). Vascular endothelial growth factor and basic fibroblast growth factor induce expression of CXCR4 on human endothelial cells: *In vivo* neovascularization induced by stromal-derived factor-1alpha. *Am. J. Pathol.* **154,** 1125–1135.

Sato, Y., Abe, M., and Takaki, R. (1990). Platelet factor 4 blocks the binding of basic fibroblast growth factor to the receptor and inhibits the spontaneous migration of vascular endothelial cells. *Biochem. Biophys. Res. Commun.* **172,** 595–600.

Schioppa, T., Uranchimeg, B., Saccani, A., Biswas, S. K., Doni, A., Rapisarda, A., Bernasconi, S., Saccani, S., Nebuloni, M., Vago, L., Mantovani, A., Melillo, G., and Sica, A. (2003). Regulation of the chemokine receptor CXCR4 by hypoxia. *J. Exp. Med.* **198,** 1391–1402.

Schrader, A. J., Lechner, O., Templin, M., Dittmar, K. E., Machtens, S., Mengel, M., Probst-Kepper, M., Franzke, A., Wollensak, T., Gatzlaff, P., Atzpodien, J., Buer, J., and Lauber, J. (2002). CXCR4/CXCL12 expression and signalling in kidney cancer. *Br. J. Cancer* **86,** 1250–1256.

Schruefer, R., Lutze, N., Schymeinsky, J., and Walzog, B. (2005). Human neutrophils promote angiogenesis by a paracrine feedforward mechanism involving endothelial interleukin-8. *Am. J. Physiol. Heart Circ. Physiol.*

Semenza, G. L. (2002). Signal transduction to hypoxia-inducible factor 1. *Biochem. Pharmacol.* **64,** 993–998.

Semenza, G. L. (2003). Targeting HIF-1 for cancer therapy. *Nat. Rev. Cancer* **3,** 721–732.

Sgadari, C., Angiolillo, A. L., Cherney, B. W., Pike, S. E., Farber, J. M., Koniaris, L. G., Vanguri, P., Burd, P. R., Sheikh, N., Gupta, G., Teruya-Feldstein, J., and Tosato, G. (1996a). Interferon-inducible protein-10 identified as a mediator of tumor necrosis *in vivo*. *Proc. Natl. Acad. Sci. USA* **93,** 13791–13796.

Sgadari, C., Angiolillo, A. L., and Tosato, G. (1996b). Inhibition of angiogenesis by interleukin-12 is mediated by the interferon-inducible protein 10. *Blood* **87,** 3877–3882.

Sgadari, C., Farber, J. M., Angiolillo, A. L., Liao, F., Teruya-Feldstein, J., Burd, P. R., Yao, L., Gupta, G., Kanegane, C., and Tosato, G. (1997). Mig, the monokine induced by interferon-gamma, promotes tumor necrosis *in vivo*. *Blood* **89,** 2635–2643.

Sharma, S., Stolina, M., Luo, J., Strieter, R. M., Burdick, M., Zhu, L. X., Batra, R. K., and Dubinett, S. M. (2000). Secondary lymphoid tissue chemokine mediates T-cell-dependent antitumor responses *In vivo* [In Process Citation]. *J. Immunol.* **164,** 4558–4563.

Sharma, S., Yang, S. C., Hillinger, S., Zhu, L. X., Huang, M., Batra, R. K., Lin, J. F., Burdick, M. D., Strieter, R. M., and Dubinett, S. M. (2003). SLC/CCL21-mediated anti-tumor responses require IFNgamma, MIG/CXCL9 and IP-10/CXCL10. *Mol. Cancer* **2,** 22.

Shellenberger, T. D., Wang, M., Gujrati, M., Jayakumar, A., Strieter, R. M., Ioannides, C., Efferson, C. L., El-Naggar, A. K., Clayman, G. L., and Frederick, M. J. (2005). BRAK/CXCL14 is a potent inhibitor of angiogenesis and is a chemotactic factor for immature dendritic cells. *Cancer Res.* (in press).

Shyamala, V., and Khoja, H. (1998). Interleukin-8 receptors R1 and R2 activate mitogen-activated protein kinases and induce c-fos, independent of Ras and Raf-1 in Chinese hamster ovary cells. *Biochemistry* **37,** 15918–15924.

Singh, R. K., Gutman, M., Radinsky, R., Bucana, C. D., and Fidler, I. J. (1994). Expression of interleukin-8 correlates with the metastatic potential of human melanoma cells in nude mice. *Cancer Res.* **54,** 3242–3247.

Smith, D. R., Polverini, P. J., Kunkel, S. L., Orringer, M. B., Whyte, R. I., Burdick, M. D., Wilke, C. A., and Strieter, R. M. (1994). Inhibition of interleukin-8 attenuates angiogenesis in bronchogenic carcinoma. *J. Exp. Med.* **179,** 1409–1415.

Soto, H., Wang, W., Strieter, R. M., Copeland, N. G., Gilbert, D. J., Jenkins, N. A., Hedrick, J., and Zlotnik, A. (1998). The CC chemokine 6Ckine binds the CXC chemokine receptor CXCR3. *Proc. Natl. Acad. Sci. USA* **95,** 8205–8210.

Staller, P., Sulitkova, J., Lisztwan, J., Moch, H., Oakeley, E. J., and Krek, W. (2003). Chemokine receptor CXCR4 downregulated by von Hippel-Lindau tumour suppressor pVHL. *Nature* **425,** 307–311.

Strieter, R. M., Polverini, P. J., Kunkel, S. L., Arenberg, D. A., Burdick, M. D., Kasper, J., Dzuiba, J., Van Damme, J., Walz, A., Marriott, D., Chan, S. Y., Roczniak, S., and Shanafelt, A. B. (1995). The functional role of the ELR motif in CXC chemokine-mediated angiogenesis. *J. Biol. Chem.* **270,** 27348–27357.

Strieter, R. M., Belperio, J. A., Arenberg, D. A., Smith, M. I., Burdick, M. D., and Keane, M. P. (2002). CXC chemokine in angiogenesis. *In* "Universies in delicate balance: Chemokines and the nervous system" (R. M. Ransohoff, K. Suzuki, A. E. I. Proudfoot, and W. F. Hickey, eds.), pp. 129–148. Amsterdam, The Netherlands., Elsevier Science B.V.

Strieter, R. M., Belperio, J. A., Phillips, R. J., and Keane, M. P. (2004a). Chemokines: Angiogenesis and metastases in lung cancer. *Novartis Found Symp.* **256,** 173–184.

Strieter, R. M., Belperio, J. A., Phillips, R. J., and Keane, M. P. (2004b). CXC chemokines in angiogenesis of cancer. *Semin. Cancer Biol.* **14,** 195–200.

Struyf, S., Burdick, M. D., Proost, P., Van Damme, J., and Strieter, R. M. (2004). Platelets release CXCL4L1, a nonallelic variant of the chemokine platelet factor-4/CXCL4 and potent inhibitor of angiogenesis. *Circ. Res.* **95,** 855–857.

Sugden, P. H., and Clerk, A. (1997). Regulation of the ERK subgroup of MAP kinase cascades through G protein-coupled receptors. *Cell Signal.* **9,** 337–351.

Tannenbaum, C. S., Tubbs, R., Armstrong, D., Finke, J. H., Bukowski, R. M., and Hamilton, T. A. (1998). The CXC chemokines IP-10 and Mig are necessary for IL-12-mediated regression of the mouse RENCA tumor. *J. Immunol.* **161,** 927–932.

Teruya-Feldstein, J., Jaffe, E. S., Burd, P. R., Kanegane, H., Kingma, D. W., Wilson, W. H., Longo, D. L., and Tosato, G. (1997). The role of Mig, the monokine induced by interferon-gamma, and IP-10, the interferon-gamma-inducible protein-10, in tissue necrosis and vascular damage associated with Epstein-Barr virus-positive lymphoproliferative disease. *Blood* **90,** 4099–4105.

White, E. S., Flaherty, K. R., Carskadon, S., Brant, A., Iannettoni, M. D., Yee, J., Orringer, M. B., and Arenberg, D. A. (2003). Macrophage migration inhibitory factor and CXC chemokine expression in non-small cell lung cancer: Role in angiogenesis and prognosis. *Clin. Cancer Res.* **9,** 853–860.

Yang, J., and Richmond, A. (2004). The angiostatic activity of interferon-inducible protein-10/CXCL10 in human melanoma depends on binding to CXCR3 but not to glycosaminoglycan. *Mol. Ther.* **9,** 846–855.

Yang, T. Y., Chen, S. C., Leach, M. W., Manfra, D., Homey, B., Wiekowski, M., Sullivan, L., Jenh, C. H., Narula, S. K., Chensue, S. W., and Lira, S. A. (2000). Transgenic expression of the chemokine receptor encoded by human herpesvirus 8 induces an angioproliferative disease resembling Kaposi's sarcoma. *J. Exp. Med.* **191,** 445–454.

Yatsunami, J., Tsuruta, N., Ogata, K., Wakamatsu, K., Takayama, K., Kawasaki, M., Nakanishi, Y., Hara, N., and Hayashi, S. (1997). Interleukin-8 participates in angiogenesis in non-small cell, but not small cell carcinoma of the lung. *Cancer Lett.* **120,** 101–108.

Yoneda, J., Kuniyasu, H., Crispens, M. A., Price, J. E., Bucana, C. D., and Fidler, I. J. (1998). Expression of angiogenesis-related genes and progression of human ovarian carcinomas in nude mice. *J. Natl. Cancer Inst.* **90,** 447–454.

Yuan, A., Pan-Chyr, Y., Chong-Jen, Y., Lee, Y., Yu-Tuang, Y., Chi-Long, C., Lee, L., Sow-Hsong, K., and Kwen-Tay, L. (1995). Tumor angiogenesis correlates with histologic type and metastasis in non-small cell lung cancer. *Am. J. Resp. Crit. Care Med.* **152,** 2157–2162.

Zetter, B. R. (1998). Angiogenesis and tumor metastasis. *Annu. Rev. Med.* **49,** 407–424.

Zhong, H., Chiles, K., Feldser, D., Laughner, E., Hanrahan, C., Georgescu, M. M., Simons, J. W., and Semenza, G. L. (2000). Modulation of hypoxia-inducible factor 1-alpha expression by the epidermal growth factor/phosphatidylinositol 3-kinase/PTEN/AKT/FRAP pathway in human prostate cancer cells: Implications for tumor angiogenesis and therapeutics. *Cancer Res.* **60,** 1541–1545.

Further Reading

American Cancer Society (2004). Cancer Facts and Figures 2004: Selected Cancers. http://www.cancer.org

CHAPTER 11

Chemokines and Chemokine Receptors in Infectious Disease

Karoll J. Cortez,* Julia O. Jackson,† and Shyam Kottilil†

*Pediatric Oncology Branch, National Cancer Institute, National Institutes of Health, Bethesda, Maryland 20892

†Laboratory of Immunoregulation, National Institute of Allergy and Infectious Diseases, National Institutes of Health, Bethesda, Maryland 20892

Current Topics in Membranes, Volume 55
1063-5823/05 $35.00
DOI: 10.1016/S1063-5823(04)55011-1

I. INTRODUCTION

Leukocytes are capable of rapid, directed movements towards the site of infection by way of chemotaxis, a property that enables them to serve as the immune system's first responders to microbes. This interaction between leukocytes and chemoattractants initiates a series of coordinated cellular and molecular events that leads to alterations in ion fluxes, integrin avidity, and transmembrane potential. These changes lead to changes in cell shape, secretion of lysosomal enzymes, production of superoxide anions, and enhanced locomotion.

The classical chemokines, like bacterial-derived N-formyl peptides, complement fragment peptides C3a and C5a, whereas lipid molecules such as leukotrienes are all chemokines that act as activators of leukocyte chemotaxis (Schiffmann *et al.*, 1975; Goldman and Goetzl, 1982; Hanahan, 1986; Gerard and Gerard, 1994). More recently, several agents in the 8–17 Kb mass range have been found to promote selective chemotaxis to leukocytes *in vitro* and contribute to the inflammatory process *in vivo* (Baggiolini *et al.*, 1994, 1997). These chemokines belong to the chemokine superfamily, which can be divided into four subgroups (CXC, CX_3C, CC, and C) according to the positioning of the first two closely paired and highly conserved cysteines of the amino-acid sequence.

The specific effects of chemokines on their target cells are mediated by members of a family of seven-transmembrane-spanning, G-protein-coupled receptors (Fig. 1) (Murphy, 1994). These receptors are part of a larger family of G-protein-coupled receptors that includes hormone receptors, neurotransmitters, paracrine substances, inflammatory mediators, proteinases, photons, and calcium ions (Watson and Arkinstall, 1994).

There are two major classes of chemokine receptors. The CC receptors bind to the CC chemokines and are named CCR1 to 9; the CXC chemokines selectively bind to the CXC receptors, termed CXCR1 to CXCR5. This group also comprises receptors for fractalkine (CX_3CR1) and lymphotactin (XCR1), as well as the Duffy antigen receptor for chemokines (DARC), which has been shown to bind promiscuously to both CXC and CC chemokines. Finally, there is a receptor termed D6, which has been renamed CCR10 by some investigators.

The significance of chemokines and their receptors in infectious diseases, which was initially guessed at because of their ability to induce leukocyte chemotaxis *in vitro*, is now firmly established. Several recent studies have shown that chemokines and their receptors play a major role not only in mediation of leukocyte trafficking to sites of antigen entry but also in T-cell polarization, and they also act as cell-entry factors for major pathogens like

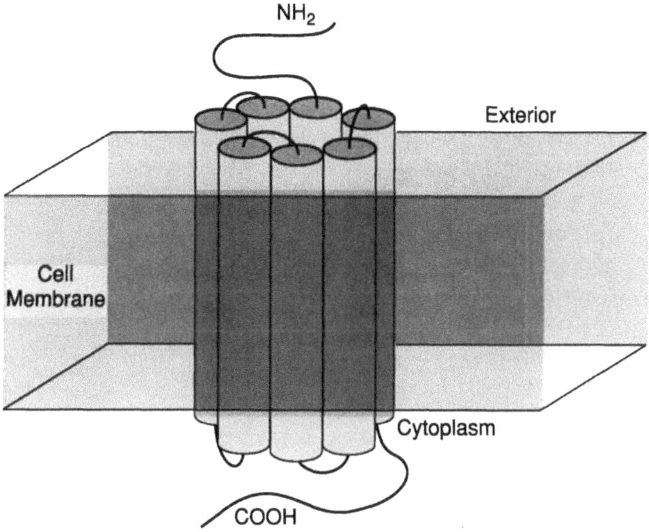

FIGURE 1 The putative CC chemokine receptor with seven-transmembrane-spanning G-protein-coupling receptors.

human immunodeficiency virus (HIV) and *Plasmodium vivax*. The latter aspect of their functional role has thereby allowed these pathogens to facilitate disease transmission and disease progression. Furthermore, most members of the *poxviridae* and *herpesviridae* families encode host chemokines, and their receptor homologues adapt them for a variety of purposes. These include enabling these viruses to evade immune defenses, dissemination of viral particles, and in certain cases angiogenesis and tumorigenesis. The discovery of virus-encoded chemokines and their receptors is a significant step towards understanding the pathogenesis of major infections as well as the development of therapeutic strategies for fighting these infections. The recent emergence of receptor antagonists and chemokine scavengers has provided insight into the role of these chemokines and their receptors in antimicrobial defense. Moreover, discovery of yet-unknown functions of viral chemokines and their receptors may unravel unrecognized functions for their host counterparts. Viral antichemokines have the potential to be exploited as therapeutic strategies against some of these pathogens. This chapter will focus on the current concepts of how host and viral chemokine systems influence the pathogenesis of infectious diseases; it will also discuss the evolving technologies for interfering with the chemokine system, particularly in HIV and other infections.

II. HUMAN CHEMOKINE RECEPTORS AS ENTRY FACTORS FOR INFECTIOUS AGENTS

In recent years, there has been an emergence of interest in the role of chemokine receptors in infectious diseases. Human chemokine receptors participate in several disease states, either by overexpression of receptors or by facilitation of viral entry into target cells to establish infection. Malaria and HIV infection have been shown to use chemokine receptors as entry receptors and coreceptors, respectively. Studies involving the genetics of these chemokine receptors and their roles in the pathogenesis of these diseases have been possible due to the existence of naturally occurring receptor gene mutations in these receptor genes that encode these entry receptors for the two pathogens.

A. Malaria

Malaria is a common infectious disease transmitted by mosquitoes infected with a protozoon, *Plasmodium*. There are four different species of *Plasmodium* that can cause malaria: *P. vivax, P. ovale, P. falciparum,* and *P. malariae*. These parasites bind to the red blood cells (RBC), causing them to undergo cell lysis and simultaneously begin to display the symptoms of malaria. Malaria is endemic to most tropical and subtropical areas of the world; however, it is most prevalent in West Africa. In the 1970s, it was noted that the Duffy antigen–negative blood group phenotype, which is common in West Africans, was associated with resistance to a major form of malaria caused by *P. vivax* (Miller *et al.*, 1976). It was also shown that an antibody to Duffy antigen could block parasite invasion into DARC-positive RBCs. It was later discovered that Duffy was a seven-transmembrane-domain chemokine-binding protein expressed on the surface of RBCs (Horuk *et al.*, 1993). The Duffy-negative phenotype is a result of a single G-to-A nucleotide substitution in a GATA-1 site within the promoter that eliminates transcription in the RBCs by producing a Gly44Asp substitution in the polypeptide chain, preventing parasite invasion into RBCs (Tournamille *et al.*, 1995). The physiological role of the Duffy receptor is not yet clearly delineated; however, it binds to both CC and CXC chemokines. Also, there is no evidence so far that Duffy-deficient individuals have altered inflammatory processes. Whether one could use chemokines as therapeutic agents that induce Duffy-receptor blockade for antimalarial therapy is currently being investigated.

B. HIV Infection

HIV uses chemokine receptors CCR5 and CXCR4 as coreceptors with CD4 molecules to mediate viral envelope–mediated fusion binding to $CD4^+$ T-lymphocytes and macrophages (Berger *et al.*, 1999). This binding will lead to subsequent membrane fusion, which precedes viral entry (Berger *et al.*, 1999). This discovery of HIV coreceptors was made in pursuit of factors that restricted specific HIV strains to different $CD4^+$ T-cells. The discovery that the same chemokines that bind to CCR5 could also suppress replication of macrophage-tropic HIV strains expedited the identification of a macrophage-tropic HIV coreceptor on CCR5 as the major HIV coreceptor (Alkhatib *et al.*, 1996; Cocchi *et al.*, 1996). Discovery of this coreceptor led to the change in nomenclature of HIV strains: they are now known as R5, X4, or dual based on their ability to bind to CCR5, CXCR4, or both, respectively. Most HIV isolates from infected individuals are of the R5 type; however, the predominant isolates seen in patients with advanced AIDS are X4 and dual, a finding that also coincides with an increased rate of $CD4^+$ T-cell destruction. Studies of primary lymphoid tissues *ex vivo* have indicated that a larger percentage of T-cells express CXCR4 than CCR5, which may allow more cells to be infected and destroyed in advanced AIDS (Margolis *et al.*, 1998). However, the factors that lead to this R5-to-X4 switch, and the reasons that HIV seems to prefer CCR5, are largely unclear.

CXCR4 is expressed on both major targets of HIV infection: $CD4^+$ T-cells and monocytes/macrophages. This leads to the question of why R5 strains are the predominant, if not exclusive, species replicating during asymptomatic infection and, in many cases, during the entire course of HIV disease (Schuitemaker *et al.*, 1991). One possible explanation is that there is a different expression of CCR5 and CXCR4 on different $CD4^+$ T-cell subsets. Phylogenetic and phenotypic analyses of HIV quasispecies replicating during the peak of viremia revealed preferential expansion of R5 viruses over X4 viruses in an individual harboring both strains at equal frequencies (Ostrowski *et al.*, 1998b). Furthermore, analysis of coreceptor expression in nonimmunized HIV+ and HIV– individuals revealed that, while the frequencies of $CXCR4^+$ $CD4^+$ and $CCR5^+$ $CD4^+$ cells in HIV-infected individuals are comparable, the majority of activated $CD4^+$ T-cells expressed CCR5 and not CXCR4 (Ostrowski *et al.*, 1998a). Of note is that CXCR4 was expressed on the majority of activated $CD4^+$ T-cells from 50% of HIV-infected individuals with advanced disease. The factors that influence the emergence of X4 HIV strains in approximately 5% of subtype-B HIV-infected individuals with advanced disease are not clearly understood. The observation that R5 HIV strains isolated from infected individuals with advanced disease are less sensitive to CC-chemokine-mediated inhibition suggests that selection pressure against CCR5 usage does occur *in vivo* (Scarlatti *et al.*, 1997). The combined selection pressures of CC chemokines and the presence of other chemokines and cytokines that support HIV replication may generate microenvironmental conditions that may play a role in the "shift" from predominantly R5 to dual or X4 HIV-strain replication (Cheng-Mayer *et al.*, 1988).

Inflammatory chemokines and cytokines can play a major role in facilitating HIV replication *in vivo* (Fig. 2). It has been shown that the level of HIV replication in peripheral blood mononuclear cell (PBMC) cultures reflects the interplay between the HIV-inhibitory activities of endogenous proinflammatory cytokines (Kinter *et al.*, 1998, 2000). In cultures of $CD4^+$ PBMCs or $CD8^+$ T-cell-depleted PBMCs, isolated from asymptomatic HIV-positive donors infected with R5 HIV, neutralization of endogenous CC chemokines was found to significantly elevate the levels of HIV replication. However, this enhancing effect was virtually abrogated by simultaneous neutralization of endogenous TNF-α, IL-1β, and IL-6, suggesting that these cytokines were important in driving viral production. The effect of the interaction between endogenous inflammatory cytokines and CC

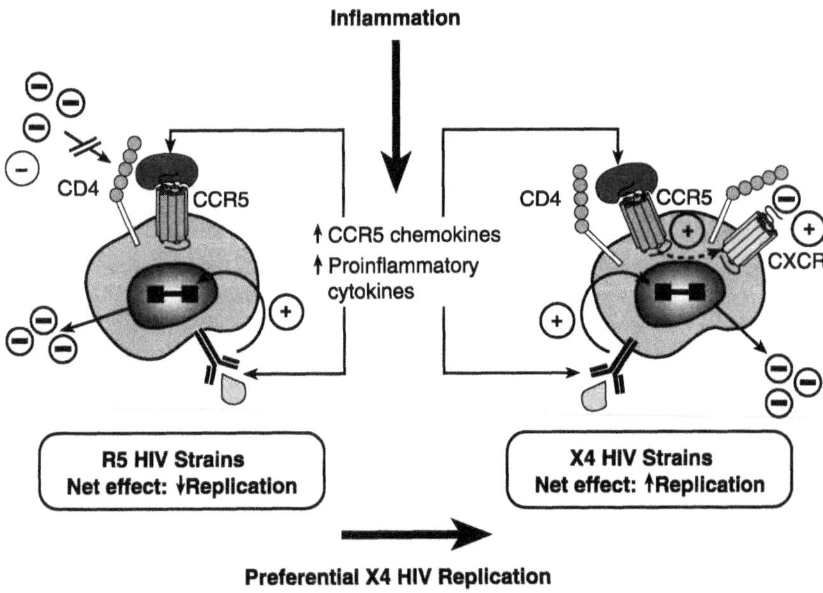

FIGURE 2 Several chemokines and cytokines generate inflammatory conditions that promote replication of X4 HIV strains and thus impact the level of HIV replication in individuals harboring primarily X4 or dual strains of HIV.

chemokines on HIV replication in *in vitro* infected macrophages has also been studied (Moriuchi *et al.*, 1998). Treatment of monocyte-derived macrophages with the bacterial endotoxin lipopolysaccharide (LPS) was found to significantly enhance X4 HIV replication, which was associated with increased expression of CXCR4 and the production of both inflammatory cytokines and CC chemokines. Antibody-mediated neutralization studies revealed that the activities of both proinflammatory cytokines and proinflammatory chemokines contributed significantly to enhancement of the replication of X4 HIV in macrophage cultures, but that these activities antagonized each other with regard to HIV replication (Fig. 2). However, results from other studies suggest that the overall effect of inflammatory cytokines in macrophages is the suppression of R5 HIV replication due to induction of high levels of CC chemokines (Verani *et al.*, 1997; Lane *et al.*, 1999). Together, these results suggest that inflammatory conditions, including the secretion of CC chemokines, result in the preferential replication of X4 over R5 strains and would, in fact. facilitate enhanced viral replication in individuals harboring the X4 virus (Fig. 2).

In contrast to early models, in which chemokines were viewed as inhibitors of HIV fusion only via a passive signal transduction–independent mechanism of competitive occupancy of HIV coreceptors (Atchison *et al.*, 1996), certain chemokine-mediated effects on viral replication have now been shown to require cellular signaling events (Fig. 3). In particular, the ability of chemokines to enhance viral replication likely involves several signal-dependent cellular events (Arthos *et al.*, 2000; Kinter *et al.*, 1998). CC-chemokine-mediated Gα1 protein signaling was first found to be involved in the enhancing effect of heterologous chemokines on X4 HIV entry in primary $CD4^+$ T-cells (Fig. 3) (Kinter *et al.*, 1996). It has been clearly shown that several CC chemokines, including both CCR5 and CCR2 ligands, directly enhance the replication of X4 HIV strains in $CD4^+$ T-cells. CC-chemokine-mediated effects were observed at the level of virus entry and were found to require intracellular signaling through G-protein coupling, as demonstrated by sensitivity to G-protein coupling inhibitor pertussis toxin. In some cases, this enhancement was found to involve protein tyrosine kinase–dependent induction of the CXCR4 gene (Dolei *et al.*, 1998). CC-chemokine-mediated enhancement of X4 entry was not associated with enhanced CXCR4 expression, but rather with increased localization of CD4 and CXCR4. Entry of R5 HIV in activated PBMCs was shown to be sensitive to inhibition by pertussis toxin, and this effect was associated with inhibition of the capping of CD4 and CCR5; no effect was observed on the entry of the T-cell-line-adapted X4 strain LAI (Lapham *et al.*, 1996; Alfano *et al.*, 1999). These studies confirm that signal transduction–dependent cocapping of CD4 and coreceptors can be mediated both by host factors (Kinter *et al.*, 2000) and by HIV itself

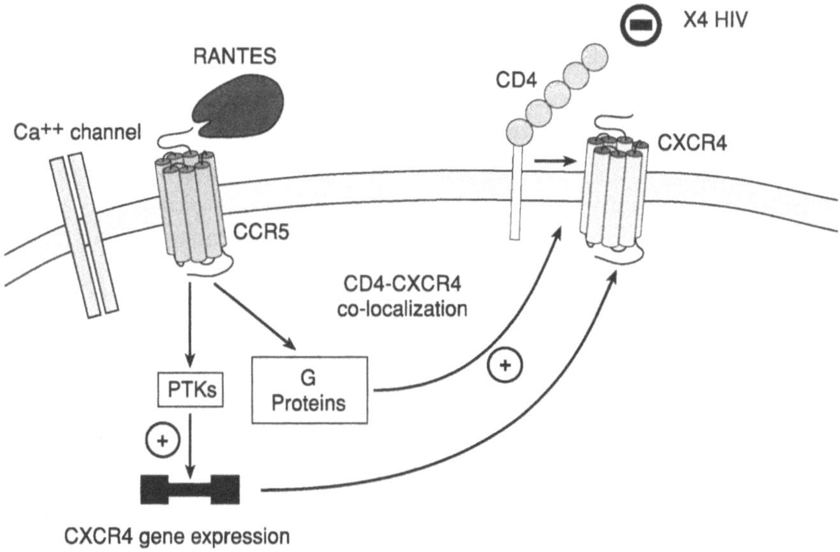

FIGURE 3 CC chemokines have been shown to enhance the entry of X4 strains of HIV into CD4 T-cell targets by signal transduction–dependent mechanisms. Signal transduction through G-protein-coupled receptors has been associated with collocalization of CD4 and CXCR4 on the surface of target cells. Signaling though protein tyrosine kinases has also been implicated in the enhancement of HIV replication via upregulation of CXCR4 cell-surface expression.

(Lapham *et al.*, 1996; Alfano *et al.*, 1999). The relative importance of host versus viral factors for HIV entry likely varies depending upon the constitutive association of CD4 and coreceptor molecules (Xiao *et al.*, 1999), the degree of CD4 dependence of the viral strain (Kabat *et al.*, 1994), and the quantity and quality of the HIV gp120 (Arthos *et al.*, 2000). In addition to the enhancing effects of chemokines on the efficiency of HIV entry, interactions with chemokine receptors have been suggested to play a role in postentry stages of viral replication (Fig. 4) (Arthos *et al.*, 2000). In this regard, it was shown that MIP-1α/CCL3 and MIP-1β/CCL4 can compensate for or rescue nonproductive infection of macrophages with certain X4 and R5 strains at the postentry level (Arthos *et al.*, 2000) (Fig. 4). The mechanisms whereby chemokines support postentry stages of HIV replication are not yet fully delineated. Several lines of evidence suggest that full reverse transcription is dependent upon actin polymerization of the cytoskeleton (Bukrinskaya *et al.*, 1998). It is likely that exposure of cells, either to chemokines or to the HIV envelope, results in activation of cytoskeletal processes necessary for early stages of HIV replication (Fig. 4).

There are several other chemokine receptors and related molecules that also could act as HIV coreceptors *in vitro* (CCR2, CCR3, CCR5, CCR8,

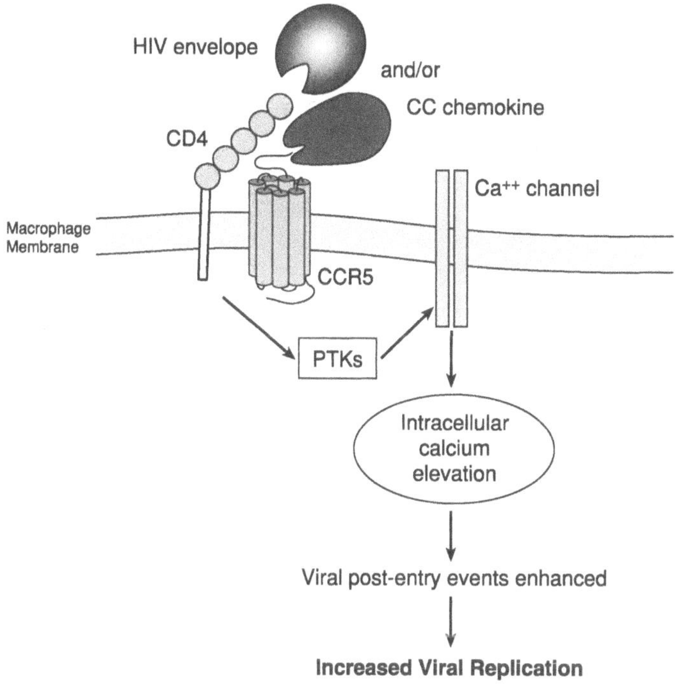

FIGURE 4 CC chemokines or HIV envelopes can bind to CCR5, leading to mobilization of intracellular calcium, and can cause progression of HIV replication through the postentry stages of HIV replication.

CXCR4, CXCR5, CXCR4, CX_3CR1, D6, etc). However, there are not enough epidemiological data to evidence a role *in vivo* for any other than perhaps CCR3, due to its expression on microglial cells (He *et al.*, 1997). The CCR5Δ32 allele has a 32-base pair deletion in the region of the ORF encoding the second extracellular domain. This causes premature termination of translation and failure to traffic to the plasma membrane, resulting in loss of both chemokine-receptor and HIV-coreceptor activity (Samson *et al.*, 1996). In addition, the CCR5Δ32 gene product may interfere with the normal functions of CCR5 or other chemokine receptors through homodimerization or heterodimerization (Rodriguez-Frade *et al.*, 1999). This allele is found in approximately 10% of all North American Caucasians but is absent in native Africans and East Asians; approximately 20% and 1% of North American Caucasians are heterozygous and homozygous, respectively. The homozygous CCR5Δ32 genotype is found at much lower frequencies than expected among HIV-positive individuals, demonstrating the importance of CCR5 in initial HIV transmission (Samson *et al.*, 1996;

Biti *et al.*, 1997). There are several anecdotal reports of isolation of X4 virus and rapid progression to AIDS in the few CCR5Δ32 homozygotes that have been identified (Biti *et al.*, 1997). These findings, along with the fact that X4 HIV correlates with the onset of AIDS, allude to the possibility that X4 strains may be more virulent.

The CCR5 promoter is highly polymorphic, which could account for the reports of wide variation in CCR5 expression. The promoter polymorphism 59029-A (PI haplotype) is a risk factor for rapid progression to AIDS in Caucasian homosexuals (McDermott *et al.*, 1998) and is associated with increased production of CCR5 (McDermott *et al.*, 1998). The 59356-T/T genotype has been associated with increased perinatal transmission of HIV in African American children of mothers who did not receive antiretroviral agents (Kostrikis *et al.*, 1999). The CCR5 promoter genotype 59353-C/C has been shown to be associated with delayed progression to AIDS (Easterbrook *et al.*, 1999). Thus, CCR5 expression is limiting for HIV infection, and small changes in the transcription or translation can result in clinical consequences.

Genetic studies have also proven a role for other chemokine receptors in HIV pathogenesis. CCR2-64I is a common single-nucleotide polymorphism (SNP) that results in a valine-to-isoleucine substitution in the second-transmembrane region of CCR2, an HIV coreceptor with limited *in vitro* activity. This mutation is associated with slower progression to AIDS (Easterbrook *et al.*, 1999). Since CCR2 and CCR5 reside within close proximity to one another on chromosome 3, it is possible that this mutation is a marker of a disease-modifying mutation in CCR5. Another possibility is that the CCR2-64I gene product has the indirect effect of heterodimerizing with CCR5 or CXCR4 in a manner different from that of wild-type CCR2 (Mellado *et al.*, 1999). However, activated $CD4^+$ T-cells from CCR2-64I individuals produced comparable levels of mRNA and protein, a fact that argues against these explanations (Mariani *et al.*, 1999).

SDF-3α is a chemokine polymorphism that has been associated with HIV pathogenesis. A common SNP in the translated region of SDF-1 mRNA, this variant was associated with a recessive effect on HIV progression (Zimmerman *et al.*, 1997). The mechanism for this effect could stabilize SDF-1 mRNA, thereby increasing the amount of SDF-1 ligand that potentially blocks HIV through CXCR4 and delaying the R5-to-X4 shift. However, these effects were not seen in other studies arguing against these mechanisms (Bai *et al.*, 1998; Arya *et al.*, 1999). Larger studies would be required to confirm this hypothesis.

Recently, two SNPs in the RANTES/CCL5 promoter called RANTES -403A and -28G have also been described (Liu *et al.*, 1999). This study has shown that this haplotype was associated with a diminished rate of $CD4^+$

T-cell-count decline in a Japanese HIV-positive hemophiliac cohort (Liu *et al.*, 1999). It has been demonstrated that the -403A polymorphism is a dominant risk factor for homosexual HIV transmission but also slows progression to AIDS at a rate comparable to that of CCR5Δ32 (McDermott *et al.*, 2000a). Thus, RANTES/CCL5 polymorphisms have effects on both HIV transmission and disease progression. Since this mutation is associated with a higher expression of RANTES/CCL5, it also suggests that RANTES/CCL5 may result in diminished progression to AIDS by blocking CCR5 entry (Nickel *et al.*, 2000).

The fractalkine receptor (CX_3CR1) has two common linked polymorphisms, called V249I and T280M, at amino-acid positions 249 and 280 (Faure *et al.*, 2000). Homozygotes for the combination of these two polymorphisms progressed faster to AIDS in the SEROCO cohort of HIV seroconverters and have less CX_3CR1 on the PBMC cell surface. However, a separate study showed that the heterozygotes of these two polymorphisms progressed at a slower rate (McDermott *et al.*, 2000b). The mechanism of action of the polymorphic variant in HIV pathogenesis is unclear but could involve modulation of CCR5 or CXCR4 activity or modulation of the HIV coreceptor or immunoregulatory activity of CX_3CR1.

Chemokine-receptor polymorphisms may also modulate response to antiretroviral therapy. In particular, Caucasians treated with several different regimens were found to be more likely to fail therapy if they lacked CCR5Δ32 (Valdez *et al.*, 1999). In ACTG 343, where everyone received the same antiretroviral treatment, those individuals homozygous for the 59029 promoter allele that lacked either CCR5Δ32 or CCR2-641 were most likely to fail therapy (O'Brien *et al.*, 2000). These findings, if confirmed, may indicate that genetic testing has prognostic significance in the treatment of HIV.

C. Human Cytomegalovirus Infection

Virus-induced alterations in the cellular expression of chemokines may play a vital role in directing the migration of leukocyte subsets to sites of infection, thereby playing a major role in immune host defense against infections. *In vitro* studies show that IL-8 expression is significantly elevated in human fibroblasts after human cytomegalovirus (HCMV) infection (Craigen *et al.*, 1997). Moreover, supernatants from HCMV-infected fibroblasts significantly enhance transendothelial migration of neutrophils, a characteristic of IL-8 (Craigen *et al.*, 1997). Since neutrophils play an important role in the dissemination of HCMV in the human body, recruitment of neutrophils is considered to enhance the spread of infection. Furthermore, exposure of fibroblasts to IL-8 augments infectious virion production and replication of cytomegalovirus (CMV), resulting in increases in levels of CMV late antigen (Murayama *et al.*, 1994). HCMV has also been shown to selectively induce the expression of CXCR1 receptor in fibroblasts (Murayama *et al.*, 1994). In addition, IL-8 has been shown to weaken the antiviral effects of interferon-α by downregulation of 2′ 5′ oligoadenylate synthetase activity, particularly against encephalomyelomyocarditis virus and polio virus (Khabar *et al.*, 1997). In summary, these findings indicate that CMV can induce IL-8 production, which, in turn, contributes to the inflammatory processes by enhancing HCMV replication directly or indirectly by counteracting the activity of interferon-α.

D. Infections of the Central Nervous System

CNS infections like viral encephalitis, bacterial meningitis, parasitic infection, and prion diseases are all associated with mortality and postinfection sequelae. The roles of chemokines and their receptors in the pathogenesis of some of these infections are summarized in the following sections and in Table I.

1. Viral Encephalitis

Viral encephalitis is a life-threatening condition caused by several viruses and can provoke an inflammatory response that leads to recruitment of leukocytes. In viral encephalitis, the inflammatory response involves activated T-cells and monocytes that are recruited into the subarachnoid space. Chemokines produced locally within the brain parenchyma play an important role in initiating the extravasation of responding leukocytes in viral encephalitis (Table I). Indeed, MCP-1/CCL-2 and IP-10/CXCL10 have been identified as the main chemokines whose expression is elevated in the cerebrospinal fluid (CSF) of patients suffering from viral meningitis (Lahrtz *et al.*, 1997).

a. CMV Encephalitis. CMV encephalitis occurs primarily in fetal development and as a common opportunistic infection of patients in the advanced stages of AIDS. There is a wide range of complications associated with CMV encephalitis, ranging from long-term cognitive deficits to death. The cellular infiltrates associated with CMV brain infection consist of macrophages (Asensio *et al.*, 1999) and T-cells, which were correlated with the production of various chemokines: IP-10/CXCL10, MCP-1/CCL2, MIP-1β/CCL4, RANTES/CCL5, and MCP-3/CCL7. Upon exposure to CMV, microglia are capable of secreting various chemokines *in vitro*, including IL-8/CXCL8, IP-10/CXCL10, MCP-1/CCL2, MIP-1β/CCL4, and RANTES/CCL5 (Asensio *et al.*, 1999). In addition, elevated levels of MCP-1/CCL2 have been detected

TABLE I
Chemokines Associated with Infectious Diseases in Human and Animal Models

Disease	Chemokine expression profiles	References
Viral Encephalitis		
Cytomegalovirus	CXCL10, CCL2, CCL4, CCL5, CCL7	Asensio *et al.*, 1999; Kielian, 2004
Herpes simplex virus	CXCL8, CXCL10, CCL2, CCL3, CCL5	Lokensgard *et al.*, 2001; Kielian, 2004
Mouse hepatitis virus	CXCL9, CXCL10, CCL2, CCL4, CCL5, CCL7	Kielian, 2004
Measles virus	CCL5, CXCL10	Patterson *et al.*, 2003
HIV encephalitis HIV-associated dementia	CCL2, CCL3, CCL4, CCL5	Sanders *et al.*, 1998; Kitai *et al.*, 2000; Garden, 2002; Kielian, 2004; Reinhart, 2003
Bacterial Meningitis		
CSF of patients	CXCL8, CCL2, CCL3, CCL4	Lahrtz *et al.*, 1997; Lahrtz *et al.*, 1998; Kielian, 2004
Listeria monocytogenes	CXCL2, CCL3, CCL4	Lahrtz *et al.*, 1998
Hemophilus influenzae	CXCL2, CCL2, CCL3, CCL5	Kielian, 2004
Streptococcus pneumoniae	CXCL2, CXCL8	Kielian, 2004; Zwijnenburg *et al.*, 2001
Brain abscess	CXCL2, CCL1, CCL2, CCL4, CCL5	Kielian, 2004
Other CNS Infections		
Toxoplasmosis	CXCL2, CXCL9, CXCL10, CCL2, CCL3, CCL4, CCL5, CCL7	Khan *et al.*, 2000, 2001; Strack *et al.*, 2002
Cryptococcus neoformans	CCL3	Huffnagle and McNeil, 1999
Creutzfeldt–Jakob Disease	CX_3CL1	Hughes *et al.*, 2002

Old nomenclature appears in parentheses: CCL1 (I-309); CCL2 (MCP-1); CCL-3 (MIP-1α); CCL4 (MIP-1β); CCL5 (RANTES); CCL7 (MCP-3); CXCL2 (GRO-β); CXCL8 (IL-8); CXCL10 (IP-10 or IFN-inducible protein 10); CXCL9 (MIG or monokine-induced by IFN-γ); CX_3CL1 (fractalkine).

in the CSF of HIV-infected individuals with CMV encephalitis, implicating MCP-1/CCL2 in macrophage/microglia recruitment and suggesting a potential mechanism to augment HIV-1 replication in the CNS (Bernasconi *et al.*, 1996).

b. Herpes Simplex Encephalitis. Herpes simplex virus (HSV) can cause acute focal necrotizing encephalitis with severe necroinflammation and edema (Hu *et al.*, 2003). The mechanisms responsible for tissue damage involve direct virus-mediated damage and indirect immediate destruction of brain parenchyma. At least one study demonstrated that MCP-1/CCL2 plays an important role in the chronic phase of HSV-2 encephalomyelitis and neutralization of MCP-1/CCL2 activity, which leads to increased severity of disease (Nakajima *et al.*, 2001). In addition, elevated levels of numerous chemokines, including MCP-1/CCL2, MIP-1α /CCL3, and RANTES/CCL5, have been detected in the CSF of patients presented with HSV-1 encephalitis; they appear to be correlated with disease severity (Rosler *et al.*, 1998). Recent evidence suggests that chemokines like IP-10/CXCL10 are capable of inhibiting HSV replication in primary neurons, providing an alternate mechanism for these mediators (Lokensgard *et al.*, 2001).

c. Measles Virus Encephalitis. Recent studies using a transgenic-mouse model of neuron-restricted measles virus infection have revealed the expression of numerous cytokines within the CNS, including RANTES/CCL5 and IP-10/CXCL10 (Patterson *et al.*, 2003). Both RANTES/CCL5 and IP-10/CXCL10 were found to collocalize with measles virus–infected neurons *in vivo*, as seen with studies performed using transgenic mice (Patterson *et al.*, 2003), indicating that these chemokines play a pivotal role in the pathogenesis of the antiviral immune response to measles virus.

d. HIV Encephalitis. CNS disease is a late complication of HIV infection, even though viral entry into the CNS is a relatively early phenomenon (Achim and Wiley, 1996). Thus, there is a significant interval between the initial establishment of CNS infection and the onset of clinically detectable disease. Nearly one-fourth of all HIV-infected patients develop cognitive, behavioral, and/or motor abnormalities that collectively are referred to as HIV-associated dementia (HAD). HIV encephalitis is a common manifestation of HAD and is characterized histologically by monocyte/macrophage infiltration into the brain and microglial nodules, formation of macrophage-derived multinucleated giant cells, and myelin pallor (Lipton *et al.*, 1994). The principal targets of HIV in the CNS are cells of mononuclear phagocytic lineage (Kure *et al.*, 1990). Several reports suggest that HIV-infected macrophages/microglia express elevated levels of numerous factors implicated in neurotoxicity, including proinflammatory cytokines (TNF-α, IL-1β, IL-6), chemokines (MIP-1β/CCL4, RANTES/CCL5), nitric oxide, and excitatory amino acids (glutamate, lysine) (Garden, 2002). These findings suggest that HIV infection in the CNS promotes neuronal injury through the release of soluble factors produced by infected macrophages/microglia. The production of CC chemokines in the infected CNS has been implicated in the recruitment of PBMCs, which serve as additional targets for HIV (Reinhart,

2003). Several reports have revealed elevated levels of the CC chemokines MCP-1/CCL2, MIP-1α/CCL3, MIP-1β/CCL4, and RANTES/CCL5 in the CSF or brain tissue of those suffering from HIV encephalitis (Sanders *et al.*, 1998). The CC chemokines MIP-1α/CCL3, MIP-1β/CCL4, and RANTES/CCL5 have been reported to prevent entry of HIV into macrophage/microglia (Kitai *et al.*, 2000), suggesting that chemokine release could be a protective response following HIV infection that competes for coreceptor binding in an attempt to contain the spread of HIV within the CNS. Conversely, enhanced CC-chemokine production during the course of HIV encephalitis may have deleterious consequences through the recruitment of additional effector cells into the CNS that could serve as additional targets for the virus for new rounds of viral replication (Reinhart, 2003). It is possible that over time, the profiles of chemokine coreceptor usage may be substantially different from those utilized by the original parental virus, due to emergence of viral variants produced during the virus swarm stage of acute HIV infection. Therefore, the release of CC chemokines by activated macrophages/microglia in the CNS may not have any effect on the infectivity of target cells by new viral variants. The overall consequences of CC-chemokine expression in the context of HIV encephalitis and progression to HAD remain to be completely defined, but they may be dictated by changes in viral phenotype and coreceptor usage throughout the evolution of disease.

2. Bacterial Meningitis

Bacterial meningitis is one of the top ten infectious causes of death worldwide, and approximately one-half of patients surviving the infection have neurological deficits and other sequelae of this disease (Nau and Bruck, 2002). Bacterial pathogens gain access into the bloodstream and then penetrate the blood-brain barrier (BBB) and replicate in the subarachnoid space. The major bacterial pathogens include *S. pneumoniae*, *N. meningitidis*, and *H. influenzae* (Townsend and Scheld, 1998). The pathological processes occurring in these patients includes cerebral edema; CSF flow blockade, leading to elevated intracranial pressures; seizures; and focal areas of ischemia and necrosis. In addition to the direct pathogen-induced damage to the brain cells, the host immune response can also induce neuron injury due to toxic effects mediated by the cytokines, chemokines, proteolytic enzymes, and superoxide anions produced locally at the site of infection (Townsend and Scheld, 1998). Neutrophils represent the main leukocyte infiltrant in bacterial meningitis. IL-8/CXCL8 is one of the predominant chemokines detected in the CSF of patients with bacterial meningitis (Lahrtz *et al.*, 1998). In addition to CXCL8, high levels of CC chemokines, including MCP-1/CCL2, MIP-1α/CCL3, and MIP-1β/CCL4, have also been demonstrated in the CSF of these patients (Lahrtz *et al.*, 1998). Altogether, these findings implicate a definitive role for CC chemokines in the pathogenesis of bacterial meningitis.

a. Listeria Monocytogenes *Meningitis.* Listeria is another causative agent of bacterial meningitis, particularly in immunocompromised individuals. Within 24 hours of Listeria infection in an experimental mouse model, over 80% of the invading cells are neutrophils, but monocytes are the predominant cellular infiltrate by 72 hours. These infiltrates are also associated with the production of the CXC and CC chemokines (Lahrtz *et al.*, 1998). Neutralization of MCP-1/CCL2 significantly reduced macrophage infiltration into the meninges, suggesting that this chemokine plays a major role in the pathogenesis of early Listeria and other bacterial meningitis.

b. Streptococcus Pneumoniae *Meningitis.* Recently, there have been major advancements in the understanding of the pathogenesis to this often fatal infection. Intranasal delivery of the pathogen led to the expression of the chemokine CXCL2, the murine functional homologue of IL-8/CXCL8 (Zwijnenburg *et al.*, 2001). This chemokine is considered to be the major factor responsible for the recruitment of neutrophils into the meninges. Bacterial meningitis is associated with higher and broader expression levels of chemokines in the CSF (Lahrtz *et al.*, 1998). This is partly explained by the high levels of immunostimulatory products associated with bacterial infections such as those in the cell wall, and secreted virulence factors, including chemokines. However, additional studies would help ascertain the importance and redundancy of individual chemokine mediators in the context of bacterial meningitis and determine whether effective therapeutic approaches can result from manipulation of the chemokine system following infection.

3. *Toxoplasma gondii* Infection

Toxoplasmosis is a worldwide parasitic infection affecting both humans and animals that is caused by an intracellular protozoon, *Toxoplasma gondii*. Natural infection with *T. gondii* results from ingestion of cysts or oocytes found in infected animal meat or fecal material. Once ingested, the parasite matures, invades the intestines, and disseminates to most organs, including the CNS (Bhopale, 2003). *T. gondii* infection induces a strong cell-mediated immune response, along with widespread expression of chemokines including GRO-β/CXCL2, MIG/CXCL9, IP-10/CXCL10, MCP-1/CCL2, MIP-1α/CCL3, MIP-1β/CCL4, RANTES/CCL5, and MCP-3/CCL7, which correlates with the coordinated recruitment of leukocytes into the CNS (Strack *et al.*, 2002). Microglia are the main source of RANTES/CCL5 and MIG/CXCL9 expression following *T. gondii* infection in the CNS, whereas astrocytes are the main producers of IP-10/CXCL10 and MCP-1/CCL2 (Strack *et al.*, 2002). Neutralization of IP-10/CXCL10 in mice following infection with *T. gondii* led to increased parasite burdens in the brain tissue of these animals, suggesting that this chemokine plays a protective antiparasite role (Khan *et al.*, 2000). Similarly, CCR1 and CXCR2 are also associated with the pathogenesis of toxoplasmosis, a fact evidenced by the enhanced parasite burdens in the brains of both CCR1 and CXCR2 knockout mice infected with *T. gondii* (Del Rio *et al.*, 2001; Khan *et al.*, 2001).

4. *Cryptococcus neoformans*

Cryptococcus neoformans is an opportunistic infection in AIDS patients acquired via the respiratory tract and disseminated to the CNS (Rodrigues *et al.*, 1999). Immunity against disseminated cryptococcosis, the disease caused by *C. neoformans*, is mediated primarily by macrophages and $CD4^+$ T-cells, hinting at the importance of CNS-derived chemokines in the effective recruitment of these effectors to contain infection. Interestingly, CCR5 knockout mice demonstrated significantly reduced monocyte infiltration into the brain, which correlated with the development of clinical signs of cryptococcal meningitis (Huffnagle *et al.*, 1999). In addition, MIP-1α/CCL3 has been shown to be another critical chemokine in cryptococcal meningitis, inasmuch as MIP-1α/CCL3 knockout mice were found to exhibit lower leukocyte recruitment and fungal clearance from the brain than were wild-type animals (Huffnaglc *et al.*, 1999).

5. Creutzfeldt–Jakob Disease

Creutzfeldt–Jakob disease (CJD) is the emerging infectious disease that presents as a chronic neurodegenerative disease. In animal models of CJD, the expression of fractalkine/CX_3CL1 was found to be increased in astrocytes, while its receptor CX_3CR1 was augmented in microglia, indicating that interactions within this chemokine–chemokine receptor pair may influence the nature of the local cytokine milieu during chronic inflammation (Hughes *et al.*, 2002).

III. VIRAL STRATEGIES TO CIRCUMVENT CHEMOKINE SYSTEMS

A unique way to address the role of chemokines and their receptors in infectious diseases is to examine the manner in which microbes interact with chemokines and their receptors. Many viruses, particularly those of the herpesvirus and poxvirus families, encode chemokines and chemokine receptor homologues that have been thought to interfere with the action of endogenous chemokines and/or allow viruses to evade host immune responses

(Murphy, 2001). A general overview is given in Table II and discussed in the next section.

A. Viral Chemokine Modulators

Human CMV, herpesvirus-6, and HIV encode proteins with chemotactic activity for human leukocytes. HCMV encodes a CXC chemokine named vCXC-1, which, when secreted, binds to CXCR2 with high affinity (Penfold *et al.*, 1999). It is a potent neutrophil chemoattractant *in vitro* that could provide a mechanism for viral infection, expansion, and dissemination as macrophages phagocytose dying neutrophils. It also induces neutrophil degranulation, which explains the pathology of retinitis, pneumonitis, encephalitis, hepatitis, carditis, and gastroenteritis in HCMV infection.

Herpesvirus 6 (HHV 6) is the etiologic agent for exanthema subitum, a common fever and rash syndrome of childhood. Rarely, HHV-6 causes bone-marrow suppression, pneumonia, or encephalitis. HHV-6 encodes a chemokine homologue named U83, which induces chemotaxis of the monocytic cell line THP-1 (Zou *et al.*, 1999). It is still unclear whether the infected cells secrete this chemokine, but because the virus typically infects monocytes, U83 is hypothesized to be important for viral dissemination.

The HIV envelope protein gp120, which is expressed on the cell surface and secreted from infected cells, can induce a calcium influx signal upon binding to CCR5 and CXCR4 (Weissman *et al.*, 1997). Gp120/CCR5 signaling causes chemotaxis of T-cells and may attract more targets for subsequent rounds of new infection. The HIV-encoded regulatory protein tat has been shown to induce the secretion of MCP-1/CCL2 expression by human astrocytes, and this has been proposed as a model for recruitment of monocytes across the BBB (Weiss *et al.*, 1999). This study also found that exposure to tat resulted in increased monocyte expression of CCR5, which would predispose patients to HIV infection. Thus, early HIV infection of the CNS promotes chemokine production, which causes recruitment of monocytes, which are then susceptible to new infection and serve as an HIV reservoir. Yet another study showed that secreted tat can induce monocyte expression of CCR5, CXCR4, and CCR3 (Hajeer *et al.*, 1999). HIV Nef protein has been shown to induce secretion of MIP-1α/CCL3 and MIP-1β/CCL4 from HIV-infected macrophages (Swingler *et al.*, 1999). Supernatants from monocytes infected with wild-type HIV were found to activate and cause chemotaxis of resting T-cells that are usually refractory to HIV infection; antibodies to MIP-1α/CCL3 and MIP-1β/CCL4 could block this, whereas infection with Nef-deleted HIV could not. Thus, these experiments provide a potential mechanism by which HIV dissemination may be facilitated by chemokine signaling.

TABLE II

Mechanisms by Which Infectious Agents May Subvert the Chemokine System

Infectious agent	Mechanism of action
Cell Entry	
Human immunodeficiency virus (HIV)	Enters cells through a variety of chemokine receptors
Plasmodium vivax	Enters cells through the Duffy antigen receptor of chemokines (DARC)
Rabbit myxoma virus	Replication regulated by CCR1, CCR5, and CXCR4
Swinepox	K2R is a chemokine-receptor homologue
Sheepox	Q2/3L is a chemokine-receptor homologue
Murine cytomegalovirus (CMV)	M33 is a chemokine-receptor homologue
Human CMV	US27, US28, and UL-78 are chemokine-receptor homologues
Herpesvirus Saimiri	ECF-3 is a functional chemokine receptor
KSHV	ORF-74 encodes a functional chemokine receptor
Spread of Virus	
Human CMV	Vcxc-1 attracts and activates neutrophils through CXCR2
Murine CMV	MCK-1 attracts monocytes and allows dissemination
Human herpesvirus-6	U83 attracts monocytes, DR1, DR6
KSHV	K4, K6
HIV	Gp120 attracts T-cells, tat induces MCP-1 attracting monocytes, tat increases CCR5, CXCR4, and CCR3 expression in monocytes; nef induces MIP-1α- and MIP-1β-attracting T-cells
Immune Evasion	
Moluscum contagiosum virus	MC148R blocks MCP-1-, MIP-1α-, RANTES-, I-309-, SDF-1-, and IL-8-induced chemotaxis
Human herpesvirus-8	VMIP-II blocks signaling of CCR1, CCR2, CCR3, CCR5, CXCR4, and US28; VMIP-1 blocks R5 HIV entry; ORF74 can induce characteristics of Kaposi's sarcoma lesions in transgenic mice
Rabbit myxoma virus	M-T7 binds to and inactivates chemokines and γ interferon; M-T1 binds to chemokines and reduces virally induced inflammation
Mouse γ herpesvirus-68	M3 binds to chemokines
Human CMV	US28 is a functional receptor for MIP-1α/β, RANTES, MCP-1, and fractalkine
	US28 sequesters chemokines
	US28 is a coreceptor for R5, X4, and primary HIV strains
Human herpes virus-6	U12 is a putative chemokine receptor
Vaccinia	B29R, 35kDa CKBP function as chemokine scavengers
Various orthopox viruses	35kDa CKBP functions as a chemokine scavenger

B. Viral Chemokine Antagonists

Herpesviruses and poxviruses also secrete chemokine homologues that act as antagonists of normal chemokine function and have been implicated in modification of the host response. One poxvirus, molluscum contagiosum virus (MCV), causes benign skin lesions that are characteristically free of surrounding inflammation. MCV encodes a chemokine homologue named MC148R that, *in vitro*, can block the chemotactic action of chemokines like MCP-1/CCL2, MCP-3/CCL7, MIP-1α/CCL3, RANTES/CCL5, and I-309/CCL1, as well as the CXC chemokine SDF-1α's human monocytes. It also blocks the chemotactic action of IL-8 on human neutrophils and SDF-1 on lymphocytes, but it has no effect on the chemotaxis of monocytes to fMet-Leu-Phe, a nonchemokine chemoattractant (Damon *et al.*, 1998). Thus, MC148R is a broad-spectrum chemokine antagonist that could block the recruitment of immune effector cells to the sites of MCV infection in the skin.

HHV-8, the causative agent of Kaposi's sarcoma, encodes several chemokine homologues. Of these, vMIP-I binds to and blocks calcium flux signaling of CCR1, CCR2, CCR5, CXCR4, and a CMV-encoded viral chemokine receptor analogue, US28 (Kledal *et al.*, 1997). It also blocks monocyte chemotaxis to RANTES/CCL5, MIP-1α/CCL3, and MIP-1β/CCL4, as well as the entry of HIV dual-tropic strain (the 89.6 strain) through CCR3, CCR5, and CXCR4 (Boshoff *et al.*, 1997). This provides a mechanism by which HHV-8 may evade the immune response and possibly inhibit HIV replication.

C. Viral Chemokine Receptors

HCMV encodes three chemokine receptor homologues, named US27, US28, and UL33. US28 is a functional chemokine receptor that binds to MIP-1α/CCL3, MIP-1β/CCL4, MCP-1/CCL2, and RANTES/CCL5, but not to CXC chemokines, when transfected into human cells (Gao and Murphy, 1994). In HCMV-infected fibroblasts, it depletes extracellular chemokines by binding to and internalizing them (Bodaghi *et al.*, 1998), thereby providing a mechanism for evading immune response. US28 also functions as an HIV coreceptor on certain cell types specific for R5, X4, and primary HIV isolates (Pleskoff *et al.*, 1997), thereby providing a mechanism for HCMV to exacerbate HIV disease.

HHV-6 and HHV-8 encode functional chemokine receptors called U12 and ORF74, respectively. ORF74 encodes a constitutively active receptor, which means that it is linked to cell proliferation *in vitro*. Moreover, ORF74 transgenic mice developed lesions characteristic of Kaposi's sarcoma, suggesting that this chemokine receptor may, in fact, be responsible for pathological lesions (Yang *et al.*, 2000). An ORF74 is syntenic with ECRF3 of the related gamma herpesvirus, herpesvirus Saimiri, a primate-restricted T-lymphotropic γ-herpes virus that causes leukemias and lymphomas in nonhuman primates (Ahuja and Murphy, 1993). ECRF3 is a functional CXC chemokine receptor when calcium influx is measured in RNA-injected frog oocytes. However, biological function in mammalian cells is still unclear.

D. Viral Chemokine Scavengers

Three structurally unrelated classes of secreted viral chemokine scavengers, which bind chemokines and prevent activation of cellular receptors, have been identified. The first is the 35kD protein class, found in diverse orthopox and leporipox viruses. Deletion of the myxoma virus 35kFa protein M-T1 increased leukocyte trafficking to infected sites but had no significant effect on mortality or viral load (Lalani *et al.*, 1999). The second class has one known member to date, M-T7, of the rabbit pathogen myxoma virus. M-T7 can scavenge both chemokines and interferon-γ and is a critical virulence factor. The third class has one member, the M3 protein of γ-herpesvirus-68, a mouse pathogen. All three of these proteins have unique sequences and bind to a broad spectrum of mammalian chemokines.

IV. STRATEGIES FOR BLOCKING THE CHEMOKINE SYSTEM IN INFECTIOUS DISEASE

This review so far has summarized the role of chemokines and chemokine receptors in infectious disease and has also provided proof of principle to target chemokine systems in disease states. The following discussion will provide a brief outline of several potential strategies that can be used to target chemokine systems in infectious disease (Table III). Viral chemokine modulators can be safe targets, as seen with Duffy and CCR5-deficient individuals. In HIV infection, identification of new targets for antiviral therapy that could suppress HIV replication is important. The effective blockade of HIV coreceptor CCR5 could be extremely useful in controlling HIV replication. Chemokines also have a number of disadvantages as potential targets. Since they are proteins, they must be administered parenterally and they tend to have very short half-lives, which means that frequent administration is required. Also, they could potentially cause undesirable adverse events due to chronic activation or blockade of their

TABLE III
Therapeutic Agents Targeting the Chemokine System

Molecule	Mechanism of action
Natural and Modified CC Chemokines	
MCP-2	Natural antagonist of MCPs and RANTES
MIP-1αP	Natural agonist of CCR5; blocks R5 HIV
RANTES (6–78)	Synthetic antagonist of CCR1, CCR3, and CCR5; blocks R5 HIV
Met-RANTES	Synthetic antagonist of CCR1, CCR3, and CCR5; blocks R5 HIV
AOP-RANTES	Synthetic antagonist of CCR1, CCR3, and CCR5; blocks R5 HIV
NNY-RANTES	Synthetic antagonist of CCR1, CCR3, and CCR5; blocks R5 HIV
MET-Cκβ7	Synthetic antagonist of CCR3; blocks eosinophil migration
BB10010	Synthetic MIP-1α variant; agonist; safe in Phase I clinical trial
Modified CXC Chemokines	
GRO-α	Synthetic antagonist of CXCR2; blocks neutrophil migration
PF4	Synthetic antagonist of CXCR2
MGSA-H19A	Synthetic antagonist of CXCR2
MGSA-E6A	Synthetic antagonist of CXCR2 and DARC, blocks *P. vivax* infection
Met-SDF-1β	Synthetic CXCR4 agonist; X4 HIV; blocks CXCR4 recycling
Antibody to Chemokines and Receptors	
α-IL-8 (K2.2, K4.3)	Inhibits IL-8 signaling in neutrophils; human monoclonal
α-CCR5 (PA14, 2D7)	Blocks R5 fusion; murine monoclonal
2D7	CCR5-directed antibody; blocks R5 HIV
PA14	CCR5-directed antibody; blocks R5 HIV
12G5	CXCR4-directed antibody; blocks X4 HIV
Peptides	
T22	Synthetic antagonist of CXCR4; blocks X4 HIV
T140	Synthetic antagonist of CXCR4; blocks X4 HIV
T134	Synthetic antagonist of CXCR4; blocks X4 HIV
ALX40-4C	Synthetic antagonist of CXCR4; blocks X4 HIV
CGP64222	Synthetic antagonist of CXCR4; blocks X4 HIV

TABLE III (*Continued*)

Molecule	Mechanism of action
V3 loop peptides	Synthetic antagonist of CXCR4; blocks X4 HIV
3bv	Synthetic antagonist of CXCR4; blocks X4 HIV
vMIP-II	Blocks both X4 and R5 strains of HIV
Small-Molecule CCR5 Antagonists	
TAK-779	Synthetic antagonist of CCR5 and CCR2b; blocks R5 HIV
NSC 651016	Synthetic antagonist of CCR1, CCR3, CCR5, and CXCR4
E913	Synthetic antagonist of CCR5; blocks R5 HIV
TAK-220	Synthetic antagonist of CCR5; blocks R5 HIV
GW873140	Synthetic antagonist of CCR5; blocks R5 HIV
SCH-C	Synthetic antagonist of CCR5; blocks R5 HIV
SCH-D	Synthetic antagonist of CCR5; blocks R5 HIV
AK 602	Synthetic antagonist of CCR5; blocks R5 HIV
UK-427857	Synthetic antagonist of CCR5; blocks R5 HIV
SB225002	Synthetic antagonist of CXCR2
BX 471	Synthetic antagonist of CCR1
RS-511336-000	Synthetic antagonist of CCR2
RS-163883-230	Synthetic antagonist of CCR3
Bindarit	Inhibits production of MCP-1 by monocytes
Small-Molecule CXCR4 Antagonists	
AMD3100	Synthetic antagonist of CXCR4; blocks X4 HIV; blocks R5-to-X4 shift
AMD3329	Synthetic antagonist of CXCR4; blocks X4 HIV
AMD 8664	Synthetic antagonist of CXCR4; blocks X4 HIV
AMD-070	Synthetic antagonist of CXCR4; blocks X4 HIV
Gene Therapy	
SDF-1α intrakine	Diminishes CXCR4 on cell surface; blocks X4 HIV
RANTES intrakine	Diminishes CCR5 on cell surface; blocks R5 HIV
CCR5 intrabody	Diminishes CCR5 on cell surface; blocks R5 HIV
CXCR4 antisense oligonucleotides	Blocks X4 HIV
CCR5 hammerhead ribozyme	Diminishes CCR5 on cell surface; blocks R5 HIV

receptor. Some studies have also shown that RANTES can, in fact, increase HIV replication (Kinter *et al.*, 1998). Moreover, targeting one receptor will probably drive HIV to use the alternative CXCR4 coreceptor, and this may lead to accelerated disease progression.

A. Chemokines and Modified Chemokines

As agonists of the CCR5 receptor, β-chemokines can inhibit HIV infectivity either by sterically blocking gp120 binding sites on the receptor or by inducing internalization of CCR5, rendering it physically inaccessible to the gp120/CD4 complex. Use of such agonists could lead to chronic receptor activation, desensitization, and internalization, but also to receptor recycling, the latter being an unacceptable property of HIV-infection blockade.

One approach is to use modified chemokines that behave primarily as receptor antagonists. $RANTES_{(9-68)}$, in which the eight N-terminal residues are removed, was the first modified chemokine to be reported to have anti-HIV activity (Arenzana-Seisdedos *et al.*, 1996). Met-RANTES, which bears an additional methionine residue at its amino-terminus, is another antagonist of CCR5 (Proudfoot *et al.*, 1999). Met-RANTES demonstrates modest antiviral activity *in vitro,* although it is weaker than RANTES itself (Simmons *et al.*, 1997). AOP-RANTES, wherein the N-terminally oxidized RANTES is derivatized with 1-aminooxy-n-pentane, behaves like a classical high-affinity receptor antagonist of CCR5 (Simmons *et al.*, 1997). It also possesses excellent *in vitro* antiviral activity against R5 strains (Simmons *et al.*, 1997). NNY-RANTES (N-nonanoyl-RANTES) has been described to have antiviral activity in *in vivo* studies using SCID/Hu/PBL mice (Mosier *et al.*, 1999). However, in those mice that had detectable viremia after 7 days of dosing, the original R5 virus has evolved to utilize CXCR4 as a co-receptor.

B. Antibodies of Chemokine Receptors

Antibodies against chemokines or their receptors offer another way to target the chemokine system. They have the advantage of having a longer half-life than peptides, but they need to be administered parenterally. Two humanized IL-8 antibodies have already been developed (Yang *et al.*, 1999). These agents are able to block IL-8 signaling completely in human neutrophils *in vitro*; they are also able to inhibit neutrophil accumulation to the injection site when given intravenously simultaneously with an intradermal injection of IL-8 in rabbits.

2D7 is a well-characterized CCR5-directed antibody whose binding has been mapped to the second extracellular domain (Wu *et al.*, 1997). *In vitro* studies the using U87/MG-$CD4^{+}$ T-cell line showed excellent antiviral activity against R5 strains (Wu *et al.*, 1997). Despite these promising results, serial passage of R5 strains in the presence of 2D7 can select for strains that are 100-fold less susceptible to the mAb than to the parental virus, although the adapted HIV strain continues to use CCR5 (Aarons *et al.*, 2001).

Another CCR5-directed mAb, PA14, has also displayed potent anti-HIV activity against R5 strains. While this mAb prevented infection of PBMCs with inhibitory potency comparable to that of RANTES (IC_{50} 17 nM), PA14 was more potent than RANTES when macrophages were used as substrates (Trkola *et al.*, 2001). Efforts are under way to prepare a humanized version of this mAb in order to initiate human clinical trials.

As in the case of CCR5, CXCR4-directed mAbs have also been shown to have anti-HIV activity. 12G5 is an anti-CXCR4 mAb, with variable effect in suppressing HIV strains, that uses CXCR4 (McKnight *et al.*, 1997). However, several studies have shown that different X4 strains of HIV have varying susceptibility to 12G5, suggesting that these HIV subtypes utilize different domains of CXCR4 and that only the 89.6 strain employs a region to which 12G5 also binds (Strizki *et al.*, 2001).

C. Peptides and Natural-Product-Derived Inhibitors of CCR5/CXCR4 Function

The extracellular domain of the CXCR4 receptor bears a net negative charge, and thus, molecules that bear a net positive charge are often well-suited to interact strongly with it. These molecules have the advantage of being easier to synthesize than chemokines; however, they are not orally bioavailable.

One example is CGP64222, originally designed as a mimic of the transactivation response-element-binding domain of the HIV tat protein (TAR) has been shown to completely block tat binding to TAR. Subsequently, it has also been shown to block an early step in the life cycle of HIV by binding to CXCR4 selectively (Daelemans *et al.*, 2000).

Another polycationic competitive inhibitor of tat, ALX40-4C, has also been shown to possess potent anti-HIV activity by binding to CXCR4 (Doranz *et al.*, 1997). Like CGP64222, it antagonizes the activity of SDF-1; however, it does not interfere with the binding of β-chemokine to CCR5, nor does it inhibit infection by R5 strains.

The V3 loop peptides (approximately 40 residues based on the V3 structure of X4 and R5X4 viruses) have been shown to exert anti-HIV activity against X4 viruses (Sakaida *et al.*, 1998). These peptides do not induce calcium mobilization upon binding the CXCR4 receptor; they function as true antagonists.

Another CXCR4 antagonist is the 18-mer, T22, which is an analog of the isolate from the hemocyte debris of American horseshoe crabs (Murakami *et al.*, 1999) and potently inhibits the infectivity of T-tropic viruses but not M-tropic viruses. Modification of T22 to improve its potency and reduce its cytotoxicity resulted in the discovery of the 14-MER T134 (Tamamura *et al.*, 1998).

A 15-mer p3bv, derived from bee venom, was extremely potent in its inhibition of SDF-1 binding to CXCR4, thereby leading to potent antiviral activity (Fenard *et al.*, 2001). Binding of p3bv does not alter the surface expression of the CXCR4 receptor, reinforcing the fact that it works as an antagonist and not by reducing the levels of receptor expression.

Viral macrophage inhibitory protein-II (vMIP-II), a viral chemokine encoded by Kaposi's sarcoma virus, displays anti-HIV activity toward the R5 and X4 viruses, which both utilize CCR3 (Kledal *et al.*, 1997). It appears to exert this inhibition via direct blockade of viral binding to the chemokine coreceptors, as demonstrated by a series of competition experiments.

D. Small-Molecule CCR5 Inhibitors

Recently, several small molecules with activity against CCR5 (Table III), which thereby inhibit R5 strains of HIV, have been discovered. TAK-779, a quaternary ammonium salt, completely inhibited RANTES-induced calcium influx in Chinese hamster ovary (CHO) cells (Baba *et al.*, 1999). At a concentration of 32nM, this molecule completely inhibited the replication of the R5 HIV strain BaL in MAGI-CCR5 cells, while having no effect on the replication of the X4 strain IIIb (Baba *et al.*, 1999). Although TAK-779 inhibits binding of R5 gp120 to CCR5, it does not block the interaction of CCR5 with anti-CCR5 antibodies that recognize the N-terminal domains (Dragic *et al.*, 2000). It is therefore plausible that the binding within the transmembrane domain induces conformational changes that disrupt binding sites on CCR5 for gp120 and RANTES, but not on the simpler domains to which antibodies react (Dragic *et al.*, 2000). *In vivo* data available for TAK-779 indicate that the compound displays 98% binding to human plasma (Baba *et al.*, 1999). Phase I study was initiated for TAK-779, but the findings of local reaction to the injection have put the study on hold.

NSC 651016 is a distamycin derivative that shows an IC_{50} between 50 uM and 200 uM on CXCR4 and CCR5 and inhibits infection by R5 and X4 viruses (Howard *et al.*, 1998). It is a potent inhibitor of chemotaxis mediated by CC chemokines and SDF-1α, and it also induces CCR5 internalization of CCR5 and CXCR4, suggesting that this activity is not driven just by simple competitive receptor antagonism.

Another compound, E913, is a high-affinity CCR5 antagonist with potent antiviral activity against R5 HIV (Maeda *et al.*, 2001). This compound also shows synergy with AMD3100, a potent CXCR4 inhibitor that inhibits the replication of dual-tropic HIV. Competition experiments indicate that E913 binds to the C-terminal half of the second extracellular loop (Maeda *et al.*, 2001).

GlaxoSmithKline, a pharmaceutical company, has recently developed a CCR5 antagonist named GW873140 that shows potent activity against HIV R5 strains, both *in vitro* and *in vivo*. It binds to CCR5 receptors and subsequently blocks the binding of HIV R5 strains to this receptor, thereby inhibiting R5 infection. Phase I/II studies show this molecule to be a potent suppressor of HIV R5 strain replication *in vivo* (Demerest *et al.*, 2004). Moreover, GW873140 was very well tolerated in both HIV-negative and HIV-positive subjects. However, there was emergence of X4 strains of HIV following the suppression of the R5 strain in some subjects who were originally harboring dual-tropic HIV strains (Westby *et al.*, 2004). The long-term clinical implications of this shift are currently unknown.

Another pharmaceutical company, Schering-Plough, has a compound named SCH 351125 (SCH-C), an oxime-piperidine that inhibits intracellular calcium influx by RANTES/CCL5, without any agonist activity (Eron and Hogan, 2002). SCH-C is highly selective for β-chemokine receptors and does not show significant affinity toward a broad array of other human receptors (Strizki *et al.*, 2001). SCH-C also exhibited antiviral activity *in vivo* after twice-daily oral dosing in SCID/Hu/Thy/Liv mice infected with R5-strain BaL (Strizki *et al.*, 2001). SCH-C has also been studied in combination with other antiviral agents *in vitro*, where their ability to suppress replication of R5 strains was tested. Synergy was observed against all viruses tested when SCH-C was combined with AZT and lamivudine (nucleoside analogs), with efavirenz (nonnucleoside analogs), with indinavir (a protease inhibitor), and with T-20 (a fusion inhibitor) (Tremblay *et al.*, 2002). Preliminary data from Phase I studies using SCH-C are now available (Eron and Hogan, 2002). The medication was well-tolerated, except for the association of QT interval prolongation in the EKG in those who received the high doses (Eron and Hogan, 2002). Also, data from some chronic HIV patients infected with R5 strains who received SCH-C showed that after ten days of dosing of 25 mg twice daily, 10 out of 12 patients experienced a drop in HIV viral load by 0.5 to 1.0 logs (Eron and Hogan, 2002).

Another product, SCH-D, was also found to have potent activity against R5 strains and was recently tested in a Phase I clinical trial. In summary, this CCR5 antagonist was found to be safe and showed potency superior to that of SCH-C in suppressing HIV R5 strains in chronic HIV-infected individuals (Schurman *et al.*, 2004).

Several other agents are under evaluation for their activity against R5 strains. These agents are summarized in Table III. Most of these agents are in the preclinical stages of development and will undergo evaluations *in vivo* as well as Phase I trials in the coming years.

E. Small-Molecule CXCR4 Antagonists

In the 1990s the bicyclams were discovered as inhibitors of T-cell-line-adapted strains of HIV, but their mechanism of action was not known until it was shown that they specifically block CXCR4 (Schols *et al.*, 1998). The prototype bicyclam, AMD3100, is a pure CXCR4 antagonist that inhibits X4 replication *in vitro*. Even though this drug showed efficacy against X4 strains and prevented an R5-to-X4 shift in long-term lymphoid tissue culture, in addition to being nontoxic, bioavailability was poor and drug-resistant strains were isolated after extended exposure to the drug. Marked synergy is observed when AMD3100 and an HIV fusion inhibitor, T20, were tested in combination against X4 strains leading to the suggestion that blocking two sequential steps in viral infection may be a particularly efficient antiviral strategy (Tremblay *et al.*, 2000). *In vivo* studies using AMD3100 in SCIDhu/Thy/Liv mice that were infected with HIV demonstrated a dose-dependent decrease in HIV p24 levels (Hendrix *et al.*, 2000). AMD3100 has also been tested in a Phase I clinical trial in healthy volunteers (Hendrix *et al.*, 2000). Unfortunately, at higher doses, cardiac arrhythmias were reported in patients leading to halting of the clinical trials. A second-generation analog, AMD8664, is currently being evaluated in Phase I clinical trials.

F. Gene Therapy-Based Approaches

Intracellular chemokines (or "intrakines") are chemokines where the normal signal peptide that facilitates secretion has been replaced with an endoplasmic reticulum (ER) retention signal (Table III). Thus, intrakines can trap chemokine receptors inside the cells. When delivered by gene-therapy vectors, an SDF-1 intrakine has been shown to reduce cell-surface CXCR4 expression and X4 HIV replication (Chen *et al.*, 1997). A modified RANTES gene is capable of yielding similar results for CCR5.

A retroviral vector that can express SDF-1 and RANTES/CCL5 intrakines, plus a selective marker, can enrich transduced human PBLs, which resist infection with X4, R5, and dual strains due to markedly reduced levels of CCR5 and CXCR4. CCR2 levels remained unaffected, and the cells proliferated normally with tetanus antigen (Bai *et al.*, 1998). One analogous antichemokine receptor approach has been to block surface expression of CCR5 (Steinberger *et al.*, 2000). This so-called "intrabody" has the potential to more precisely target just one chemokine receptor at a time. Other methods are being developed to genetically disrupt the chemokine genes directly with triplex DNA, cleaving enzymes, promoter-modifying agents, and antisense oligonucleotides. One study demonstrated that an antisense oligonucleotide directed to the CXCR4 gene translational initiation region was able to reduce the replication of X4 viral strain in COS cells stably expressing CD4 and CXCR4 (Kusunoki *et al.*, 1999). All these techniques could be used on hematopoietic cells, which, after transduction and purification, could be infused into patients to diminish HIV replication. Theoretically, the transduced cells could last for many months, providing the possibility of prolonged benefit. A potential problem with RANTES intrakine is that RANTES can also bind to CCR1 and CCR3 (in addition to CCR5), and one would expect reduction of these two other high-affinity receptors also. Hence, these agents need to be tested carefully in nonhuman primate models for potential adverse events prior to the initiation of human trials.

G. Strategies for Selective Targeting of HIV-Infected Cells

The gp120/41 molecule must be expressed in the outer membrane of cells to allow the budding of infectious virus and is thus a characteristic marker of cellular infection. Through the encapsulation of drugs or toxins in liposomes coated with CCR5 or CXCR4 and CD4, it may be possible to selectively target gp120/41-expressing cells. This carries the obvious advantage of hitting HIV where it abides and offers hope of minimizing adverse drug events.

Two studies have used pseudotyped vesicular stomatitis virus (VSV) and rabies virus with CD4 and CXCR4 in place of their normal envelope proteins (Mebatsion *et al.*, 1997; Schnell *et al.*, 1997). This system targets a cytolytic virus to HIV-infected cells. While the rabies-virus pseudotype was only able to undergo a single round of infection, the VSV pseudotype could continue to propagate in culture, destroying HIV-infected cells.

V. FUTURE DIRECTIONS

Recent advances in the understanding of the interactions between chemokine systems and infectious agents have added a great deal to our understanding of how microbes interact with host immune systems. Use of animal models and human genetic variants has unraveled several new members of the chemokine system that are critical for normal homeostasis. Investigators have begun several approaches that target different components of chemokine systems, including the development of vaccines and adjuvants. Certain chemokines, like MCP-1/CCL2, have immunoregulatory properties that have the potential to boost or shape the immune response to vaccination (Gu *et al.*, 2000). Furthermore, several chemokines, like IL-4, IL-12, and IFN-γ, have adjuvant properties. The current approved adjuvant, alum, favors only a humoral immune response. Hence, chemokines could provide an option for favoring a T-helper-1 response, which may greatly improve the efficacy of vaccines directed against intracellular pathogens. Furthermore, it is possible that epitopes of gp120/41 that are not exposed to the immune system because of protein folding are transiently revealed during the fusion process. By fixation of these HIV envelope CD4 coreceptor structures with formaldehyde, whole-cell vaccines in transgenic mouse models have been shown to result in broadly reactive neutralizing antibodies. If the relevant epitopes can be defined, this may allow better HIV vaccines to be developed in the future.

It is largely unknown whether the effects seen with infectious conditions are due to a direct interaction between the pathogen and its host or to an indirect effect seen as a consequence of the overwhelming immune activation associated with the infectious process. In order to better understand the pathogenesis of infection, one needs to understand the roles of infectious agents and immune activation in the induction of chemokines and chemokine receptors. In this regard, a recent study demonstrated that the expression of CCR5 on the surface of natural killer cells is probably due to immune activated states rather than to the infection itself, as similar levels of CCR5 receptor upregulation were associated with HIV viremia and inflammatory conditions like Wegener's granulomatosis (Kottilil *et al.*, 2004). Further studies should unravel the common pathways of interaction between inflammatory cells and chemokine systems in infectious diseases and other inflammatory conditions.

One of the problems in using chemokine-based therapies for infectious diseases like HIV is the emergence of resistance or development of alternate receptor usage, as evident in the case of CCR5 and CXCR4 antagonists (Fig. 5A) (Schols, 2004; Westby *et al.*, 2004). Development of therapeutic strategies that target both coreceptors for HIV would preclude this problem (Fig. 5B). As seen in this diagram, one could potentially target both CCR5 and CXCR4; each agent thereby would individually block the respective tropic virus (Fig. 5B). Hence, such a combination could prevent replication of mono- and dual-tropic HIV viruses *in vivo*. Furthermore, several of these small molecules demonstrated synergy with already-existing antiretroviral agents *in vitro* and could soon be part of the antiretroviral armamentarium used to control HIV infection.

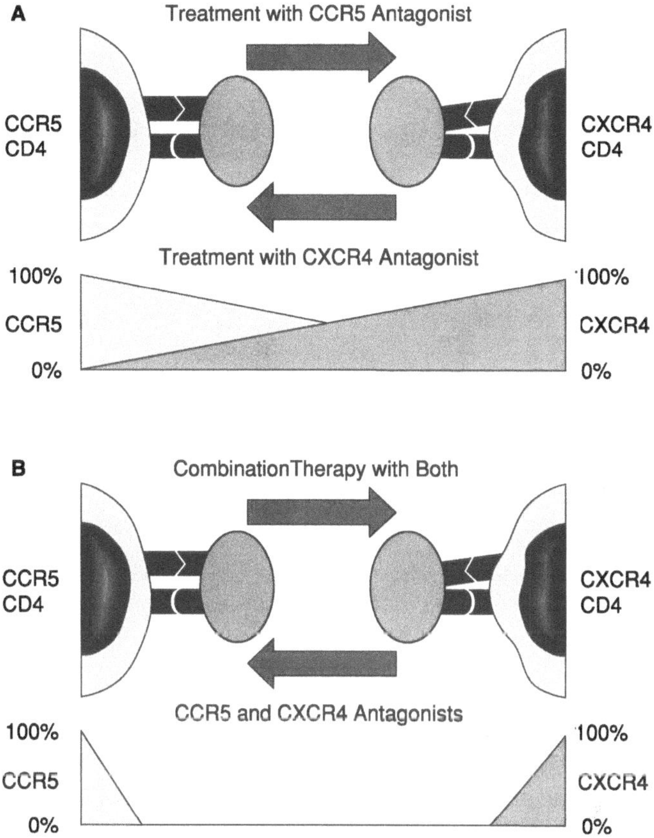

FIGURE 5 Panel A represents the clinical scenario in which a CCR5 antagonist or CXCR4 antagonist is used alone in the treatment of HIV infection. As seen in the graph at the bottom of the figure, such treatments are always associated with escape of the mutant strains of HIV with alternate receptor usage. A novel way to address this particular pattern of resistance is to use both CCR5 and CXCR4 antagonists in combinations so as to suppress the development of mutant strains with alternate receptor usage.

Several studies have also undertaken the task of exploring the effects of CCR5 and CXCR4 receptor engagement in the pathogenesis of HIV infection (Cicala *et al.*, 2002). Such studies could unravel novel potential targets against which new therapeutics could be developed. Knowledge of how pathogens evade host immune responses will ultimately lead to the identification of potential targets. It is expected that several of these chemokine-based therapies will provide clinicians with new ways to augment host response and circumvent infections. Such treatments aimed at novel targets within the chemokine system will have applications in the management of inflammatory conditions as well.

Acknowledgments

We would like to thank John Weddle for help with the manuscript preparation.

References

Aarons, E. J., Beddows, S., Willingham, T., Wu, L., and Koup, R. A. (2001). Adaptation to blockade of human immunodeficiency virus type 1 entry imposed by the anti-CCR5 monoclonal antibody 2D7. *Virology* **287,** 382–390.

Achim, C. L., and Wiley, C. A. (1996). Inflammation in AIDS and the role of the macrophage in brain pathology. *Curr. Opin. Neurol.* **9,** 221–225.

Ahuja, S. K., and Murphy, P. M. (1993). Molecular piracy of mammalian interleukin-8 receptor type B by herpesvirus saimiri. *J. Biol. Chem.* **268,** 20691–20694.

Alfano, M., Schmidtmayerova, H., Amella, C. A., Pushkarsky, T., and Bukrinsky, M. (1999). The B-oligomer of pertussis toxin deactivates CC chemokine receptor 5 and blocks entry of M-tropic HIV-1 strains. *J. Exp. Med.* **190,** 597–605.

Alkhatib, G., Combadiere, C., Broder, C. C., Feng, Y., Kennedy, P. E., Murphy, P. M., and Berger, E. A. (1996). CC CKR5: A RANTES, MIP-1alpha, MIP-1beta receptor as a fusion cofactor for macrophage-tropic HIV-1. *Science* **272,** 1955–1958.

Arenzana-Seisdedos, F., Virelizier, J. L., Rousset, D., Clark-Lewis, I., Loetscher, P., Moser, B., and Baggiolini, M. (1996). HIV blocked by chemokine antagonist. *Nature* **383,** 400.

Arthos, J., Rubbert, A., Rabin, R. L., Cicala, C., Machado, E., Wildt, K., Hanbach, M., Steenbeke, T. D., Swofford, R., Farber, J. M., and Fauci, A. S. (2000). CCR5 signal transduction in macrophages by human immunodeficiency virus and simian immunodeficiency virus envelopes. *J. Virol.* **74,** 6418–6424.

Arya, S. K., Ginsberg, C. C., Davis-Warren, A., and D'Costa, J. (1999). *In vitro* phenotype of SDF1 gene mutant that delays the onset of human immunodeficiency virus disease *in vivo*. *J. Hum. Virol.* **2,** 133–138.

Asensio, V. C., Kincaid, C., and Campbell, I. L. (1999). Chemokines and the inflammatory response to viral infection in the central nervous system with a focus on lymphocytic choriomeningitis virus. *J. Neurovirol.* **5,** 65–75.

Atchison, R. E., Gosling, J., Monteclaro, F. S., Franci, C., Digilio, L., Charo, I. F., and Goldsmith, M. A. (1996). Multiple extracellular elements of CCR5 and HIV-1 entry: Dissociation from response to chemokines. *Science* **274,** 1924–1926.

Baba, M., Nishimura, O., Kanzaki, N., Okamoto, M., Sawada, H., Iizawa, Y., Shiraishi, M., Aramaki, Y., Okonogi, K., Ogawa, Y., Meguro, K., and Fujino, M. (1999). A small-molecule, nonpeptide CCR5 antagonist with highly potent and selective anti-HIV-1 activity. *Proc. Natl. Acad. Sci. USA* **96,** 5698–5703.

Baggiolini, M., Dewald, B., and Moser, B. (1994). Interleukin-8 and related chemotactic cytokines–CXC and CC chemokines. *Adv. Immunol.* **55,** 97–179.

Baggiolini, M., Dewald, B., and Moser, B. (1997). Human chemokines: An update. *Annu. Rev. Immunol.* **15,** 675–705.

Bai, X., Chen, J. D., Yang, A. G., Torti, F., and Chen, S. Y. (1998). Genetic co-inactivation of macrophage- and T-tropic HIV-1 chemokine coreceptors CCR-5 and CXCR-4 by intrakines. *Gene. Ther.* **5,** 984–994.

Berger, E. A., Murphy, P. M., and Farber, J. M. (1999). Chemokine receptors as HIV-1 coreceptors: Roles in viral entry, tropism, and disease. *Annu. Rev. Immunol.* **17,** 657–700.

Bernasconi, S., Cinque, P., Peri, G., Sozzani, S., Crociati, A., Torri, W., Vicenzi, E., Vago, L., Lazzarin, A., Poli, G., and Mantovani, A. (1996). Selective elevation of monocyte chemotactic protein-1 in the cerebrospinal fluid of AIDS patients with cytomegalovirus encephalitis. *J. Infect. Dis.* **174,** 1098–1101.

Bhopale, G. M. (2003). Pathogenesis of toxoplasmosis. *Comp. Immunol. Microbiol. Infect. Dis.* **26,** 213–222.

Biti, R., Ffrench, R., Young, J., Bennetts, B., Stewart, G., and Liang, T. (1997). HIV-1 infection in an individual homozygous for the CCR5 deletion allele. *Nat. Med.* **3,** 252–253.

Bodaghi, B., Jones, T. R., Zipeto, D., Vita, C., Sun, L., Laurent, L., Arenzana-Seisdedos, F., Virelizier, J. L., and Michelson, S. (1998). Chemokine sequestration by viral chemoreceptors as a novel viral escape strategy: Withdrawal of chemokines from the environment of cytomegalovirus-infected cells. *J. Exp. Med.* **188,** 855–866.

Boshoff, C., Endo, Y., Collins, P. D., Takeuchi, Y., Reeves, J. D., Schweickart, V. L., Siani, M. A., Sasaki, T., Williams, T. J., Gray, P. W., Moore, P. S., Chang, Y., and Weiss, R. A. (1997). Angiogenic and HIV-inhibitory functions of KSHV-encoded chemokines. *Science* **278,** 290–294.

Bukrinskaya, A., Brichacek, B., Mann, A., and Stevenson, M. (1998). Establishment of a functional human immunodeficiency virus type 1 (HIV-1) reverse transcription complex involves the cytoskeleton. *J. Exp. Med.* **188,** 2113–2125.

Chen, J. D., Bai, X., Yang, A. G., Cong, Y., and Chen, S. Y. (1997). Inactivation of HIV-1 chemokine co-receptor CXCR-4 by a novel intrakine strategy. *Nat. Med.* **3,** 1110–1116.

Cheng-Mayer, C., Seto, D., Tateno, M., and Levy, J. A. (1988). Biologic features of HIV-1 that correlate with virulence in the host. *Science* **240,** 80–82.

Cicala, C., Arthos, J., Selig, S. M., Dennis, G., Jr., Hosack, D. A., Van Ryk, D., Spangler, M. L., Steenbeke, T. D., Khazanie, P., Gupta, N., Yang, J., Daucher, M., Lempicki, R. A., and Fauci, A. S. (2002). HIV envelope induces a cascade of cell signals in non-proliferating target cells that favor virus replication. *Proc. Natl. Acad. Sci. USA* **99,** 9380–9385.

Cocchi, F., DeVico, A. L., Garzino-Demo, A., Cara, A., Gallo, R. C., and Lusso, P. (1996). The V3 domain of the HIV-1 gp120 envelope glycoprotein is critical for chemokine-mediated blockade of infection. *Nat. Med.* **2,** 1244–1247.

Craigen, J. L., Yong, K. L., Jordan, N. J., MacCormac, L. P., Westwick, J., Akbar, A. N., and Grundy, J. E. (1997). Human cytomegalovirus infection up-regulates interleukin-8 gene expression and stimulates neutrophil transendothelial migration. *Immunology* **92,** 138–145.

Daelemans, D., Schols, D., Witvrouw, M., Pannecouque, C., Hatse, S., van Dooren, S., Hamy, F., Klimkait, T., de Clercq, E., and VanDamme, A. M. (2000). A second target for the peptoid Tat/transactivation response element inhibitor CGP64222: Inhibition of human immunodeficiency virus replication by blocking CXC-chemokine receptor 4-mediated virus entry. *Mol. Pharmacol.* **57,** 116–124.

Damon, I., Murphy, P. M., and Moss, B. (1998). Broad spectrum chemokine antagonistic activity of a human poxvirus chemokine homolog. *Proc. Natl. Acad. Sci. USA* **95,** 6403–6407.

Del Rio, L., Bennouna, S., Salinas, J., and Denkers, E. Y. (2001). CXCR2 deficiency confers impaired neutrophil recruitment and increased susceptibility during Toxoplasma gondii infection. *J. Immunol.* **167,** 6503–6509.

Demerest, J., Sparks, A. K., Shachoy-Clark, S., Schell, K., O'Mara, S. F., Shibayama, S., and Piscitelli, S. (2004). Single and dose escalation study to investigate the safety, pharmacokinetics, and receptor binding of GW873140, a novel CCR5 antagonist, in healthy subjects. *In* "11th Conference on Retroviruses and Opportunistic Infections," Abstract #139, San Francisco.

Dolei, A., Biolchini, A., Serra, C., Curreli, S., Gomes, E., and Dianzani, F. (1998). Increased replication of T-cell-tropic HIV strains and CXC-chemokine receptor-4 induction in T cells treated with macrophage inflammatory protein (MIP)-1alpha, MIP-1beta and RANTES beta-chemokines. *AIDS* **12,** 183–190.

Doranz, B. J., Grovit-Ferbas, K., Sharron, M. P., Mao, S. H., Goetz, M. B., Daar, E. S., Doms, R. W., and O'Brien, W. A. (1997). A small-molecule inhibitor directed against the chemokine receptor CXCR4 prevents its use as an HIV-1 coreceptor. *J. Exp. Med.* **186,** 1395–1400.

Dragic, T., Trkola, A., Thompson, D. A., Cormier, E. G., Kajumo, F. A., Maxwell, E., Lin, S. W., Ying, W., Smith, S. O., Sakmar, T. P., and Moore, J. P. (2000). A binding pocket for a small molecule inhibitor of HIV-1 entry within the transmembrane helices of CCR5. *Proc. Natl. Acad. Sci. USA* **97,** 5639–5644.

Easterbrook, P. J., Rostron, T., Ives, N., Troop, M., Gazzard, B. G., and Rowland-Jones, S. L. (1999). Chemokine receptor polymorphisms and human immunodeficiency virus disease progression. *J. Infect. Dis.* **180,** 1096–1105.

Eron, J. J., and Hogan, C. M. (2002). Entry and fusion inhibitors: An update. *The PRN Notebook* **7,** 16–22.

Faure, S., Meyer, L., Costagliola, D., Vaneensberghe, C., Genin, E., Autran, B., Delfraissy, J. F., McDermott, D. H., Murphy, P. M., Debre, P., Theodorou, I., and Combadiere, C. (2000). Rapid progression to AIDS in HIV+ individuals with a structural variant of the chemokine receptor CX3CR1. *Science* **287,** 2274–2277.

Fenard, D., Lambeau, G., Maurin, T., Lefebvre, J. C., and Doglio, A. (2001). A peptide derived from bee venom-secreted phospholipase A2 inhibits replication of T-cell tropic HIV-1 strains via interaction with the CXCR4 chemokine receptor. *Mol. Pharmacol.* **60,** 341–347.

Gao, J. L., and Murphy, P. M. (1994). Human cytomegalovirus open reading frame US28 encodes a functional beta chemokine receptor. *J. Biol. Chem.* **269,** 28539–28542.

Garden, G. A. (2002). Microglia in human immunodeficiency virus-associated neurodegeneration. *Glia* **40,** 240–251.

Gerard, C., and Gerard, N. P. (1994). C5A anaphylatoxin and its seven transmembrane-segment receptor. *Annu. Rev. Immunol.* **12,** 775–808.

Goldman, D. W., and Goetzl, E. J. (1982). Specific binding of leukotriene B4 to receptors on human polymorphonuclear leukocytes. *J. Immunol.* **129,** 1600–1604.

Gu, L., Tseng, S., Horner, R. M., Tam, C., Loda, M., and Rollins, B. J. (2000). Control of TH2 polarization by the chemokine monocyte chemoattractant protein-1. *Nature* **404,** 407–411.

Hajeer, A. H., al Sharif, F., and Ollier, W. E. (1999). A polymorphism at position -403 in the human RANTES promoter. *Eur. J. Immunogenet.* **26,** 375–376.

Hanahan, D. J. (1986). Platelet activating factor: A biologically active phosphoglyceride. *Annu. Rev. Biochem.* **55,** 483–509.

He, J., Chen, Y., Farzan, M., Choe, H., Ohagen, A., Gartner, S., Busciglio, J., Yang, X., Hofmann, W., Newman, W., Mackay, C. R., Sodroski, J., and Gabuzda, D. (1997). CCR3 and CCR5 are co-receptors for HIV-1 infection of microglia. *Nature* **385,** 645–649.

Hendrix, C. W., Flexner, C., MacFarland, R. T., Giandomenico, C., Fuchs, E. J., Redpath, E., Bridger, G., and Henson, G. W. (2000). Pharmacokinetics and safety of AMD-3100, a novel antagonist of the CXCR-4 chemokine receptor, in human volunteers. *Antimicrob. Agents. Chemother.* **44,** 1667–1673.

Horuk, R., Chitnis, C. E., Darbonne, W. C., Colby, T. J., Rybicki, A., Hadley, T. J., and Miller, L. H. (1993). A receptor for the malarial parasite Plasmodium vivax: The erythrocyte chemokine receptor. *Science* **261,** 1182–1184.

Howard, O. M., Korte, T., Tarasova, N. I., Grimm, M., Turpin, J. A., Rice, W. G., Michejda, C. J., Blumenthal, R., and Oppenheim, J. J. (1998). Small molecule inhibitor of HIV-1 cell fusion blocks chemokine receptor-mediated function. *J. Leukoc. Biol.* **64,** 6–13.

Hu, Z., Yang, Q., Bo, X., and Li, J. (2003). Herpes simplex encephalitis. *Lancet* **362,** 280.

Huffnagle, G. B., and McNeil, L. K. (1999). Dissemination of C. neoformans to the central nervous system: Role of chemokines, Th1 immunity and leukocyte recruitment. *J. Neurovirol.* **5,** 76–81.

Huffnagle, G. B., McNeil, L. K., McDonald, R. A., Murphy, J. W., Toews, G. B., Maeda, N., and Kuziel, W. A. (1999). Cutting edge: Role of C-C chemokine receptor 5 in organ-specific and innate immunity to *Cryptococcus neoformans*. *J. Immunol.* **163,** 4642–4646.

Hughes, P. M., Botham, M. S., Frentzel, S., Mir, A., and Perry, V. H. (2002). Expression of fractalkine (CX3CL1) and its receptor, CX3CR1, during acute and chronic inflammation in the rodent CNS. *Glia* **37,** 314–327.

Kabat, D., Kozak, S. L., Wehrly, K., and Chesebro, B. (1994). Differences in CD4 dependence for infectivity of laboratory-adapted and primary patient isolates of human immunodeficiency virus type 1. *J. Virol.* **68,** 2570–2567.

Khabar, K. S., Al-Zoghaibi, F., Al-Ahdal, M. N., Murayama, T., Dhalla, M., Mukaida, N., Taha, M., Al-Sedairy, S. T., Siddiqui, Y., Kessie, G., and Matsushima, K. (1997). The alpha chemokine, interleukin-8, inhibits the antiviral action of interferon alpha. *J. Exp. Med.* **186,** 1077–1085.

Khan, I. A., MacLean, J. A., Lee, F. S., Casciotti, L., DeHaan, E., Schwartzman, J. D., and Luster, A. D. (2000). IP-10 is critical for effector T cell trafficking and host survival in Toxoplasma gondii infection. *Immunity* **12,** 483–494.

Khan, I. A., Murphy, P. M., Casciotti, L., Schwartzman, J. D., Collins, J., Gao, J. L., and Yeaman, G. R. (2001). Mice lacking the chemokine receptor CCR1 show increased susceptibility to Toxoplasma gondii infection. *J. Immunol.* **166,** 1930–1937.

Kielian, T. (2004). Microglia and chemokines in infectious diseases of the nervous system: Views and reviews. *Front. Biosci.* **9,** 732–750.

Kinter, A., Arthos, J., Cicala, C., and Fauci, A. S. (2000). Chemokines, cytokines and HIV: A complex network of interactions that influence HIV pathogenesis. *Immunol. Rev.* **177,** 88–98.

Kinter, A., Catanzaro, A., Monaco, J., Ruiz, M., Justement, J., Moir, S., Arthos, J., Oliva, A., Ehler, L., Mizell, S., Jackson, R., Ostrowski, M., Hoxie, J., Offord, R., and Fauci, A. S. (1998). CC-chemokines enhance the replication of T-tropic strains of HIV-1 in CD4(+) T cells. Role of signal transduction. *Proc. Natl. Acad. Sci. USA* **95,** 11880–11885.

Kinter, A. L., Ostrowski, M., Goletti, D., Oliva, A., Weissman, D., Gantt, K., Hardy, E., Jackson, R., Ehler, L., and Fauci, A. S. (1996). HIV replication in CD4+ T cells of HIV-infected individuals is regulated by a balance between the viral suppressive effects of endogenous beta-chemokines and the viral inductive effects of other endogenous cytokines. *Proc. Natl. Acad. Sci. USA* **93,** 14076–14081.

Kitai, R., Zhao, M. L., Zhang, N., Hua, L. L., and Lee, S. C. (2000). Role of MIP-1beta and RANTES in HIV-1 infection of microglia: Inhibition of infection and induction by IFNbeta. *J. Neuroimmunol.* **110,** 230–239.

Kledal, T. N., Rosenkilde, M. M., Coulin, F., Simmons, G., Johnsen, A. H., Alouani, S., Power, C. A., Luttichau, H. R., Gerstoft, J., Clapham, P. R., Clark-Lewis, I., Wells, T. N., and Schwartz, T. W. (1997). A broad-spectrum chemokine antagonist encoded by Kaposi's sarcoma-associated herpesvirus. *Science* **277,** 1656–1659.

Kostrikis, L. G., Neumann, A. U., Thomson, B., Korber, B. T., McHardy, P., Karanicolas, R., Deutsch, L., Huang, Y., Lew, J. F., McIntosh, K., Pollack, H., Borkowsky, W., Spiegel,

H. M., Palumbo, P., Oleske, J., Bardeguez, A., Luzuriaga, K., Sullivan, J., Wolinsky, S. M., Koup, R. A., Ho, D. D., and Moore, J. P. (1999). A polymorphism in the regulatory region of the CC-chemokine receptor 5 gene influences perinatal transmission of human immunodeficiency virus type 1 to African-American infants. *J. Virol.* **73,** 10264–10271.

Kottilil, S., Shin, K., Planta, M., McLaughlin, M., Hallahan, C. W., Ghany, M., Chun, T. W., Sneller, M. C., and Fauci, A. S. (2004). Expression of chemokine and inhibitory receptors on natural killer cells: Effect of immune activation and HIV viremia. *J. Infect. Dis.* **189,** 1193–1198.

Kure, K., Weidenheim, K. M., Lyman, W. D., and Dickson, D. W. (1990). Morphology and distribution of HIV-1 gp41-positive microglia in subacute AIDS encephalitis. Pattern of involvement resembling a multisystem degeneration. *Acta. Neuropathol. (Berl)* **80,** 393–400.

Kusunoki, A., Wada, A., Kurosaki, N., Kimura, T., Takai, K., Yamamoto, N., and Takaku, H. (1999). Antisense oligodeoxynucleotide complementary to CXCR4 mRNA block replication of HIV-1 in COS cells. *Nucleosides. Nucleotides.* **18,** 1705–1708.

Lahrtz, F., Piali, L., Nadal, D., Pfister, H. W., Spanaus, K. S., Baggiolini, M., and Fontana, A. (1997). Chemotactic activity on mononuclear cells in the cerebrospinal fluid of patients with viral meningitis is mediated by interferon-gamma inducible protein-10 and monocyte chemotactic protein-1. *Eur. J. Immunol.* **27,** 2484–2489.

Lahrtz, F., Piali, L., Spanaus, K. S., Seebach, J., and Fontana, A. (1998). Chemokines and chemotaxis of leukocytes in infectious meningitis. *J. Neuroimmunol.* **85,** 33–43.

Lalani, A. S., Masters, J., Zeng, W., Barrett, J., Pannu, R., Everett, H., Arendt, C. W., and McFadden, G. (1999). Use of chemokine receptors by poxviruses. *Science* **286,** 1968–1971.

Lane, B. R., Markovitz, D. M., Woodford, N. L., Rochford, R., Strieter, R. M., and Coffey, M. J. (1999). TNF-alpha inhibits HIV-1 replication in peripheral blood monocytes and alveolar macrophages by inducing the production of RANTES and decreasing C-C chemokine receptor 5 (CCR5) expression. *J. Immunol.* **163,** 3653–3661.

Lapham, C. K., Ouyang, J., Chandrasekhar, B., Nguyen, N. Y., Dimitrov, D. S., and Golding, H. (1996). Evidence for cell-surface association between fusin and the CD4-gp120 complex in human cell lines. *Science* **274,** 602–605.

Lipton, S. A., Yeh, M., and Dreyer, E. B. (1994). Update on current models of HIV-related neuronal injury: Platelet-activating factor, arachidonic acid and nitric oxide. *Adv. Neuroimmunol.* **4,** 181–188.

Liu, H., Chao, D., Nakayama, E. E., Taguchi, H., Goto, M., Xin, X., Takamatsu, J. K., Saito, H., Ishikawa, Y., Akaza, T., Juji, T., Takebe, Y., Ohishi, T., Fukutake, K., Maruyama, Y., Yashiki, S., Sonoda, S., Nakamura, T., Nagai, Y., Iwamoto, A., and Shioda, T. (1999). Polymorphism in RANTES chemokine promoter affects HIV-1 disease progression. *Proc. Natl. Acad. Sci. USA* **96,** 4581–4585.

Lokensgard, J. R., Hu, S., Sheng, W., vanOijen, M., Cox, D., Cheeran, M. C., and Peterson, P. K. (2001). Robust expression of TNF-alpha, IL-1beta, RANTES, and IP-10 by human microglial cells during nonproductive infection with herpes simplex virus. *J. Neurovirol.* **7,** 208–219.

Maeda, K., Yoshimura, K., Shibayama, S., Habashita, H., Tada, H., Sagawa, K., Miyakawa, T., Aoki, M., Fukushima, D., and Mitsuya, H. (2001). Novel low molecular weight spirodiketopiperazine derivatives potently inhibit R5 HIV-1 infection through their antagonistic effects on CCR5. *J. Biol. Chem.* **276,** 35194–35200.

Margolis, L. B., Glushakova, S., Grivel, J. C., and Murphy, P. M. (1998). Blockade of CC chemokine receptor 5 (CCR5)-tropic human immunodeficiency virus-1 replication in human lymphoid tissue by CC chemokines. *J. Clin. Invest.* **101,** 1876–1880.

Mariani, R., Wong, S., Mulder, L. C., Wilkinson, D. A., Reinhart, A. L., LaRosa, G., Nibbs, R., O'Brien, T. R., Michael, N. L., Connor, R. I., Macdonald, M., Busch, M., Koup, R. A., and Landau, N. R. (1999). CCR2-64I polymorphism is not associated with altered CCR5 expression or coreceptor function. *J. Virol.* **73,** 2450–2459.

McDermott, D. H., Beecroft, M. J., Kleeberger, C. A., Al-Sharif, F. M., Ollier, W. E., Zimmerman, P. A., Boatin, B. A., Leitman, S. F., Detels, R., Hajeer, A. H., and Murphy, P. M. (2000a). Chemokine RANTES promoter polymorphism affects risk of both HIV infection and disease progression in the Multicenter AIDS Cohort Study. *AIDS* **14,** 2671–2678.

McDermott, D. H., Colla, J. S., Kleeberger, C. A., Plankey, M., Rosenberg, P. S., Smith, E. D., Zimmerman, P. A., Combadiere, C., Leitman, S. F., Kaslow, R. A., Goedert, J. J., Berger, E. A., O'Brien, T. R., and Murphy, P. M. (2000b). Genetic polymorphism in CX3CR1 and risk of HIV disease. *Science* **290,** 2031.

McDermott, D. H., Zimmerman, P. A., Guignard, F., Kleeberger, C. A., Leitman, S. F., and Murphy, P. M. (1998). CCR5 promoter polymorphism and HIV-1 disease progression. Multicenter AIDS Cohort Study (MACS). *Lancet* **352,** 866–870.

McKnight, A., Wilkinson, D., Simmons, G., Talbot, S., Picard, L., Ahuja, M., Marsh, M., Hoxie, J. A., and Clapham, P. R. (1997). Inhibition of human immunodeficiency virus fusion by a monoclonal antibody to a coreceptor (CXCR4) is both cell type and virus strain dependent. *J. Virol.* **71,** 1692–1696.

Mebatsion, T., Finke, S., Weiland, F., and Conzelmann, K. K. (1997). A CXCR4/CD4 pseudotype rhabdovirus that selectively infects HIV-1 envelope protein-expressing cells. *Cell* **90,** 841–847.

Mellado, M., Rodriguez-Frade, J. M., Vila-Coro, A. J., de Ana, A. M., and Martinez, A. C. (1999). Chemokine control of HIV-1 infection. *Nature* **400,** 723–724.

Miller, L. H., Mason, S. J., Clyde, D. F., and McGinniss, M. H. (1976). The resistance factor to Plasmodium vivax in blacks. The Duffy-blood-group genotype, FyFy. *N. Engl. J. Med.* **295,** 302–304.

Moriuchi, M., Moriuchi, H., Turner, W., and Fauci, A. S. (1998). Exposure to bacterial products renders macrophages highly susceptible to T-tropic HIV-1. *J. Clin. Invest.* **102,** 1540–1550.

Mosier, D. E., Picchio, G. R., Gulizia, R. J., Sabbe, R., Poignard, P., Picard, L., Offord, R. E., Thompson, D. A., and Wilken, J. (1999). Highly potent RANTES analogues either prevent CCR5-using human immunodeficiency virus type 1 infection *in vivo* or rapidly select for CXCR4-using variants. *J. Virol.* **73,** 3544–3550.

Murakami, T., Zhang, T. Y., Koyanagi, Y., Tanaka, Y., Kim, J., Suzuki, Y., Minoguchi, S., Tamamura, H., Waki, M., Matsumoto, A., Fujii, N., Shida, H., Hoxie, J. A., Peiper, S. C., and Yamamoto, N. (1999). Inhibitory mechanism of the CXCR4 antagonist T22 against human immunodeficiency virus type 1 infection. *J. Virol.* **73,** 7489–7496.

Murayama, T., Kuno, K., Jisaki, F., Obuchi, M., Sakamuro, D., Furukawa, T., Mukaida, N., and Matsushima, K. (1994). Enhancement human cytomegalovirus replication in a human lung fibroblast cell line by interleukin-8. *J. Virol.* **68,** 7582–7585.

Murphy, P. M. (1994). The molecular biology of leukocyte chemoattractant receptors. *Annu. Rev. Immunol.* **12,** 593–633.

Murphy, P. M. (2001). Viral exploitation and subversion of the immune system through chemokine mimicry. *Nat. Immunol.* **2,** 116–122.

Nakajima, H., Kobayashi, M., Pollard, R. B., and Suzuki, F. (2001). Monocyte chemoattractant protein-1 enhances HSV-induced encephalomyelitis by stimulating Th2 responses. *J. Leukoc. Biol.* **70,** 374–380.

Nau, R., and Bruck, W. (2002). Neuronal injury in bacterial meningitis: Mechanisms and implications for therapy. *Trends. Neurosci.* **25,** 38–45.

Nickel, R. G., Casolaro, V., Wahn, U., Beyer, K., Barnes, K. C., Plunkett, B. S., Freidhoff, L. R., Sengler, C., Plitt, J. R., Schleimer, R. P., Caraballo, L., Naidu, R. P., Levett, P. N., Beaty, T. H., and Huang, S. K. (2000). Atopic dermatitis is associated with a functional mutation in the promoter of the C-C chemokine RANTES. *J. Immunol.* **164,** 1612–1616.

O'Brien, T. R., McDermott, D. H., Ioannidis, J. P., Carrington, M., Murphy, P. M., Havlir, D. V., and Richman, D. D. (2000). Effect of chemokine receptor gene polymorphisms on the response to potent antiretroviral therapy. *AIDS* **14,** 821–826.

Ostrowski, M. A., Justement, S. J., Catanzaro, A., Hallahan, C. A., Ehler, L. A., Mizell, S. B., Kumar, P. N., Mican, J. A., Chun, T. W., and Fauci, A. S. (1998a). Expression of chemokine receptors CXCR4 and CCR5 in HIV-1-infected and uninfected individuals. *J. Immunol.* **161,** 3195–3201.

Ostrowski, M. A., Krakauer, D. C., Li, Y., Justement, S. J., Learn, G., Ehler, L. A., Stanley, S. K., Nowak, M., and Fauci, A. S. (1998b). Effect of immune activation on the dynamics of human immunodeficiency virus replication and on the distribution of viral quasispecies. *J. Virol.* **72,** 7772–7784.

Patterson, C. E., Daley, J. K., Echols, L. A., Lane, T. E., and Rall, G. F. (2003). Measles virus infection induces chemokine synthesis by neurons. *J. Immunol.* **171,** 3102–3109.

Penfold, M. E., Dairaghi, D. J., Duke, G. M., Saederup, N., Mocarski, E. S., Kemble, G. W., and Schall, T. J. (1999). Cytomegalovirus encodes a potent alpha chemokine. *Proc. Natl. Acad. Sci. USA* **96,** 9839–9844.

Pleskoff, O., Treboute, C., Brelot, A., Heveker, N., Seman, M., and Alizon, M. (1997). Identification of a chemokine receptor encoded by human cytomegalovirus as a cofactor for HIV-1 entry. *Science* **276,** 1874–1878.

Proudfoot, A. E., Buser, R., Borlat, F., Alouani, S., Soler, D., Offord, R. E., Schroder, J. M., Power, C. A., and Wells, T. N. (1999). Amino-terminally modified RANTES analogues demonstrate differential effects on RANTES receptors. *J. Biol. Chem.* **274,** 32478–32485.

Reinhart, T. A. (2003). Chemokine induction by HIV-1: Recruitment to the cause. *Trends. Immunol.* **24,** 351–353.

Rodrigues, M. L., Alviano, C. S., and Travassos, L. R. (1999). Pathogenicity of Cryptococcus neoformans: Virulence factors and immunological mechanisms. *Microbes. Infect.* **1,** 293–301.

Rodriguez-Frade, J. M., Vila-Coro, A. J., de Ana, A. M., Albar, J. P., Martinez, A. C., and Mellado, M. (1999). The chemokine monocyte chemoattractant protein-1 induces functional responses through dimerization of its receptor CCR2. *Proc. Natl. Acad. Sci. USA* **96,** 3628–3633.

Rosler, A., Pohl, M., Braune, H. J., Oertel, W. H., Gemsa, D., and Sprenger, H. (1998). Time course of chemokines in the cerebrospinal fluid and serum during herpes simplex type 1 encephalitis. *J. Neurol. Sci.* **157,** 82–89.

Sakaida, H., Hori, T., Yonezawa, A., Sato, A., Isaka, Y., Yoshie, O., Hattori, T., and Uchiyama, T. (1998). T-tropic human immunodeficiency virus type 1 (HIV-1)-derived V3 loop peptides directly bind to CXCR-4 and inhibit T-tropic HIV-1 infection. *J. Virol.* **72,** 9763–9770.

Samson, M., Libert, F., Doranz, B. J., Rucker, J., Liesnard, C., Farber, C. M., Saragosti, S., Lapoumeroulie, C., Cognaux, J., Forceille, C., Muyldermans, G., Verhofstede, C., Burtonboy, G., Georges, M., Imai, T., Rana, S., Yi, Y., Smyth, R. J., Collman, R. G., Doms, R. W., Vassart, G., and Parmentier, M. (1996). Resistance to HIV-1 infection in caucasian individuals bearing mutant alleles of the CCR-5 chemokine receptor gene. *Nature* **382,** 722–725.

Sanders, V. J., Pittman, C. A., White, M. G., Wang, G., Wiley, C. A., and Achim, C. L. (1998). Chemokines and receptors in HIV encephalitis. *AIDS* **12,** 1021–1026.

Scarlatti, G., Tresoldi, E., Bjorndal, A., Fredriksson, R., Colognesi, C., Deng, H. K., Malnati, M. S., Plebani, A., Siccardi, A. G., Littman, D. R., Fenyo, E. M., and Lusso, P. (1997). *In vivo* evolution of HIV-1 co-receptor usage and sensitivity to chemokine-mediated suppression. *Nat. Med.* **3,** 1259–1265.

Schiffmann, E., Showell, H. V., Corcoran, B. A., Ward, P. A., Smith, E., and Becker, E. L. (1975). The isolation and partial characterization of neutrophil chemotactic factors from *Escherichia coli*. *J. Immunol.* **114,** 1831–1837.

Schnell, M. J., Johnson, J. E., Buonocore, L., and Rose, J. K. (1997). Construction of a novel virus that targets HIV-1-infected cells and controls HIV-1 infection. *Cell* **90,** 849–857.

Schols, D., Proost, P., Struyf, S., Wuyts, A., De Meester, I., Scharpe, S., Van Damme, J., and De Clercq, E. (1998). CD26-processed RANTES(3-68), but not intact RANTES, has potent anti-HIV-1 activity. *Antiviral. Res.* **39,** 175–187.

Schols, D., Vermeire, K., Hatse, S., Princen, K., De Clerq, E., Calandra, G., Fricker, S., Nelson, K., Labrecque, J., Bogucki, D., Zhou, Y., Skerlj, R., and Bridger, G. (2004). *In vitro* anti-HIV activity profile of AMD887, a novel CCR5 antagonist, in combination with the CXCR4 inhibitor AMD070. *In* "11th Conference on Retroviruses and Opportunistic Infections," Abstract #539, San Francisco.

Schuitemaker, H., Kootstra, N. A., de Goede, R. E., de Wolf, F., Miedema, F., and Tersmette, M. (1991). Monocytotropic human immunodeficiency virus type 1 (HIV-1) variants detectable in all stages of HIV-1 infection lack T-cell line tropism and syncytium-inducing ability in primary T-cell culture. *J. Virol.* **65,** 356–363.

Schurman, D., Nougarede, R. R., Reynes, J., Fatkenheur, G., Raffi, F., Michelet, C., Tarral, A., Hoffman, C., Kiunke, J., Spenger, H., banLier, J., Sansome, S., Jackson, M., and Laughlin, M. (2004). SCH-D: Antiviral activity of a CCR5 receptor antagonist. *In* "11th Conference on Retroviruses and Opportunistic Infections," Abstract #140LB, San Francisco.

Simmons, G., Clapham, P. R., Picard, L., Offord, R. E., Rosenkilde, M. M., Schwartz, T. W., Buser, R., Wells, T. N., and Proudfoot, A. E. (1997). Potent inhibition of HIV-1 infectivity in macrophages and lymphocytes by a novel CCR5 antagonist. *Science* **276,** 276–279.

Steinberger, P., Andris-Widhopf, J., Buhler, B., Torbett, B. E., and Barbas, C. F., 3rd. (2000). Functional deletion of the CCR5 receptor by intracellular immunization produces cells that are refractory to CCR5-dependent HIV-1 infection and cell fusion. *Proc. Natl. Acad. Sci. USA.* **97,** 805–810.

Strack, A., Schluter, D., Asensio, V. C., Campbell, I. L., and Deckert, M. (2002). Regulation of the kinetics of intracerebral chemokine gene expression in murine Toxoplasma encephalitis: Impact of host genetic factors. *Glia* **40,** 372–377.

Strizki, J. M., Xu, S., Wagner, N. E., Wojcik, L., Liu, J., Hou, Y., Endres, M., Palani, A., Shapiro, S., Clader, J. W., Greenlee, W. J., Tagat, J. R., McCombie, S., Cox, K., Fawzi, A. B., Chou, C. C., Pugliese-Sivo, C., Davies, L., Moreno, M. E., Ho, D. D., Trkola, A., Stoddart, C. A., Moore, J. P., Reyes, G. R., and Baroudy, B. M. (2001). SCH-C (SCH 351125), an orally bioavailable, small molecule antagonist of the chemokine receptor CCR5, is a potent inhibitor of HIV-1 infection *in vitro* and *in vivo*. *Proc. Natl. Acad. Sci. USA* **98,** 12718–12723.

Swingler, S., Mann, A., Jacque, J., Brichacek, B., Sasseville, V. G., Williams, K., Lackner, A. A., Janoff, E. N., Wang, R., Fisher, D., and Stevenson, M. (1999). HIV-1 Nef mediates lymphocyte chemotaxis and activation by infected macrophages. *Nat. Med.* **5,** 997–1003.

Tamamura, H., Xu, Y., Hattori, T., Zhang, X., Arakaki, R., Kanbara, K., Omagari, A., Otaka, A., Ibuka, T., Yamamoto, N., Nakashima, H., and Fujii, N. (1998). A low-molecular-weight inhibitor against the chemokine receptor CXCR4: A strong anti-HIV peptide T140. *Biochem. Biophys. Res. Commun.* **253,** 877–882.

Tournamille, C., Colin, Y., Cartron, J. P., and Le Van Kim, C. (1995). Disruption of a GATA motif in the Duffy gene promoter abolishes erythroid gene expression in Duffy-negative individuals. *Nat. Genet.* **10,** 224–228.

Townsend, G. C., and Scheld, W. M. (1998). Infections of the central nervous system. *Adv. Intern. Med.* **43,** 403–447.

Tremblay, C. L., Giguel, F., Kollmann, C., Guan, Y., Chou, T. C., Baroudy, B. M., and Hirsch, M. S. (2002). Anti-human immunodeficiency virus interactions of SCH-C (SCH 351125), a CCR5 antagonist, with other antiretroviral agents *in vitro*. *Antimicrob. Agents Chemother.* **46,** 1336–1339.

Tremblay, C. L., Kollmann, C., Giguel, F., Chou, T. C., and Hirsch, M. S. (2000). Strong *in vitro* synergy between the fusion inhibitor T-20 and the CXCR4 blocker AMD-3100. *J. Acquir. Immune. Defic. Syndr.* **25,** 99–102.

Trkola, A., Ketas, T. J., Nagashima, K. A., Zhao, L., Cilliers, T., Morris, L., Moore, J. P., Maddon, P. J., and Olson, W. C. (2001). Potent, broad-spectrum inhibition of human immunodeficiency virus type 1 by the CCR5 monoclonal antibody PRO 140. *J. Virol.* **75,** 579–588.

Valdez, H., Purvis, S. F., Lederman, M. M., Fillingame, M., and Zimmerman, P. A. (1999). Association of the CCR5delta32 mutation with improved response to antiretroviral therapy. *JAMA* **282,** 734.

Verani, A., Scarlatti, G., Comar, M., Tresoldi, E., Polo, S., Giacca, M., Lusso, P., Siccardi, A. G., and Vercelli, D. (1997). C-C chemokines released by lipopolysaccharide (LPS)-stimulated human macrophages suppress HIV-1 infection in both macrophages and T cells. *J. Exp. Med.* **185,** 805–816.

Watson, S., and Arkinstall, S. (1994). "The G-protein linked receptor." Academic Press, London.

Weiss, J. M., Nath, A., Major, E. O., and Berman, J. W. (1999). HIV-1 Tat induces monocyte chemoattractant protein-1-mediated monocyte transmigration across a model of the human blood-brain barrier and up-regulates CCR5 expression on human monocytes. *J. Immunol.* **163,** 2953–2959.

Weissman, D., Rabin, R. L., Arthos, J., Rubbert, A., Dybul, M., Swofford, R., Venkatesan, S., Farber, J. M., and Fauci, A. S. (1997). Macrophage-tropic HIV and SIV envelope proteins induce a signal through the CCR5 chemokine receptor. *Nature* **389,** 981–985.

Westby, M. W. J., Huang, W., James, I., Abel, S., Petropoulos, C., Perros, M., and van der Ryst, E. (2004). Reversible predominance of CXCR4 utilizing variants in a non-responsive dual-tropic Patient Receiving the CCR5 Antagonist UK-427857. *In* "11th Conference on Retroviruses and Opportunistic Infections," Abstract #538, San Francisco.

Wu, L., LaRosa, G., Kassam, N., Gordon, C. J., Heath, H., Ruffing, N., Chen, H., Humblias, J., Samson, M., Parmentier, M., Moore, J. P., and Mackay, C. R. (1997). Interaction of chemokine receptor CCR5 with its ligands: Multiple domains for HIV-1 gp120 binding and a single domain for chemokine binding. *J. Exp. Med.* **186,** 1373–1381.

Xiao, X., Wu, L., Stantchev, T. S., Feng, Y. R., Ugolini, S., Chen, H., Shen, Z., Riley, J. L., Broder, C. C., Sattentau, Q. J., and Dimitrov, D. S. (1999). Constitutive cell surface association between CD4 and CCR5. *Proc. Natl. Acad. Sci. USA* **96,** 7496–7501.

Yang, T. Y., Chen, S. C., Leach, M. W., Manfra, D., Homey, B., Wiekowski, M., Sullivan, L., Jenh, C. H., Narula, S. K., Chensue, S. W., and Lira, S. A. (2000). Transgenic expression of the chemokine receptor encoded by human herpesvirus 8 induces an angioproliferative disease resembling Kaposi's sarcoma. *J. Exp. Med.* **191,** 445–454.

Yang, X. D., Corvalan, J. R., Wang, P., Roy, C. M., and Davis, C. G. (1999). Fully human anti-interleukin-8 monoclonal antibodies: Potential therapeutics for the treatment of inflammatory disease states. *J. Leukoc. Biol.* **66,** 401–410.

Zimmerman, P. A., Buckler-White, A., Alkhatib, G., Spalding, T., Kubofcik, J., Combadiere, C., Weissman, D., Cohen, O., Rubbert, A., Lam, G., Vaccarezza, M., Kennedy, P. E., Kumaraswami, V., Giorgi, J. V., Detels, R., Hunter, J., Chopek, M., Berger, E. A., Fauci, A. S., Nutman, T. B., and Murphy, P. M. (1997). Inherited resistance to HIV-1 conferred by an inactivating mutation in CC chemokine receptor 5: Studies in populations with contrasting clinical phenotypes, defined racial background, and quantified risk. *Mol. Med.* **3,** 23–36.

Zou, P., Isegawa, Y., Nakano, K., Haque, M., Horiguchi, Y., and Yamanishi, K. (1999). Human herpesvirus 6 open reading frame U83 encodes a functional chemokine. *J. Virol.* **73,** 5926–5933.

Zwijnenburg, P. J., van der Poll, T., Florquin, S., van Deventer, S. J., Roord, J. J., and van Furth, A. M. (2001). Experimental pneumococcal meningitis in mice: A model of intranasal infection. *J. Infect. Dis.* **183,** 1143–1146.

CHAPTER 12

New Therapies Targeting Chemokine Receptors: Can Changing the Way Cells Traffic be Used to Treat Human Disease?

Linda Bendall and Kenneth Bradstock

Westmead Millennium Institute, University of Sydney, and Department of Haematology, Westmead Hospital Westmead, New South Wales, Australia 2145

I. INTRODUCTION

A growing understanding in recent years of the biology of chemokines and their cellular receptors has resulted in the identification of the central importance of these molecules in many physiological and pathophysiological

Current Topics in Membranes, Volume 55 1063-5823/05 $35.00

DOI: 10.1016/S1063-5823(04)55112-8

processes, as well as revealing a number of potential mechanisms by which chemokine networks could be targeted for therapeutic benefit. Interfering with chemokine-receptor function has the potential not only to alter the trafficking of cells through the body and change the sites of cell accumulation, but also to affect more fundamental cellular processes of growth and survival. As a result, this potentially powerful strategy for the development of new therapeutic interventions is likely to have an impact on a diverse range of pathological conditions, from vascular disease, infections, and inflammatory disorders to cancer-cell growth and metastasis.

The ability to specifically target key chemokine–receptor interactions, and selectively disrupt these to fulfil therapeutic objectives, is dependent on two issues: a comprehensive knowledge of the pathobiology of the system, and the availability of chemical compounds that are highly specific and that fulfil basic criteria for pharmacological success, such as stability, binding affinity, and bioavailability.

The interest in blocking chemokine receptors leapt into prominence in 1996 with the realization that antagonists of chemokine receptors had potential as therapeutics for HIV treatment (Balter, 1996). By 1997 there were over 14 separate developers of chemokine receptor antagonists studying a range of strategies, including antibodies, modified chemokines, and small molecule antagonists. The first useful chemokine receptor blockers were antibodies and mutated chemokines, but in less than a decade, a vast array of structurally diverse small molecule antagonists dominated research, and several were moving into human clinical trials in a variety of diseases. The first small-molecule chemokine-receptor antagonists were directed at CXCR2 and CCR1 and were developed in 1998 by SmithKline Beecham (now GlaxoSmithKline) and Berlex Biosciences, respectively (Hesselgesser *et al.*, 1998; White *et al.*, 1998). Gradually, the structural requirements for specific and effective chemokine receptor antagonists were understood. Today there are large numbers of structurally distinct compounds registered with patent agencies directed at a diverse range of receptors, including CCR1, CCR2, CCR3, CCR5, CCR10, CXCR2, CXCR3, and CXCR4. The final obstacle to the use of chemokine antagonists in the clinic has been the poor pharmacokinetic and toxicity profiles of antagonists with high potency. This raises issues about dosing schedules in relation to oral bioavailability and the therapeutic window of the drugs. Despite this, a number of drugs, including antagonists of CCR1, CCR3, CCR5, and CXCR4, have entered Phase I clinical trials, and a smaller number are now in Phase II trials. This has occurred despite the lack of full knowledge of the complexity of the underlying biology and its perturbations in disease.

In this chapter, we describe current drugs that can antagonize the function of chemokine receptors, primarily by preventing ligand binding, and detail their investigation in preclinical studies and clinical trials. We discuss the potential directions of research in this exciting area of medical science.

II. AVAILABLE DRUGS THAT MODULATE CHEMOKINE/RECEPTOR FUNCTION

A. *CCR1 Antagonists*

1. Xanthene Carboxamides

The xanthene carboxamides were initially identified as antagonists of CCR1 and CCR3. However, the initial compound studied, although effective, had poor bioavailability (Sabroe *et al.*, 2000). This compound was developed into selective CCR1 antagonists; different agents were developed for use in mouse and human systems (Naya *et al.*, 2001a). These compounds continue to be developed, but further refinement appears to be required to produce potent antagonists that are orally available (Naya *et al.*, 2003a).

2. BX471

BX471 was identified from a drug screen as a highly selective antagonist of CCR1 (Liang *et al.*, 2000a). It is orally active and has shown efficacy in animal models of multiple sclerosis, organ transplant, and lupus nephritis (Horuk *et al.*, 2001a,b; Anders *et al.*, 2004). BX471 entered Phase I clinical trial in 2002 for multiple sclerosis, and Phase II clinical trials have apparently been initiated (Elices, 2002). However, published outcomes for these are not available.

3. CP-481,715

CP-481,715 was developed more recently and effectively blocks a range of CCL3-induced cellular responses (Gladue *et al.*, 2003). A number of other compounds have recently been developed from CP-481,715 with greater potency and improved metabolic stability, but these have yet to be tested *in vivo* (Brown *et al.*, 2004; Kath *et al.*, 2004a,b). Structures and potency information are given in Fig. 1. In addition, another CCR1 antagonist, MLN3897, has been produced by Millennium Pharmaceuticals, but its relationship to published antagonists is not known because its structure is currently not publicly available.

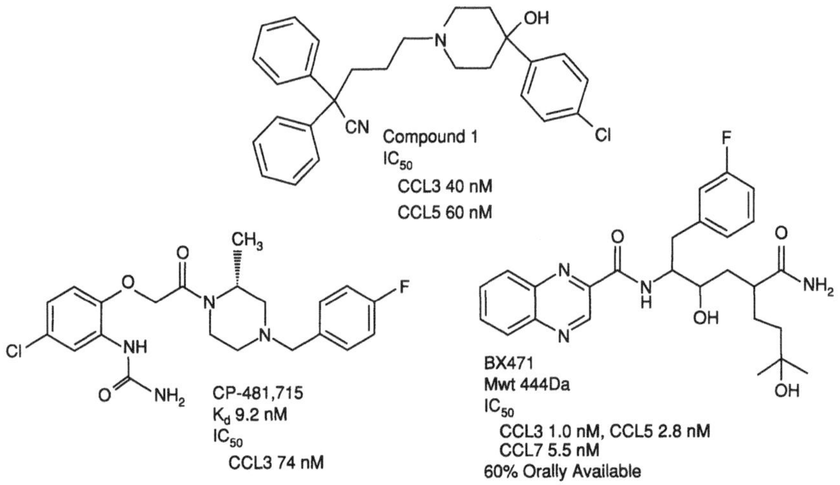

FIGURE 1 CCR1 antagonists.

B. CCR2 Antagonists

Small-molecule inhibitors of CCR2 have not been widely reported; **RS-504393** is the only significant player. It has a K_d of 89 nM and an IC_{50} for the chemotactic response to 3 nM CCL2 of 330 nM. Despite this relatively low potency, it has been used successfully in animal models of ischemia reperfusion injury and kidney fibrosis (Furuichi *et al.*, 2003; Kitagawa *et al.*, 2004). An indirect blockade of CCR2 has also been reported using **propagermanium** via its interaction with a GPI-linked protein associated with CCR2 (Yokochi *et al.*, 2001).

C. CCR3 Antagonists

1. N-(Ureidoalkyl)-Benzyl-Piperidines

These were identified by Bristol-Myers Squibb as potent CCR3 antagonists with IC_{50}s in the low nanomolar range (De Lucca *et al.*, 2002; Wacker *et al.*, 2002). However, they displayed considerable cross-reactivity with serotonin receptors, particularly $5HT_{2A}$, suggesting that serious adverse side effects arising from interactions with the central nervous system may occur. Recently, modifications of these compounds have been developed with improved specificity and IC_{50}s for CCR3 as low as 1 nM, but results from *in vivo* experiments have yet to be published (Varnes *et al.*, 2004).

2. 2-(Benzothiazolylthio)acetamides

These were developed from a CCR1 and CCR3 antagonist (Naya *et al.*, 2001a,b). These newly developed compounds have a specificity of greater than 200-fold for CCR3 over CCR1, and IC_{50}s in the low nanomolar range (Naya *et al.*, 2003b), although *in vivo* data are not yet available.

3. A-122057 and A122058

These were identified from a screen of over 23,000 compounds. They have IC_{50}s of 600 and 975 nM, respectively. Surprisingly, they inhibited CCL11-induced calcium mobilization in transfected cells with IC_{50}s of 165 and 195 nM, respectively. Despite the lower potency of these inhibitors, they have been shown to inhibit CCL11-induced peritoneal eosinophilia (Warrior *et al.*, 2003). Two additional compounds, 766994 (from GlaxoSmithKline) and DPC-168 (from Bristol-Myers Squibb), are available, but they have not been published and their structures are unavailable.

D. CCR4 Antagonists

Very few CCR4 antagonists have been developed to date, and none have been reported as having been used *in vivo*. The lead published compound is a trisubstituted thiazolidinone with a K_d of 140 nM and an IC_{50} for CCL22-induced chemotaxis of 2 μM (Allen *et al.*, 2004).

E. CCR5 Antagonists

The first CCR5 small-molecule antagonist developed was **TAK-779**, which inhibits signaling through CCR5 at nanomolar concentrations (Baba *et al.*, 1999; Iizawa *et al.*, 2003). Further study led to the development of **TAK-220**, which has similar *in vitro* activity, preventing HIV replication in the low nanomolar range, and is orally available, making it an excellent candidate for clinical trials (Iizawa *et al.*, 2003; Tremblay *et al.*, 2003). **SCH-C, SCH-D, UK-427,857**, and the spirodiketopiperazine derivatives, such as **AK602**, each block *in vitro* viral replication in the nanomolar range, making them feasible potential therapeutic agents (Strizki *et al.*, 2001). Most are highly specific for CCR5, although SCH-C also blocks CCR2 at high doses (Strizki *et al.*, 2001). UK-427,857 and AK602 are of particular interest. UK-427,857 does not block chemokine binding to CCR5, suggesting that there is a separation of chemotactic and HIV coreceptor function in CCR5 (Dorr *et al.*, 2003). AK602 has similar properties, being a potent inhibitor of HIV entry and CCL3 activity but having no effect on CCL4- or CCL5-induced responses (Maeda *et al.*, 2004). Each of these compounds is orally available

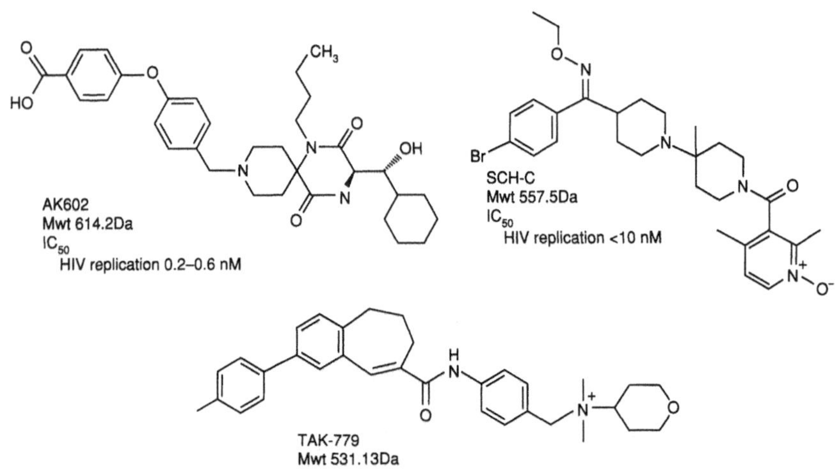

FIGURE 2 CCR5 antagonists.

and demonstrates acceptable pharmacokinetics. The structures of representative CCR5 antagonists are shown in Fig. 2. New compounds continue to be developed, some with greater potency (i.e., IC_{50}s down to 0.7 nM), but this has not, as yet, been associated with favorable pharmacokinetics (Shankaran *et al.*, 2004; Thoma *et al.*, 2004). In addition, a number of agents have recently been detected in natural products; these include anibamine, a novel pyridine quaternary alkaloid, Ophiobolin C, and 19,20-epoxycytochalasin Q. These block gp120 binding with IC_{50}s of 1, 40, and 60 μM, respectively (Jayasuriya *et al.*, 2004). The potential of these compounds has yet to be fully investigated.

F. CXCR2 Antagonists

1. SB225002

SB225002 was among the first small-molecule chemokine receptor antagonists described, and it has been used in animal models to inhibit IL-8-induced neutrophil margination (White *et al.*, 1998). A modification of this compound, with an IC_{50} for blockade of CXCL8 binding of 9.3 nM, attenuates leukocyte infiltration into the synovium in models of acute and chronic arthritis in rabbits (Podolin *et al.*, 2002).

2. N,N-diarylurea

GlaxoSmithKline has further developed the **N,N-diarylurea**–based antagonists, and orally available compounds with IC_{50}s of as low as 7 nM have been published (Jin *et al.*, 2004; Widdowson *et al.*, 2004).

G. CXCR3 Antagonists

To date, no small-molecule inhibitors of CXCR3 have been published, although murine CXCR3 can be blocked by the CCR5 antagonist TAK-779 (Gao *et al.*, 2003). Despite this absence of published data, T-487 has entered Phase II clinical trial by ChemoCentryx for the treatment of psoriasis. The drug is orally available and was well tolerated in Phase I trials, according to company reports.

H. CXCR4 Antagonists

1. Histone Deacetylase Inhibitors

This category includes drugs such as suberoylanilide hydroxamic acid (SAHA) and the butyrate prodrug tributyrin. Although these drugs are not classic antagonists of CXCR4, they reduce CXCR4 mRNA expression and cellular responses to CXCL12 (Jourdan *et al.*, 1998).

2. ALX40-4C

ALX40-4C is a small peptide designed as an inhibitor of Tat-TAR binding, but it was identified as an inhibitor of initial viral–target cell interactions because of its association with CXCR4 (Doranz *et al.*, 1997). More recently, ALX40-4C was found to block HIV entry via the APJ receptor. This receptor has recently been shown to act as a coreceptor for HIV strains derived from the brain (Shieh *et al.*, 1998; Albright *et al.*, 1999), where APJ is normally expressed (O'Dowda *et al.*, 1993). The ligands for APJ can be chemotactic in certain cells, and their binding to APJ is inhibited by ALX40-4C. Interestingly, ALX40-4C was the first chemokine-receptor antagonist to be tested in Phase I/II clinical trials (Doranz *et al.*, 2001), and it was well tolerated.

3. Bicyclams

The bicyclams were originally discovered as impurities in the large-scale preparations of the macrocyclic polyamine cyclam (1, 4, 8, 11-tetraazacyclo-tetradecane) and were found to be potent and selective inhibitors of the X4 strains of HIV (De Clercq, 2003). These antagonists, these AMD3100 and AMD070, inhibit CXCL12 binding and signaling through CXCR4 and are able to block the binding of the functionally blocking CXCR4 monoclonal antibody 12G5 at nanomolar concentrations (De Clercq, 2003). AMD070 has the added advantage of being absorbed orally.

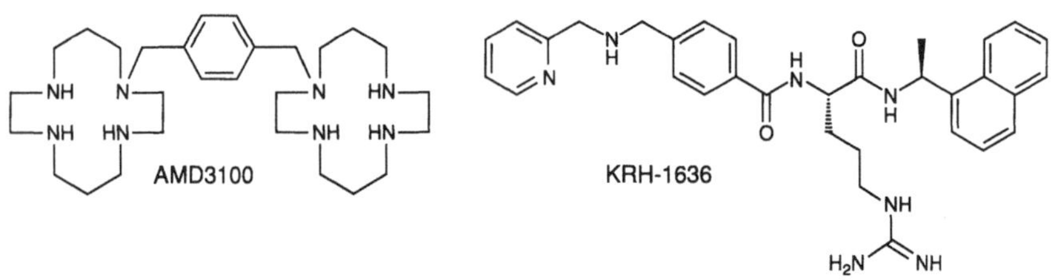

FIGURE 3 CXCR4 antagonists.

4. KRH-1636

KRH-1636 has antiviral properties similar to those of AMD3100, but it can be absorbed from the duodenum, making it potentially orally available, and it is able to reduce viral loads in SCID mice xenografted with human lymphocytes (Ichiyama *et al.*, 2003). Structures of selected CXCR4 antagonists are shown in Fig. 3.

III. CLINICAL USES FOR CHEMOKINE-RECEPTOR ANTAGONISTS

Some of the small-molecule antagonists that were just described, in addition to a number of less-well-described compounds, have entered clinical trials for a number of indications. A summary of currently listed clinical trials is given in Table I.

A. Mobilization and Transplantation of Hematopoietic Stem Cells

Until the early 1990s, the primary source of hematopoietic stem cells (HSCs) for transplantation purposes was the bone marrow. Currently however, the great majority of transplants are performed using hematopoietic progenitor cells (HPCs) or HSCs collected from the peripheral blood following mobilization from the bone marrow. Mobilization is achieved by the administration of cytokines, most commonly granulocyte colony-stimulating factor (G-CSF) (Sato *et al.*, 1994), alone or in combination with chemotherapy (To *et al.*, 1984, 1997). The use of mobilized peripheral blood HPCs/HSCs is associated with more rapid engraftment, decreased morbidity, and reduced costs in comparison to bone-marrow transplantation, and all of these factors have contributed to the decline in the use of bone marrow as a source of HPCs/HSCs (To *et al.*, 1997). Expression of CXCR4 by hematopoietic progenitor cells and production of CXCL12 by bone-marrow stromal cells and osteoblasts (Lapidot, 2001; Netelenbos

TABLE I

Evaluation of Chemokine Receptors in Clinical Trials

Receptor	Drug	Sponsor	Phase of trial	Indication
CCR1	BX471 (ZK-811752)	Schering AG/Berlex Biosciences	Phase II	Multiple sclerosis
			Phase I/II	Alzheimer's disease (diagnostic tool)
	CP-481715	Pfizer, Inc.	Phase II	Rheumatoid arthritis
	MLN-3897	Millennium pharmaceuticals/Aventis SA	Phase I	Rheumatoid arthritis, Multiple sclerosis, psoriasis
CCR2	MLN-1202 (Antibody)	Millennium pharmaceuticals	Phase I	Rheumatoid arthritis
CCR3	766994	GlaxoSmithKline	Phase II	Asthma, allergic rhinitis
	DPC-168	Bristol-Myers Squibb	Phase I	Asthma
CCR5	GW873140 (ONO-4128)	Ono Pharmaceutical Co., Ltd./Glaxo SmithKline	Phase II	HIV
	UK-427857	Pfizer	Phase I	HIV
	SCH-C (SCH-351125)	Schering-Plough	Phase I	HIV
	SCH-D (SCH-417690)	Schering-Plough	Phase I	HIV
CXCR4	AMD3100	AnorMED	Phase II	Stem cell mobilization for Myeloma, Non-Hodgkin's lymphoma, Cancers/Tumors unspecified
	AMD070	AnorMED	Phase I	HIV
CXC unspecified	Traficet-EN	ChemoCentryx	Phase I	Inflammatory bowel disease, Crohn's disease, ulcerative colitis

et al., 2003) are important for the homing of HPCs to the bone marrow, both during ontogeny and in the adult, and for the retention of these cells within this organ (Kawabata *et al.*, 1999; Lapidot, 2001). Degradation of bone-marrow CXCL12 and CXCR4 by neutrophil enzymes is believed to be an important factor behind G-CSF and chemotherapy-induced mobilization (Petit *et al.*, 2002; Levesque *et al.*, 2003). Indeed, perturbation of the CXCL12/CXCR4 axis in isolation *in vivo* can lead to the mobilization of HSCs from the bone marrow. This has been observed after upregulation of CXCL12 in the plasma, through either adenoviral transduction of CXCL12 or displacement of HS-bound CXCL12 with the sulfated polysaccharide fucoidan (Hattori *et al.*, 2001; Sweeney *et al.*, 2002). More importantly, *in vivo* administration of the small-molecule CXCR4 antagonist AMD3100 results in rapid mobilization of HPCs and has led to this compound being investigated as a potential mobilization agent. Studies by Liles *et al.* (2003) have demonstrated that a single dose of AMD3100 (80 μg/kg subcutaneously) results in an increase in peripheral-blood $CD34^+$ HPC to levels appropriate for collection for transplantation. AMD3100 also synergizes with G-CSF to achieve a greater level of mobilization than either agent alone (Liles *et al.*, 2002). Furthermore, in a murine model mimicking the heterogeneous responses of patients to G-CSF mobilisation, AMD3100 enhanced mobilization of myeloid progenitors in mice that were known to be poor mobilizers (Broxmeyer *et al.*, 2002). The positive effects of disrupting the CXCL12/CXCR4 axis on HSC mobilization strongly suggest that there is a role for current small-molecule CXCR4 inhibitors as mobilization agents. Considering the not-insignificant side effects of cytokine-mediated mobilization in normal volunteer HPC donors, the use of a single dose and a same-day collection regimen has substantial appeal. It is possible that compounds such as AMD3100 will replace cytokines like G-CSF as the preferred mobilizing agents in normal donors in the near future. AMD3100 was successful in Phase I and is currently in Phase II clinical trials for the collection of stem cells from patients receiving treatment for multiple myeloma and non-Hodgkin's lymphoma (Devine *et al.*, 2004).

Recent studies have suggested that stem cells obtained from the bone marrow or from mobilized peripheral blood can increase cardiac recovery following infarction (Kang *et al.*, 2004; Wollert *et al.*, 2004). However, there are conflicting reports regarding the safety of G-CSF to this high-risk group of patients, making alternative mobilization agents appealing (Hill *et al.*, 2003; Kang *et al.*, 2004). AnorMED, a pharmaceutical company, plans to initiate trials using AMD3100 to mobilize stem cells for the treatment of heart-attack patients in the near future.

B. Inhibition of Cancer Metastasis and Growth

Chemokines have been shown to be involved in all stages of cancer development, including neoplastic transformation, tumor growth, immune evasion, metastasis, and angiogenesis. This has resulted in great interest in the utilization of chemokine receptor antagonists in new therapeutic strategies. Modulators of chemokine function have the potential to inhibit tumor-promoting leukocyte infiltrates, metastasis, angiogenesis, and even tumor-cell proliferation. Oncogenic events, such as the overexpression of NFKB, loss of tumor-suppressor gene expression, or viral infection, can increase or induce the expression of chemokine receptors, such as CXCR4 or the CXCR2 homologue ORF74 from the herpesvirus-8, in malignant cells (Yang *et al.*, 2000; Helbig *et al.*, 2003; Staller *et al.*, 2003). Local environmental factors, such as hypoxia or inflammatory cytokines, can also modulate chemokine-receptor expression (Bachelder *et al.*, 2002; Hall and Korach, 2003).

Testing of this strategy is still in its infancy; few *in vivo* studies have been reported. Administration of a CCL5 antagonist based on CCL5, met-CCL5, to mice bearing breast tumors produced modest inhibition of tumor growth, which was associated with reduced leukocyte infiltration (Robinson *et al.*, 2003). This is consistent with reports associating increased CCL5 expression with increased lymphocyte infiltration in non-small-cell lung cancer. However, in that study, the increased lymphocyte infiltration was a positive predictor of survival (Moran *et al.*, 2002), suggesting that CCL5 expression was desirable.

To date, most studies have involved inhibition of CXCR4. CXCR4 was among the first chemokine receptors implicated in cancer metastasis and has now been implicated in the metastatic spread of an impressive list of malignancies, including breast, ovarian, lung, and prostate cancer and T-cell lymphoma (Koshiba *et al.*, 2000; Zeelenberg *et al.*, 2001; Bachelder *et al.*, 2002; Scotton *et al.*, 2002; Taichman *et al.*, 2002; Phillips *et al.*, 2003; Sun *et al.*, 2003). Studies using monoclonal antibodies to block receptor function or intrakines to downmodulate surface expression have demonstrated the predicted reduction in metastasis in breast cancer, non-Hodgkin's lymphoma, non-small-cell lung carcinoma, and melanoma (Muller *et al.*, 2001; Bertolini *et al.*, 2002; Phillips *et al.*, 2003). The CXCR4 peptide antagonist, T140, and its newer analogue, TN14003, have similarly inhibited breast cancer metastasis (Murakami *et al.*, 2002; Liang *et al.*, 2004), while the related peptides T22 and TE14011 inhibited lung metastasis of melanoma cells in murine models (Tamamura *et al.*, 2003; Takenaga *et al.*, 2004). In colon cancer, metastasis was not directly prevented by inhibition of CXCR4, but the growth of metastases was prevented (Zeelenberg *et al.*, 2003). Similarly, the small-molecule CXCR4 antagonist AMD3100 inhibited growth of glioblastoma and medulloblastoma xenografts (Rubin *et al.*, 2003). These results are consistent with studies demonstrating the proliferative effects of CXCL12 on a number of cancer-cell types, including glioblastomas (Sehgal *et al.*, 1998; Barbero *et al.*, 2003), colon cancers, chronic B-cell leukemia, acute lymphoblastic leukaemia (Nishii *et al.*, 1999; Burger *et al.*, 2000; Juarez *et al.*, 2003), non-Hodgkin's lymphoma (Bertolini *et al.*, 2002), small-cell lung cancer (Kijima *et al.*, 2002), glioma (Zhou *et al.*, 2002; Barbero *et al.*, 2003), and breast and ovarian cancer (Scotton *et al.*, 2001; Hall and Korach, 2003). Surprisingly, AMD3100 reduced growth of lymphomas *in vivo* by 40–60% when administered thrice weekly but enhanced tumor growth when given by continuous infusion (Paul *et al.*, 2002). One potential confounder is the role of CXCR4 in the immune response, where it has been reported to have both positive and negative effects (Nomura *et al.*, 2001; Zou *et al.*, 2001; Dunussi-Joannopoulos *et al.*, 2002). Obviously, further preclinical studies are required to clarify the mechanisms of action of chemokine antagonists *in vivo* and to optimize treatment protocols. To date, no small-molecule chemokine receptor antagonists have been registered for clinical trials for the direct treatment of cancers; however, AMD3100 is in Phase II clinical trials for the mobilization of hematopoietic stem cells for the treatment of myeloma and non-Hodgkin's lymphoma.

The CXCR4 antagonists perhaps at the forefront in cancer treatment are the histone deacetylase inhibitors SAHA and butyrate. Butyrate and SAHA can significantly downregulate CXCR4 protein expression in acute lymphoblastic leukemia cells by up to 60% after 12 hours of treatment at millimolar and micromolar concentrations, respectively (Crazzolara *et al.*, 2002). At these concentrations inhibition was found to be specific to CXCR4 and not due to cytotoxic effects of either agent (Crazzolara *et al.*, 2002). Therefore, it is likely that these agents will have effects on malignant cell behavior relating to inhibition of the CXCL12/CXCR4 axis in addition to their primary mechanism of action. These agents are currently in clinical trials for their antitumor activities; the results of Phase I studies are already available (Marshall *et al.*, 2002; Sandor *et al.*, 2002). Although dose-limiting toxicity was observed, tolerated doses achieved plasma levels likely to be biologically active, and researchers observed increased histone acetylation in patient-derived peripheral blood mononuclear cells. One of 37 patients showed a partial response. Considering that the doses required to achieve inhibition of CXCR4 expression are well below those required for cytotoxicity, these drugs may provide a dual role in the treatment of cancers.

Many other chemokines and chemokine receptors have been implicated in the pathophysiology of malignancy, including CXCR2 and its ligands,

CXCL1, CXCL2, CXCL3, and CXCL8, in tumor-cell proliferation (Richmond and Thomas, 1986; Schadendorf *et al.*, 1993; Owen *et al.*, 1997; Miyamoto *et al.*, 1998; Fujisawa *et al.*, 2000; Takamori *et al.*, 2000), CCR7, CCR3, CCR4, CCR9, CCR10, and CXCR3 in metastasis (Arenberg *et al.*, 2001; Ishida *et al.*, 2003; Kleinhans *et al.*, 2003; Kawada *et al.*, 2004; Letsch *et al.*, 2004), and the CXCR2 ligands CXCL1, CXCL2, CXCL3, and CXCL8 in angiogenesis (Luan *et al.*, 1997; Gupta *et al.*, 1998; Moore *et al.*, 1999; Veltri *et al.*, 1999). However, studies using specific small-molecule antagonists of these chemokine receptors have not been performed in tumor models. Antagonists exist for many of these receptors, which raises the possibility of studying the effect of blocking these as-yet-untested receptors alone or potentially in combination.

C. *Treatment of Inappropriate Inflammatory Responses*

Regulation of the immune response probably represents the largest market for chemokine-receptor antagonists, and a large number of compounds are currently under development, some in clinical trials, for the treatment of these diseases. However, there are some major problems facing the development of drugs in this category. These include identifying the chemokines/chemokine receptors that are most important for the pathology of a diverse range of diseases. It is clear from a review of the current literature that the role of chemokines in immune disease is far from clear or simple. Potential redundancy in the chemokine network may also cause major problems in the obtainment of effective blockade of any single component of leukocyte trafficking or activation. Disparity between human and animal chemokine receptors means that drugs that would be effective on the human protein may have limited efficacy on animal versions of the proteins, making it difficult (if not impossible) to study the drugs in animal models. A prominent example of this is BX471, which is 100-fold less potent on murine CCR1 than on the human form of the receptor (Anders *et al.*, 2002). This phenomenon appears to affect particular chemically related series of compounds, and another CCR1 antagonist series based on xanthene carboxamide is equally effective in human and murine systems (Naya *et al.*, 2001a), while yet another series is completely inactive in the murine system (Liang *et al.*, 2000b).

Proof-of-principle studies using protein antagonists such as met-CCL5 have shown that inhibition of chemokine receptors improves outcome in animal models of atherosclerosis, graft rejection, and viral infection (Bonville *et al.*, 2004; Veillard *et al.*, 2004; Yun *et al.*, 2004). Blockade of chemokine function has also been explored as a strategy to treat inflammatory diseases like asthma, arthritis, and multiple sclerosis in animal models of disease. The potential efficacy of this strategy for atherosclerosis was further supported by a transgenic model in which conditional expression of M3, a herpesvirus protein that binds and inhibits chemokine activity, resulted in reduced plaque size (Pyo *et al.*, 2004). Despite problems with species specificity of some small-molecule antagonists, animal models have demonstrated potential efficacy of these agents in a number of inflammatory diseases, including rheumatoid arthritis, multiple sclerosis, and transplant rejection.

1. Arthritis

Rheumatoid arthritis is a chronic inflammatory disease characterized by an infiltration of Th1 cells, neutrophils, and (most notably) monocytes in the synovial joint, resulting in cartilage destruction and bone remodeling. There are good animal models of this disease, and it is probably the most extensively studied inflammatory disease. Many chemokines and their receptors are upregulated in arthritis, including CCL3, CCL4, CCL5, CCL20, CXCL8, CXCL9, CXCL10, CCR1, CCR2, and CCR5 (Hosaka *et al.*, 1994; Robinson *et al.*, 1995; Chabaud *et al.*, 2001; Patel *et al.*, 2001; Shahrara *et al.*, 2003), and some of these have been shown to act as chemoattractants for leukocytes within the synovial fluid within arthritic joints (Koch *et al.*, 1994; Volin *et al.*, 1998; Shadidi *et al.*, 2002; Ruth *et al.*, 2003). The importance of each of these chemokines in the pathogenesis of rheumatoid arthritis is not clear, although small-molecule targeting of chemokine receptors for the treatment of this disease is underway.

Inhibition of CXCR4 using TN14003 (the T140 derivative) or AMD3100 significantly reduced joint inflammation and the severity of the disease in mouse models of arthritis (Matthys *et al.*, 2001; Tamamura *et al.*, 2004), a finding that is consistent with the believed role of CXCR4 and CXCL12 in the retention of T-cells within the inflamed synovial compartment (Buckley *et al.*, 2000; Nanki *et al.*, 2000). In contrast to the phenotype of the CCR5−/− mouse, which is not protected from collagen-induced arthritis (Quinones *et al.*, 2004), administration of the CCR5 antagonist, TAK-779, significantly reduced leukocyte infiltration and improved outcome in animal models (Yang *et al.*, 2002). This result is, however, consistent with a protective effect of the CCR5Δ32 polymorphism in this disease (Pokorny *et al.*, 2004). Similarly to results using the CXCR2 antagonist CXCL1 (8-73) (Gong *et al.*, 2004) and those obtained using the CXCR2−/− mouse (Brown *et al.*, 2003), blockade of CXCR2 using SB225002 derivatives reduced the intensity of acute and chronic arthritis in rabbit models (Podolin *et al.*, 2002).

In humans, an unspecified CCR1 antagonist has been evaluated in a Phase I clinical trial in patients with chronic arthritis, without severe toxicity and with significantly reduced leukocyte infiltration, including macrophages and

T-cells (Haringman *et al.*, 2003). The CCR1 antagonist, MLN3897, entered Phase I trials in January 2004, and, although results are not yet publicly available, this agent has advanced to Phase II clinical trials. CCR1 and its ligands have been implicated in joint pain and swelling in animals and humans, providing good preclinical support for these studies (Plater-Zyberk *et al.*, 1997; Tak *et al.*, 1997; Barnes *et al.*, 1998). The roles of CCR2 and its ligand CCL2 in arthritis are not clear; animal knockout models and antibody blockade studies have produced conflicting results (Akahoshi *et al.*, 1993; Ogata *et al.*, 1997; Brown *et al.*, 2003; Quinones *et al.*, 2004). Despite this, the anti-CCR2 antibody, MLN1202, has also entered its Phase II clinical trial for the treatment of arthritis. Disappointingly, a monoclonal antibody to CXCL8 proved to be ineffective in a Phase II clinical trial of patients with rheumatoid arthritis, despite CXCL8 being heavily implicated in this disease by animal models and by human laboratory studies (Endo *et al.*, 1991; Harada *et al.*, 1994; Kraan *et al.*, 2001). Overall, inhibition of six individual chemokine receptors has demonstrated efficacy in preclinical studies. One of these has failed in human trials, although the reasons underlying this result are not clear and may be due to factors relating to *in vivo* efficacy rather than to lack of relevance of the ligand being manipulated. The respective efficacies of the remainder await verification, although some have had encouraging early results.

2. Other Immune Diseases

Little work has been done using small-molecule antagonists for the treatment of other immune diseases. CCR1–/– mice have greatly reduced graft rejection, and the CCR1 antagonist BX471 has been shown to decrease graft rejection in animal models of heart and kidney transplantation, an effect that was further enhanced by combination with the immunosuppressive drug cyclosporin (Gao *et al.*, 2000; Horuk *et al.*, 2001a,b). However, this drug has not yet entered clinical trials for this indication. In addition to having reduced graft rejection, CCR1–/– mice are also resistant to experimental immune encephalitis (Rottman *et al.*, 2000). CCR1 antagonists have produced encouraging results in rodent models of multiple sclerosis, abrogating clinical and histopathological signs of disease (Eltayeb *et al.*, 2003). The lead candidate from Berlex Biosciences/Schering AG, BX471, has entered Phase II clinical trial for the treatment of this disease, and patents have been lodged for new formulations of this product. In asthma, AMD3100 is the first small-molecule chemokine-receptor antagonist to show modification of disease progression in animal models (Lukacs *et al.*, 2002), a finding consistent with earlier antibody-blocking experiments (Gonzalo *et al.*, 2000). CCR3 and its ligand CCL11 have been widely implicated in allergic inflammation in *in vitro* studies (Gonzalo *et al.*, 1996; Ying *et al.*, 1997), but gene knockout studies have not provided clear evidence for the importance of this axis (Rothenberg *et al.*, 1997; Humbles *et al.*, 2002; Schuh *et al.*, 2002). Interestingly, small-molecule antagonists of CCR3, including DCP-168, have apparent efficacy in murine and monkey models of asthma, but this data has not been published in mainstream literature. The CCR3 antagonists, 766994 from GlaxoSmithKline and DPC168 from Bristol-Myers Squibb, are orally available and are in Phase II and Phase I clinical trials (respectively) for the treatment of this disease. Finally, Traficet-EN, an unspecified CXC chemokine receptor antagonist of unknown structure, has entered Phase I clinical trials for the treatment of inflammatory bowel disease, Crohn's disease, and ulcerative colitis. Considering the available *in vitro* and animal data on these diseases (Yuan *et al.*, 2001), it is likely that CXCR3 is the target of this agent.

Many potential chemokine-receptor targets have yet to be tested in inflammatory diseases, including CXCR2 and CXCR4 in arthritis (Buckley *et al.*, 2000; Nanki *et al.*, 2000; Brown *et al.*, 2003); CCR5 and CCR2 in multiple sclerosis (Fife *et al.*, 2000; Izikson *et al.*, 2000; Tran *et al.*, 2000; Gaupp *et al.*, 2003; Silversides *et al.*, 2004); CCR3, CCR4, CCR6, CCR10, and CXCR3 in psoriasis (Gerber *et al.*, 1997; Yawalkar *et al.*, 1999; Homey *et al.*, 2000, 2002; Flier *et al.*, 2001; Reiss *et al.*, 2001); CCR9 in Crohn's disease (Papadakis *et al.*, 2001); and CCR2 in ulcerative colitis (Andres *et al.*, 2000). In addition, CCR2 and CX3CR1 have potential roles in atherosclerosis (Roque *et al.*, 2002; Combadiere *et al.*, 2003; Lesnik *et al.*, 2003) and CCR1 and CCR2 in viral infection (Gao *et al.*, 1997; Kurihara *et al.*, 1997; Warmington *et al.*, 1999; Blease *et al.*, 2000; Sato *et al.*, 2000; Khan *et al.*, 2001), while graft rejection may be reduced by antagonists of CXCR3, CCR1, and CCR5 (Gao *et al.*, 2000, 2001; Hancock *et al.*, 2000, 2001; Baker *et al.*, 2003). Antagonists are available for most of these and are in clinical trials for other indications. It is likely that, once safety has been demonstrated in one setting, these may go directly into Phase I/II clinical trials. Overall, in diseases of the immune system the number of potential intervention sites is extensive, and the design of optimal treatment protocols is likely to take some time.

D. HIV Treatment

Treatment of HIV infection using chemokine-receptor antagonists has been the most intensively studied of all potential therapeutic uses of this nature. Use of chemokine-receptor antagonists for the treatment of HIV is much simpler than for the treatment of inflammatory disease or malignancy, because essentially only two receptors are involved, CCR5 and CXCR4.

HIV entry into a cell requires the fusion of its viral envelope with the cell membrane. This normally occurs through an initial binding of the viral envelope protein, gp120, to CD4, followed by a second interaction with a coreceptor, primarily chemokine receptor CXCR4 or CCR5 (Greene *et al.*, 2002). HIV strains are mostly restricted to the use of CCR5 (R5-tropic) or CXCR4 (X4-tropic), although some strains can utilize both coreceptors (Alkhatib *et al.*, 1996; Dragic *et al.*, 1996; Feng *et al.*, 1996; Berger *et al.*, 1998). R5 viruses are usually transmitted between individuals and infect both macrophages and primary T-cells, while X4 strains develop in about 50% of infected individuals and are usually associated with disease progression (Connor *et al.*, 1997); X4 strains infect T-cells and T-cell lines but not macrophages (Deng *et al.*, 1996). Individuals homozygous for defective CCR5 are highly resistant to HIV infection, underscoring the importance of this receptor in HIV transmission (Huang *et al.*, 1996; Michael *et al.*, 1997). As a result, inhibitors of coreceptor engagement have been intensively investigated as therapeutics for HIV infection.

Initially, peptide derivatives of the CCR5 chemokines, CCL3, CCL4, and CCL5, were investigated as potential therapeutics, but their poor bioavailability and their induction of receptor signaling are likely to prevent further development (Ylisastigui *et al.*, 1998; Detheux *et al.*, 2000). However, small-molecule inhibitors have provided a more promising approach. The first non-peptide small-molecule inhibitor was TAK-779 (Baba *et al.*, 1999; Iizawa *et al.*, 2003), but it induced local inflammation upon parental administration, preventing further development. However, a follow-up compound, TAK-220, which has similar in *in vitro* activity, is orally available, making it an excellent candidate for clinical trials (Iizawa *et al.*, 2003; Tremblay *et al.*, 2003).

SCH-C displayed potent antiviral activity in NOD/SCID mice xenografted with human liver and thymus, and it reduced viral load by up to 1.5 logs in clinical trials. Toxicity is likely to prevent further development of this agent (Reynes *et al.*, 2002), although clinical trials of this agent are still listed as being active. However, the next-generation compound, SCH-D, which has greater *in vitro* and *in vivo* potency, has performed better in clinical trials, with improved pharmacokinetics and no toxicities observed to date (Schurmann *et al.*, 2004). UK-427,857 caused no severe adverse effects in healthy volunteers and reduced the viral load in infected individuals by 1.4 logs (Abel *et al.*, 2003; Napier *et al.*, 2003). AK602 (ONO4128/GW873140) also blocks viral replication in animal models (Maeda *et al.*, 2003) and entered Phase I clinical trials in July 2003 with GlaxoSmithKline, with no significant adverse events documented.

The emergence of X4 strains is associated with accelerated progression to AIDS, so the development of CXCR4 antagonists has also been sought for HIV treatment. Similarly to the evolution of R5-specific drugs, chemokine-derived peptides have been investigated and discarded. However, a number of peptide inhibitors, based on the Polyphemusin II protein of the American horseshoe crab, have been brought to *in vivo* studies using animal models and have further potential for development (Murakami *et al.*,1997; Takenaga *et al.*, 2004). However, it is the small-molecule inhibitors that hold the greatest promise. ALX40-4C, AMD3100, AMD070, and KRH-1636 inhibit *in vitro* viral replication of X4 strains in the nanomolar range (Doranz *et al.*, 1997; Schols *et al.*, 1997; Ichiyama *et al.*, 2003). ALX40-4C and AMD3100 have been used in Phase I/II clinical trials in asymptomatic HIV patients and normal volunteers, respectively (Hendrix *et al.*, 2000). Both agents were well tolerated and showed no grade II toxicity (Doranz *et al.*, 2001). ALX40-4C did not produce consistent reductions in viral load, but only 12 of the 40 enrolled patients harbored X4 viral strains (Doranz *et al.*, 1997). The absence of an elevation in white-cell count, despite peak serum concentrations that clearly exceeded the IC_{50} for blockade of CXCL12, raises questions relating to *in vivo* effectiveness of the CXCR4 blockade by ALX40-4C. Interestingly, this agent may experience a renaissance in regard to its role in blocking the APJ receptor in the CNS (Albright *et al.*, 1999). In Phase II clinical trials, AMD3100 treatment resulted in a significant reduction in viral loads in patients harboring X4 strains (Schols *et al.*, 2002); however, cardiac toxicity and insufficient efficacy have resulted in its withdrawal from further development for the treatment of HIV (Scozzafava *et al.*, 2002). More recently, the orally available AMD070 demonstrated anti-HIV activity, and it will be investigated for its clinical potential (Schols *et al.*, 2003). It is currently in Phase I clinical trials with AnorMED.

HIV treatment will require long-term use of chemokine receptor antagonists. Potential detrimental effects resulting from the long-term disruption of hematopoiesis, immune-cell trafficking, and other normal physiological roles of the natural ligands for these receptors remain to be seen. HIV is a highly polymorphic virus existing as number of genetically related variants or quasispecies, which are subject to rapid evolution (Meyerhans *et al.*, 1989; Domingo, 1992; Holland *et al.*, 1992). Previous experience with other anti-HIV agents has shown that the highly polymorphic nature of the virus is a primary stumbling block to achieving control of disease progression (Mansky, 2002). Switching of viral strains from X4 to R5 and vice versa has already been observed in the presence of specific inhibitors (Este *et al.*, 1999; Mosier *et al.*, 1999). Of more concern is the emergence of viral strains resistant to antagonists in the absence of altered coreceptor usage (Schols *et al.*, 1998; Kanbara *et al.*, 2001; Trkola *et al.*, 2002). Encouragingly, these emerging resistant strains appear to have reduced replicative and pathogenic potential (Armand-Ugon *et al.*, 2003). In order to effectively block HIV entry in patients, it may be necessary to use more than one antagonist of

CCR5 (and/or CXCR4, depending on tropism) in combination or in parallel to minimize and control escape variants of the virus. One advantage for the treatment of R5-tropic HIV is the ability to block HIV entry without inhibiting the natural ligands of CCR5, thereby minimizing the potential for side effects, including the potential for immunosuppression.

IV. DISCUSSION

The strategy of pharmacological targeting of chemokine receptors is clearly a rapidly evolving area of experimental therapeutics, where a large number of new compounds have been made available over the last few years for human clinical trials. Predictably, the outcome of these studies so far has been mixed, predominantly because of toxicity observed following extended use. Considering that most of the indications for the use of these compounds are chronic conditions, this remains a major issue to be resolved. However, new and improved antagonists are continually being developed, and it seems likely that this problem will be overcome. The requirement for only short-term administration for stem-cell mobilization has positioned this utilization of chemokine inhibitors as the most successful to date and also the most likely to enter routine clinical use in the near future. The exclusive relationship of CXCL12 and CXCR4 has made targeting this axis relatively straightforward, and the biology of stem-cell mobilization is arguably better understood than that of immune or malignant cell regulation. The pathophysiology of HIV is also relatively well understood, and the *in vitro* results in HIV infection are now being translated into clinical trials in the therapy of AIDS. Undoubtedly, further positive therapeutic results will be reported from clinical trials of compounds that interact with other receptors.

There are, however, a number of caveats that require consideration. There is still only a partial understanding of the complex biology of the chemokine family and its receptors, a situation contributed to by the considerable redundancy of binding of various structurally related chemokines by many of the known receptors. The role of chemokines in various disease states is incompletely known, and the consequences of receptor blockade are similarly not fully established, partly due to the difficulty of producing suitable animal models.

These problems make interpretation of a number of clinical trials (particularly studies with negative results) difficult. Further issues that contribute to uncertainty include a lack of knowledge about optimal scheduling and delivery of some of the receptor antagonists and about the toxicities that may result from long-term blockade of key chemokine-receptor pathways in patients receiving chronic therapy with pharmacological inhibitors. Another important issue that has emerged recently is the strong commercial focus in the clinical research development of certain chemokine-receptor antagonists. Greater access by the scientific community to the results of clinical studies being conducted by pharmaceutical companies would be of significant benefit to research in chemokine therapeutics.

Future progress in this area is likely to be dependent on a number of factors. First, development of more pharmacologically effective inhibitors of certain chemokine receptors, and establishment of robust animal models for their evaluation, will be critical for progress in some situations. Second, more research on delivery and scheduling of inhibitors will be highly valuable in optimizing the use of existing drugs. Next, it is naïve to believe that any individual agent will be able to solve difficult therapeutic problems, particularly in complex diseases such as cancer. Chemokine-receptor antagonists with demonstrable effects as single agents are likely to be then subjected to further evaluation in combination with other established therapies, such as anticancer cytotoxic drugs, or with other chemokine inhibitors, to identify possible therapeutic synergies. A new phase of clinical evaluation is therefore foreseeable, as these exciting new compounds are incorporated into the treatment of a diverse range of human diseases.

References

Abel, S., Van der Ryst, E., Muirhead, G. J., Rosario, M., and Edgington, A. (2003). Pharmacokinetics of single and multiple oral doses of UK-427,857-a novel CCR5 antagonist in healthy volumteers. 10th Conference on Retroviruses and Opportunistic Infections Boston, MA.

Akahoshi, T., Wada, C., Endo, H., Hirota, K., Hosaka, S., Takagishi, K., Kondo, H., Kashiwazaki, S., and Matsushima, K. (1993). Expression of monocyte chemotactic and activating factor in rheumatoid arthritis. Regulation of its production in synovial cells by interleukin-1 and tumor necrosis factor. *Arthritis. Rheum.* **36,** 762–771.

Albright, A., Shieh, J., Itoh, T., Lee, B., Pleasure, D., O'Connor, M., Doms, R., and Gonzalez-Scarano, F. (1999). Microglia express CCR5, CXCR4, and CCR3, but of these, CCR5 is the principal coreceptor for human immunodeficiency virus type 1 dementia isolates. *J. Virol.* **73,** 205–213.

Alkhatib, G., Combadiere, C., Broder, C. C., Feng, Y., Kennedy, P. E., Murphy, P. M., and Berger, E. A. (1996). CC CKR5: A RANTES, MIP-1alpha, MIP-1beta receptor as a fusion cofactor for macrophage-tropic HIV-1. *Science* **272,** 1955–1958.

Allen, S., Newhouse, B., Anderson, A., Fauber, B., Allen, A., Chantry, D., Eberhardt, C., Odingo, J., and Burgess, L. (2004). Discovery and SAR of trisubstituted thiazolidinones as CCR4 antagonists. *Bioorg. Med. Chem. Lett.* **14,** 1619–1624.

Anders, H., Vielhauer, V., Frink, M., Linde, Y., Cohen, C., Blattner, S., Kretzler, M., Strutz, F., Mack, M., Grone, H., Onuffer, J., Horuk, R., Nelson, P., and Schlondorff, D. (2002). A chemokine receptor CCR-1 antagonist reduces renal fibrosis after unilateral ureter ligation. *J. Clin. Invest.* **109,** 251–259.

Anders, H., Belemezova, E., Eis, V., Segerer, S., Vielhauer, V., Perez de Lema, G., Kretzler, M., Cohen, C., Frink, M., Horuk, R., Hudkins, K., Alpers, C., Mampaso, F., and Schlondorff, D. (2004). Late onset of treatment with a chemokine receptor CCR1 antagonist prevents progression of lupus nephritis in MRL-Fas(lpr) mice. *J. Am. Soc. Nephrol.* **15,** 1504–1513.

Andres, P., Beck, P., Mizoguchi, E., Mizoguchi, A., Bhan, A., Dawson, T., Kuziel, W., Maeda, N., MacDermott, R., Podolsky, D., and Reinecker, H. (2000). Mice with a selective deletion of the CC chemokine receptors 5 or 2 are protected from dextran sodium sulfate-mediated colitis: Lack of CC chemokine receptor 5 expression results in a NK1.1+ lymphocyte-associated Th2-type immune response in the intestine. *J. Immunol.* **164,** 6303–6312.

Arenberg, D., Zlotnick, A., Strom, S., Burdick, M., and Strieter, R. (2001). The murine CC chemokine, 6C-kine, inhibits tumor growth and angiogenesis in a human lung cancer SCID mouse model. *Cancer Immunol. Immunother.* **49,** 587–592.

Armand-Ugon, M., Quinones-Mateu, M., Gutierez, A., Barretina, J., Blanco, J., Schols, D., De Clercq, E., Clotet, B., and Este, J. (2003). Reduced fitness of HIV-1 resistant to CXCR4 antagonists. *Antiviral. Ther.* **8,** 1–8.

Baba, M., Nishimura, O., Kanzaki, N., Okamoto, M., Sawada, H., Iizawa, Y., Shiraishi, M., Aramaki, Y., Okonogi, K., Ogawa, Y., Meguro, K., and Fujino, M. (1999). A small-molecule, non-peptide CCR5 antagonist with highly potent and selective anti-HIV activity. *Proc. Nat. Acad. Sci. USA* **96,** 5698–5703.

Bachelder, R. E., Wendt, M. A., and Mercurio, A. M. (2002). Vascular endothelial growth factor promotes breast carcinoma invasion in an autocrine manner by regulating the chemokine receptor CXCR4. *Cancer Res.* **62,** 7203–7206.

Baker, M., Chen, X., Rotramel, A., Nelson, J., Lu, B., Gerard, C., Kanwar, Y., and Kaufman, D. (2003). Genetic deletion of chemokine receptor CXCR3 or antibody blockade of its ligand IP-10 modulates posttransplantation graft-site lymphocytic infiltrates and prolongs functional graft survival in pancreatic islet allograft recipients. *Surgery* **134,** 126–133.

Balter, M. (1996). New hope in HIV disease. *Science* **274,** 1988–1989.

Barbero, S., Bonavia, R., Bajetto, A., Porcile, C., Pirani, P., Ravetti, J., Zona, G., Spaziante, R., Florio, T., and Schettini, G. (2003). Stromal cell-derived factor 1alpha stimulates human glioblastoma cell growth through the activation of both extracellular signal-regulated kinases 1/2 and Akt. *Cancer Res.* **63,** 1969–1974.

Barnes, D., Tse, J., Kaufhold, M., Owen, M., Hesselgesser, J., Strieter, R., Horuk, R., and Perez, H. (1998). Polyclonal antibody directed against human RANTES ameliorates disease in the Lewis rat adjuvant-induced arthritis model. *J. Clin. Invest.* **101,** 2910–2919.

Berger, E. A., Doms, R. W., Fenyo, E. M., Korber, B. T., Littman, D. R., Moore, J. P., Sattentau, Q. J., Schuitemaker, H., Sodroski, J., and Weiss, R. A. (1998). A new classification for HIV-1. *Nature* **391,** 240.

Bertolini, F., Dell'Agnola, C., Mancuso, P., Rabascio, C., Burlini, A., Monestiroli, S., Gobbi, A., Pruneri, G., and Martinelli, G. (2002). CXCR4 neutralization, a novel therapeutic approach for non-Hodgkin's lymphoma. *Cancer Res.* **62,** 3106–3112.

Blease, K., Mehrad, B., Standiford, T., Lukacs, N., Gosling, J., Boring, L., Charo, I., Kunkel, S., and Hogaboam, C. (2000). Enhanced pulmonary allergic responses to Aspergillus in CCR2-/- mice. *J. Immunol.* **165,** 2603–2611.

Bonville, C., Lau, V., De Leon, J., Gao, J., Easton, A., Rosenberg, H., and Domachowske, J. (2004). Functional antagonism of chemokine receptor CCR1 reduces mortality in acute pneumovirus infection *in vivo*. *J. Virol.* **78,** 7984–7989.

Brown, C., Blaho, V., and Loiacono, C. (2003). Susceptibility to experimental Lyme arthritis correlates with KC and monocyte chemoattractant protein-1 production in joints and requires neutrophil recruitment via CXCR2. *J. Immunol.* **171,** 893–901.

Brown, M., Avery, M., Brissette, W., Chang, J., Colizza, K., Conklyn, M., DiRico, A., Gladue, R., Kath, J., Krueger, S., Lira, P., Lillie, B., Lundquist, G., Mairs, E., McElroy, E., McGlynn, M., Paradis, T., Poss, C., Rossulek, M., Shepard, R., Sims, J., Strelevitz, T., Truesdell, S., Tylaska, L., Yoon, K., and Zheng, D. (2004). Novel CCR1 antagonists with improved metabolic stability. *Bioorg. Med. Chem. Lett.* **14,** 2175–2179.

Broxmeyer, H. E., Hangoc, G., Cooper, S., Li, X., Bridger, G., and Clapp, D. W. (2002). AMD3100, an antagonist of CXCR4 and mobilizer of myeloid progenitor cells, is a potent mobilizer of competitive repopulating long term marrow self-renewing stem cells in mice. *In* "American Society of Hematology Annual Meeting," Vol. 100: "Blood." Philadelphia, Pennsylvania, USA.

Buckley, C., Amft, N., Bradfield, P., Pilling, D., Ross, E., Arenzana-Seisdedos, F., Amara, A., Curnow, S., Lord, J., Scheel-Toellner, D., and Salmon, M. (2000). Persistent induction of the chemokine receptor CXCR4 by TGF-beta 1 on synovial T cells contributes to their accumulation within the rheumatoid synovium. *J. Immunol.* **165,** 3423–3429.

Burger, J., Tsukada, N., Burger, M., Zvaifler, N., Dell'Aquila, M., and Kipps, T. (2000). Blood-derived nurse-like cells protect chronic lymphocytic leukemia B cells from spontaneous apoptosis through stromal cell-derived factor-1. *Blood* **96,** 2655–2663.

Chabaud, M., Page, G., and Miossec, P. (2001). Enhancing effect of IL-1, IL-17, and TNF-alpha on macrophage inflammatory protein-3alpha production in rheumatoid arthritis: Regulation by soluble receptors and Th2 cytokines. *J. Immunol.* **167,** 6015–6020.

Combadiere, C., Potteaux, S., Gao, J., Esposito, B., Casanova, S., Lee, E., Debre, P., Tedgui, A., Murphy, P., and Mallat, Z. (2003). Decreased atherosclerotic lesion formation in CX3CR1/apolipoprotein E double knockout mice. *Circulation* **107,** 1009–1016.

Connor, R., Sheridan, K., Ceradini, D., Choe, S., and Landau, N. (1997). Change in co-receptor use correlates with disease progression in HIV-infected individuals. *J. Exp. Med.* **185,** 621–628.

Crazzolara, R., Johrer, K., Johnstone, R., Greil, R., Ofler, R., Meister, B., and Bernhard, D. (2002). Histone deacetylase inhibitors potently repress CXCR4 chemokine receptor expression and function in acute lymphoblastic leukaemia. *Br. J. Haematol.* **119,** 965–969.

De Clercq, E. (2003). The bicyclam AMD3100 story. *Nat. Rev. Drug Discov.* **2,** 581–587.

De Lucca, G., Kim, U., Johnson, C., Vargo, B., Welch, P., Covington, M., Davies, P., Solomon, K., Newton, R., Trainor, G., Decicco, C., and Ko, S. (2002). Discovery and structure-activity relationship of N-(ureidoalkyl)-benzyl-piperidines as potent small molecule CC chemokine receptor-3 (CCR3) antagonists. *J. Med. Chem.* **45,** 3794–3804.

Deng, H., Liu, R., Ellmeier, W., Choe, S., Unutmaz, D., Burkhart, M., Di Marzio, P., Marmon, S., Sutton, R. E., Hill, C. M., Davis, C. B., Peiper, S. C., Schall, T. J., Littman, D. R., and Landau, N. R. (1996). Identification of a major co-receptor for primary isolates of HIV-1. *Nature* **381,** 661–666.

Detheux, M., Standker, L., Vakili, J., Munch, J., Forssmann, U., Aderman, K., Pohlmann, S., Vassart, G., Kirchoff, F., Parmentier, M., and Forssmann, W. (2000). Natural proteolytic processing of hemofiltrate CC chemokine 1 generates a potent CC chemokine receptor (CCR)1 and CCR5 agonist with anti-HIV properties. *J. Exp. Med.* **192,** 1501–1508.

Devine, S., Flomenberg, N., Vesole, D., Liesveld, J., Weisdorf, D., Badel, K., Calandra, G., and DiPersio, J. (2004). Rapid mobilization of CD34+ cells following administration of the CXCR4 antagonist AMD3100 to patients with multiple myeloma and non-Hodgkin's lymphoma. *J. Clin. Oncol.* **22,** 1095–1102.

Domingo, E. (1992). Genetic variation and quasi-species. *Curr. Opin. Genet. Dev.* **2,** 61–63.

Doranz, B. J., Grovit-Ferbas, K., Sharron, M. P., Mao, S. H., Goetz, M. B., Daar, E. S., Doms, R. W., and O'Brien, W. A. (1997). A small-molecule inhibitor directed against the chemokine receptor CXCR4 prevents its use as an HIV-1 coreceptor. *J. Exp. Med.* **186,** 1395–1400.

Doranz, B. J., Filion, L. G., Diaz-Mitoma, F., Sitar, D. S., Sahai, J., Baribaud, F., Orsini, M. J., Benovic, J. L., Cameron, W., and Doms, R. W. (2001). Safe use of the CXCR4 inhibitor ALX40-4C in humans. AIDS Res Hum Retroviruses. *AIDS Res. Hum. Retroviruses* **17,** 475–486.

Dorr, P., Macartney, M., Rickett, G., Smith-Burchnell, C., Dobbs, S., Mori, J., Griffin, P., Lok, J., Irvin, R., Westby, M., Hitchcock, C., Stammen, B., Price, D., Armour, D., Wood, A., and Perros, M. (2003). UK-427,857, a novel small molecule HIV entry inhibitor is a specific antagonist of the chemokine receptor CCR5. *In* "10th Conference on Retroviruses and Opportunistic Infections." Boston, MA.

Dragic, T., Litwin, V., Allaway, G. P., Martin, S. R., Huang, Y., Nagashima, K. A., Cayanan, C., Maddon, P. J., Koup, R. A., Moore, J. P., and Paxton, W. A. (1996). HIV-1 entry into CD4+ cells is mediated by the chemokine receptor CC-CKR-5. *Nature* **381,** 667–673.

Dunussi-Joannopoulos, K., Zuberek, K., Runyon, K., Hawley, R. G., Wong, A., Erickson, J., Herrmann, S., and Leonard, J. P. (2002). Efficacious immunomodulatory activity of the chemokine stromal cell-derived factor 1 (SDF-1): Local secretion of SDF-1 at the tumor site serves as T-cell chemoattractant and mediates T-cell-dependent antitumor responses. *Blood* **100,** 1551–1558.

Elices, M. (2002). Bx-471 Berlex. *Curr. Opin. Investig. Drugs* **3,** 865–869.

Eltayeb, S., Sunnemark, D., Berg, A., Nordvall, G., Malmberg, A., Lassmann, H., Wallstrom, E., Olsson, T., and Ericsson-Dahlstrand, A. (2003). Effector stage CC chemokine receptor-1 selective antagonism reduces multiple sclerosis-like rat disease. *J. Neuroimmunol.* **142,** 75–85.

Endo, H., Akahoshi, T., Takagishi, K., Kashiwazaki, S., and Matsushima, K. (1991). Elevation of interleukin-8 (IL-8) levels in joint fluids of patients with rheumatoid arthritis and the induction by IL-8 of leukocyte infiltration and synovitis in rabbit joints. *Lymphokine Cytokine Res.* **10,** 245–252.

Este, J., Cabrera, C., Blanco, J., Gutierrez, A., Bridger, G., Henson, G., Clotet, B., Schols, D., and De Clercq, E. (1999). Shift of clinical human immunodeficiency virus type 1 isolates from X4 to R5 and prevention of emergence of the syncythium-inducing phenotype by blockade of CXCR4. *J. Virol.* **73,** 5577–5585.

Feng, Y., Broder, C. C., Kennedy, P. E., and Berger, E. A. (1996). HIV-1 entry cofactor: Functional cDNA cloning of a seven-transmembrane, G protein-coupled receptor. *Science* **272,** 872–877.

Fife, B., Huffnagle, G., Kuziel, W., and Karpus, W. (2000). CC chemokine receptor 2 is critical for induction of experimental autoimmune encephalomyelitis. *J. Exp. Med.* **92,** 899–906.

Flier, J., Boorsma, D., van Beek, P., Nieboer, C., Stoof, T., Willemze, R., and Tensen, C. (2001). Differential expression of CXCR3 targeting chemokines CXCL10, CXCL9, and CXCL11 in different types of skin inflammation. *J. Pathol.* **194,** 398–405.

Fujisawa, N., Sakao, Y., Hayashi, S., Hadden, Wr., Harmon, C., and Miller, E. (2000). Alpha-chemokine growth factors for adenocarcinomas; a synthetic peptide inhibitor for alpha-chemokines inhibits the growth of adenocarcinoma cell lines. *J. Cancer Res. Clin. Oncol.* **26,** 19–26.

Furuichi, K., Wada, T., Iwata, Y., Kitagawa, K., Kobayashi, K., Hashimoto, H., Ishiwata, Y., Asano, M., Wang, H., Matsushima, K., Takeya, M., Kuziel, W., Mukaida, N., and Yokoyama, H. (2003). CCR2 signaling contributes to ischemia-reperfusion injury in kidney. *J. Am. Soc. Nephrol.* **14,** 2503–2515.

Gao, J., Wynn, T., Chang, Y., Lee, E., Broxmeyer, H., Cooper, S., Tiffany, H., Westphal, H., Kwon, C., and Murphy, P. (1997). Impaired host defense, hematopoiesis, granulomatous inflammation and type 1-type 2 cytokine balance in mice lacking CC chemokine receptor 1. *J. Exp. Med.* **185,** 1959–1968.

Gao, W., Topham, P., King, J., Smiley, S., Csizmadia, V., Lu, B., Gerard, C., and Hancock, W. (2000). Targeting of the chemokine receptor CCR1 suppresses development of acute and chronic cardiac allograft rejection. *J. Clin. Invest.* **105,** 35–44.

Gao, W., Faia, K., Csizmadia, V., Smiley, S., Soler, D., King, J., Danoff, T., and Hancock, W. (2001). Beneficial effects of targeting CCR5 in allograft recipients. *Transplantation* **72,** 1199–1205.

Gao, P., Zhou, X., Yashiro-Ohtani, Y., Yang, Y., Sugimoto, N., Ono, S., Nakanishi, T., Obika, S., Imanishi, T., Egawa, T., Nagasawa, T., Fujiwara, H., and Hamaoka, T. (2003). The unique target specificity of a nonpeptide chemokine receptor antagonist: Selective blockade of two Th1 chemokine receptors CCR5 and CXCR3. *J. Leukoc. Biol.* **73,** 273–280.

Gaupp, S., Pitt, D., Kuziel, W., Cannella, B., and Raine, C. (2003). Experimental autoimmune encephalomyelitis (EAE) in CCR2(-/-) mice: Susceptibility in multiple strains. *Am. J. Pathol.* **162,** 139–150.

Gerber, B., Zanni, M., Uguccioni, M., Loetscher, M., Mackay, C., Pichler, W., Yawalkar, N., Baggiolini, M., and Moser, B. (1997). Functional expression of the eotaxin receptor CCR3 in T lymphocytes co-localizing with eosinophils. *Curr. Biol.* **7,** 836–843.

Gladue, R., Tylaska, L., Brissett, W., Lira, P., Kath, J., Poss, C., Brown, M., Paradis, T., Conklyn, M., Ogborne, K., McGlynn, M., Lillie, B., DiRico, A., Mairs, E., McElroy, E., Martin, W., Stock, I., Shepard, R., Showell, H., and Neote, K. (2003). CP-481,715, a potent and selective CCR1 antagonist with potential therapeutic implications for inflammatory diseases. *J. Biol. Chem.* **278,** 40473–40480.

Gong, J., Yan, R., Waterfield, J., and Clark-Lewis, I. (2004). Post-onset inhibition of murine arthritis using combined chemokine antagonist therapy. *Rheumatology (Oxford)* **43,** 39–42.

Gonzalo, J., Lloyd, C., Kremer, L., Finger, E., Martinez-A, C., Siegelman, M., Cybulsky, M., and Gutierrez-Ramos, J. (1996). Eosinophil recruitment to the lung in a murine model of allergic inflammation. The role of T cells, chemokines, and adhesion receptors. *J. Clin. Invest.* **98,** 2332–2345.

Gonzalo, J., Lloyd, C., Peled, A., Delaney, T., Coyle, A., and Gutierrez-Ramos, J. (2000). Critical involvement of the chemotactic axis CXCR4/stromal cell-derived factor-1 alpha in the inflammatory component of allergic airway disease. *J. Immunol.* **165,** 499–508.

Greene, W., and Peterlin, B. (2002). Charting HIV's remarkable voyage through the cell: Basic science as a passport to future therapy. *Nat. Med.* **8,** 673–680.

Gupta, S., Lysko, P., Pillarisetti, K., Ohlstein, E., and Stadel, J. (1998). Chemokine receptors in human endothelial cells. Functional expression of CXCR4 and its transcriptional regulation by inflammatory cytokines. *J. Biol. Chem.* **273,** 4282–4287.

Hall, J. M., and Korach, K. S. (2003). Stromal cell-derived factor 1, a novel target of estrogen receptor action, mediates the mitogenic effects of estradiol in ovarian and breast cancer cells. *Mol. Endocrinol.* **17,** 792–803.

Hancock, W., Lu, B., Gao, W., Csizmadia, V., Faia, K., King, J., Smiley, S., Ling, M., Gerard, N., and Gerard, C. (2000). Requirement of the chemokine receptor CXCR3 for acute allograft rejection. *J. Exp. Med.* **192,** 1515–1520.

Hancock, W., Gao, W., Csizmadia, V., Faia, K., Shemmeri, N., and Luster, A. (2001). Donor-derived IP-10 initiates development of acute allograft rejection. *J. Exp. Med.* **193,** 975–980.

Harada, A., Sekido, N., Akahoshi, T., Wada, T., Mukaida, N., and Matsushima, K. (1994). Essential involvement of interleukin-8 (IL-8) in acute inflammation. *J. Leukoc. Biol.* **56,** 559–564.

Haringman, J., Kraan, M., Smeets, T., Zwinderman, K., and Tak, P. (2003). Chemokine blockade and chronic inflammatory disease: Proof of concept in patients with rheumatoid arthritis. *Ann. Rheum. Dis.* **62,** 715–721.

Hattori, K., Heissig, B., Tashiro, K., Honjo, T., Tateno, M., Shieh, J., Hackett, N., Quitoriano, M., Crystal, R., Rafii, S., and Moore, M. (2001). Plasma elevation of stromal cell-derived factor-1 induces mobilization of mature and immature hematopoietic progenitor and stem cells. *Blood* **97,** 3354–3360.

Helbig, G., Christopherson, K., Bhat-Nakshatri, P., Kumar, S., Kishimoto, H., Miller, K., Broxmeyer, H., and Nakshatri, H. (2003). NF-kappaB promotes breast cancer cell migration and metastasis by inducing the expression of the chemokine receptor CXCR4. *J. Biol. Chem.* **278,** 21631–21638.

Hesselgesser, J., Ng, H., Liang, M., Zheng, W., May, K., Bauman, J., Monahan, S., Islam, I., Wei, G., Ghannam, A., Taub, D., Rosser, M., Snider, R., Morrissey, M., Perez, H., and Horuk, R. (1998). Identification and characterization of small molecule functional antagonists of the CCR1 chemokine receptor. *J. Biol. Chem.* **273,** 15687–15692.

Hendrix, C., Flexner, C., MacFarland, R., Giandomenico, C., Fuchs, E., Redpath, E., Bridger, G., and Henson, G. (2000). Pharmacokinetics and safety of AMD-3100, a novel antagonist of the CXCR-4 chemokine receptor, in human volunteers. *Antimicrob. Agents Chemother.* **44,** 1667–1673.

Hill, J., Paul, J., and Powell, T. (2003). Efficacy and risk of granulocyte macrophage colony stimulating factor administration in patients with severe coronary artery disease: Circulation, vol. 108, p. 478, Supplement 4 Vols.

Holland, J. J., De La, Torre, J. C., and Steinhauer, D. A. (1992). RNA virus populations as quasispecies. *Curr. Top. Microbiol. Immunol.* **176,** 1–20.

Homey, B., Dieu-Nosjean, M., Wiesenborn, A., Massacrier, C., Pin, J., Oldham, E., Catron, D., Buchanan, M., Muller, A., deWaal Malefyt, R., Deng, G., Orozco, R., Ruzicka, T., Lehmann, P., Lebecque, S., Caux, C., and Zlotnik, A. (2000). Up-regulation of macrophage inflammatory protein-3 alpha/CCL20 and CC chemokine receptor 6 in psoriasis. *J. Immunol.* **164,** 6621–6632.

Homey, B., Alenius, H., Muller, A., Soto, H., Bowman, E., Yuan, W., McEvoy, L., Lauerma, A., Assmann, T., Bunemann, E., Lehto, M., Wolff, H., Yen, D., Marxhausen, H., To, W., Sedgwick, J., Ruzicka, T., Lehmann, P., and Zlotnik, A. (2002). CCL27-CCR10 interactions regulate T cell-mediated skin inflammation. *Nat. Med.* **8,** 157–165.

Horuk, R., Clayberger, C., Krensky, A., Wang, Z., Grone, H., Weber, C., Weber, K., Nelson, P., May, K., Rosser, M., Dunning, L., Liang, M., Buckman, B., Ghannam, A., Ng, H., Islam, I., Bauman, J., Wei, G., Monahan, S., Xu, W., Snider, R., Morrissey, M., Hesselgesser, J., and Perez, H. (2001a). A non-peptide functional antagonist of the CCR1 chemokine receptor is effective in rat heart transplant rejection. *J. Biol. Chem.* **276,** 4199–4204.

Horuk, R., Shurey, S., Ng, H., May, K., Bauman, J., Islam, I., Ghannam, A., Buckman, B., Wei, G., Xu, W., Liang, M., Rosser, M., Dunning, L., Hesselgesser, J., Snider, R., Morrissey, M., Perez, H., and Green, C. (2001b). CCR1-specific non-peptide antagonist: Efficacy in a rabbit allograft rejection model. *Immunol. Lett.* **76,** 193–201.

Hosaka, S., Akahoshi, T., Wada, C., and Kondo, H. (1994). Expression of the chemokine superfamily in rheumatoid arthritis. *Clin. Exp. Immunol.* **97,** 451–457.

Huang, Y., Paxton, W., Wolinsky, S., Neumann, A., Zhang, L., He, T., Kang, S., Ceradini, D., Jin, Z., Yazdanbakhsh, K., Kunstman, K., Erickson, D., Dragon, E., Landau, N., Phair, J., Ho, D., and Koup, R. (1996). The role of a mutant CCR5 allele in HIV-1 transmission and disease progression. *Nat. Med.* **2,** 1240–1243.

Humbles, A., Lu, B., Friend, D., Okinaga, S., Lora, J., Al-Garawi, A., Martin, T., Gerard, N., and Gerard, C. (2002). The murine CCR3 receptor regulates both the role of eosinophils and mast cells in allergen-induced airway inflammation and hyperresponsiveness. *Proc. Nat. Acad. Sci. USA* **99,** 1479–1484.

Ichiyama, K., Yokoyama-Kumakura, S., Tanaka, Y., Tanaka, R., Hirose, K., Bannai, K., Edamatsu, T., Yanaka, M., Niitani, Y., Miyano-Kurosaki, N., Takaku, H., Koyanagi, Y., and Yamamoto, N. (2003). A duodenally absorbable CXC chemokine receptor 4 antagonist, KRH-1636, exhibits a potent and selective anti-HIV-1 activity. *Proc. Nat. Acad. Sci. USA* **100,** 4185–4190.

Iizawa, Y., Knzaki, N., Takashima, K., Miyake, H., Tagawal, Y., Sugihara, Y., and Baba, M. (2003). Activity of TAK-220, a small molecule CCR5 antagonist. *In* "10th Conference on Retroviruses and Opportunistic Infections," pp. Abstract 562. Boston, MA.

Ishida, T., Utsunomiya, A., Iida, S., Inagaki, H., Takatsuka, Y., Kusumoto, S., Takeuchi, G., Shimizu, S., Ito, M., Komatsu, H., Wakita, A., Eimoto, T., Matsushima, K., and Ueda, R. (2003). Clinical significance of CCR4 expression in adult T-cell leukemia/lymphoma: Its close association with skin involvement and unfavorable outcome. *Clin. Cancer Res.* **9,** 3625–3634.

Izikson, L., Klein, R., Charo, I., Weiner, H., and Luster, A. (2000). Resistance to experimental autoimmune encephalomyelitis in mice lacking the CC chemokine receptor (CCR)2. *J. Exp. Med.* **192,** 1075–1080.

Jayasuriya, H., Herath, K., Ondeyka, J., Polishook, J., Bills, G., Dombrowski, A., Springer, M., Siciliano, S., Malkowitz, L., Sanchez, M., Guan, Z., Tiwari, S., Stevenson, D., Borris, R., and Singh, S. (2004). Isolation and structure of antagonists of chemokine receptor (CCR5). *J. Nat. Prod.* **67,** 1036–1038.

Jin, Q., Nie, H., McCleland, B., Widdowson, K., Palovich, M., Elliott, J., Goodman, R., Burman, M., Sarau, H., Ward, K., Nord, M., Orr, B., Gorycki, P., and Busch-Petersen, J. (2004). Discovery of potent and orally bioavailable N,N'-diarylurea antagonists for the CXCR2 chemokine receptor. *Bioorg. Med. Chem. Lett.* **14,** 4375–4378.

Jourdan, P., Abbal, C., Noraz, N., Hori, T., Uchiyama, T., Vendrell, J. P., Bousquet, J., Taylor, N., Pene, J., Yssel, H., and Nora, N. (1998). IL-4 induces functional cell-surface expression of CXCR4 on human T cells. *J. Immunol.* **160,** 4153–4157.

Juarez, J., Bradstock, K., Gottlieb, D., and Bendall, L. (2003). Effects of inhibitors of the chemokine receptor CXCR4 on acute lymphoblastic leukemia cells *in vitro*. *Leukemia* **17,** 1294–1300.

Kanbara, K., Sato, S., Tanuma, J., Tamamura, H., Kanamoto, T., Mochizuki, K., Fujii, N., and Nakashima, H. (2001). Biological and genetic characterization of a human immunodeficiency virus strain resistant to CXCR4 antagonist T134. *AIDS Res. Hum. Retroviruses* **17,** 615–622.

Kang, H., Kim, H., Zhang, S., Park, K., Cho, H., Koo, B., Kim, Y., Soo Lee, D., Sohn, D., Han, K., Oh, B., Lee, M., and Park, Y. (2004). Effects of intracoronary infusion of peripheral blood stem-cells mobilised with granulocyte-colony stimulating factor on left ventricular systolic function and restenosis after coronary stenting in myocardial infarction: The MAGIC cell randomised clinical trial. *Lancet* **363,** 751–756.

Kath, J., Di Rico, A., Gladue, R., Martin, W., McElroy, E., Stock, I., Tylaska, L., and Zheng, D. (2004a). The discovery of structurally novel CCR1 antagonists derived from a hydroxyethylene peptide isostere template. *Bioorg. Med. Chem. Lett.* **14,** 2163–2167.

Kath, J., Brissette, W., Brown, M., Conklyn, M., DiRico, A., Dorff, P., Gladue, R., Lillie, B., Lira, P., Mairs, E., Martin, W., McElroy, E., McGlynn, M., Paradis, T., Poss, C., Stock, I., Tylaska, L., and Zheng, D. (2004b). Potent small molecule CCR1 antagonists. *Bioorg. Med. Chem. Lett.* **14,** 2169–2173.

Kawabata, K., Ujikawa, M., Egawa, T., Kawamoto, H., Tachibana, K., Iizasa, H., Katsura, Y., Kishimoto, T., and Nagasawa, T. (1999). A cell-autonomous requirement for CXCR4 in long-term lymphoid and myeloid reconstitution. *Proc. Natl. Acad. Sci. USA* **96,** 5663–5667.

Kawada, K., Sonoshita, M., Sakashita, H., Takabayashi, A., Yamaoka, Y., Manabe, T., Inaba, K., Minato, N., Oshima, M., and Taketo, M. (2004). Pivotal role of CXCR3 in melanoma cell metastasis to lymph nodes. *Cancer Res.* **64,** 4010–4017.

Khan, I., Murphy, P., Cascotti, L., Schwartzman, J., Collins, J., Gao, J., and Yeaman, G. (2001). Mice lacking the chemokine receptor CCR1 show increased susceptibility to Toxoplasma gondii infection. *J. Immunol.* **166,** 1930–1937.

Kijima, T., Maulik, G., Ma, P. C., Tibaldi, E. V., Turner, R. E., Rollins, B., Sattler, M., Johnson, B. E., and Salgia, R. (2002). Regulation of cellular proliferation, cytoskeletal function, and signal transduction through CXCR4 and c-Kit in small cell lung cancer cells. *Cancer Res.* **62,** 6304–6311.

Kitagawa, K., Wada, T., Furuichi, K., Hashimoto, H., Ishiwata, Y., Asano, M., Takeya, M., Kuziel, W., Matsushima, K., Mukaida, N., and Yokoyama, H. (2004). Blockade of CCR2 ameliorates progressive fibrosis in kidney. *Am. J. Pathol.* **165,** 237–246.

Kleinhans, M., Tun-Kyi, A., Gilliet, M., Kadin, M., Dummer, R., Burg, G., and Nestle, F. (2003). Functional expression of the eotaxin receptor CCR3 in CD30+ cutaneous T-cell lymphoma. *Blood* **101,** 1487–1493.

Koch, A., Kunkel, S., Harlow, L., Mazarakis, D., Haines, G., Burdick, M., Pope, R., and Strieter, R. (1994). Macrophage inflammatory protein-1 alpha. A novel chemotactic cytokine for macrophages in rheumatoid arthritis. *J. Clin. Invest.* **93,** 921–928.

Koshiba, T., Hosotani, R., Miyamoto, Y., Ida, J., Tsuji, S., Nakajima, S., Kawaguchi, M., Kobayashi, H., Doi, R., Hori, T., Fujii, N., and Imamura, M. (2000). Expression of stromal cell-derived factor 1 and CXCR4 ligand receptor system in pancreatic cancer: A possible role for tumor progression. *Clin. Cancer Res.* **6,** 3530–3535.

Kraan, M., Patel, D., Haringman, J., Smith, M., Weedon, H., Ahern, M., Breedveld, F., and Tak, P. (2001). The development of clinical signs of rheumatoid synovial inflammation is associated with increased synthesis of the chemokine CXCL8 (interleukin-8). *Arthritis Res.* **3,** 65–71.

Kurihara, T., Warr, G., Loy, J., and Bravo, R. (1997). Defects in macrophage recruitment and host defense in mice lacking the CCR2 chemokine receptor. *J. Exp. Med.* **186,** 1757–1762.

Lapidot, T. (2001). Mechanism of human stem cell migration and repopulation of NOD/SCID and B2mnull NOD/SCID mice. The role of SDF-1/CXCR4 interactions. *Ann. N. Y. Acad. Sci.* **938,** 83–95.

Lesnik, P., Haskell, C., and Charo, I. (2003). Decreased atherosclerosis in CX3CR1-/- mice reveals a role for fractalkine in atherogenesis. *J. Clin. Invest.* **111,** 333–340.

Letsch, A., Keilholz, U., Schadendorf, D., Assfalg, G., Asemissen, A., Thiel, E., and Scheibenbogen, C. (2004). Functional CCR9 expression is associated with small intestinal metastasis. *J. Invest. Dermatol.* **122,** 685–690.

Levesque, J., Hendy, J., Takamatsu, Y., Simmons, P., and Bendall, L. (2003). Disruption of the CXCR4/CXCL12 chemotactic interaction during hematopoietic stem cell mobilization induced by GCSF or cyclophosphamide. *J. Clin. Invest.* **111,** 187–196.

Liang, M., Mallari, C., Rosser, M., Ng, H., May, K., Monahan, S., Bauman, J., Islam, I., Ghannam, A., Buckman, B., Shaw, K., Wei, G., Xu, W., Zhao, Z., Ho, E., Shen, J., Oanh, H., Subramanyam, B., Vergona, R., Taub, D., Dunning, L., Harvey, S., Snider, R., Hesselgesser, J., Morrissey, M., and Perez, H. (2000a). Identification and characterization of a potent, selective, and orally active antagonist of the CC chemokine receptor-1. *J. Biol. Chem.* **275,** 19000–19008.

Liang, M., Rosser, M., Ng, H., May, K., Bauman, J., Islam, I., Ghannam, A., Kretschmer, P., Pu, H., Dunning, L., Snider, R., Morrissey, M., Hesselgesser, J., Perez, H., and Horuk, R. (2000b). Species selectivity of a small molecule antagonist for the CCR1 chemokine receptor. *Eur. J. Pharmacol.* **389,** 41–49.

Liang, Z., Wu, T., Lou, H., Yu, X., Taichman, R., Lau, S., Nie, S., Umbreit, J., and Shim, H. (2004). Inhibition of breast cancer metastasis by selective synthetic polypeptide against CXCR4. *Cancer Res.* **64,** 4302–4308.

Liles, W., Rodger, E., Broxmeyer, H., Srour, E., Dehner, C., Badel, K., Calandra, G., Christensen, J., Wood, B., Price, T., and Dale, D. (2002). Mobilization and collection of CD34+ progenitor cells from normal human volunteers with AMD-3100, a CXCR4 antagonist, and G-CSF. *In* "American Society for Hematology," Vol. 100, p. 404a, 11 Vols. Philadelphia.

Liles, W., Broxmeyer, H., Rodger, E., Wood, B., Hubel, K., Cooper, S., Hangoc, G., Bridger, G., Henson, G., Calandra, G., and Dale, D. (2003). Mobilization of hematopoietic progenitor cells in healthy volunteers by AMD3100, a CXCR4 antagonist. *Blood* **102,** 2728–2730.

Luan, J., Shattuck-Brandt, R., Haghnegahdar, H., Owen, J., Strieter, R., Burdick, M., Nirodi, C., Beauchamp, D., Johnson, K., and Richmond, A. (1997). Mechanisms and biological significance of constitutive expression of MGSA/GRO chemokines in malignant melanoma tumor progression. *J. Leukoc. Biol.* **62,** 588–597.

Lukacs, N., Berlin, A., Schols, D., Skerlj, R., and Bridger, G. (2002). AMD3100, a CxCR4 antagonist, attenuates allergic lung inflammation and airway hyperreactivity. *Am. J. Pathol.* **160,** 1353–1360.

Maeda, K., Nakata, H., Miyakawa, T., Ogata, H., Koh, Y., Shibayama, S., Sagawa, K., Takaoka, Y., Moravek, J., Koyanagi, H., and Mitsuya, H. (2003). AK602: A novel HIV-specific spirodiketopiperazine CCR5 inhibitor potent against a wide spectrum of R5-HIV. 10th Conference on Retroviruses and Opportunistic Infections Boston, MA.

Maeda, K., Nakata, H., Koh, Y., Miyakawa, T., Ogata, H., Takaoka, Y., Shibayama, S., Sagawa, K., Fukushima, D., Moravek, J., Koyanagi, Y., and Mitsuya, H. (2004). Spirodiketopiperazine-based CCR5 inhibitor which preserves CC-chemokine/CCR5 interactions and exerts potent activity against R5 human immunodeficiency virus type 1 *in vitro*. *J. Virol.* **78,** 8654–8662.

Mansky, L. (2002). HIV mutagenesis and the evolution of antiretroviral drug resistance. *Drug Resist. Updat.* **5,** 219–223.

Marshall, J., Rizvi, N., Kauh, J., Dahut, W., Figuera, M., Kang, M., Figg, W., Wainer, I., Chaissang, C., Li, M., and Hawkins, M. (2002). A phase I trial of depsipeptide (FR901228) in patients with advanced cancer. *J. Exp. Ther. Oncol.* **2,** 325–332.

Matthys, P., Hatse, S., Vermeire, K., Wuyts, A., Bridger, G., Henson, G. W., De Clercq, E., Billiau, A., and Schols, D. (2001). AMD3100, a potent and specific antagonist of the stromal cell-derived factor-1 chemokine receptor CXCR4, inhibits autoimmune joint inflammation in IFN-gamma receptor-deficient mice. *J. Immunol.* **167,** 4686–4692.

Meyerhans, A., Cheynier, R., Albert, J., Seth, M., Kwok, S., Sninsky, J., Morfeldt-Manson, L., Asjo, B., and Wain-Hobson, S. (1989). Temporal fluctuations in HIV quasispecies *in vivo* are not reflected by sequential HIV isolations. *Cell* **58,** 901–910.

Michael, N., Chang, G., Louie, L., Mascola, J., Dondero, D., Birx, D., and Sheppard, H. (1997). The role of viral phenotype and CCR-5 gene defects in HIV-1 transmission and disease progression. *Nat. Med.* **3,** 338–340.

Miyamoto, M., Shimizu, Y., Okada, K., Kashii, Y., Higuchi, K., and Watanabe, A. (1998). Effect of IL-8 on the production of tumor associated substances and autocrine growth of human liver and pancreatic cancer cells. *Cancer Immunol. Immunother.* **47,** 47–57.

Moore, B., Arenberg, D., Stoy, K., Morgan, T., Addison, C., Morris, S., Glass, M., Wilke, C., Xue, Y., Sitterding, S., Kunkel, S., Burdick, M., and Strieter, R. (1999). Distinct CXC chemokines mediate tumorigenicity of prostate cancer cells. *Am. J. Pathol.* **154,** 1503–1512.

Moran, C., Arenberg, D., Huang, C., Giordano, T., Thomas, D., Misek, D., Chen, G., Iannettoni, M., Orringer, M., Hanash, S., and Beer, D. (2002). RANTES expression is a predictor of survival in stage 1 lung adenocarcinoma. *Clin. Cancer Res.* **8,** 3803–3812.

Mosier, D., Picchio, G., Gulizia, R., Sabbe, R., Piognard, P., Picard, L., Offord, R., Thompson, D., and Wilken, J. (1999). Highly potent RANTES analogues either prevent CCR5-usinghuman immunodeficiency type-1 infection *in vivo* or rapidly select for CXCR4-using variants. *J. Virol.* **73,** 3544–3550.

Muller, A., Homey, B., Soto, H., Ge, N., Catron, D., Buchanan, M., McClanahan, T., Murphy, E., Yuan, W., Wagner, S., Barrerak, J., Mohark, A., Verastegui, E., and Zlotnik, A. (2001). Involvement of chemokine receptors in breast cancer metastasis. *Nature* **410,** 50–56.

Murakami, T., Nakajima, T., Koyanagi, Y., Tachibana, K., Fujii, N., Tamamura, H., Yoshida, N., Waki, M., Matsumoto, A., Yoshie, O., Kishimoto, T., Yamamoto, N., and Nagasawa, T. (1997). A small molecule CXCR4 inhibitor that blocks T cell line-tropic HIV-1 infection. *J. Exp. Med.* **186,** 1389–1393.

Murakami, T., Maki, W., Cardones, A., Fang, H., Tun Kyi, A., Nestle, F., and Hwang, S. (2002). Expression of CXC chemokine receptor-4 enhances the pulmonary metastatic potential of murine B16 melanoma cells. *Cancer Res.* **62,** 7328–7334.

Nanki, T., Hayashida, K., El Gabalawy, H., Suson, S., Shi, K., Girschick, H., Yavuz, S., and Lipsky, P. (2000). Stromal cell-derived factor-1-CXC chemokine receptor 4 interactions play a central role in CD4+ T cell accumulation in rheumatoid arthritis synovium. *J. Immunol.* **165,** 6590–6598.

Napier, C., Dorr, P., Gladue, R., Halliday, R., Leishman, D., Machin, I., Mitchell, R., Nedderman, A., Perros, M., Roffey, S., Walker, D., and Webster, R. (2003). The preclinical pharmacokinetics and safety pharmacology of the anti-HIV CCR5 antagonist, UK-427,857. *In* "10th Conference on Retroviruses and Opportunistic Infections." Boston, MA.

Naya, A., Sagara, Y., Ohwaki, K., Saeki, T., Ichikawa, D., Iwasawa, Y., Noguchi, K., and Ohtake, N. (2001a). Design, synthesis, and discovery of a novel CCR1 antagonist. *J. Med. Chem.* **44,** 1429–1435.

Naya, A., Kobayashi, K., Ishikawa, M., Ohwaki, K., Saeki, T., Noguchi, K., and Ohtake, N. (2001b). Discovery of a novel CCR3 selective antagonist. *Bioorg. Med. Chem. Lett.* **11,** 1219–1223.

Naya, A., Ishikawa, M., Matsuda, K., Ohwaki, K., Saeki, T., Noguchi, K., and Ohtake, N. (2003a). Structure-activity relationships of xanthene carboxamides, novel CCR1 receptor antagonists. *Bioorg. Med. Chem.* **11,** 875–884.

Naya, A., Kobayashi, K., Ishikawa, M., Ohwaki, K., Saeki, T., Noguchi, K., and Ohtake, N. (2003b). Structure-activity relationships of 2-(benzothiazolylthio)acetamide class of CCR3 selective antagonist. *Chem. Pharm. Bull.* **51,** 697–701.

Netelenbos, T., van den Born, J., Kessler, F., Zweegman, S., Merle, P., van Oostveen, J., Zwaginga, J., Huijgens, P., and Drager, A. (2003). Proteoglycans on bone marrow endothelial cells bind and present SDF-1 towards hematopoietic progenitor cells. *Leukemia* **17,** 175–184.

Nishii, K., Katayama, N., Miwa, H., Shikami, M., Masuya, M., Shiku, H., and Kita, K. (1999). Survival of human leukaemic B-cell precursors is supported by stromal cells and cytokines: Association with the expression of bcl-2 protein. *Br. J. Haematol.* **105,** 701–710.

Nomura, T., Hasegawa, H., Kohno, M., Sasaki, M., and Fujita, S. (2001). Enhancement of anti-tumor immunity by tumor cells transfected with the secondary lymphoid tissue chemokine EBI-1-ligand chemokine and stromal cell-derived factor-1alpha chemokine genes. *Int. J. Cancer* **91,** 597–606.

O'Dowda, B., Heiberb, M., Chanb, A., Hengd, H., Tsuid, L.-C, Kennedye, J., Shid, X., Petronise, A., Georgeb, S., and Nguyena, T. (1993). A human gene that shows identity with the gene encoding the angiotensin receptor is located on chromosome 11. *Gene* **136,** 355–360.

Ogata, H., Takeya, M., Yoshimura, T., Takagi, K., and Takahashi, K. (1997). The role of monocyte chemoattractant protein-1 (MCP-1) in the pathogenesis of collagen-induced arthritis in rats. *J. Pathol.* **182,** 106–114.

Owen, J., Strieter, R., Burdick, M., Haghnegahdar, H., Nanney, L., Shattuck-Brandt, R., and Richmond, A. (1997). Enhanced tumor-forming capacity for immortalized melanocytes expressing melanoma growth stimulatory activity/growth-regulatory cytokine beta and gamma proteins. *Int. J. Cancer* **73,** 94–103.

Papadakis, K., Prehn, J., Moreno, S., Cheng, L., Kouroumalis, E., Deem, R., Breaverman, T., Ponath, P., Andrew, D., Green, P., Hodge, M., Binder, S., and Targan, S. (2001). CCR9-positive lymphocytes and thymus-expressed chemokine distinguish small bowel from colonic Crohn's disease. *Gastroenterology* **121,** 246–254.

Patel, D., Zachariah, J., and Whichard, L. (2001). CXCR3 and CCR5 ligands in rheumatoid arthritis synovium. *Clin. Immunol.* **98,** 39–45.

Paul, S., Mancuso, P., Rabascio, C., Gobbi, A., Capillo, M., Pruneri, G., Martinelli, G., Fricker, S., Bridger, G., and Bertolini, F. (2002). *In vitro* and preclinical activity of the novel AMD3100 CXCR4 antagonist in lymphoma models. *In* "American Society of Hematology Annual Meeting," Vol. 100: "Blood." Philadelphia, Pennsylvania, USA.

Petit, I., Szyper-Kravitz, M., Nagler, A., Lahav, M., Peled, A., Habler, L., Ponomaryov, T., Taichman, R., Arenzana-Seisdedos, F., Fujii, N., Sandbank, J., Zipori, D., and Lapidot, T. (2002). G-CSF induces stem cell mobilization by decreasing bone marrow SDF-1 and up-regulating CXCR4. *Nat. Immunol.* **3,** 687–694.

Phillips, R. J., Burdick, M. D., Lutz, M., Belperio, J. A., Keane, M. P., and Strieter, R. M. (2003). The stromal derived factor-1/CXCL12-CXC chemokine receptor 4 biological axis in non-small cell lung cancer metastases. *Am. J. Respir. Crit. Care Med.* **167,** 1676–1686.

Plater-Zyberk, C., Hoogewerf, A., Proudfoot, A., Power, C., and Wells, T. (1997). Effect of a CC chemokine receptor antagonist on collagen induced arthritis in DBA/1 mice. *Immunol. Lett.* **57,** 117–120.

Podolin, P., Bolognese, B., Foley, J., Schmidt, D., Buckley, P., Widdowson, K., Jin, Q., White, J., Lee, J., Goodman, R., Hagen, T., Kajikawa, O., Marshall, L., Hay, D., and Sarau, H. (2002). A potent and selective nonpeptide antagonist of CXCR2 inhibits acute and chronic models of arthritis in the rabbit. *J. Immunol.* **169,** 6435–6444.

Pokorny, V., McQueen, F., Yeoman, S., Merriman, M., Merriman, T., Harrison, A., Highton, J., and McLean, L. (2004). Evidence for negative association of the chemokine receptor CCR5 d32 polymorphism with rheumatoid arthritis. *Ann. Rheum. Dis.* (In press).

Pyo, R., Jensen, K., Wiekowsk, M., Manfra, D., Alcami, A., Taubman, M., and Lira, S. (2004). Inhibition of intimal hyperplasia in transgenic mice conditionally expressing the chemokine-binding protein M3. *Am. J. Pathol.* **164,** 2289–2297.

Quinones, M., Ahuja, S., Jimenez, F., Schaefer, J., Garavito, E., Rao, A., Chenaux, G., Reddick, R., Kuziel, W., and Ahuja, S. (2004). Experimental arthritis in CC chemokine receptor 2-null mice closely mimics severe human rheumatoid arthritis. *J. Clin. Invest.* **113,** 856–866.

Reiss, Y., Proudfoot, A., Power, C., Campbell, J., and Butcher, E. (2001). CC chemokine receptor (CCR)4 and the CCR10 ligand cutaneous T cell-attracting chemokine (CTACK) in lymphocyte trafficking to inflamed skin. *J. Exp. Med.* **194,** 1541–1547.

Reynes, J., Rouzier, R., Kanouni, T., Baillat, V., Baroudy, B., and Keung, A. (2002). SCH-C: Safety and antiviral effects of a CCR5 receptor antagonist in HIV-1 infected subjects. *In* "9th Conference on Retroviruses and Opportunistic Infections," Abstract 1.

Richmond, A., and Thomas, H. (1986). Purification of melanoma growth stimulatory activity. *J. Cell Physiol.* **129,** 375–384.

Robinson, E., Keystone, E., Schall, T., Gillett, N., and Fish, E. (1995). Chemokine expression in rheumatoid arthritis (RA): Evidence of RANTES and macrophage inflammatory protein (MIP)-1 beta production by synovial T cells. *Clin. Exp. Immunol.* **101,** 398–407.

Robinson, S., Scott, K., Wilson, J., Thompson, R., Proudfoot, A., and Balkwill, F. (2003). A chemokine receptor antagonist inhibits experimental breast tumor growth. *Cancer Res.* **63,** 8360–8365.

Roque, M., Kim, W., Gazdoin, M., Malik, A., Reis, E., Fallon, J., Badimon, J., Charo, I., and Taubman, M. (2002). CCR2 deficiency decreases intimal hyperplasia after arterial injury. *Arterioscler. Thromb. Vasc. Biol.* **22,** 554–559.

Rothenberg, M., MacLean, J., Pearlman, E., Luster, A., and Leder, P. (1997). Targeted disruption of the chemokine eotaxin partially reduces antigen-induced tissue eosinophilia. *J. Exp. Med.* **185,** 785–790.

Rottman, J., Slavin, A., Silva, R., Weiner, H., Gerard, C., and Hancock, W. (2000). Leukocyte recruitment during onset of experimental allergic encephalomyelitis is CCR1 dependent. *Eur. J. Immunol.* **30,** 2372–2377.

Rubin, J., Kung, A., Klein, R., Chan, J., Sun, Y., Schmidt, K., Kieran, M., Luster, A., and Segal, R. (2003). A small-molecule antagonist of CXCR4 inhibits intracranial growth of primary brain tumors. *Proc. Natl. Acad. Sci. USA* **100,** 13513–13518.

Ruth, J., Shahrara, S., Park, C., Morel, J., Kumar, P., Qin, S., and Koch, A. (2003). Role of macrophage inflammatory protein-3alpha and its ligand CCR6 in rheumatoid arthritis. *Lab. Invest.* **83,** 579–588.

Sabroe, I., Peck, M., Van Keulen, B., Jorritsma, A., Simmons, G., Clapham, P., Williams, T., and Pease, J. (2000). A small molecule antagonist of chemokine receptors CCR1 and CCR3. Potent inhibition of eosinophil function and CCR3-mediated HIV-1 entry. *J. Biol. Chem.* **275,** 25985–25992.

Sandor, V., Bakke, S., Robey, R., Kang, M., Blagosklonny, M., Bender, J., Brooks, R., Piekarz, R., Tucker, E., Figg, W., Chan, K., Goldspiel, B., Fojo, A., Balcerzak, S., and Bates, S. (2002). Phase I trial of the histone deacetylase inhibitor, depsipeptide (FR901228, NSC 630176), in patients with refractory neoplasms. *Clin. Cancer Res.* **8,** 718–728.

Sato, N., Sawada, K., Takahashi, T., Mogi, Y., Asano, S., Koike, T., and Sekiguchi, S. (1994). A time course study for optimal harvest of peripheral blood progenitor cells by granulocyte colony-stimulating factor in healthy volunteers. *Exp. Hematol.* **22,** 973–978.

Sato, N., Ahuja, S., Quinones, M., Kostecki, V., Reddick, R., Melby, P., Kuziel, W., and Ahuja, S. (2000). CC chemokine receptor (CCR)2 is required for langerhans cell migration and localization of T helper cell type 1 (Th1)-inducing dendritic cells. Absence of CCR2 shifts the Leishmania major-resistant phenotype to a susceptible state dominated by Th2 cytokines, b cell outgrowth, and sustained neutrophilic inflammation. *J. Exp. Med.* **192,** 205–218.

Schadendorf, D., Moller, A., Algermissen, B., Worm, M., Sticherling, M., and Czarnetzki, B. (1993). IL-8 produced by human malignant melanoma cells *in vitro* is an essential autocrine grwoth factor. *J. Immunol.* **151,** 2267–2275.

Schols, D., Este, J., Henson, G., and De Clercq, E. (1997). Bicyclams, a class of potent anti-HIV agents, are targeted at the HIV coreceptor fusin/CXCR-4. *Antiviral Res.* **35,** 147–156.

Schols, D., Este, J., Cabrera, C., and De Clercq, E. (1998). T-cell-line-tropic human immunodeficiency virus type 1 that is made resistant to stromal cell-derived factor 1alpha conatins mutations in the envelope gp120 but does not show a switch in coreceptor use. *J. Virol.* **72,** 4032–4037.

Schols, D., Claes, S., De Clercq, E., Hendrix, C., Bridger, G., Calandra, G., *et al.* (2002). Group A-HS. AMD3100, a CXCR4 antagonist, reduced HIV viral load and X4 virus levels in humans. *In* "9th Conference on Retroviruses and Opportunistic Infections," Vol. 2, p. 53. Seattle, WA.

Schols, D., Claes, S., Hatse, S., Princen, K., Vermeire, K., De Clercq, E., Skerlj, R., Bridger, G., and Calandra, G. (2003). Anti-HIV activity profile of AMD070, an orally bioavailable CXCR4 antagonist. *In* "10th Conference on Retroviruses and Opportunistic Infections," p. 562. (Abstract), Bostan, MA.

Schuh, J., Blease, K., Kunkel, S., and Hogaboam, C. (2002). Eotaxin/CCL11 is involved in acute, but not chronic, allergic airway responses to Aspergillus fumigatus. *Am. J. Physiol. Lung. Cell. Mol. Physiol.* **283,** L198–L204.

Schurmann, D., Rouzier, R., Nougarede, R., Reynes, J., Fatkenhever, G., Raffi, F., Michelet, C., Tarral, A., Haffman, C., Kiunke, J., Sprenger, H., vanLier, J., Sansune, A., Jackson, M., and Laughlin, M. (2004). SCH-D: Antiviral activity of a CCR5 receptor antagonist. *In* "11th Conference on Retroviruses and Opportunistic Infections." San Francisco, CA.

Scotton, C., Wilson, J., Milliken, D., Stamp, G., and Balkwill, F. (2001). Epithelial cancer cell migration: A role for chemokine receptors? *Cancer Res.* **61,** 4961–4965.

Scotton, C. J., Wilson, J. L., Scott, K., Stamp, G., Wilbanks, G. D., Fricker, S., Bridger, G., and Balkwill, F. R. (2002). Multiple actions of the chemokine CXCL12 on epithelial tumor cells in human ovarian cancer. *Cancer Res.* **62,** 5930–5938.

Scozzafava, A., Mastrolorenzo, A., and Supuran, C. (2002). Non-peptidic chemokine receptors antagonists as emerging anti-HIV agents. *Enzyme. Inhib. Med. Chem.* **17,** 69–76.

Sehgal, A., Keener, C., Boynton, A., Warrick, J., and Murphy, G. (1998). CXCR4, a chemokine receptor, is overexpressed in and required for proliferation of glioblastoma tumor cells. *J. Surg. Oncol.* **69,** 99–104.

Shadidi, K., Thompson, K., Henriksen, J., Natvig, J., and Aarvak, T. (2002). Association of antigen specificity and migratory capacity of memory T cells in rheumatoid arthritis. *Scand. J. Immunol.* **55,** 274–283.

Shahrara, S., Amin, M., Woods, J., Haines, G., and Koch, A. (2003). Chemokine receptor expression and *in vivo* signaling pathways in the joints of rats with adjuvant-induced arthritis. *Arthritis. Rheum.* **48,** 3568–3583.

Shankaran, K., Donnelly, K., Shah, S., Guthikonda, R., MacCoss, M., Mills, S., Gould, S., Malkowitz, L., Siciliano, S., Springer, M., Carella, A., Carver, G., Hazuda, D., Holmes, K., Kessler, J., Lineberger, J., Miller, M., Emini, E., and Schleif, W. (2004). Syntheses and SAR studies of 4-(heteroarylpiperdin-1-yl-methyl)-pyrrolidin-1-yl-acetic acid antagonists of the human CCR5 chemokine receptor. *Bioorg. Med. Chem. Lett.* **14,** 3419–3424.

Shieh, J., Albright, A., Sharron, M., Gartner, S., Strizki, J., Doms, R., and Gonzalez-Scarano, F. (1998). Chemokine receptor utilization by human immunodeficiency virus type 1 isolates that replicate in microglia. *J. Virol.* **72,** 4243–4249.

Silversides, J., Heggarty, S., McDonnell, G., Hawkins, S., and Graham, C. (2004). Influence of CCR5 delta32 polymorphism on multiple sclerosis susceptibility and disease course. *Mult. Scler.* **10,** 149–152.

Staller, P., Sulitkova, J., Lisztwan, J., Moch, H., Oakeley, E. J., and Krek, W. (2003). Chemokine receptor CXCR4 downregulated by von Hippel-Lindau tumour suppressor pVHL. *Nature* **425,** 307–311.

Strizki, J., Xu, S., Wagner, N., Wojcik, L., Liu, J., Hou, Y., Endres, M., Palani, A., Shapiro, S., Clader, J., Greenlee, W., Tagat, J., McCombie, S., Cox, K., Fawzi, A., Chou, C., Pugliese-Sivo, C., Davies, L., Moreno, M., Ho, D., Trkola, A., Stoddart, C., Moore, J., Reyes, G., and Baroudy, B. (2001). SCH-C (SCH351125), an orally bioavailable, small molecule antagonist of the chemokine CCR5, is a potent inhibitor of HIV-1 infection *in vitro* and *in vivo*. *Proc. Nat. Acad. Sci. USA* **98,** 12718–12723.

Sun, Y. X., Wang, J., Shelburne, C. E., Lopatin, D. E., Chinnaiyan, A. M., Rubin, M. A., Pienta, K. J., and Taichman, R. S. (2003). Expression of CXCR4 and CXCL12 (SDF-1) in human prostate cancers (PCa) *in vivo*. *J. Cell. Biochem.* **89,** 462–473.

Sweeney, E., Lortat-Jacob, H., Priestley, G., Nakamoto, B., and Papayannopoulou, T. (2002). Sulfated polysaccharides increase plasma levels of SDF-1 in monkeys and mice: Involvement in mobilization of stem/progenitor cells. *Blood* **99,** 44–51.

Taichman, R., Cooper, C., Keller, E., Pienta, K., Taichman, N., and McCauley, L. (2002). Use of the stromal cell-derived factor-1/CXCR4 pathway in prostate cancer metastasis to bone. *Cancer Res.* **62,** 1832–1837.

Tak, P., Smeets, T., Daha, M., Kluin, P., Meijers, K., Brand, R., Meinders, A., and Breedveld, F. (1997). Analysis of the synovial cell infiltrate in early rheumatoid synovial tissue in relation to local disease activity. *Arthritis. Rheum.* **40,** 217–225.

Takamori, H., Oades, Z., Hoch, O., Burger, M., and Schraufstatter, I. (2000). Autocrine growth effect of IL-8 and GROalpha on a human pancreatic cancer cell line. *Pancreas* **21,** 52–56.

Takenaga, M., Tamamura, H., Hiramatsu, K., Nakamura, N., Yamaguchi, Y., Kitagawa, A., Kawai, S., Nakashima, H., Fujii, N., and Igarashi, R. (2004). A single treatment with microcapsules containing a CXCR4 antagonist suppresses pulmonary metastasis of murine melanoma. *Biochem. Biophys. Res. Commun.* **320,** 226–232.

Tamamura, H., Hori, A., Kanzaki, N., Hiramatsu, K., Mizumoto, M., Nakashima, H., Yamamoto, N., Otaka, A., and Fujii, N. (2003). T140 analogs as CXCR4 antagonists identified as anti-metastatic agents in the treatment of breast cancer. *FEBS Lett.* **550,** 79–83.

Tamamura, H., Fujisawa, M., Hiramatsu, K., Mizumoto, M., Nakashima, H., Yamamoto, N., Otaka, A., and Fujii, N. (2004). Identification of a CXCR4 antagonist, a T140 analog, as an anti-rheumatoid arthritis agent. *FEBS Lett.* **569,** 99–104.

Thoma, G., Nuninger, F., Schaefer, M., Akyel, K., Albert, R., Beerli, C., Bruns, C., Francotte, E., Luyten, M., MacKenzie, D., Oberer, L., Streiff, M., Wagner, T., Walter, H., Weckbecker, G., and Zerwes, H. (2004). Orally bioavailable competitive CCR5 antagonists. *J. Med. Chem.* **47,** 1939–1955.

To, L., Haylock, D., Kimber, R., and Juttner, C. (1984). High levels of circulating haemopoietic stem cells in very early remission from acute non-lymphoblastic leukaemia and their collection and cryopreservation. *Br. J. Haematol.* **58,** 399–410.

To, L., Haylock, D., Simmons, P., and Juttner, C. (1997). The biology and clinical uses of blood stem cells. *Blood* **89,** 2233–2258.

Tran, E., Kuziel, W., and Owens, T. (2000). Induction of experimental autoimmune encephalomyelitis in C57BL/6 mice deficient in either the chemokine macrophage inflammatory protein-1alpha or its CCR5 receptor. *J. Immunol.* **30,** 1410–1415.

Tremblay, C., Giguel, F., Hicks, J., Chou, T., Lizawa, Y., Sugihara, Y., and Hirsch, M. (2003). TAK-220, a novel small molecule inhibitor of CCR5 has favorable anti-HIV interactions with other antiretrovirals *in vitro*. *In* "10th Conference on Retroviruses and Opportunistic Infections," pp. Abstract 562. Boston, MA.

Trkola, A., Kuhmann, S., Strizki, J., Maxwell, E., Ketas, T., Morgan, T., Pugach, P., Xu, S., Wojcik, L., Tagat, J., Palani, A., Shapiro, S., Clader, J., McCombie, S., Reyes, G., Baroudy, B., and Moore, J. (2002). HIV-1 escape from a small molecule, CCR5-specific entry inhibitor does not involve CXCR4 use. *Proc. Nat. Acad. Sci. USA* **99,** 395–400.

Varnes, J., Gardner, D., Santella, Jr., Duncia, J., Estrella, M., Watson, P., Clark, C., Ko, S., Welch, P., Covington, M., Stowell, N., Wadman, E., Davies, P., Solomon, K., Newton, R., Trainor, G., Decicco, C., and Wacker, D. (2004). Discovery of N-propylurea 3-benzylpiperidines as selective CC chemokine receptor-3 (CCR3) antagonists. *Bioorg. Med. Chem. Lett.* **14,** 1645–1649.

Veillard, N., Kwak, B., Pelli, G., Mulhaupt, F., James, R., Proudfoot, A., and Mach, F. (2004). Antagonism of RANTES receptors reduces atherosclerotic plaque formation in mice. *Circ. Res.* **94,** 253–261.

Veltri, R., Miller, M., Zhao, G., Ng, A., Marley, G., Wright, G. J., Vessella, R., and Ralph, D. (1999). Interleukin 8 serum levels in patients with benign prostatic hyperplasia and prostate cancer. *Urology* **53,** 139–147.

Volin, M., Shah, M., Tokuhira, M., Haines, G., Woods, J., and Koch, A. (1998). RANTES expression and contribution to monocyte chemotaxis in arthritis. *Clin. Immunol. Immunopathol.* **89,** 44–53.

Wacker, D., Santella, Jr., Gardner, D., Varnes, J., Estrella, M., De Lucca, G., Ko, S., Tanabe, K., Watson, P., Welch, P., Covington, M., Stowell, N., Wadman, E., Davies, P., Solomon, K., Newton, R., Trainor, G., Friedman, S., Decicco, C., and Duncia, J. (2002). CCR3 antagonists: A potential new therapy for the treatment of asthma. Discovery and structure-activity relationships. *Bioorg. Med. Chem. Lett.* **12,** 1785–1789.

Warmington, K., Boring, L., Ruth, J., Sonstein, J., Hogaboam, C., Curtis, J., Kunkel, S., Charo, I., and Chensue, S. (1999). Effect of C-C chemokine receptor 2 (CCR2) knockout on type-2 (schistosomal antigen-elicited) pulmonary granuloma formation: Analysis of cellular recruitment and cytokine responses. *Am. J. Pathol.* **154,** 1407–1416.

Warrior, U., McKeegan, E., Rottinghaus, S., Garcia, L., Traphagen, L., Grayson, G., Komater, V., McNally, T., Helfrich, R., Harris, R., Bell, R., and Burns, D. (2003). Identification and characterization of novel antagonists of the CCR3 receptor. *J. Biomol. Screen* **8,** 324–331.

White, J., Lee, J., Young, P., Hertzberg, R., Jurewicz, A., Chaikin, M., Widdowson, K., Foley, J., Martin, L., Griswold, D., and Sarau, H. (1998). Identification of a potent, selective non-peptide CXCR2 antagonist that inhibits interleukin-8-induced neutrophil migration. *J. Biol. Chem.* **273,** 10095–10098.

Widdowson, K., Elliott, J., Veber, D., Nie, H., Rutledge, M., McCleland, B., Xiang, J., Jurewicz, A., Hertzberg, R., Foley, J., Griswold, D., Martin, L., Lee, J., White, J., and Sarau, H. (2004). Evaluation of potent and selective small-molecule antagonists for the CXCR2 chemokine receptor. *J. Med. Chem.* **47,** 1319–1321.

Wollert, K., Meyer, G., Lotz, J., Ringes-Lichtenberg, S., Lippolt, P., Breidenbach, C., Fichtner, S., Korte, T., Hornig, B., Messinger, D., Arseniev, L., Hertenstein, B., Ganser, A., and Drexler, H. (2004). Intracoronary autologous bone-marrow cell transfer after myocardial infarction: The BOOST randomised controlled clinical trial. *Lancet* **364,** 141–148.

Yang, T., Chen, S., Leach, M., Manfra, D., Homey, B., Wiekowski, M., Sullivan, L., Jenh, C., Narula, S., Chensue, S., and Lira, S. (2000). Transgenic expression of the chemokine receptor encoded by human herpiesvirus 8 induces an angioproliferative disease resembling Karposi's sarcoma. *J. Exp. Med.* **191,** 445–454.

Yang, Y., Mukai, T., Gao, P., Yamaguchi, N., Ono, S., Iwaki, H., Obika, S., Imanishi, T., Tsujimura, T., Hamaoka, T., and Fujiwara, H. (2002). A non-peptide CCR5 antagonist inhibits collagen-induced arthritis by modulating T cell migration without affecting anti-collagen T cell responses. *Eur. J. Immunol.* **32,** 2124–2132.

Yawalkar, N., Uguccioni, M., Scharer, J., Braunwalder, J., Karlen, S., Dewald, B., Braathen, L., and Baggiolini, M. (1999). Enhanced expression of eotaxin and CCR3 in atopic dermatitis. *J. Invest. Dermatol.* **113,** 43–48.

Ying, S., Robinson, D., Meng, Q., Rottman, J., Kennedy, R., Ringler, D., Mackay, C., Daugherty, B., Springer, M., Durham, S., Williams, T., and Kay, A. (1997). Enhanced expression of eotaxin and CCR3 mRNA and protein in atopic asthma. Association with airway hyperresponsiveness and predominant co-localization of eotaxin mRNA to bronchial epithelial and endothelial cells. *Eur. J. Immunol.* **27,** 3507–3516.

Ylisastigui, L., Vizzavona, J., Drakopoupou, E., Painavoine, P., Calvo, C., Parmentier, M., Gluckman, J., Vita, C., and Benjouad, A. (1998). Synthetic full-length and truncated RANTES inhibit HIV-1 infection of primary macrophages. *AIDS* **12,** 977–984.

Yokochi, S., Hashimoto, H., Ishiwata, Y., Shimokawa, H., Haino, M., Terashima, Y., and Matsushima, K. (2001). An anti-inflammatory drug, propagermanium, may target GPI-anchored proteins associated with an MCP-1 receptor, CCR2. *J. Interferon. Cytokine. Res.* **21,** 389–398.

Yuan, Y., ten Hove, T., The, F., Slors, J., van Deventer, S., and te Velde, A. (2001). Chemokine receptor CXCR3 expression in inflammatory bowel disease. *Inflamm. Bowel Dis.* 7(4), 281–286.

Yun, J., Whiting, D., Fischbein, M., Banerji, A., Irie, Y., Stein, D., Fishbein, M., Proudfoot, A., Laks, H., Berliner, J., and Ardehali, A. (2004). Combined blockade of the chemokine receptors CCR1 and CCR5 attenuates chronic rejection. *Circulation* **109,** 932–937.

Zeelenberg, I. S., Ruuls-Van Stalle, L., and Roos, E. (2001). Retention of CXCR4 in the endoplasmic reticulum blocks dissemination of a T cell hybridoma. *J. Clin. Invest.* **108,** 269–277.

Zeelenberg, I. S., Ruuls-Van, Stalle,L, and Roos, E. (2003). The chemokine receptor CXCR4 is required for outgrowth of colon carcinoma micrometastases. *Cancer Res.* **63,** 3833–3839.

Zhou, Y., Larsen, P. H., Hao, C., and Yong, V. W. (2002). CXCR4 is a major chemokine receptor on glioma cells and mediates their survival. *J. Biol. Chem.* **277,** 49481–49487.

Zou, W., Machelon, V., Coulomb-L'Hermin, A., Borvak, J., Nome, F., Isaeva, T., Wei, S., Krzysiek, R., Durand-Gasselin, I., Gordon, A., Pustilnik, T., Curiel, D. T., Galanaud, P., Capron, F., Emilie, D., and Curiel, T. J. (2001). Stromal-derived factor-1 in human tumors recruits and alters the function of plasmacytoid precursor dendritic cells. *Nat. Med.* **7,** 1339–1346.

Index

N

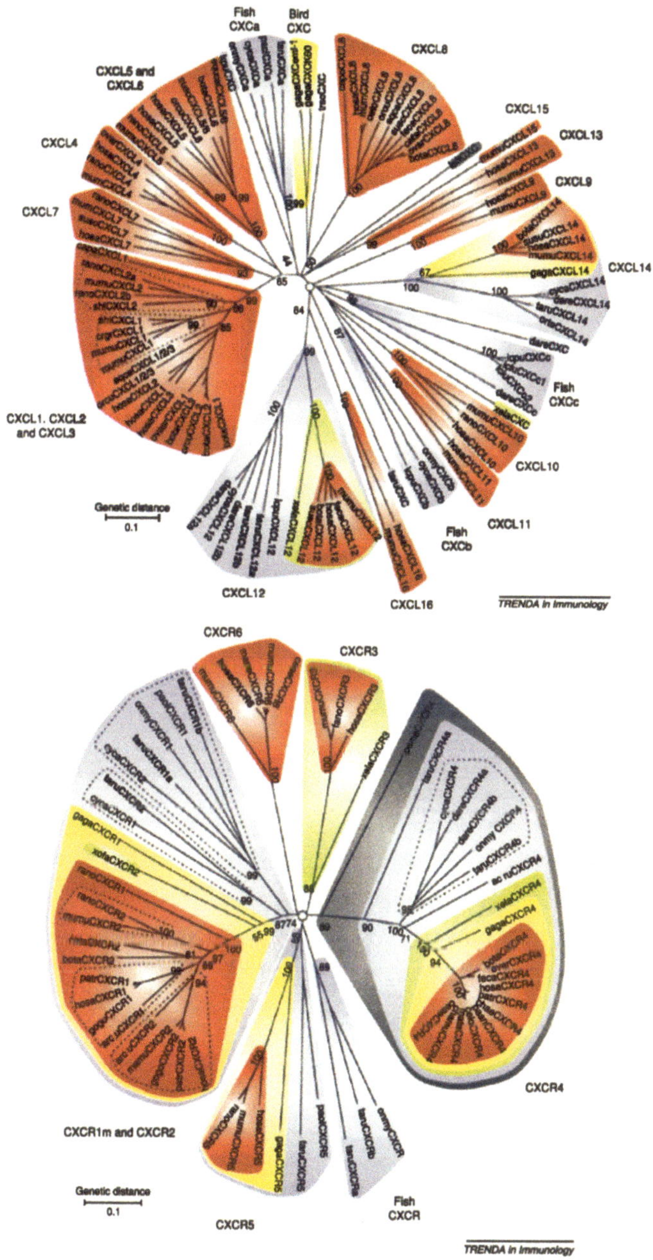

CHAPTER 2, FIGURE 1 Phylogenetic analysis of the CXC chemokine ligands (A) and receptors (B). Colors represent major vertebrate divisions: red, mammalian; yellow, avian; green, amphibian; lilac, teleost fish; and grey, agnathan. Human and pufferfish sequences are indicated in bold. Subclusters (statistically reliable clusters within clusters) are indicated by broken lines. The abbreviations for the organisms are as follows: bota, *Bos taurus* (cow); cafa, *Canis familiaris* (dog); ceae, *Cercopithecus aethiops* (vervet monkey); capo, *Cavia porcellus* (guinea pig); ceto, *Cercocebus torquatus atys* (mangabey); chae, *Chlorocebus aethiops* (African green monkey); crgr, *Cricetulus griseus* (Chinese hamster); cyca, *Cyprinus carpio* (carp); dare, *Danio rerio* (zebrafish); eqca, *Equus caballus* (horse); feca, *Felis cattus* (cat); gaga, *Gallus gallus* (chicken); gogo, *Gorilla gorilla* (gorilla); hosa, *Homo sapiens* (human); icfu, *Ictalurus furcatus* (blue catfish); icpu, *Ictalurus punctatus* (channel catfish); lafl, *Lampetra fluviatilis* (lamprey); mamu, *Macaca mulatta* (rhesus macaque); mane, *Macaca nemestrina* (pig-tailed macaque); mumu, *Mus musculus* (mouse); onmy, *Oncorhynchus mykiss* (rainbow trout); orcu, *Oryctolagus cuniculus* (rabbit); orla, *Oryzias latipes* (medaka); ovar, *Ovis aries* (sheep); paan, *Papio anubis* (olive baboon); paol, *Paralichtys olivaceous* (flounder); patr, *Pan troglodytes* (chimpanzee); pema, *Petromyzon marinus* (sea lamprey); rano, *Rattus norvegicus* (rat); sihi, *Sigmodon hispidus* (hispid cotton rat); susc, *Sus scrofa* (pig); taru, *Takifugu rubripes* (pufferfish); trsc, *Triakis scyllium* (shark); xela, *Xenopus laevis* (African clawed toad). Figure and parts of legend reprinted from Huising *et al.* (2003) with permission.

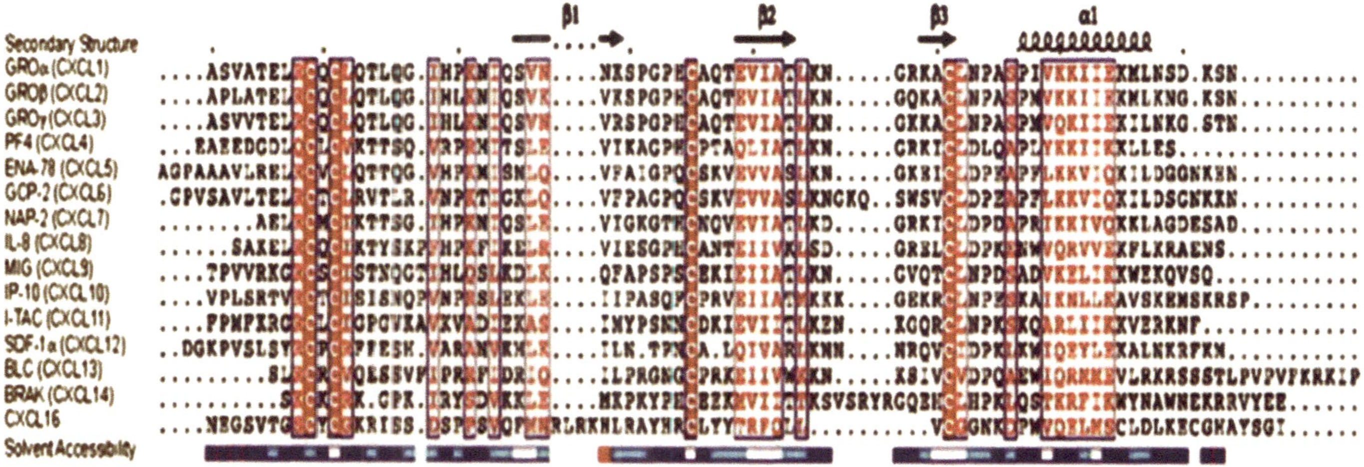

CHAPTER 2, FIGURE 2 Protein sequence alignment of the human CXC chemokine family. Structural cysteines are shown with a red background, and homologous residues in the family are shown in red. Secondary structural character is designated at the top of the diagram, while solvent accessibility is shown at the bottom. Dark blue indicates completely solvent exposed side chains, while light blue indicates partially exposed residues, and white amino acids are buried. Reprinted from Fernandez and Lolis (2002) with permission.

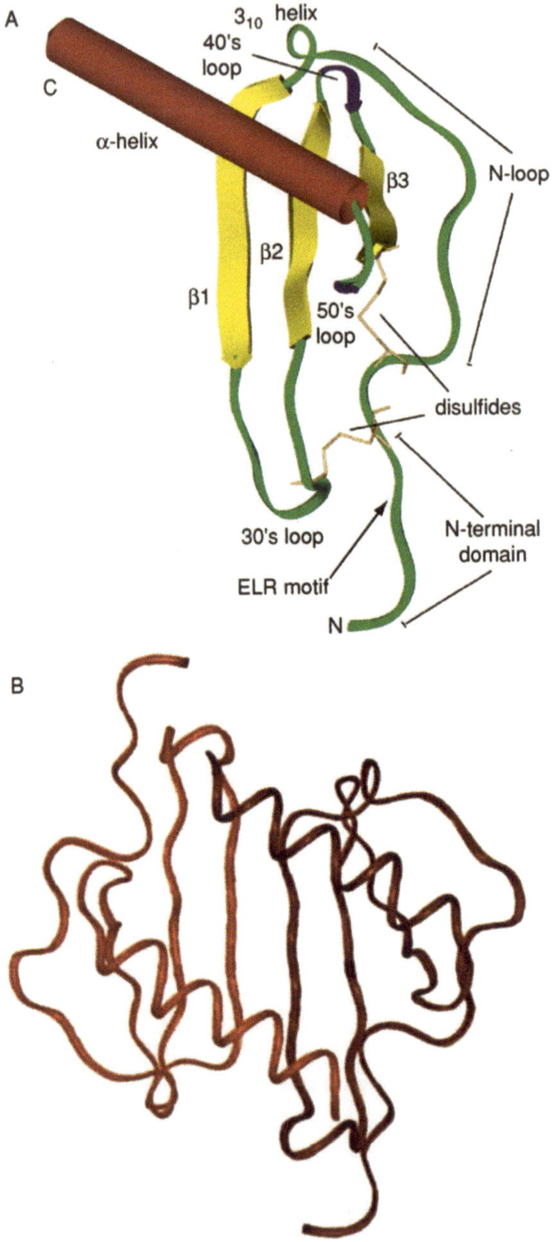

CHAPTER 2, FIGURE 3 The degenerate structure of CXC chemokines in monomeric form (A) and the dimeric structure of CXCL8 (B). The secondary structural motifs discussed in the text are labeled in A. In B, it can be seen that dimerization occurs along the $\beta 1$ strand of the molecules. Reprinted from Fernandez and Lolis (2002) with permission.

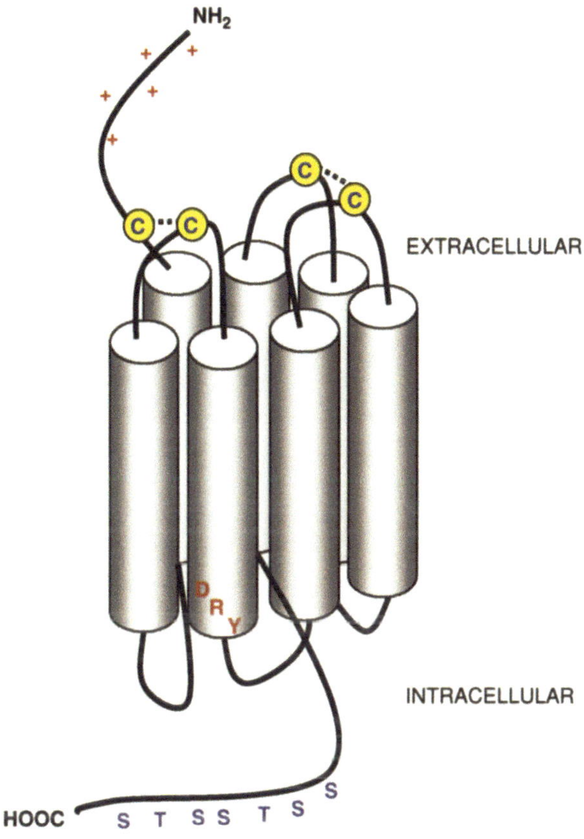

CHAPTER 3, FIGURE 1 This cartoon illustrates the secondary structure typical of a CC chemokine receptor. The seven-transmembrane (TM) helices in the bundle are depicted as cylinders and are held together by disulphide bonding of conserved cysteine residues (yellow). The N-terminus is negatively charged and binds the predominantly basic chemokine, while the intracellular C-terminus is rich in serine and threonine residues, some of which undergo phosphorylation following receptor activation. The DRY motif of TM helix 3 is also illustrated.

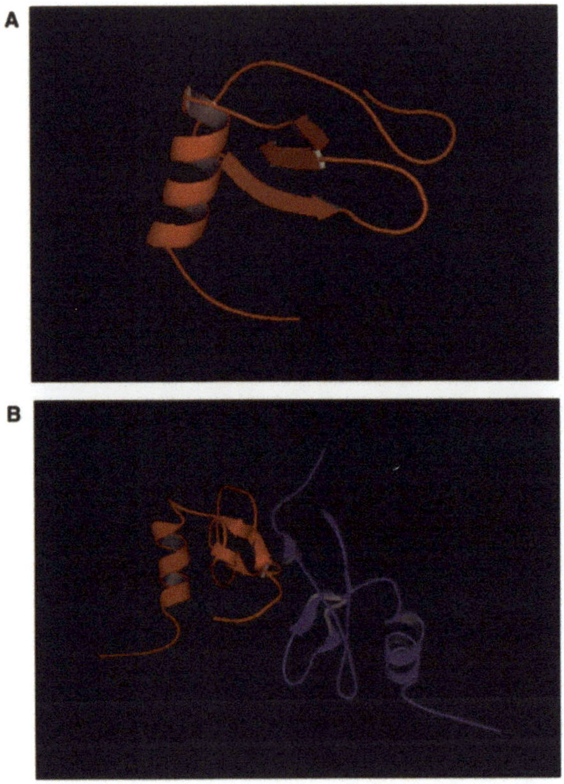

CHAPTER 3, FIGURE 2 Panel A shows the secondary structural "Greek Key" motif of CC chemokines, as typified by CCL2. Three antiparallel β-pleated sheets overlay a C-terminal, α-helical domain. Panel B shows the elongated CCL2 dimer, which is facilitated by interactions of the chemokine N-termini. The CCL2 monomers making up the dimer are depicted in blue and red.

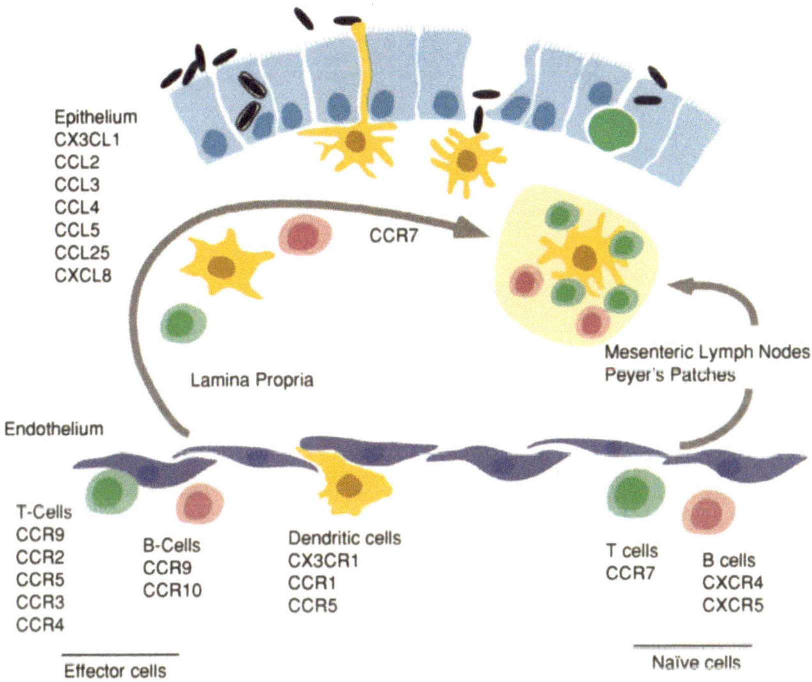

CHAPTER 6, FIGURE 1 Binding of chemokines to their receptors expressed by T-cells, B-cells, and dendritic cells directs them to the intestinal lamina propria or to mesenteric lymph nodes and Peyer's patches.

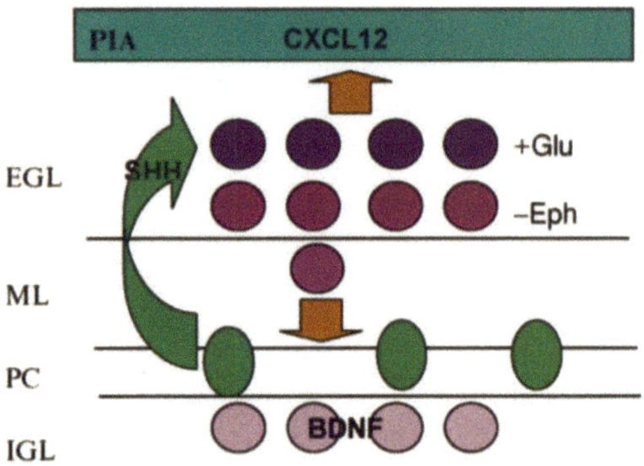

CHAPTER 7, FIGURE 1 CXCL12 is critical for the normal formation of the cerebellum. CXCR4-expressing granule precursor cells (GPCs; dark purple circles) migrate from the rhombic lip to form the external granule cell layer (EGL) in proximity to CXCL12-expressing pia mater. CXCR4 signaling responses are enhanced by glutamate receptor activation (+Glu) (Klein *et al.*, 2001). Purkinje cell (PC)–derived Sonic hedgehog (SHH) induces the proliferation of GPCs in the EGL, which is enhanced by CXCL12. Migration of GPCs out of the EGL to form the internal granule cell layer (IGL) is believed to occur through alterations in CXCL12 chemoattraction achieved, in part, by decreased expression of CXCR4 by maturing GPCs (lighter purple circles), increased expression of ephrin ligands and receptors (-Eph), which antagonize CXCR4 responses to CXCL12, and by the increased expression of brain-derived neurotrophic factor (BDNF) by mature granule cell neurons (lavender ovals), which enhance the movement of GPCs into the IGL. ML, molecular layer containing Purkinje cell and granule cell neuronal axons and dendrites.

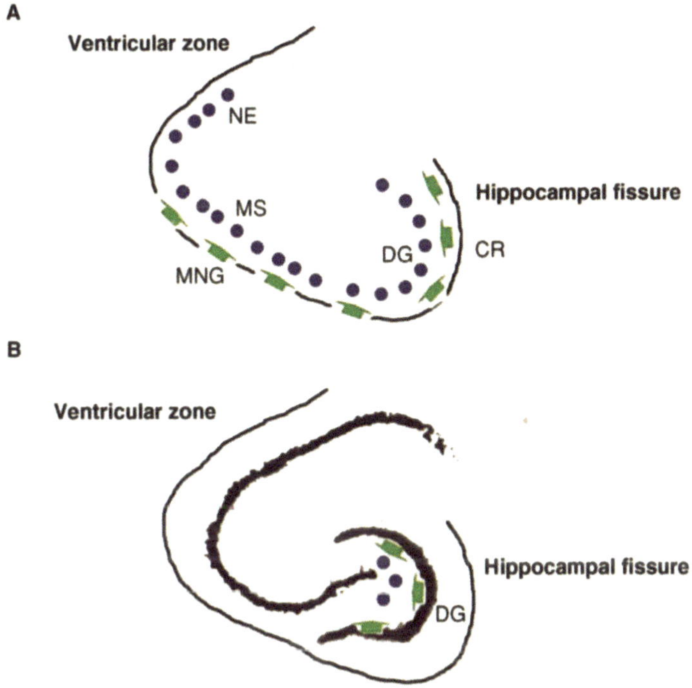

CHAPTER 7, FIGURE 2 CXCL12 and CXCR4 are critical for the normal formation of the dentate gyrus (DG). (A) Dentate granule cells (blue circles) first arise in the dentate neuroepithelium (NE), a typical subventricular primary neural germinal zone. The dentate migratory stream (MS) arises from this and constitutes a secondary proliferative zone in which granule precursor cells can be seen to migrate and divide. Finally, a tertiary proliferative zone forms in the hilus of the DG. During embryonic and early postnatal development, granule cells are expressed by the meninges (MNG) as well as by Cajal–Retzius cells (CR) lining the hippocampal fissure. Targeted deletion of CXCR4 results in the mislocalization of dentate granule cells and a reduction in their numbers. During early hippocampal development, CXCL12 may also modulate neurotransmission in developing hippocampal circuits, further affecting hippocampal patterning. (B) The pattern of CXCL12 and CXCR4 expression is different beyond the postnatal period. Now CXCL12 is highly expressed by cells populating the blades of the DG (green arrows), and proliferating cells within the dentate hilus can be seen to express CXCR4 (blue circles).

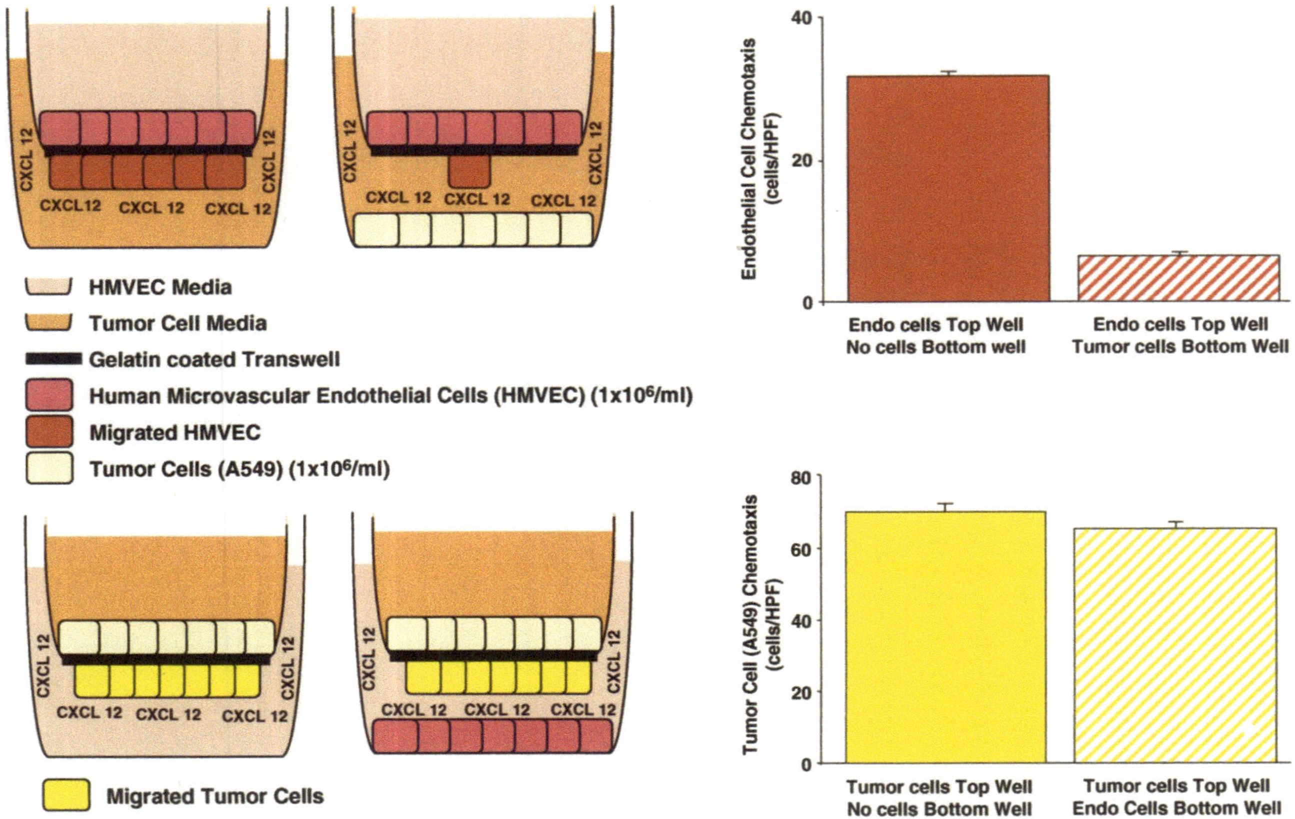

CHAPTER 10, FIGURE 3 Transwell chemotaxis assay analogous to the tumor microenvironment. Chemotaxis of tumor and endothelial cells expressing CXCR4 in response to CXCL12. The presence of the tumor outcompetes endothelial cells for CXCL12-mediated cellular migration.

FSC
www.fsc.org
MIX
Papier | Fördert gute Waldnutzung
FSC® C083411

Kolibri 360 GmbH
Ferdinand-Jühlke-Straße 7
99095 Erfurt, Deutschland
produktsicherheit@kolibri360.de